PHYSICS
FOR THE
BIOLOGICAL SCIENCES

A TOPICAL APPROACH TO BIOPHYSICAL CONCEPTS

Fourth Edition

F.R. Hallett
J.L. Hunt

E.L. McFarland
G.H. Renninger

R.H. Stinson
D.E. Sullivan

THOMSON
NELSON

Australia   Canada   Mexico   Singapore   Spain   United Kingdom   United States

THOMSON

NELSON

Physics for the Biological Sciences:
A Topical Approach to Biophysical Concepts
Fourth Edition

by F.R. Hallett , J.L. Hunt, R.H. Stinson, Ernie
McFarland, George Renninger, and D. Sullivan

Editorial Director and Publisher:
Evelyn Veitch

Executive Editor:
Joanna Cotton

Acquisitions Editor:
Brad Lambertus

Marketing Manager:
Janet Piper

Developmental Editor:
Glen Herbert

Production Coordinator:
Helen Jager Locsin

Creative Director:
Angela Cluer

Cover Design:
Katherine Strain

Photo Credits:
Front cover, left to right:
PhotoDisc; Lawrence
Lawry/PhotoDisc; David Chasey/
PhotoDisc; Back cover, left to
right: Bruce Heinemann/PhotoDisc;
Janis Christie/PhotoDisc; Spike
Mafford/PhotoDisc

Printer:
Webcom

COPYRIGHT © 2003 by Nelson,
a division of Thomson Canada
Limited.

Printed and bound in Canada
3   4   05   04

For more information contact
Nelson, 1120 Birchmount Road,
Toronto, Ontario, M1K 5G4.
Or you can visit our Internet site at
http://www.nelson.com

ALL RIGHTS RESERVED. No part of
this work covered by the copyright
hereon may be reproduced,
transcribed, or used in any form or
by any means—graphic, electronic,
or mechanical, including
photocopying, recording, taping,
Web distribution, or information
storage and retrieval systems—
without the written permission of
the publisher.

For permission to use material
from this text or product, contact
us by
Tel 1-800-730-2214
Fax 1-800-730-2215
www.thomsonrights.com

Every effort has been made to
trace ownership of all copyrighted
material and to secure permission
from copyright holders. In the
event of any question arising as to
the use of any material, we will be
pleased to make the necessary
corrections in future printings.

**National Library of Canada
Cataloguing in Publication Data**

Physics for the biological
sciences : a topical approach to
biophysical concepts / F.R. Hallett
... [et al.]. -- 4th ed.

Includes index.
ISBN 0-17-622451-3

1. Biophysics. I. Hallett, F.R.
(Frederick Ross), 1942–

QH505.H25 2003   571.4
C2002-905735-3

Preface

The life sciences have undergone vast changes in the last quarter-century. At the beginning of this period, the role of DNA was understood and the basics of protein structures were being worked out, but the bulk of biological understanding was at the level of taxonomy, genetics, population dynamics and microbiology. Many developments have changed this situation, perhaps the most important being the ubiquity of powerful computing. With this tool it has become possible to descend to the level of molecules and to understand them in terms of fundamental physical and chemical processes. Consequently, it is more important than ever that life scientists have a firm foundation in the physical sciences.

This book is the product of 30 years of experience, innovation, and advances in pedagogy. In 1972, Hallett, Stinson, Graham and Speight produced the first edition of this textbook for life-science students. The book was a radical change from the usual introductory physics text, which has a presentation determined by the perceived needs of physical-science and engineering students. Instead, the objective of their book was to educate the student in the fundamentals of physics (both classical and quantum) that were most appropriate for understanding phenomena in the life sciences.

The topic of sound and acoustics was centred on human hearing and bat navigation. The study of optics (ray, wave and quantum) was motivated by the human visual system; of mechanics was centred around locomotion, material strength, and scaling; and of electricity led to an understanding of the neural spike. Thirty years of use has justified the choice of topics and the engagement of life-science students with these and other examples.

The present edition contains updated content, more sample problems, and many more end-of-chapter exercises and problems. A chapter on magnetism has been added. The book is more geared to "student-centred learning," that is, it is more self-contained and comprehensive so that students can learn the material more easily by just using the textbook.

Who Are the Intended Students?

This book is written for first-year university students in life sciences and environmental sciences. The students are expected to have some background from high-school physics and must have good skills in algebra and trigonometry.

$$\frac{dy}{dx}$$

Although calculus is used occasionally in derivations, the students are not expected to use calculus in problem solutions. Where calculus is used in the text, it is contained within a grey background and identified by a dy/dx icon. The calculus derivations may be skipped by the reader with no loss in continuity.

Order of Chapters

The material in the book can be covered in two semesters, but the chapters do not have to be followed in order. Depending on students' background, lecturers may wish to begin with kinematics (Chapter 7) and cover the second half of the book in one semester, followed by the first half in a second semester. Indeed, at the University of Guelph, the life-science physics courses follow the chapters in order, whereas the environmental-science physics courses cover the second half of the book first.

Notation, Symbols, and Units

Vector quantities such as force and acceleration are indicated in boldface type with an arrow above, such as $\vec{F}$ and $\vec{a}$. The magnitude of a vector is in non-boldface type, e.g., F is the magnitude of the vector $\vec{F}$. Similarly, a scalar quantity is shown in non-boldface type, such as T for temperature. Vector components are indicated by x and y subscripts on the non-boldface symbol for the vector; for example, F_x and F_y are the x and y components of vector $\vec{F}$. Appendix 3 provides a complete list of all symbols and their meanings.

SI units (Système International d'Unités) are used exclusively in this book.

Exercises and Problems

Each chapter has an extensive selection of end-of-chapter exercises and problems. The exercises are straightforward application of the principles discussed in the chapter. The problems are more challenging and typically require two or more nontrivial steps.

Worldwide Web Resources

As discussed in further detail in Appendix 2, the following Web site contains links to useful tutorials in mathematics and physics, interactive Web-based activities from around the world, and, if applicable, corrections to the textbook:

http://www.physics.uoguelph.ca/biophysics/

FRH, JLH, ELM, GHR, RHS, DES
Guelph, Ontario
November 2002

Following the first printing of the third edition of this text, Prof. R.H. Stinson passed away after a brief illness. The remaining authors wish to acknowledge the career of this fine gentleman. His life was one of devotion and charity, and his dedication to education and the welfare of students was an example to everyone who knew him. In his retirement he continued his lifelong work by volunteer teaching in Malawi, a nation for which he developed a strong affection. He is greatly missed.

Acknowledgements

The authors wish to thank the following people for their contributions in producing this book:

Lenore Latta, whose meticulous editing and thoughtful questions to the authors have resulted in a book that will be much more useful to students;

Catherine and Gerard Pacey for the attractive layout and their many hours working with all the equations;

Carol Croft for her careful work on the mechanics chapters;

Brad Lambertus and staff at Nelson-Thompson for their guidance and assistance;

The Department of Physics at the University of Guelph for its support of this project.

Credits

The authors would like to thank these sources for permission to reprint the following images:

Figure 2-18 compiled from *Bats: Biology and Behaviour* by John Altringham (New York: Oxford University Press, 1996).

Figure 3-11 from *Comparative Anatomy of the Eye* by Jack H. Prince (Springfield, IL: Charles C. Thomas, 1956). Courtesy of Charles C. Thomas, Publisher, Ltd., Springfield, Illinois.

Figure 5-1 compiled from *The Senses* by H.B. Barlow and J.D. Mollon (Cambridge: Cambridge University Press, 1982). Reprinted with permission of Cambridge University Press.

Figure 5-4 compiled from *The Senses* by H.B. Barlow and J.D. Mollon (Cambridge: Cambridge University Press, 1982) and from *Photoreceptors, Their Role in Vision* by Z. Ete (New York: Cambridge University Press, 1982). Reprinted with the permission of Cambridge University Press.

The image in Box 9-2 from *Nature as Constructor* by Klaus Wunderlich and Wolfgang Gloede (New York: Arco Publishing, 1981).

Figure 10-11 from *Life's Devices* by Steven Vogel (Princeton, NJ: Princeton University Press, 1988).

Figure 12-8 from *McDonald's Blood Flow in Arteries: Theoretic, Experimental and Clinical Principles* by Wilmer W. Nichols and Michael F. O'Rourke (Philadelphia: Lea & Febiger, 1990).

Figure 13-11 from *Random Walks in Biology* by Howard C. Berg (Princeton: Princeton University Press, 1983).

Contents

Chapter 4   Absorption and Emission of Light by Molecules

Chapter 5   Quantum Nature of Vision

Chapter 12   Fluid Dynamics

Chapter 13   Thermal Motion of Molecules

1 Vibrations and Waves

1.1 Introduction

Vibrations and waves are among the most important types of motion found in nature. The movement of an insect's wing during flight and the oscillations of the tympanic membrane of the ear are examples of vibrational motions of interest in biology. Wave phenomena occur in studies of sound, light, and the energy and arrangement of electrons in atoms and molecules. In this chapter some of the terms and mathematics used to describe vibrational and wave motion are presented. Many applications to living systems will be discussed in later chapters.

1.2 Simple Harmonic Motion

Many objects in nature, ranging from atoms in molecules to fluttering leaves on tree branches, are subject, when displaced slightly from their equilibrium position, to a force that is proportional to the displacement from equilibrium. Such objects, if displaced slightly from equilibrium and then released, will undergo back-and-forth oscillatory motion, known as *simple harmonic motion* (SHM). A simple, although somewhat academic, model of such a system is shown in Fig. 1-1. A mass m rests on a horizontal surface on which friction is negligible. The mass is attached to a spring S that has a mass which is very small compared to m.

Figure 1-1

Mass-spring system undergoing simple harmonic motion.

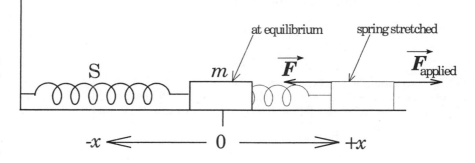

When the mass is at the equilibrium position ($x = 0$), the spring exerts no force on it, and if the mass is released from rest at equilibrium, it remains there. If the mass is displaced to the right (+x-direction) or to the left (–x-direction), the spring is stretched or compressed and exerts a force $\vec{F}$ (see Box 1-1) on the mass in the direction opposite to the direction of the displacement, i.e., the force is directed back toward $x = 0$. This force is referred to as a *restoring force* since it tends to push the mass back toward equilibrium. For small displacements, it is found that the magnitude of $\vec{F}$ is proportional to the displacement x.[1] Because of this proportionality, a graph of F vs. x is linear, and the force is called a "linear" restoring force.

Box 1-1   Vector and Scalar Notation

Vectors are quantities that have both magnitude and direction. Examples include force, velocity, acceleration, and electric field. *Scalar* quantities have only magnitude; time, temperature, and speed are examples. In this book, vectors are shown in boldface type with an arrow above, such as $\vec{F}$ for force. Scalars appear in non-boldface type, such as t for time.

The *magnitude of a vector*, which is always a positive quantity (or zero), is also displayed in non-boldface type without an arrow above. Hence, the magnitude of the force vector $\vec{F}$ is F.

Vectors are often expressed in terms of their *components* in x- and y- (and possibly z-) directions. These components, which can be positive, negative (or zero), are shown in non-boldface type with subscripts. Thus, the x- and y-components of force $\vec{F}$ are F_x and F_y.

The proportionality between the restoring force and the displacement is commonly expressed as

$$F_x = -kx \qquad\qquad \text{[1-1]}$$

---

1. The proportionality between F and x is often called "Hooke's law," although it is really an experimental observation rather than a law of nature. This "law" was first stated by the English scientist Robert Hooke (1635–1703).

where k is a positive constant of proportionality known as the *force constant* of the system. This constant respresents the force per unit displacement of the spring. A stiff spring has a lage force constant, and a soft spring has a small force constant.

In the equation $F_x = -kx$, the SI units of F_x and x are newton (N) and metre (m), respectively, and hence the SI unit of k is newton per metre (N/m).

Figure 1-2

The linear restoring force for simple harmonic motion.

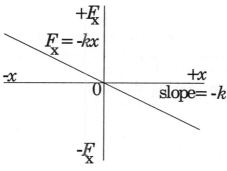

The correctness of Eq. [1-1] can be confirmed as follows. When the mass in Fig. 1-1 is displaced to the right of equilibrium, x is positive and, from Eq. [1-1], F_x is negative, indicating that the restoring force $\vec{F}$ is (correctly) to the left. When the mass is displaced to the left, x is negative and F_x is then positive, indicating a force to the right. Equation [1-1] is expressed graphically in Fig. 1-2; notice the linearity of the graph.

Elastic Potential Energy

The system of Fig. 1-1 may be described in terms of its potential energy, expressed in SI units as joules (J). Suppose a force is applied such that the mass m is displaced from equilibrium at constant velocity.

Figure 1-3

As the spring is stretched a distance *x* from equilibrium, the elastic potential energy stored is ½*kx²*.

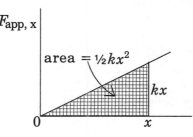

Since the velocity is constant, the acceleration $\vec{a}$ of the mass is zero, and from Newton's second law of motion ($\vec{F}_{net} = m\vec{a}$), the net force must be zero. This means that the applied force $\vec{F}_{app}$ must be equal in magnitude to the force exerted by the spring, but in the opposite direction, i.e., $F_{app,x} = +kx$ (Fig. 1-3).

As the mass moves, work is done by the applied force. If the force were constant, then the work W done could be calculated from $W = F_{app,x}\,\Delta x$. However, the applied force is proportional to x, which is variable. The work done in this case can be expressed as the area under the graph of $F_{app,x}$ vs. x. Hence, the work done in stretching the spring from 0 to x is simply the area of the triangle in Fig. 1-3, which is $\frac{1}{2}(x)(kx) = \frac{1}{2}kx^2$.

This work W can also be written as a definite integral[2]

$$W = \int_0^x F_{app,x}\,dx = \int_0^x kx\,dx = \tfrac{1}{2}kx^2$$

This work is positive for elongation of the spring ($x > 0$) and for compression ($x < 0$), and may be considered as *elastic potential energy* U stored in the spring. Thus, for a linear restoring force, the elastic potential energy of the system is given by

$$U = \tfrac{1}{2}kx^2 \qquad\qquad \textbf{[1-2]}$$

More generally, elastic potential energy refers to energy that is stored in any object as a result of twisting, stretching, or compressing. It may be converted to other forms of energy. For example, if the spring in Fig. 1-1 is released when away from equilibrium, elastic potential energy is converted to kinetic energy as the mass moves.

Angular Frequency, Period, and Frequency of SHM

If the mass of Fig. 1-1 is pulled aside from equilibrium to a position $x = A$ and released, how will its position x vary with time t? In a truly frictionless system, the mass will oscillate indefinitely between the positions $x = A$ and $x = -A$. The position x is a sinusoidal function of t, and can be expressed as

$$x = A\sin(\omega t) \qquad\qquad \textbf{[1-3]}$$

where ω is a constant[3] related to the force constant k and the mass m. The quantity A is the *amplitude* of the oscillation, which is the maximum distance that the mass moves away from the equilibrium position.

In the next few paragraphs the sinusoidal nature of the motion is justified, and ω is determined in terms of k and m. The starting point is $F_x = -kx$ (Eq. [1-1]). Newton's second law of motion can be used to replace F_x with ma_x, where a_x is the x-component of the acceleration. Hence, $ma_x = -kx$, or

---

2. As stated in the Preface, calculus is not required for students using this textbook. However, for students who do have a background in calculus, any comments or derivations using calculus are shown with a gray background and identified with the dy/dx icon.

3. ω is the lowercase Greek letter omega.

$$a_x = -\frac{k}{m}x \qquad\qquad \textbf{[1-4]}$$

Equation [1-4] indicates that, for a linear restoring force ($F_x = -kx$), the acceleration is equal to a negative constant ($-k/m$) times the position x. Acceleration is the time derivative of velocity, which in turn is the time derivative of position. Hence, acceleration is the second derivative of position, and Eq. [1-4] can be written as

$$\frac{d^2x}{dt^2} = -\frac{k}{m}x \qquad\qquad \textbf{[1-5]}$$

Equation [1-5] stipulates a condition that must be satisfied by the position function x for simple harmonic motion: its second derivative must equal a negative constant times the function itself. This condition is satisfied if position is a sinusoidal function of time (Eq. [1-3]), as will now be demonstrated. The left-hand side of Eq. [1-5] will be used. The second derivative of $x = A\sin(\omega t)$ with respect to time is to be determined. The first derivative of $x = A\sin(\omega t)$ gives velocity:

$$v_x = \frac{dx}{dt} = \frac{d}{dt}A\sin(\omega t) = A\omega\cos(\omega t) \qquad\qquad \textbf{[1-6]}$$

The time derivative of this velocity is the acceleration:

$$a_x = \frac{d^2x}{dt^2} = \frac{dv_x}{dt} = -A\omega^2\sin(\omega t) \qquad\qquad \textbf{[1-7]}$$

Since $x = A\sin(\omega t)$ (Eq. [1-3]), Eq. [1-7] can be rewritten as:

$$\frac{d^2x}{dt^2} = -\omega^2(A\sin(\omega t)) = -\omega^2 x \qquad\qquad \textbf{[1-8]}$$

Equation [1-8] confirms that the second derivative of the sinusoidal function $x = A\sin(\omega t)$ equals a negative constant times the function. A comparison of the right-hand sides of Eqs. [1-8] and [1-5] shows that the constants ω^2 and k/m must be equal, or:

$$\omega = \sqrt{\frac{k}{m}} \qquad\qquad \textbf{[1-9]}$$

What has been shown is that if the acceleration and position of an object are related by Eq. [1-4], $a_x = -(k/m)x$, then the position x can be written as a function of time t as $x = A\sin(\omega t)$, with ω given by Eq. [1-9]. Many types of oscillations in biological systems, such as the the vibrations of the tympanic membrane in the human ear, can be characterized very well by this mathematical representation.

The constant ω is called the *angular frequency* of the oscillation, and its SI unit is radians per second (rad/s). A discussion of the radian as a unit of angular measure is presented in Box 1-2. Since angular frequency is normally expressed in radians per second, the product ωt, for example in $x = A\sin(\omega t)$, has units of radians. It is important to remember to set your calculator to "radians mode" when evaluating trigonometric functions involving ωt.

Box 1-2   Radians — A Unit of Angular Measure

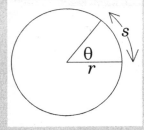

In the accompanying figure, the angle θ subtends an arc of length s in a circle of radius r. The angle θ in *radians* is defined as the ratio of the arc length to the radius, that is, $\theta = s/r$. Since both s and r are distances measured in, say, metres, then θ has no real units and is said to be *dimensionless*. (See Appendix 2 for information on dimensional analysis.) However, the unit of radian (abbreviated "rad") is often written for an angle defined in this way to distinguish it from an angle measured in degrees. If an angle has no unit stated, then radian is the default unit. Hence, sin(4) means the sine of four radians; sin(4°) is the sine of four degrees.

If the angle θ is 360°, then the arc length s is one circumference of the circle, that is, $s = 2\pi r$. In radians, the angle is then $\theta = s/r = 2\pi r/r = 2\pi$. Thus, the conversion between radians and degrees is:

$$2\pi \text{ radians} = 360°, \text{ or}$$
$$\pi \text{ radians} = 180°$$

Dividing both sides of the above equation by π gives 1 radian = 57.3°.

Since angles are dimensionless quantities, the unit of radian (or degree) is included in units where appropriate and omitted where it is not needed. The two equations labeled as [1-10] serve to illustrate this point. Suppose that a period T is known (in seconds) and ω is calculated using $\omega = 2\pi/T$. The unit for ω is thus seconds^{-1}; however, the unit of radian is then inserted, giving radians per second. Conversely, if ω is known in radians per second, and T is calculated using $T = 2\pi/\omega$, the unit for T would be seconds per radian, but the radian is omitted, giving seconds.

The angular frequency is closely related to the *period*, T, of the oscillation, which is the time for one complete oscillation or cycle of the particle. For example, one period is the time to travel from $x = A$ to $x = -A$ and back to $x = A$, as shown in Fig. 1-4. During a time interval of one period, the argument (ωt) of the sine function in $x = A \sin(\omega t)$ must increase by 2π radians in order that the value of x return to the value it had at the beginning of the period. This statement can be expressed as an equation: $\Delta(\omega t) = 2\pi$. Because ω is constant, this equation can be written as $\omega \Delta t = 2\pi$. Since the time interval is one period, $\Delta t = T$, and thus, $\omega T = 2\pi$. This relation is usually expressed as:

$$\omega = \frac{2\pi}{T} \quad \text{or} \quad T = \frac{2\pi}{\omega} \qquad \textbf{[1-10]}$$

Figure 1-4

Displacement of a simple harmonic oscillator as a function of time.

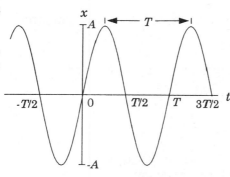

The number of complete oscillations per second is the *frequency*, *f*, and is given by the reciprocal of the period *T*:

$$f = \frac{1}{T} \quad \text{or} \quad T = \frac{1}{f} \qquad \textbf{[1-11]}$$

For example, if the period of an oscillation is ¼ s, the particle will make four complete oscillations per second, and the frequency *f* is 4 s^{-1}. The unit of frequency, s^{-1}, is given the name hertz (Hz).[4] Thus, the frequency in the above case is normally written as 4 Hz, although it could also be written as 4 oscillations/s or 4 vibrations/s.

Angular frequency ω can be related to frequency *f* by combining Eqs. [1-10] and [1-11]:

$$\omega = 2\pi f \qquad \textbf{[1-12]}$$

Since the angular frequency ω for a mass *m* attached to a spring having a force constant *k* is equal to $(k/m)^{1/2}$ (Eq. [1-9]), the expressions for period *T* and frequency *f* given in Eqs. [1-10] and [1-11] can be expressed in terms of *k* and *m* as:

$$T = 2\pi\sqrt{\frac{m}{k}} \quad \text{and} \quad f = \frac{1}{2\pi}\sqrt{\frac{k}{m}} \qquad \textbf{[1-13]}$$

---

4.   The German physicist Heinrich Hertz (1857–1894) was one of the original workers in the field of electromagnetic waves.

EXAMPLE 1-1

The vibration of the tympanic membrane in the ear is essentially simple harmonic motion. A tympanic membrane having a mass of 2.4×10^{-5} kg is vibrating with a frequency of 550 Hz.

(a) What is the force constant associated with the tympanic membrane?

(b) What are the angular frequency and period of the vibrations?

SOLUTION

(a) Values of m and f are given, and k is required. Equation [1-13] relates these three variables:

$$f = \frac{1}{2\pi}\sqrt{\frac{k}{m}}$$

Squaring this equation, and rearranging to solve for k gives:

$$k = 4\pi^2 f^2 m$$
$$= 4\pi^2 (550 \text{ Hz})^2 (2.4 \times 10^{-5} \text{kg})$$
$$= 2.9 \times 10^2 \text{ N/m}$$

(b) Angular frequency ω is related to frequency f by Eq. [1-12]:

$$\omega = 2\pi f = 2\pi(550 \text{ Hz}) = 3.5 \times 10^3 \text{ rad/s}$$

Since the frequency is given, the period can be calculated from Eq. [1-11]:

$$T = \frac{1}{f} = \frac{1}{550 \text{ Hz}} = 1.8 \times 10^{-3} \text{s}$$

Position, Velocity and Acceleration in SHM

An equation for position as a function of time has already been presented as Eq. [1-3] for simple harmonic motion:

$$x = A\sin(\omega t)$$

Since $\omega = 2\pi/T$ (Eq. [1-10]), the above expression can be rewritten as:

$$x = A\sin\left(\frac{2\pi t}{T}\right) \qquad \textbf{[1-14]}$$

Alternatively, the argument of the sine function can be written in terms of frequency f by using $T = 1/f$ (Eq [1-11]) to give:

$$x = A\sin(2\pi ft) \qquad \textbf{[1-15]}$$

Whichever of the above three equations is used for position, the various arguments of the sine function—ωt, $2\pi t/T$, and $2\pi ft$—are all angles expressed in radians. The quantities ω, $2\pi/T$, and $2\pi f$ may be considered as multiplicative factors that convert the variable time t to an angle that increases linearly with t.

Of course, $t = 0$ can be chosen to be at any point in the cyclic motion of the particle. The general graph of x vs. t will still be sinusoidal as in Fig. 1-4, but with the sine curve shifted anywhere up to ½ period ($T/2$) to the left or right of the time origin, with the exact shift depending on just where $t = 0$ is chosen to be. The equation of such a shifted sine curve is:

$$x = A\sin(\omega t + \delta) \qquad \textbf{[1-16]}$$

where δ is a dimensionless constant[5] called the *phase angle*. The exact value of δ depends on the point in the cycle where t is chosen to be zero, i.e., it depends on the amount of shift in the sine curve. It is normally simplest to use the $t = 0$ position shown in Fig. 1-4, in which case $\delta = 0$ and the sine function is unshifted.

Another simple choice is to let $t = 0$ when the particle is at $x = A$. Substituting $t = 0$ and $x = A$ into Eq. [1-16] gives $A = A\sin(\delta)$, from which $\sin(\delta) = 1$, and $\delta = \pi/2$. The graph of x vs. t is then the sine curve of Fig. 1-4 shifted ¼ period to the left, which is the same as a

---

5. δ is the lowercase Greek letter *delta*.

standard cosine curve. In this case, the analytical relation between x and t is $x = A\sin(\omega t + \pi/2)$, which can also be written $x = A\cos(\omega t)$.

Objects that oscillate with the same period and pass through $x = 0$ at the same time, traveling in the same direction, are said to oscillate *in phase* since their phase angles must be identical. However, the amplitudes can be different, as shown for objects 1 and 2 in Fig. 1-5.

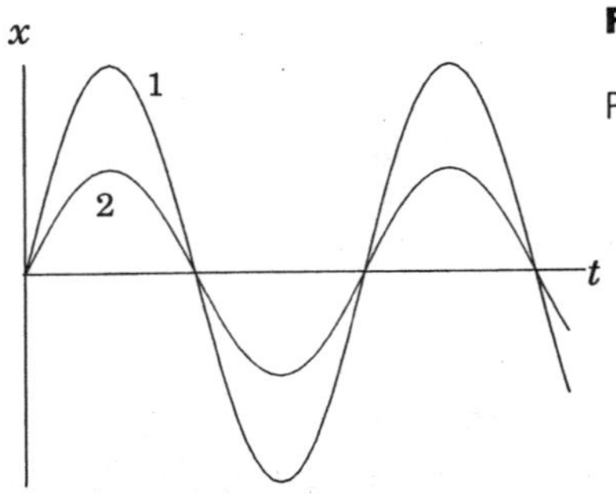

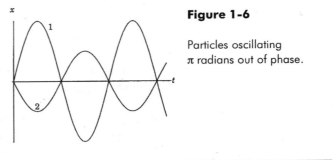

Figure 1-5

Particles oscillating in phase.

If one object is at its maximum positive position when the other is at its negative maximum, as in Fig. 1-6, their phase angles differ by π radians or 180°, and they are said to oscillate 180° or π radians out of phase. All phase differences from 0 to $\pm\pi$ radians are possible.

Figure 1-6

Particles oscillating π radians out of phase.

As an object oscillates back and forth, its velocity changes with time t. If position is written as $x = A\sin(\omega t)$ (Eq. [1-3]), then it has already been shown, using calculus, that the velocity is $v_x = A\omega\cos(\omega t)$ (Eq. [1-6]). This sinusoidally varying velocity is largest ($\pm A\omega$) at the times when the object passes through the equilibrium position. You already know this from experience—the motion of a person swinging on a playground swing is approximately simple harmonic, and the person travels fastest when passing through the central part. The maximum *speed* (magnitude of velocity) is:

$$v_{max} = A\omega \text{ (at the equilibrium position)} \qquad \textbf{[1-17]}$$

The velocity is zero at those times when the object is farthest from equilibrium, that is, when $x = \pm A$; on a swing, you come to rest instantaneously at the end points of your swing.

The acceleration also varies sinusoidally with time: $a_x = -A\omega^2\sin(\omega t)$ (Eq. [1-7]). Notice that the negative sign in this equation means that the acceleration and position are π radians out of phase with each other. The acceleration is zero as the object passes through $x = 0$ and is largest ($\pm A\omega^2$) at the extreme positions $x = \mp A$ (when x is negative, the acceleration is positive and vice versa.)

This is in agreement with the linear restoring force relationship $F_x = -kx$ and Newton's second law of motion $F_x = ma_x$, since acceleration a_x is zero where $F_x = 0$ (at $x = 0$), and where $|F_x|$ is a maximum the acceleration has a maximum magnitude, given by:

$$a_{max} = A\omega^2 \text{ (at the end points } x = \pm A) \qquad \text{[1-18]}$$

EXAMPLE 1-2

The wind is causing a leaf to vibrate, undergoing simple harmonic motion with a period of 0.40 s and an amplitude of 1.2 cm. At time $t = 0$ the leaf is passing through its equilibrium position ($x = 0$) and is traveling in the $+x$-direction (eastward).

(a) What is the equation relating x (in metres) and t (in seconds)?

(b) Sketch a graph of x vs. t.

(c) At time $t = 0.72$ s, how far is the leaf from the equilibrium position? In which direction?

SOLUTION

(a) Since $x = 0$ at $t = 0$ and the velocity (slope of x vs. t graph) is positive, then x is related to t by a sine function with no phase shift (Eq. [1-3]):

$$x = A \sin(\omega t)$$

Since the period T is given, use $\omega = 2\pi/T$ (Eq. [1-10]) to rewrite the above equation (or simply start with Eq. [1-14]):

$$x = A \sin\left(\frac{2\pi t}{T}\right)$$

Since x is requested in metres, the amplitude must be converted to metres:

$$A = 1.2 \text{ cm} \times \frac{1 \text{ m}}{100 \text{ cm}} = 0.012 \text{ m}$$

Then Eq. [1-14] becomes:

$$x = (0.012) \sin\left(\frac{2\pi t}{0.40}\right) = (0.012) \sin(5.0\pi t)$$

where x is in metres and t is in seconds.

Figure 1-7

Solution for Example 1-2(b).

(b) Figure 1-7 shows the sinusoidal function that at $t = 0$ has $x = 0$ and a positive velocity (i.e., slope), and for which $A = 0.012$ m and $T = 0.40$ s.

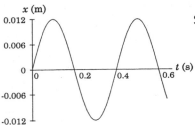

(c) The equation relating x and t was developed in part (a); it is rewritten here with all the units included:

$$x = (0.012 \text{ m}) \sin[(5.0 \, \pi \, \text{s}^{-1})t]$$

Substitution of $t = 0.72$ s gives:

$$\begin{aligned}
x &= (0.012 \text{ m}) \sin[(5.0 \, \pi \, \text{s}^{-1})(0.72 \text{ s})] \\
&= (0.012 \text{ m}) \sin(11.3) \\
&= (0.012 \text{ m})(-0.951) \\
&= -0.011 \text{ m}
\end{aligned}$$

Thus,[6] the leaf is 0.011 m from the equilibrium position, and since the $+x$-direction was stated to be eastward, the negative sign indicates that the leaf is to the west of equilibrium.

EXAMPLE 1-3

A rubber raft is bobbing up and down beside a dock, undergoing simple harmonic motion. The equation $x = 0.30 \sin(5.0t + 0.40)$ describes the simple harmonic motion of the raft, where x represents the vertical position relative to equilibrium, and all distances are in metres and times in seconds. What are the amplitude, frequency, and period of the motion?

---

6. Recall that sin(11.3) means the sine of 11.3 radians. Notice also that the intermediate results (11.3 and −0.951) have three significant digits to avoid round-off error.

SOLUTION

By comparing $x = 0.30 \sin(5.0t + 0.40)$ with $x = A \sin(\omega t + \delta)$ (Eq. [1-16]), it is apparent that the amplitude A is 0.30 m and the angular frequency ω is 5.0 rad/s. Equation [1-12] can then be used to determine the frequency f:

$$\omega = 2\pi f \qquad \therefore f = \frac{\omega}{2\pi} = \frac{5.0 \text{ rad/s}}{2\pi} = 0.80 \text{ Hz}$$

To calculate the period, Eq. [1-11] is used:

$$T = \frac{1}{f} = \frac{1}{0.80 \text{ Hz}} = 1.3 \text{ s}$$

Simple Harmonic Motion and Walking

As a person walks, the foot swings back and forth with a motion that is similar to the simple harmonic motion of a pendulum. The equations already developed for simple harmonic motion allow the maximum speed and acceleration of the foot to be calculated. The speed with which the foot moves during walking is surprisingly large when compared to the forward speed of the body as a whole.

Figure 1-8

Analysis of walking using a simple pendulum.

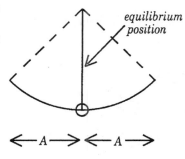

The motion of a simple pendulum (Fig. 1-8) is familiar to everyone. The pendulum swings about an equilibrium position with a period T, which is the time for one complete cycle. The amplitude A can be approximated as the maximum horizontal distance from the equilibrium position of the pendulum bob. The pendulum comes momentarily to rest at the extremities of the swing, and has its maximum speed when it crosses the equilibrium position. The acceleration at any time is roughly proportional to the distance from the equilibrium position. In walking, every swing of the moving foot is equivalent to one-half cycle of simple harmonic motion because the foot starts from rest, is accelerated forward at a rate roughly proportional to its distance from the centre of the swing, passes the stationary foot at high speed, and comes to rest as it touches the ground.

EXAMPLE 1-4

A man who is walking briskly takes 120 steps per minute, each step being 0.80 m in length. What are the maximum speed and maximum acceleration of the foot?

SOLUTION

Equations for the maximum speed and acceleration during simple harmonic motion were developed in the previous subsection (Eq. [1-17] and [1-18]):

$$v_{max} = A\omega \quad \text{and} \quad a_{max} = A\omega^2$$

As indicated in Fig. 1-8, the amplitude A is equal to one-half of the step length, i.e., $A = 0.40$ m. The angular frequency ω is not given; what is provided instead is the information that there are 120 steps per minute, i.e., each step takes 0.50 s. Since one step is only one-half of a complete oscillation, the period T is the time for taking <u>two</u> steps. Thus, $T = 1.0$ s. It is easiest now to proceed by rewriting Eqs. [1-17] and [1-18] in terms of T instead of ω, by substituting $\omega = 2\pi/T$ (Eq. [1-10]). Thus, the maximum speed is

$$v_{max} = \frac{2\pi A}{T} = \frac{2\pi(0.40 \text{ m})}{1.0 \text{ s}} = 2.5 \text{ m/s}$$

Note that this speed is large compared to the forward speed of the body as a whole, which is

$$v = \frac{\text{distance}}{\text{time}} = \frac{120(0.80 \text{ m})}{60 \text{ s}} = 1.6 \text{ m/s}$$

The maximum acceleration from Eq. [1-18] is

$$a_{max} = \frac{4\pi^2 A}{T^2} = \frac{4\pi^2(0.40 \text{ m})}{(1.0 \text{ s})^2} = 16 \text{ m/s}^2$$

Note that this acceleration is also large—approximately 1.6 times the acceleration due to gravity.

Damped and Forced Harmonic Motion and Resonance

In any real macroscopic oscillating system there is some friction. Thus, such systems do not oscillate indefinitely if simply pulled aside and released. The friction causes the amplitude to decrease with time, although it has very little effect on the period. This *damped harmonic motion* is illustrated by the position vs. time graph in Fig. 1-9. If the friction is very large relative to the restoring force, the particle will not oscillate at all but will simply return slowly to its equilibrium point, as shown in Fig. 1-10. It is only in atomic systems that friction is absent and that simple harmonic motion continues indefinitely.

The simple harmonic oscillations that have been discussed so far are free oscillations of a system at the frequency f determined by the mechanical properties of mass m and force constant k, as given by $f = (1/2\pi)(k/m)^{1/2}$ (Eq. [1-13]). However, it is possible to attempt to drive a system into oscillation at any frequency by applying a driving force that varies periodically with time. The resulting oscillations are referred to as *"forced" harmonic motion*.

Figure 1-9

Damped harmonic motion with small damping.

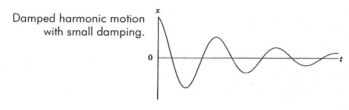

Figure 1-10

Damped harmonic motion with large damping.

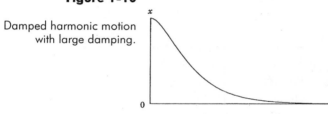

It is found that it is quite difficult to force a system to vibrate with large amplitude except at or very near its free oscillation frequency, i.e., the frequency given by Eq. [1-13]. If the system is driven at its free oscillation frequency, oscillations of very large amplitude can easily be built up. This phenomenon is called *resonance,* and the free oscillation frequency is often referred to as the *resonant frequency*. A child's swing is a common example. It has a natural oscillating or swinging frequency and it is hard to push it back and forth at any other frequency. However, gentle pushes applied at the natural oscillating frequency can easily cause the swing to swing back and forth with a very large amplitude.

1.3 **Traveling Waves**

We are all familiar with waves traveling across the surface of water. In such waves, the disturbance moves forward across the water although the water particles move approximately up and down (except where the water is very shallow) at one location. The water does not move forward although the wave or disturbance does. This may be verified by watching a floating object bob up and down at one location as waves move forward beneath it. A *traveling wave* may be defined as a disturbance which moves through space carrying energy without the bulk forward movement of matter.

Sound is another example of wave motion and is discussed in more detail in Chapter 2. Sound waves differ from water waves in one important way. In water waves, the particles oscillate in a direction (up and down) perpendicular to the direction (horizontal) that the wave travels. Such waves are called *transverse waves*. On the other hand, in a sound wave, the particles oscillate back and forth in the same direction as the wave as a whole travels. Such waves are called *longitudinal waves*.

Light consists of (transverse) oscillations of electric and magnetic fields, and is therefore called an *electromagnetic wave*. These waves are discussed in detail in Chapter 3. They differ from sound and water waves in that the oscillations of material particles are not involved. Consequently, they can readily carry energy through space that is completely void of material particles. For example, light from the sun travels through empty space between the sun and Earth.

Equation of a Traveling Wave

Suppose that a wave is traveling along the $+x$-axis. The symbol y will be used to designate the size of the wave disturbance (i.e., the displacement from equilibrium) at a position x along the wave at a time t. In a water wave, y would represent the vertical displacement ($+$up, $-$down) of the particle from its equilibrium position (the position it would have if the water were perfectly flat). The displacement y varies with both x and t.

The relation of y to x and t may be illustrated graphically, as shown in Fig. 1-11. In these graphs and in the subsequent discussion it is assumed that the waves are sinusoidal in shape. Although other shapes are commonly

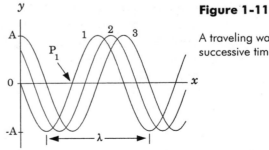

Figure 1-11

A traveling wave at three successive times (1, 2, 3).

found in nature, the sinusoidal waves are the simplest and most important. It can be shown that any periodic wave can be represented as the superposition of sinusoidal waves, and hence a sinusoidal wave will be considered as the fundamental model wave.

The curve labeled "1" in Fig. 1-11 is a graph of y versus x at one particular instant in time ($t = 0$, say). The wave is sinusoidal in shape and repeats itself after a particular length called the *wavelength* of the wave. The wavelength is the distance between two successive *crests* or two successive *troughs*. The usual symbol for wavelength is the lowercase Greek letter λ *(lambda)*, and the SI unit for wavelength is the metre (m).

If the wave is traveling in the $+x$-direction, with the crests moving along with some speed v, then the graph of y versus x will change with time, as the crests move to the right along the x-axis. Curve "2" of Fig. 1-11 represents the wave at some short time after $t = 0$ (curve "1"), and curve "3" occurs a short time later than "2." Each of these curves is essentially a snapshot of the wave at one particular instant of time.

As the sinusoidal wave moves along, the disturbance y at any particular point at a fixed value of position x (such as point P_1 in Fig. 1-11) oscillates in simple harmonic motion about its equilibrium position. The variation of y with time for point P_1 is given in Fig. 1-12. It is important to recognize the difference between what is being plotted in Fig. 1-11 and 1-12. Each curve in Fig. 1-11 is a graph of y vs. x for a constant value of t, whereas Fig. 1-12 shows a graph of y vs. t for a constant value of x. The graphs of y vs. t for other constant x-values would be similar to Fig. 1-12, but shifted along the time axis.

Figure 1-12

Simple harmonic motion of a particle at constant position x along a wave.

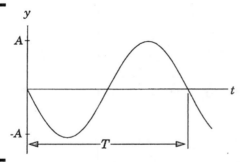

As shown in Fig. 1-12, the disturbance oscillates through one complete cycle in time T, which is the *period of the wave*. The reciprocal of the period is the *frequency of the wave, f.* During one period T, the wave moves forward one wavelength λ, so the frequency of the wave is also the number of wavelengths which pass a given point per second. For example, if the period is ⅓ s, one wavelength takes ⅓ s to pass any given constant x-point. Therefore, in one second three wavelengths pass any given point, and the wave frequency is $3 \text{ s}^{-1} = 3$ Hz.

The sinusoidal relationship between the disturbance y, the position x and the time t, illustrated graphically by the various curves of Fig. 1-11 and 1-12, can also be expressed analytically, i.e., by an equation. What is required is an equation in which y is a sinusoidal function of x for constant t, and y is also a sinusoidal function of t for constant x. Such an equation is:

$$y = A\sin\left(\frac{2\pi}{T}t - \frac{2\pi}{\lambda}x\right) \qquad \textbf{[1-19]}$$

This equation corresponds to the wave shown in Fig. 1-11. At time $t = 0$, Eq. [1-19] becomes $y = A\sin(-2\pi x/\lambda)$, which is graphed as curve "1" in Fig. 1-11. As t takes on increasing positive values, the wave moves to the right, i.e., in the $+x$-direction (curves "2" and "3").

For a wave traveling in the $-x$-direction, the equation is:

$$y = A\sin\left(\frac{2\pi}{T}t + \frac{2\pi}{\lambda}x\right) \qquad \textbf{[1-20]}$$

Notice that the only difference between Eqs. [1-19] and [1-20] is the sign preceding $2\pi x/\lambda$. The quantity $2\pi/\lambda$ in these equations is the *wave number* or *wave vector* and is usually represented by the symbol k. (Note that the wave vector k is not the same as the force constant k discussed earlier in this chapter.)

$$k = \frac{2\pi}{\lambda} \qquad \textbf{[1-21]}$$

Since λ is a length, k is an inverse length, and its SI unit is metre^{-1} (m^{-1}). When multiplied by the position x, the wave vector generates a dimensionless quantity ($2\pi x/\lambda$) which may be considered as an angle in radians.

Equations [1-19] and [1-20] for traveling waves may be written in numerous other forms. In the most useful form, $2\pi/\lambda$ is replaced by the wave vector k, and $2\pi/T$ is replaced by the angular frequency ω:

$$y = A\sin(\omega t \pm kx) \qquad \textbf{[1-22]}$$

In this text, Eq. [1-22] will be used as the standard equation for a traveling wave.[7]

7.　The traveling wave equation can also be written with the x and t terms in reverse order. For example, $y = A\sin(kx - \omega t)$ represents a wave traveling in the $+x$-direction just as does Eq. [1-19]. The difference is simply in the choice of the wave's position at $t = 0$.

Wave Speed

The speed at which a traveling wave moves can be expressed in terms of wavelength and period. Speed is distance divided by time, and since it takes one period T for the wave to travel a distance of one wavelength λ, the wave speed v is just λ/T. Since $f = 1/T$ (Eq. [1-11]), the speed can also be written as $v = \lambda f$. Thus,

$$v = \frac{\lambda}{T} = \lambda f \qquad \textbf{[1-23]}$$

The speed of a wave depends largely on the properties of the medium through which it moves. Sound waves travel at a particular speed in air at a given temperature but at a different speed in water. In addition, the wave speed may vary slightly with the wave frequency.

The frequency and period of a wave are determined by the wave source. Once a wave is "launched," its frequency normally remains constant, but as the wave passes from one medium to another, the speed and hence the wavelength change (in accordance with Eq. [1-23]).

EXAMPLE 1-5

A wave moves along a string in the $+x$-direction with speed of 8.0 m/s, a frequency of 4.0 Hz, and an amplitude of 0.050 m. What are the

(a) wavelength? (b) wave number?

(c) period? (d) angular frequency?

(e) Determine an equation for this wave.

(f) What is the value of y for the point at $x = \frac{3}{4}$ m at time $t = \frac{1}{8}$ s?

SOLUTION

The given parameters for this wave are speed $v = 8.0$ m/s, frequency $f = 4.0$ Hz, and amplitude $A = 0.050$ m.

(a) Since $v = \lambda f$ (Eq. [1-23]),
     wavelength $\lambda = v/f = (8.0 \text{ m/s})/(4.0 \text{ Hz}) = 2.0$ m.

(b) Using Eq. [1-21],
     wave number $k = 2\pi/\lambda = (2\pi)/(2.0 \text{ m}) = \pi \text{ m}^{-1}$.

(c) From Eq. [1-11], period $T = 1/f = 1/(4.0 \text{ Hz}) = 0.25$ s.

(d) Equation [1-12] gives angular frequency
$\omega = 2\pi f = 8\pi$ rad/s.

(e) A general equation for a wave traveling in the
+x-direction is given by Eq. [1-22]:

$$y = A \sin(\omega t - kx)$$

Since A, ω, and k have already been determined, this equation
becomes

$$y = 0.050 \sin(8\pi t - \pi x)$$

where y and x are in metres and t is in seconds.

(f) To determine the value of y for given values of t and x, the
equation determined in part (e) is used. It is rewritten
here with all the units included:

$$y = (0.050 \text{ m}) \sin[(8\pi \text{ rad/s})t - (\pi \text{ m}^{-1})x]$$

Substitution of $x = \frac{3}{4}$ m and time $t = \frac{1}{8}$ s gives:

$$
\begin{aligned}
y &= (0.050 \text{ m}) \sin[(8\pi \text{ rad/s})(\tfrac{1}{8} \text{ s}) - (\pi \text{ m}^{-1})(\tfrac{3}{4} \text{ m})] \\
&= (0.050 \text{ m}) \sin[(\pi - 3\pi/4)] \\
&= (0.050 \text{ m}) \sin(\pi/4) \\
&= 0.035 \text{ m}
\end{aligned}
$$

EXAMPLE 1-6

The equation of a traveling wave is $y = 4.0 \sin(2.0t + 3.0x)$
where y and x are in metres and t in seconds. What are the
wave's (a) amplitude? (b) wavelength? (c) period and
frequency? (d) velocity (speed and direction)?

SOLUTION

By comparison with the general wave equation
$y = A \sin(\omega t + kx)$ (Eq. [1-22]), it follows that:

(a) amplitude $A = 4.0$ m

(b) wave vector $k = 3.0 \text{ m}^{-1} = 2\pi/\lambda$ (Eq. [1-21])

Thus, wavelength $\lambda = 2\pi/(3.0\ \text{m}^{-1}) = 2.1\ \text{m}$

(c)   angular frequency $\omega = 2.0\ \text{rad/s} = 2\pi/T$ (Eq. [1-10])

Hence, $T = 2\pi/(2.0\ \text{rad/s}) = \pi\ \text{s}$

frequency $f = 1/T$ (Eq. [1-11]), and thus $f = 1/(\pi\ \text{s}) = 0.32\ \text{Hz}$

(d)   speed $v = \lambda f$ (Eq. [1-23]),
      and hence $v = (2.1\ \text{m})(0.32\ \text{Hz}) = 0.67\ \text{m/s}$

As discussed in the paragraphs describing Eqs. [1-19] to [1-22], a wave expressed by an equation of the form $y = A\sin(\omega t + kx)$ is traveling in the negative x-direction. (If the equation is $y = A\sin(\omega t - kx)$, the wave is traveling in the positive x-direction.)

Thus, the velocity is 0.67 m/s in the negative x-direction.

EXAMPLE 1-7

A traveling wave is described by the equation $y = 3\sin[(\pi/2)t - \pi x]$, where all distances are in metres and time is in seconds. Sketch the following graphs for this wave:

(a) y vs. x at time $t = 0$

(b) y vs. t at position $x = 0$

(c) y vs. x at time $t = 0.8$ s.

SOLUTION

(a)   Substitution of $t = 0$ into the equation for this wave gives:

$$y = 3\sin(-\pi x) = -3\sin(\pi x)$$

Figure 1-13

Solution for Example 1-7(a).

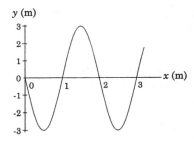

This is the equation of a wave with amplitude $A = 3$ m and wave number $k = \pi\ \text{m}^{-1}$. The wavelength is $\lambda = 2\pi/k = 2\pi/(\pi\ \text{m}^{-1}) = 2$ m. The negative sign in the equation indicates that as the wave is plotted from $x = 0$ to $x > 0$, a trough is encountered first. This wave is sketched in Fig. 1-13. Notice that the repeat unit is one wavelength (2 m).

(b) Substitution of $x = 0$ into the equation for the wave gives:

$$y = 3 \sin[(\pi/2)t]$$

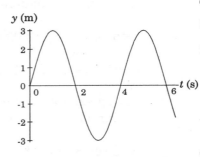

Figure 1-14

Solution for Example 1-7(b).

The amplitude is still 3 m, and the angular frequency is $\omega = (\pi/2)$ rad/s. Hence, the period is $T = 2\pi/\omega = 2\pi/[(\pi/2)\text{ rad/s}] = 4$ s. The graph of y vs. t is shown in Fig. 1-14. Note that the repeat unit is one period (4 s).

(c) The equation obtained by substituting $t = 0.8$ s into the equation for the wave is:

$$y = 3 \sin[(\pi/2)(0.8) - \pi x]$$
$$= 3 \sin(0.4\pi - \pi x)$$

This equation could be plotted by substituting several values of x and calculating the corresponding y-values. However, there are other ways that are less laborious. Since the graph of y vs. x at $t = 0$ has already been plotted in part (a), it is easiest simply to determine how far the wave has moved from its position at $t = 0$ in the given time of 0.8 s.

First, calculate what fraction of a period the time of 0.8 s represents. Since the period is 4 s, then 0.8 s is (0.8 s)/(4 s) = 0.2 (or 1/5) of a period. In this time, the wave will advance $0.2\,\lambda$ in the $+x$-direction. (Recall that in one period, the wave advances one wavelength.) Since one wavelength is 2 m, then $0.2\,\lambda$ is 0.4 m. Hence, the

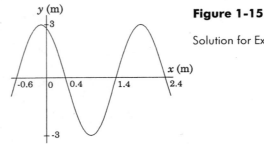

Figure 1-15

Solution for Example 1-7(c).

required graph is similar to the one at $t = 0$ (same amplitude and wavelength), except that the wave is shifted 0.4 m in the $+x$-direction. This graph is shown in Fig. 1-15; compare it with the wave at $t = 0$ (Fig. 1-13).

1.4 Standing Waves

The last type of wave to be considered is a *standing wave*, or *stationary wave*. These waves are fairly common, although perhaps not always obvious. They are found, for example, on guitar strings, in organ pipes and sometimes in buildings. Standing waves are important in many biological systems, such as in the auditory canals of animals.

A standing wave is produced when two waves of the same wavelength travel in opposite directions through the same medium. Very often one wave is simply the reflection of the other wave from a surface. The two waves interfere with each other, and by the *principle of wave superposition*, the resultant wave displacement y is simply the sum of the displacements, y_1 and y_2, of the two interfering waves. Thus, when a crest from one wave is at the same position as the crest from the other wave, the waves are said to undergo *constructive interference*, and the result is a "supercrest" having a displacement that is the sum of the displacements of the two crests. Similarly, when two troughs interfere, constructive interference produces a "supertrough."

Where a crest interferes with a trough, the two waves undergo *destructive interference*, and the result is a small displacement. If the crest and trough have the same size, then destructive interference produces a complete cancellation of the crest and trough, giving a net displacement of zero.

Equation for a Standing Wave

An equation for a standing wave can be developed by adding the equations for two waves traveling in opposite directions. Consider one wave to be traveling in the $+x$-direction, and the second wave (traveling in the $-x$-direction) to be the reflection of the first wave from a surface. Represent the first wave by the standard equation:

$$y_1 = A \sin(\omega t - kx) \qquad \textbf{[1-24]}$$

When a traveling wave is reflected from a surface at which the net vibration must be zero (for example, at a fixed end of a string), it undergoes a phase change of π radians. Thus, the wave changes not only in direction, but also in phase. Mathematically, the change in direction can be accommodated by changing the negative sign in the argument $(\omega t - kx)$ of the sine function in Eq. [1-24] to a positive sign; the phase change is indicated by introducing a negative sign in front of the amplitude A. Hence, the equation of the reflected wave is

$$y_2 = -A \sin(\omega t + kx) \qquad \textbf{[1-25]}$$

These two waves add together to produce the standing wave. Thus,

$$y = y_1 + y_2 = A\,[\sin(\omega t - kx) - \sin(\omega t + kx)] \qquad \textbf{[1-26]}$$

This equation can be simplified by using the trigonometric identity:

$$\sin(\theta - \phi) - \sin(\theta + \phi) = -2\cos(\theta)\sin(\phi)$$

In our case, $\theta = \omega t$ and $\phi = kx$, hence Eq. [1-26] becomes

$$y = A\,[-2\cos(\omega t)\sin(kx)]$$

or

$$y = -\,[2A\cos(\omega t)]\sin(kx) \qquad \textbf{[1-27]}$$

Equation [1-27] represents a wave that has a constant wavelength $\lambda = 2\pi/k$. The quantity in brackets $[2A\cos(\omega t)]$ can be considered as an amplitude that varies with time. The maximum amplitude is $2A$, and the minimum amplitude is zero. Such a wave is shown in Fig. 1-16 for a case where the period ($T = 2\pi/\omega$) is 12 s. The shape of the wave is shown second by second over the whole period. At time $t = 0$ s, $\cos(\omega t) = 1$, and Eq. [1-27] becomes $y = -2A\sin(kx)$; hence, the wave at this time is a negative sine function with an amplitude of $2A$. The curve labeled "12" in Fig. 1-16 represents the function at $t = 0$ s, $t = 12$ s, $t = 24$ s, etc. The other curves represent subsequent positions at times $t = 1$ s, 2 s, 3 s, etc.

In Fig. 1-16, notice that there are positions where there is zero displacement at all times (*nodes*), and other positions where the maximum displacement occurs at various times (*antinodes*). Note also that the distance from one node to the next node (or from antinode to antinode) is one-half wavelength ($\lambda/2$).

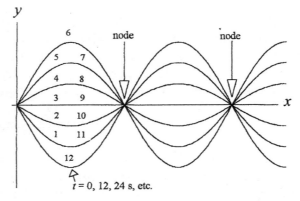

Figure 1-16

A standing wave.

EXAMPLE 1-8

A traveling wave is represented by the equation
$y_1 = 0.10 \sin(2\pi t - 1.57x)$, with all distances in metres and time
in seconds. The wave reflects from a surface at which the net
vibration must be zero.

(a) What is the equation of the reflected wave?

(b) What is the equation of the resulting standing wave?

(c) Sketch the standing wave at time $t = 0.30$ s.

SOLUTION

(a) As discussed in the paragraph leading to Eq. [1-25], the
equation of the reflected wave is
$y_2 = -0.10 \sin(2\pi t + 1.57x)$

(b) The equation of the standing wave is
$y = -0.20 \cos(2\pi t) \sin(1.57x)$, as shown in the discussion
leading to Eq. [1-27].

(c) At $t = 0.30$ s, the standing wave equation becomes

Figure 1-17

Solution for Example 1-8(c).

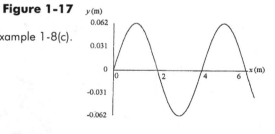

$$y = -0.20 \cos(0.60\pi) \sin(1.57x)$$
$$= -(0.20)(-0.309) \sin(1.57x)$$
$$= 0.062 \sin(1.57x)$$

Hence, at $t = 0.30$ s, the standing wave
is a sine function having an amplitude
of 0.062 m and a wavelength of
$\lambda = 2\pi/k = 2\pi/(1.57 \text{ m}^{-1}) = 4.00$ m.
This wave is sketched in Fig. 1-17.

Standing Waves on a String

A string fixed at both ends, such as a guitar string, provides a good example of standing waves. Because each end of the string is fixed and hence unable to vibrate, there must be a node there. A number of possible standing-wave patterns have a node at each end. The simplest pattern has no nodes between the end ones, and is shown in the upper drawing of Fig. 1-18. Other patterns have 1, 2, 3, or more nodes between the end nodes; two examples are shown in the bottom two drawings of Fig. 1-18.

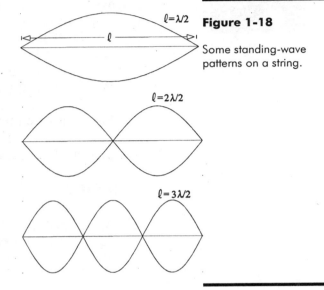

Figure 1-18

Some standing-wave patterns on a string.

When a guitar string is plucked, all possible standing-wave patterns with nodes at the ends are produced. Normally the simplest one has the largest amplitude and corresponds to the musical note (fundamental tone, or first harmonic) that is heard. The other patterns that are present are the overtones or higher musical harmonics.

Standing waves on a string fixed at both ends occur only for waves of specific wavelengths. Since the distance from one node to the next node is $\lambda/2$, the length ℓ of the string must equal an integral number n of half-wavelengths:

$$\ell = \frac{n\lambda}{2} \text{ where } n = 1, 2, 3, \dots \qquad \textbf{[1-28]}$$

EXAMPLE 1-9

The length of the high-"E" string on a guitar is 0.627 m.

(a) What are the three longest wavelengths of standing waves on this string?

(b) The frequency of the fundamental tone (first harmonic) of the high-"E" string is 329.6 Hz. What is the wave speed along this string?

SOLUTION

(a) Rearranging Eq. [1-28] to solve for wavelength:

$$\lambda = \frac{2\ell}{n} \text{ where } n = 1, 2, 3, \ldots$$

The three longest wavelengths will occur for the three smallest values of n (1, 2, and 3). Substituting $n = 1$ gives $\lambda = 2\ell = 1.25$ m. Similarly, when n equals 2 or 3, the values of λ are 0.627 m or 0.418 m, respectively.

(b) The fundamental tone corresponds to the simplest standing-wave pattern, i.e., the one with the longest wavelength (1.25 m, from part (a)). Since the frequency is given, Eq. [1-23] can be used to determine the speed:

$$v = \lambda f = (1.25 \text{ m})(329.6 \text{ Hz}) = 412 \text{ m/s}$$

Standing Waves in Other Objects

In an organ pipe that is closed at one end and open at the other end, there is negligible sound vibration (thus, a node) at the closed end, and maximum vibration (an antinode) at the open end. The simplest standing-wave pattern that satisfies these conditions is shown in Fig. 1-19; one-quarter wavelength ($\lambda/4$) is contained in the length of the tube. The next simplest patterns contain odd multiples of quarter-wavelengths in the tube: $3\lambda/4$, $5\lambda/4$, etc. Hence,

Figure 1-19

Standing wave in an organ pipe.

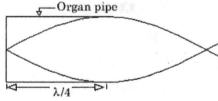

Organ pipe

$\lambda/4$

$$\ell = \frac{n\lambda}{4} \text{ where } n = 1, 3, 5, \ldots \qquad \textbf{[1-29]}$$

An interesting biological analogy to an organ pipe closed at one end is the auditory canal in animals, discussed further in Chapter 2.

Figure 1-20

Standing wave on a ring.

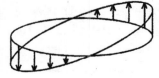

If standing waves are produced on a ring, such as in a bell, then the standing waves must join smoothly on themselves after traveling around one circumference. Thus, one circumference must equal a whole number of wavelengths. With the radius of the ring written as r, this condition is:

$$2\pi r = n\lambda \text{ where } n = 1, 2, 3, \ldots \qquad \textbf{[1-30]}$$

Figure 1-20 shows the standing-wave pattern around a ring for $n = 1$.

1.5   Beats

One last concept in waves and wave motion which should not be overlooked is the process of beating by waves of two or more frequencies. This effect can be readily experienced in sound waves, and is of value in tuning pianos, guitars, and other instruments. The series of clicks produced by the cricket is a sound which is thought to arise from the beating of two higher-frequency sounds.

To see how this beating process occurs, imagine that two sound waves of slightly different frequency are being received by your ear. These two waves are shown in Fig. 1-21a,b as graphs of displacement y vs. time t at some constant position x. Since your ear responds to the total sound displacement arriving at the eardrum, you are really experiencing the sum of these two waves, which is shown in Fig. 1-21c. The slight difference in frequencies produces constructive interference (loud sound) and destructive interference (faint sound) at regular intervals. As shown below, the *beat frequency*, which is the frequency at which the loud sounds occur, is equal to the frequency difference between the waves. The production of beats can have biological importance, as discussed in Box 1-3.

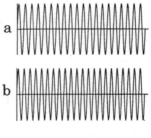

Figure 1-21

Two waves (a, b) of slightly different frequencies add to form a beat pattern (c).

Box 1-3   How Penguins Use Beats

Imagine being an emperor penguin trying to identify family members in a dense, noisy colony with up to 10 birds per square metre. The birds emit two distinct calls simultaneously, generating a unique beat pattern that serves as a vocal identity code that permits easy identification of parents and chicks. A penguin can produce two calls at once because its voicebox is in two parts.

To derive an expression for the beat frequency, start with traveling wave equations (Eq. [1-22]) for the two waves:

$$y_1 = A\sin(\omega_1 t - k_1 x)$$

$$y_2 = A\sin(\omega_2 t - k_2 x)$$

Notice that the waves have different angular frequencies (and hence different frequencies) and different wave numbers. Both waves are traveling in the $+x$-direction. To determine the resultant wave in Fig. 1-21c, what is needed is the sum of y_1 and y_2 as a function of time, at a constant position x. To make the mathematics as easy as possible, the constant x-value is chosen to be zero. (This choice does not affect the final expression for beat frequency.) Adding y_1 and y_2, with $x = 0$, gives

$$y_1 + y_2 = A[\sin(\omega_1 t) + \sin(\omega_2 t)]$$

This wave equation can be rewritten, using the trig identity:

$$\sin(\alpha) + \sin(\beta) = 2\,\sin\left(\frac{\alpha+\beta}{2}\right)\cos\left(\frac{\alpha-\beta}{2}\right)$$

Thus,

$$y_1 + y_2 = 2A\sin\left(\frac{(\omega_1+\omega_2)t}{2}\right)\cos\left(\frac{(\omega_1-\omega_2)t}{2}\right)$$

The above equation shows that the resultant wave has two dependencies on time. There is a sine portion having an angular frequency of $\frac{1}{2}(\omega_1 + \omega_2)$, i.e., the average angular frequency of the two waves. This sine-function oscillation corresponds to the closely spaced crests and troughs in Fig. 1-21c. Superimposed on this sine oscillation is a cosine function that has an angular frequency of $\frac{1}{2}(\omega_1 - \omega_2)$. The cosine oscillation corresponds to the more gradual rise and fall in the wave pattern in Fig. 1-21c.

The resultant sound wave is loudest (i.e., a beat is heard) when the cosine function has a value of $+1$ or -1, which occurs when its argument $[\frac{1}{2}(\omega_1 - \omega_2)t]$ equals 0, $\pm\,\pi$, $\pm\,2\pi$, etc. Stated another way, beats occur at time intervals Δt for which $\frac{1}{2}(\omega_1 - \omega_2)\Delta t = \pi$. Solving this equation for Δt, which is the *beat period*, gives

$$\Delta t = \frac{2\pi}{\omega_1 - \omega_2}$$

Using $\omega = 2\pi f$ (Eq. [1-12]) to rewrite angular frequencies in terms of frequencies:

$$\Delta t = \frac{2\pi}{2\pi(f_1 - f_2)} = \frac{1}{f_1 - f_2}$$

The beat frequency f_B is the reciprocal of the beat period Δt, and hence:

$$f_B = f_1 - f_2 \qquad \qquad \textbf{[1-31]}$$

Thus, as stated earlier in this section, the beat frequency equals the difference in the frequencies of the two waves.

Exercises

1-1   A spider web acts as an elastic membrane that vibrates when an insect first becomes trapped. An insect of mass 1.5 g is caught in a spider web that has an effective force constant of 7.2 N/m. What are the frequency and period of the resulting vibration?

1-2   After the insect of Exercise 1-1 hits the web, the web stretches a maximum distance of 3.0 mm. How much elastic potential energy is stored in the web at this position?

1-3   When the insect of the previous two questions first hits the web, assume that position $x = 0$ and time $t = 0$, and that the insect is traveling in the $+x$-direction. What is the equation expressing x as a function of t for the resulting SHM?

1-4   A box of candies of mass 0.50 kg, resting on a horizontal surface, is attached to a horizontal spring having a force constant of 4.8 N/m. The box is pulled 10.0 cm away from the equilibrium position ($x = 0$), and released from rest at time $t = 0$. Neglect friction.

   (a)   What are the amplitude and angular frequency of the resulting SHM?

   (b)   If the release position $x > 0$, what is the equation relating x and t?

   (c)   What is the maximum speed of the box? Where does this speed occur?

   (d)   What is the maximum magnitude of the acceleration of the box? Where does this acceleration occur?

   (e)   Sketch the graph of x vs. t.

1-5   A boat is bobbing up and down in a lake. Its vertical position x above the flat surface of the water is given by $x = 0.60 \sin(3.0t + 4.2)$, where distances are in metres and time t is in seconds.

   (a)   What is the boat's position at $t = 0$? at $t = 2.8$ s?

(b)   What is the period of the boat's oscillation?

(c)   What is the boat's maximum speed during the oscillation?

1-6   Consider the traveling wave: $y = 2.0 \sin(4.0t + \pi x)$, where x and y are measured in metres, and t is measured in seconds. What are the

(a)   amplitude? (b) wavelength? (c) period?

(d)   velocity (magnitude and direction)?

1-7   Show that the traveling wave equation $y = A \sin(\omega t - kx)$ can be written in the alternative forms:

(a) $y = A \sin[2\pi(ft - x/\lambda)]$ (b) $y = A \sin[k(vt - x)]$

Where A = amplitude
ω = angular frequency
k = wave number
f = frequency
λ = wavelength
v = speed

1-8   A wave moves along a string in the $+x$-direction at a speed of 8.33 cm/s with a period of 0.24 s and an amplitude of 10 cm.

(a)   Determine the frequency, angular frequency, and wavelength of the wave.

(b)   What is the equation of the wave?

Sketch the following graphs:

(c) y vs. x, at $t = 0$ s

(d) y vs. x, at $t = 0.16$ s

(e) y vs. t, at $x = 0.50$ cm.

1-9   The wave speed along the "B" string of a guitar is 322 m/s; the frequency of the fundamental tone of this string is 247 Hz.

(a)   How long is the string?

(b)   What are the two next highest frequencies of standing waves in this string?

1-10   The function $y = -0.70 \cos(4.0t) \sin(\pi x)$ describes a standing wave. If t is measured in seconds, and y and x in metres, determine the wave's

(a)   amplitude   (b)   period   (c)   wavelength.

1-11 Sketch the profile of the standing wave given by
$y = 3 \cos(2t) \sin(x)$ at times $t = 0$ and $t = T/4$, where T is the period of
oscillation (t in seconds, x and y in metres).

1-12 Two wave trains of the same frequency, speed, and amplitude are
traveling in opposite directions along a string:

$$y_1 = 2 \sin(3t - x)$$

$$y_2 = -2 \sin(3t + x)$$

where y_1, y_2, and x are in centimetres, and t is in seconds. Write the
equation of the resultant standing wave, and plot the three wave forms
at times $t = 0$, $T/4$, $T/2$, $3T/4$, and T, where T is the period.

1-13 Two tuning forks are set side-by-side and struck. One fork has a
frequency of 300 Hz. The other's frequency is unknown. Five beats per
second are heard. The same fork of unknown frequency is now struck
along with a fork of frequency 290 Hz. Five beats per second are heard
again. What is the unknown frequency?

Problem

1-14 A raft is bobbing up and down beside a dock with an amplitude of 0.40 m
and a period of 4.0 s. The equilibrium position of the raft is 0.45 m
below the dock. It is safe and comfortable to step onto the raft only
when it is no more than 0.35 m below the dock. How much time do
people have to step on to the raft safely during each period of the raft's
oscillation?

Answers

1-1 11 Hz, 0.091 s

1-2 3.2×10^{-5} J

1-3 $x = (3.0 \times 10^{-3}$ m$) \sin[(69$ rad/s$)t]$

1-4 (a) 10.0 cm, 3.1 rad/s
    (b) $x = (10.0$ cm$) \cos[(3.1$ rad/s$)t]$
    (c) 0.31 m/s, at $x = 0$
    (d) 0.96 m/s^2, at $x = \pm 10.0$ cm
    (e)

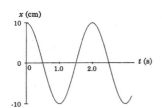

(d)

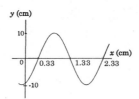

(e)

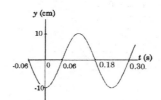

1-5 (a) -0.52 m, $+0.020$ m
    (b) 2.1 s
    (c) 1.8 m/s

1-6 (a) 2.0 m
    (b) 2 m
    (c) 1.6 s ($\pi/2$ s)
    (d) 1.3 m/s in $-x$-direction

1-8 (a) 4.2 Hz, 26 rad/s, 2.0 cm
    (b) $y = (10$ cm$) \sin[(26$ rad/s$)t - (\pi$ cm$^{-1})x]$
    (c)

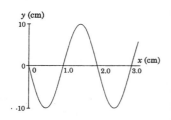

1-9 (a) 0.652 m
    (b) 494 Hz, 741 Hz

1-10 (a) 0.70 m
    (b) 1.6 s ($\pi/2$ s)
    (c) 2.0 m

1-11

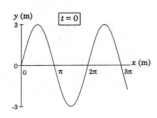

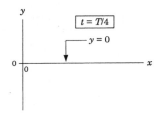

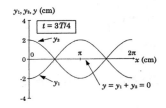

1-12 $y = -4 \cos(3t) \sin(x)$

1-13 295 Hz

1-14 1.7 s

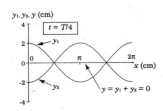

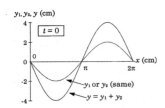

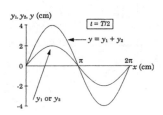

2 Sound, Hearing and Echolocation

2.1 Introduction

Sound is the result of *acoustic waves* propagating through a medium, such as air or water. Humans and many other species rely on the detection of sound, that is, on hearing, to provide information about the world around them. Hearing involves not only detecting how "strong" the sound is, but also what frequencies are present in the acoustic waves, and from what direction the waves are coming. All of this information may indicate the source of the sound, and what, if any, action the organism should take. The organism's ears carry out the task of extracting this information from the acoustic waves impinging on it. The emphasis in this chapter will be on human ears and hearing. Other mammals, bats for example, use sound and their ears for *echolocation*. Some species of bat can estimate the relative speed between them and their prey by using the *Doppler effect*. The physical basis of these uses of sound will also be discussed in this chapter.

2.2 The Nature of Sound

Sound (acoustic) waves occur in substances which are *elastic*. These substances can be compressed by an external force, and they recover from the compression once the force is removed. The molecules of these substances can be thought of as masses connected to one another by springs, so that if the external force moves one of the molecules, it will in turn cause its neighbours to move, each at a slightly later time (Fig. 2-1). After the force is removed, the molecules return toward their original positions.

Figure 2-1

Masses connected by springs.

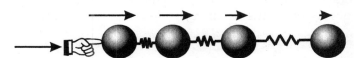

Sound waves are *longitudinal waves*, unlike waves on a string or on the surface of the ocean, which are transverse. This means that the displacements of the molecules of the medium in which the sound takes place are along the direction of propagation of the wave.

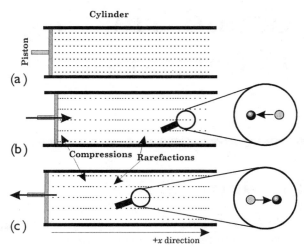

Figure 2-2

Generation of sound waves in a cylinder by a piston moving in SHM: (a) Piston not moving. All molecules at their equilibrium positions; (b) Piston moving, generating a traveling sound wave. Magnification showing molecules (black dot) in this region displaced to the left (in −x-direction) of their equilibrium positions (grey dot); (c) Piston moving. Magnification showing molecules in this region displaced to the right (in +x-direction) of their equilibrium positions.

Imagine a piston at the left end of a cylinder filled with air which can be moved back and forth in the horizontal direction in simple harmonic motion (SHM), as illustrated in Fig. 2-2. When the piston is not moving (as in Fig. 2-2a), the air molecules can be considered to be at rest,[1] and a molecule located at some particular distance x to the right of the piston is said to be at its *equilibrium position x*. When the piston is moving in SHM, the air molecules just to the right of the piston (located at $x = 0$) are moved from their equilibrium positions by the same amount as the piston. When the piston moves to the right (as in Fig. 2-2b) it decreases the volume available to the molecules next to it and so they move closer together (*compression*). When it moves to the left (Fig. 2-2c), it increases the volume available to the molecules next to it and so they move further apart, thus increasing the distance between them and the molecules farther to the right (*rarefaction*).

The elastic nature of air allows the compressions and rarefactions to move into the air in the cylinder to the right of the piston, so that, as the piston executes SHM, a sequence of compressions separated by rarefactions moves to the right through the cylinder. The piston moving in SHM generates a longitudinal sine wave in the displacement of the air molecules in the cylinder; that is, the piston is the source of a sound wave propagating to the right. The distance between consecutive compressions or rarefactions is the wavelength λ of the wave; the compressions occur at any point inside the cylinder with frequency f.

---

1. The molecules in any substance move about randomly within the substance at temperatures above 0 kelvin (K). For the purpose of describing waves in the substance, these random motions can be ignored.

The mathematical description of sound waves is the same as that for waves on a string or on the surface of water (Eq. [1-22]). Consider a molecule (equilibrium position x) vibrating about x sinusoidally in time with amplitude A and angular frequency $\omega = 2\pi f$ in a sound wave with wave vector $k = 2\pi/\lambda$. The position of the molecule at time t relative to its equilibrium position x is the *displacement* of the molecule[2] $\xi(t)$ (Fig. 2-3):

$$\xi(t) = A\sin(\omega t - kx) \qquad \text{[2-1]}$$

With respect to the source of the sound, the position $x(t)$ of the molecule at time t is just

$$\begin{aligned} x(t) &= x + \xi(t) \\ &= x + A\sin(\omega t - kx) \end{aligned} \qquad \text{[2-2]}$$

so that at each distance x from the piston, the molecules vibrate about their equilibrium position in SHM with the same angular frequency and amplitude, but with phases which depend on their equilibrium positions x.

Figure 2-3

Displacement $\xi(t)$ of molecules from their equilibrium positions in a traveling sound wave at one instant of time t.

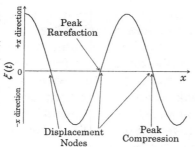

The speed v of propagation of the sound waves equals the product of the frequency f and wavelength λ (refer to Eq. [1-23]: $v = f\lambda$). This speed depends on the substance in which the sound is generated and on other quantities, such as the temperature of the substance. Values for the speed of propagation in various substances are given in Table 2-1, along with the densities of the substances.

Table 2-1   Density and Speed of Sound

Substance	Density ρ (kg/m³)	Speed v (m/s)
Dry air (17°C at standard pressure)	1.217	340
Water (20°C)	9.982×10^2	1.41×10^3
Seawater (20°C)	10.25×10^2	1.54×10^3

2. ξ is the lowercase Greek letter *xi*.

How does this mathematical expression (Eq. [2-2]) give the alternating compressions and rarefactions shown in Fig. 2-2, and what relation do they have to the peak displacements of the molecules? Suppose that, at a particular point x in the air at a particular time t, the molecules to the left of x are moving towards x (i.e., to the right, because the sine function in Eq. [2-2] for those molecules is positive), while molecules to the right of x are also moving towards x at time t (i.e., to the left, because the sine is negative for them). At x at time t, then, the air molecules are crowded together and thus have a higher density at x than if no wave were present. The air around x is said to be compressed, and there is a peak in the compression of the air molecules at x. Rarefaction of the air (lower density) occurs near a point at which molecules to the left are moving to the left, away from the point (the sine is negative for them), while those on the right are moving to the right, also away from the point (the sine is positive for them). At the point itself there is a peak rarefaction. Peak compressions and rarefactions thus occur at the *displacement nodes of the traveling wave*, that is, at places where the displacement of the molecules from their equilibrium positions is zero. Figure 2-3 summarizes the relation between the displacements of the air molecules and the compressions and rarefactions.

Changes in density are accompanied by changes in pressure.[3] A region of compression in a sound wave is thus also a region of high pressure, relative to the pressure in the medium when no sound waves are present. Correspondingly, a region of rarefaction in a sound wave is a region of pressure which is lower than that when no sound wave is present in the medium. Points at which peak compression occurs are points of maximum pressure, while points at which peak rarefaction occurs are points of minimum pressure. Between a point of maximum pressure and a point of minimum pressure in the wave is a point at which the pressure is equal to the pressure in the medium when no wave is present. Such a point is called a *pressure node*. Sound waves can thus be pictured as displacement waves, density waves, or pressure waves. Of course, they are all of these.

The maximum change in pressure can be related to the amplitude of the sound wave by thinking about what causes the medium to move back and forth as the sound wave passes through it. The simple harmonic motion of the medium at position x is described by Eq. [2-1], which gives the displacement from equilibrium ξ at x at time t. Any mass moving in SHM has an acceleration, and so there must be a force acting on the medium to accelerate it. Assume that a sound wave is traveling in the positive x-direction, and consider the mass m contained

3.   Pressure, which is the force perpendicular to the surface of an object per unit area of the surface, is discussed in Chapter 11.

Figure 2-4

Forces acting on a
mass *m* of air.

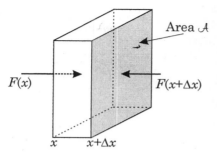

in a slab of the medium shown in Fig. 2-4. The mass is just the density ρ of the medium times the volume of the slab. The volume $\mathcal{V}$ equals the area $\mathcal{A}$ of the slab times the slab's thickness Δx ($\mathcal{V} = \mathcal{A}\,\Delta x$). This mass has an acceleration a in the positive x-direction; the acceleration is the rate of change of its velocity v, and its velocity v in turn is just the rate of change of its displacement ξ. In other words, for mass m in the slab,

$$ma = \rho\mathcal{A}\Delta x\; a = \rho\mathcal{A}\Delta x\; dv/dt = \rho\mathcal{A}\Delta x\; d^2\xi/dt^2$$

The total force acting on the slab equals ma. The force $F(x)$ on the left face of the slab is the pressure at x in the medium times the area $\mathcal{A}$ of the face, and this force is in the positive x-direction. Similarly, the force on the right face is the pressure at $x+\Delta x$ times the area $\mathcal{A}$ of the right face; it is in the negative x-direction. The pressure equals the air pressure when there is no wave present in the medium plus the change in pressure ΔP associated with the sound wave. Thus, the total force acting on the slab in the positive x-direction at time t is equal to

$$\Delta P(x)\mathcal{A} - \Delta P(x+\Delta x)\mathcal{A}$$

and this is equal to ma:

$$\rho\mathcal{A}\Delta x\; d^2\xi/dt^2 = \Delta P(x)\mathcal{A} - \Delta P(x+\Delta x)\mathcal{A}$$

Both sides of this equation can be divided by the volume $\mathcal{V}$ of the slab ($= \mathcal{A}\Delta x$), with the following result:

$$\rho d^2\xi/dt^2 = -\,(\Delta P(x+\Delta x) - \Delta P(x))/\Delta x$$

As the thickness Δx of the slab is allowed to approach zero, the right-hand side approaches

$$-d\Delta P/dx$$

i.e., the negative of the derivative of ΔP with respect to position x. The relationship between the back-and-forth motion (described by ξ) and the pressure difference ΔP is given by

$$\rho d^2\xi/dt^2 = -d\Delta P/dx \qquad\qquad \text{[2-3]}$$

which is the *equation of motion* of the medium.

The relationship between the maximum change in pressure and the amplitude in simple harmonic motion is obtained by applying the equation of motion when a sound wave passes through the medium. The velocity and acceleration in simple harmonic motion (see Eq. [2-1]) are given by

$$v = d\xi/dt = d(A \sin(\omega t - kx))/dt = A\omega \cos(\omega t - kx) \qquad \textbf{[2-4]}$$

and

$$d^2\xi/dt^2 = dv/dt = -A\omega^2 \sin(\omega t - kx) = -\omega^2 \xi \qquad \textbf{[2-5]}$$

The discussion about the relation between the location of peak compression (peak pressure) and location of a displacement node (see Fig. 2-3) indicates that the pressure ΔP and displacement ξ are 90° out of phase in the position x. Thus, instead of the sine function, the pressure change depends on the cosine function:

$$\Delta P = \Delta P_{Max} \cos(\omega t - kx)$$

The maximum change in pressure in the pressure wave is denoted by ΔP_{Max}. Consequently,

$$-d\Delta P/dx = -d(\Delta P_{Max} \cos(\omega t - kx))/dx = -\Delta P_{Max} k \sin(\omega t - kx) \qquad \textbf{[2-6]}$$

Substitution of Eq. [2-5] and [2-6] into Eq. [2-3] yields

$$\Delta P_{Max} = \rho(\omega^2/k)A$$

Since the speed of propagation of sound v is equal to ω/k (see Eqs. [1-12], [1-21], and [1-23]),

$$\Delta P_{Max} = \rho v \omega A \qquad \textbf{[2-7]}$$

Equation [2-7] says that the maximum change in pressure in a sound wave is proportional to the amplitude A, which is the maximum change in displacement. The proportionality involves the product ρv, the *acoustic resistance* R_A of the medium. This quantity is useful in understanding the function of the middle ear (see Box 2-2).

The combination ωA occurring in Eq. [2-7] is just the maximum velocity v_{Max} in simple harmonic motion (see Eq. [2-4]). Equation [2-7] is thus equivalent to the relation

$$\Delta P_{Max} = R_A v_{Max} \qquad \textbf{[2-8]}$$

In other words, the peak change in pressure in a sound wave is equal to the peak velocity in the simple harmonic motion of the medium times the acoustic resistance. This relationship is reminiscent of Ohm's law in electricity (Chapter 15).

EXAMPLE 2-1

A sound wave travels in seawater (20°C) in the $+x$-direction. The frequency of the wave is 1000 Hz. Suppose that at $t = 0$ s there is a peak compression at $x = 0$ m. Locate other peak compressions and rarefactions in x.

SOLUTION

Peak compressions are spaced multiples of a wavelength λ apart. Peak rarefactions are located between peak compressions and are separated from them by $\lambda/2$. From Eq. [1-23] the wavelength of this wave $\lambda = v/f = 1.54 \times 10^3$ m·s^{-1}/1000 Hz = 1.54 m (v from Table 2-1). Thus, peak compressions are located at $x = \ldots, -4.62, -3.08, -1.54, 0, +1.54, +3.08, +4.62$ m, $\ldots$, while peak rarefactions are at $x = \ldots, -3.85, -2.31, -0.77, +0.77, +2.31, +3.85, \ldots$.

EXAMPLE 2-2

A sound wave in air (17°C at standard pressure) with frequency 300 Hz and amplitude equal to 0.1 μm travels in the $+x$-direction. A peak compression passes the point $x = 0$ m at $t = 0$ s.

(a) Plot the displacement ξ as a function of x at $t = 0$ s.

(b) Where is that peak compression 0.001 s later?

SOLUTION

(a) Refer to Chapter 1 for methods of plotting traveling waves. The wavelength of the sound $\lambda = v/f = 340$ m·s^{-1}/300 Hz = 1.13 m. As shown in Fig. 2-3, a peak compression occurs at a displacement node to the left of which the displacement ξ is to the right, i.e., positive (the sine function is positive), and to the right of which the displacement ξ is to the left, i.e., negative (the sine function is negative). At $t = 0$ then, ξ is as shown in Fig. 2-5.

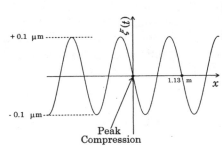

Figure 2-5

Solution for Example 2-2(a).

(b) Plot ξ as a function of x at $t = 0.001$ s.

In 0.001 s, each part of the wave has moved a distance $d = vt = 340$ m·s⁻¹ × 0.001 s = 0.34 m to the right, and so the peak compression which was at $x = 0$ m at $t = 0$ s is at $x = 0.34$ m at $t = 0.001$ s. Thus ξ as a function of x at $t = 0.001$ s is as shown in Fig. 2-6.

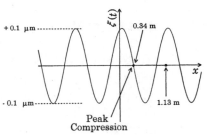

Figure 2-6

Solution for Example 2-2(a).

EXAMPLE 2-3

A sound wave in dry air at 17°C has frequency $f = 1000$ Hz and amplitude $A = 1.0 \times 10^{-11}$ m. Calculate the maximum pressure difference in the air due to the wave.

SOLUTION

Equation [2-7] relates the maximum pressure difference ΔP_{Max} to the density of air and the speed of propagation at this temperature (see Table 2-1):

$$\Delta P_{\text{Max}} = \rho v \omega A$$

$$= 1.2 \text{ kg·m}^{-3} \times 340 \text{ m·s}^{-1} \times 2\pi \times 1000 \text{ Hz} \times 1.0 \times 10^{-11} \text{ m}$$

$$= 2.6 \times 10^{-5} \text{ N·m}^{-2}$$

The unit of pressure (force per unit area) is the Newton per square metre; this unit is called the *pascal*[4] (P). As a point of reference, normal atmospheric pressure at sea level is approximately equal to 1.0×10^5 P, so that this maximum pressure difference is only about **260 billionths** of the pressure of the atmosphere.

2.3   Acoustic Resonance

Standing sound waves can be set up in a medium, just as standing transverse waves can exist on a string, as discussed in Chapter 1. The music produced by wind instruments and organ pipes is the result of standing waves in the air inside the circular cylindrical tubes forming these instruments. All wind instruments (flutes, clarinets, oboes, etc.) are based on tubes with circular cross sections and varying lengths, which are open to the atmosphere at both ends. The pipes of the organ can be open at both ends, or open at one end and closed at the other end. The latter are called *stopped pipes*.

Consider the prototype of every wind instrument, a circular tube of length ℓ shown in Fig. 2-7. Because the tube is open at both ends, the air pressure at the ends[5] is equal to atmospheric pressure, and thus every standing wave in the tube must have a pressure node at each end, since the pressure at a pressure node is just the pressure of the atmosphere. Since a pressure node is associated with no compression or rarefaction of the air, where there is maximum displacement, this is equivalent to saying that the standing wave must have a displacement antinode at each end.

Figure 2-7

The first two harmonics in a pipe open at both ends.

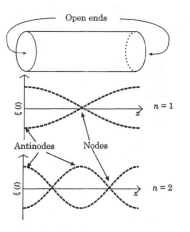

The simplest standing wave which can be set up in such a tube is one with a displacement antinode (pressure node) at each end and a displacement node (pressure antinode) halfway along the tube (Fig. 2-7). The length of the tube is thus equal to one half-wavelength;

----

4.   Blaise Pascal (1623–1662), French physicist.
5.   Strictly speaking, the pressure node is slightly beyond the ends of the tube by a distance which depends on the diameter of the tube. This effect will be ignored here.

it is the *first harmonic* of the tube (harmonic number $n = 1$). This standing wave thus has a wavelength equal to twice the length of the tube (see Fig. 2-7):

$$\lambda_1 = 2\ell$$

Similarly, the *second harmonic* has a displacement antinode at each end and one halfway along the tube, so that its wavelength is equal to the length of the tube:

$$\lambda_2 = \ell$$

A standing wave with harmonic number n has one displacement antinode at each end and n nodes spaced equally along the length of the tube. Its wavelength is given by

$$\lambda_n = \frac{2\ell}{n} \quad \text{with } n = 1, 2, 3, \ldots \qquad \textbf{[2-9]}$$

Compare this with the case of standing waves on a string discussed in Chapter 1. The frequencies of the harmonics can be calculated from Eq. [2-9] using the relation $v = f\lambda$:

$$f_n = \frac{nv}{2\ell} \qquad \textbf{[2-10]}$$

The harmonic structure of the sound produced by a stopped organ pipe, which is open at one end and closed at other end (Fig. 2-8), differs from that of the wind instruments and organ pipes open at both ends. Since sound waves are longitudinal, the air molecules cannot move from their equilibrium position at the closed end, since they cannot move into the material closing the end of the tube. A displacement node must therefore be present at the closed end of the tube. Just as in the case of the wind instruments, a standing wave in an organ pipe will have a displacement antinode at the open end of the tube. Here, then, the simplest standing wave has a displacement node at the closed end and a displacement antinode at the open end, with no other nodes or antinodes in between. The distance between a node and a neighbouring antinode is one quarter-wavelength. The first harmonic in a stopped organ pipe thus has a wavelength which equals four times the length of the tube:

$$\lambda_1 = 4\ell$$

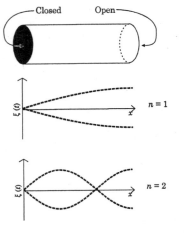

Figure 2-8

The first two harmonics in a pipe closed at one end.

The second harmonic has an additional displacement node within the tube so that three-quarters of a wave is in the tube, i.e., the length of the tube is 3/4 of the wavelength, or

$$\lambda_2 = \frac{4\ell}{3}$$

The nth harmonic has $n - 1$ displacement nodes within the tube, together with one at the closed end and a displacement antinode at the open end, so that its wavelength is given by

$$\lambda_n = \frac{4\ell}{(2n-1)} \qquad n = 1, 2, 3, \ldots \qquad \textbf{[2-11]}$$

The frequencies of the harmonics are thus

$$f_n = \frac{(2n-1)v}{4\ell} \qquad \textbf{[2-12]}$$

EXAMPLE 2-4

(a) How long does a hollow cylinder open at both ends have to be to make a Pan pipe with fundamental frequency (first harmonic) at 256 Hz in air at 17°C and standard pressure?

(b) At what frequency will its second harmonic be?

SOLUTION

(a) Using Eq. [2-10] (the frequency of the harmonics of the open–open tube): for the first harmonic, $f_1 = v/(2\ell)$, so that 256 Hz = 340 m·s⁻¹/(2ℓ). Solve for ℓ to find ℓ = 0.66 m.

(b) Using Eq. [2-10] gives $f_2 = 2v/(2\ell)$, so that $f_2 = 2f_1$ = 512 Hz. When a frequency is twice another frequency, it is said to be an *octave* higher than the lower frequency.

EXAMPLE 2-5

(a) How long does a hollow cylinder have to be to make an organ pipe (open at one end and closed at the other) with a fundamental frequency of 128 Hz in air at 17°C and standard pressure?

(b) What is the frequency of its second harmonic?

SOLUTION

(a) For $n = 1$ for the open–closed tube, Eq. [2-12] gives
$f_1 = (2 \times 1 - 1)v/(4\ell) = v/(4\ell)$. Thus, $128 \text{ Hz} = 340 \text{ m·s}^{-1}/(4\ell)$,
so that $\ell = 0.66$ m. This is the same length as in
Example 2-4, but the fundamental frequency is a factor of
2 (an octave) lower here because one end is closed.

(b) For $n = 2$, Eq. [2-12] gives $f_2 = (2 \times 2 - 1)v/(4\ell) = 3v/(4\ell)$
$= 3f_1 = 384$ Hz.

Standing waves in tubes are examples of *acoustic resonances*.
Acoustic resonances can occur in other situations in which standing
sound waves can be set up. A bare room with all windows and doors
closed is an example of an acoustic resonator. Standing waves can be
set up between opposite walls, each of which acts like the closed end of
a stopped organ pipe. A sound wave generated by a source in the room
will sound particularly loud if the wavelength matches one of the
possible standing waves which can be set up in the room.

2.4   Energy, Power, and Intensity

Sound waves arise as the result of some source, like the piston in
Fig. 2-2, causing a disturbance in the medium in which the sound wave
propagates. It requires energy (SI unit is joules [J]) to move the
medium, and that energy comes from the source. This energy E is
carried away from the source in the form of the wave. The rate at which
the source delivers energy to the medium is called the power P of the
source; this is the power carried away by the wave. If the rate at which
energy is produced by the source is constant, the power is simply the
ratio of the total energy E produced in time t divided by the time t:

$$P = \frac{E}{t} \qquad\qquad \textbf{[2-13]}$$

The SI unit of power is joules per second, which is a watt (W).

EXAMPLE 2-6

A sound amplifier and loudspeaker system can produce a peak of 20 W of acoustic power. How much acoustic energy will this system put into the air if its output is kept at 20 W for 1.0 h?

SOLUTION

From Eq. [2-13], the acoustic energy $E = Pt = 20$ W $\times$ 1.0 h $= 20$ J/s $\times$ 60 min $\times$ 60 s$\cdot$min$^{-1} = 7.2 \times 10^4$ J

The energy delivered to the medium by the source goes into the energy of the molecules as they execute SHM in the wave. The total energy of a molecule $E_{molecule}$ executing SHM is just the sum of its kinetic energy due to its speed v:

$$K = \tfrac{1}{2} m_{molecule}\, v^2$$

and its potential energy (Eq. [1-2])

$$U = \tfrac{1}{2}\, kx^2$$

The kinetic and potential energies oscillate between 0 and their maximum values during the molecule's motion. When the potential energy is at its maximum ($\tfrac{1}{2} kA^2$, with A the amplitude of SHM), the kinetic energy is 0 and *vice versa*. The total energy $E_{molecule}$, however, is constant during the motion. Thus, for example, $E_{molecule}$ equals the maximum value of the molecule's kinetic energy, when the molecule is traveling at its maximum speed v_{max}. According to Eq. [1-17], $v_{max} = A\omega$, so that

$$E_{molecule} = \tfrac{1}{2} m_{molecule}\, A^2 \omega^2 \qquad \textbf{[2-14]}$$

This describes the energy carried by the wave as the energy per molecule. An alternative description is in terms of the energy per unit volume, or *energy density* $E_\circ$ (in joules per cubic metre [J/m^3]) of the wave. The energy density is equal to the energy per molecule multiplied by the number of molecules per unit volume $n_\circ$:

$$E_{\backsim} = E_{\text{molecule}} n_{\backsim}$$
$$= \tfrac{1}{2} m_{\text{molecule}} A^2 \omega^2 n_{\backsim} \qquad \textbf{[2-15]}$$
$$= \tfrac{1}{2} \rho A^2 \omega^2$$

where ρ,[6] the mass density of the medium, equals $m_{\text{molecule}} \, n_{\backsim}$.

The energy density is carried by the wave at the speed of propagation of the wave v. The power (energy per unit time, Eq. [2-13]) carried by the wave across unit area is called the *intensity (I)* of the wave ($I = P/\mathcal{A}$) and has the SI unit of watts per square metre (W/m^2). The intensity can be calculated using the energy density of the wave. Imagine a surface with area $\mathcal{A}$ perpendicular to the direction of the propagation of the wave. Suppose that a peak compression arrives at that surface at time = 0 s (Fig. 2-9). At time t later, the compression has moved a distance $d = vt$ to the right of the surface (Fig. 2-9), forming a volume $\mathcal{V} = \mathcal{A}d$. A total energy E_{total} has been transported into this volume by the sound wave equal to

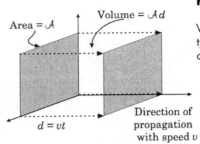

Figure 2-9

Volume swept out by a traveling wave front of area $\mathcal{A}$.

$$E_{\text{total}} = E_{\backsim} \mathcal{V}$$
$$= E_{\backsim} \mathcal{A} v t$$

The intensity I is the energy crossing the surface per unit area of the surface per unit time

$$I = E_{\text{total}}/(\mathcal{A}\, t)$$

so that

$$I = E_{\backsim} v \qquad \textbf{[2-16]}$$

Equation [2-16] says that the intensity I of the wave is the energy density times the speed of propagation. Using this equation, the intensity can be expressed in terms of the amplitude of oscillation and the angular frequency by substituting Eq. [2-15] into Eq. [2-16] to find

$$I = \tfrac{1}{2}\rho \, A^2 \omega^2 v \qquad \textbf{[2-17]}$$

6. ρ is the lowercase Greek letter *rho*.

In terms of the frequency $f = \omega/2\pi$ (Eq. [1-12]), the intensity I is given by

$$I = 2\pi^2 \rho v f^2 A^2 \qquad \textbf{[2-18]}$$

The relationship between the maximum change in pressure ΔP_{Max} and the amplitude A (Eq. [2-7]) can be used in Eq. [2-17] to express the intensity of the sound wave in terms of ΔP_{Max}:

$$I = \Delta P_{Max}^2 /(2\rho v) \qquad \textbf{[2-19]}$$

EXAMPLE 2-7

For a sound of frequency 1000 Hz, the intensity at the threshold of hearing in humans is 1.0×10^{-12} W/m². What is the amplitude of motion of the molecules in air (17°C; standard pressure) in this case?

SOLUTION

Equation [2-18] relates intensity I, amplitude A, and frequency f. From Table 2-1, the density ρ of air at the stated conditions is 1.2 kg/m³, to two significant figures, and the speed v of sound propagation is 340 m/s. Thus,

1.0×10^{-12} W/m² $= 2\pi^2 \times$ 1.2 kg/m³ $\times$ 340 m/s $\times$ (1000 Hz)² $\times A^2$

Since 1 W = 1 J/s and 1 J = 1 kg·m²/s², 1 W/m² = 1 kg/s³. Therefore,

1.0×10^{-12} kg/s³ $= 8.1 \times 10^9$ kg·m^{-2}·s$^{-3} \times A^2$

Thus, the amplitude of oscillation $A = 1.1 \times 10^{-11}$ m at the threshold of hearing, which is a small fraction of the diameter of the smallest atom.

EXAMPLE 2-8

Sounds are produced in water with various frequencies. The sounds are detected by an underwater microphone which reveals that all the sounds have the same intensity. Which of the graphs in Fig. 2-10 shows the qualitative dependence of the amplitude A of the sound on the frequency f of the sound?

SOLUTION

Since the intensities of all of the underwater sounds detected are equal, Eq. [2-18] implies that $f^2 A^2 = $ constant, or $A = $ constant$/f$. The answer is Fig. 2-10b.

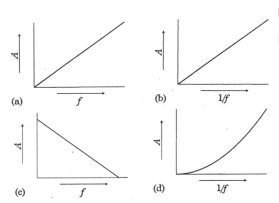

Figure 2-10

Example 2-8.

2.5 Dependence of Intensity on Distance

In situations such as that shown in Fig. 2-2, the sound wave travels in one direction only, in the *x*-direction in that particular case. Such waves are called *plane waves*, and the intensity of such a wave is a constant. In other cases, the wave moves in more than one direction away from the source, and the intensity depends on the distance from the source. An important special case occurs when the source pushes and pulls on the air equally in all directions, as a "breathing" balloon will do, whose radius increases and decreases regularly in time. If the radius of the balloon varies sinusoidally in time about an equilibrium value, the balloon is the source of a sinusoidal sound wave which is spherically symmetric.[7] These are called *spherical waves*. In spherical sound waves, the points at which peak compressions or peak rarefactions occur form spherical surfaces centred on the source.

The intensity of a spherical wave depends on the distance from the source. Recall that the intensity I of a wave is the power which the wave carries across a surface of unit area. The total power carried by the wave P_{wave} across a spherical surface a distance r from the centre is just

$$P_{\text{wave}} = I\,(4\pi r^2) \qquad \qquad \textbf{[2-20]}$$

because the surface area of a sphere of radius r is equal to $4\pi r^2$. It will be assumed in this text that the wave is not *attenuated*, i.e., that there

---

7.   The wave seems to emanate from a point at the centre of the balloon.

are no energy losses due to friction between air molecules (viscosity, discussed in Chapter 12) or from any other cause. Thus, the power P_{wave} carried across the spherical surface a distance r from the source is exactly equal to the power P_{source} delivered to the medium by the source. In this case, the intensity I at a distance r from a spherical source is given by

$$I = \frac{P_{source}}{4\pi r^2} \qquad \qquad \textbf{[2-21]}$$

The intensity I is inversely proportional to the square of the distance, or $Ir^2 = constant$. For this reason, I is said to "obey the *inverse square law*" (Fig. 2-11). Equation [2-21] can be used as a good approximation in cases in which the source and the waves it produces are not spherically symmetric, as long as the distance r is very large (> 10 times) compared to the size of the source.

Figure 2-11

Basis of the inverse square law.

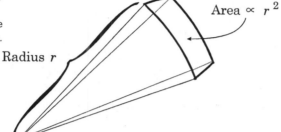

Radius r

Area $\propto r^2$

EXAMPLE 2-9

An underwater source of sound produces spherical waves in a lake. The intensity of the sound 5.0 m from the source is 7.0×10^{-3} W/m².

(a) What is the intensity 2.0 m from the source?

(b) What is the total acoustic power of the source?

SOLUTION

(a) Equation [2-21] says that the intensity I is inversely proportional to r^2. In this example, the intensity I_1 is known at distance r_1 from the source, and it is required to calculate I_2 at a distance r_2 from the source. The inverse square law relates I_1 and I_2 in the following way: $I_2 r_2^2 = I_1 r_1^2$.
Here, $I_2 =$ intensity at 2.0 m, and $I_1 = 7.0 \times 10^{-3}$ W/m² at 5.0 m. Thus, $I_2 \times (2.0 \text{ m})^2 = 7.0 \times 10^{-3}$ W/m² $\times (5.0 \text{ m})^2$, which gives $I_2 = 7.0 \times 10^{-3}$ W/m² $\times (25/4) = 4.4 \times 10^{-2}$ W/m².

(b) From Eq. [2-21],
$$P_{source} = (4\pi r^2)I = 4\pi \times (5\text{ m})^2 \times 7.0 \times 10^{-3} \text{ W/m}^2 = 2.2 \text{ W}.$$

Note that one could calculate I_2 in part (a) by first finding P_{source} and then using Eq. [2-21] with $r = r_2$. The method used here emphasizes the essence of the inverse square law, namely, that if the distance increases by a factor, the intensity decreases by that factor squared.

2.6   The Ear and Hearing

The ear is the organ which detects sound waves and produces neural signals which are interpreted by the brain, giving the organism an internal *(psychophysical)* representation *(perception)* of the physical characteristics of the sound. For example, the intensity of the sound is perceived as *loudness* and the frequency as *pitch*. The pitch of a pure tone is a familiar perception, and some humans have no difficulty in associating the perception with a particular frequency of sound *(perfect pitch)*, while most humans can easily determine the relation of one pitch to another.

Humans can also gauge the loudness of a sound and its relation to the loudness of another sound. It turns out that humans respond to a large range of intensities, from a threshold intensity of about 1.0×10^{-12} W/m^2 to intensities causing pain or damage to the ear (above about 1.0×10^{-2} W/m^2). This is a range covering 10 orders of magnitude.

How is this possible? In the 19th century, Weber[8] found that the ear distinguishes very well between low-intensity sounds, but does not respond well to differences in high-intensity sound. For low-intensity sounds, the just-noticeable difference in intensity between two sounds is small; for high-intensity sounds this difference is large. That the ear should respond this way makes sense from the perspective of self-preservation: it may be very important to notice if a weak sound is just a little louder than it was, but less important with sounds which are loud. These studies led to the conclusion that perceived loudness increases not as the intensity, but rather approximately as the *logarithm* of the intensity, so that loudness is roughly proportional to $\log(I)$ (see Box 2-1).

---

8.   Ernst Heinrich Weber (1795–1878), German physiologist. The discovery was popularized by German physicist Gustav Theodor Fechner (1801–1887), and is known as the *Weber-Fechner law*.

On the basis of this observation, the *intensity level* scale was developed. The intensity level of a sound of intensity *I*, denoted by *IL*, is defined as

$$IL = 10 \log(I/I_0) \qquad \textbf{[2-22]}$$

Box 2-1 The Weber-Fechner Law and the Intensity Level Scale

The intensities of two sounds of the same frequency sounded one after the other can be distinguished if the intensities differ sufficiently. The minimum difference in the intensities which can be distinguished is δI (the just-noticeable difference). The results of Weber's psychophysical research led him to the conclusion that δI is approximately proportional to the average intensity *I* of the two sounds, that is, that the higher the intensities of the sounds, the larger the difference between the intensities must be for them to be distinguished:

$$\delta I \propto I, \text{ or } \delta I / I \approx \text{constant.}$$

Weber's research thus suggested that the perception of intensity depends on intensity in such a way that the just-noticeable difference in the intensity of two sounds divided by the average intensity of the two sounds is a constant. In other words, the internal scale by which humans gauge intensity cannot be arbitrarily finely divided: the minimum difference in loudness is approximately a constant, different from zero, which is characteristic of the human sensory system of hearing.

The *intensity level scale* is based on Weber's observations: the intensity level *IL* is chosen to be a function of intensity with the property that

$$dIL = dI / I,$$

where the differential "d" represents an infinitesimal difference. In this scale, then, just-noticeable differences in intensity correspond to an approximately constant difference in the intensity levels of the two sounds, independent of how high the intensities of the sounds are. What function of intensity *I* has this property? If the equation relating d*IL* to d*I* is rewritten as

(cont'd.)

$$\mathrm{d}IL/\mathrm{d}I = 1/I,$$

it becomes clear that IL is the function of intensity whose derivative with respect to the intensity is just the reciprocal of the intensity. The function which has that property is the logarithm, that is,

$$IL = 10 \log(I/I_0),$$

when the intensity level is measured in decibels.

The reference intensity I_0 is taken to be $1.0 \times 10^{-12}\,\mathrm{W/m^2}$, which is approximately the threshold of human hearing at 1000 Hz (Fig. 2-12). Although the logarithm is dimensionless, and hence has no unit, the value of the intensity level is quoted in *decibels*[9] (db).

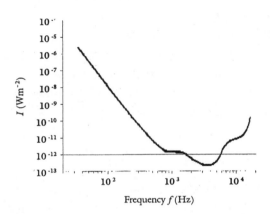

Figure 2-12

Threshold of human hearing as a function of frequency.

EXAMPLE 2-10

A sound has an intensity $I = 2.0 \times 10^{-5}\,\mathrm{W/m^2}$. What is the intensity level IL of this sound?

SOLUTION

According to Eq. [2-22], the intensity level

$IL = 10 \log(I/I_0)$
$\quad = 10 \log(2.0 \times 10^{-5}\,\mathrm{W \cdot m^{-2}}/1.0 \times 10^{-12}\,\mathrm{W \cdot m^{-2}}) = 73\ \mathrm{db}$

---

9.   The decibel = 0.1 bel, an older unit named after Scottish-Canadian inventor Alexander Graham Bell, who patented the telephone in 1876.

Two sounds of different intensities I_1 and I_2 can be compared quantitatively by calculating either the ratio of their intensities I_1/I_2 or their intensity level difference $IL_1 - IL_2$. According to Eq. [2-22], $IL_1 = 10 \log(I_1/I_0)$ and $IL_2 = 10 \log(I_2/I_0)$. The intensity level difference is

$$IL_1 - IL_2 = 10 \log(I_1/I_0) - 10 \log(I_2/I_0)$$

Recall that log a – log b = log(a/b), Hence,

$$IL_1 - IL_2 = 10 \log(I_1/I_2) \qquad \textbf{[2-23]}$$

EXAMPLE 2-11

A sound has an intensity I_1 which is twice the intensity I_2 of a second sound. What is the difference in their intensity levels?

SOLUTION

According to Eq. [2-23],

$$IL_1 - IL_2 = 10 \log(I_1/I_2) = 10 \log(2) \doteq 3.0 \text{ db}$$

How the Human Ear Works

The human ear detects the frequencies present in sound and the intensity of sound through an elaborate structure. The organ can be divided into three sections: the *outer (external) ear*; the *middle ear*; and the *inner ear*. The sensory cells which produce the first neural signals as a result of sound impinging on the outer ear are located in the inner ear.

As illustrated in Fig. 2-13, the outer ear consists of the *auricle*, which helps guide sound waves into the *auditory canal* and helps distinguish the direction of the source of the sound. The auditory canal is the small air-filled cylinder connected to the auricle and terminated by the *eardrum* or *tympanic membrane*. The canal is about 2.5 cm long in the adult human. It is an acoustic resonator of the same type as the organ pipe closed at one end, and so its first harmonic has a wavelength which is 4 times its length, i.e., $\lambda_1 = 10$ cm (Eq. [2-11]), which corresponds to a frequency of about 3.4 kHz. The resonance can be observed as a minimum in the threshold of hearing at this frequency (see Fig. 2-12).

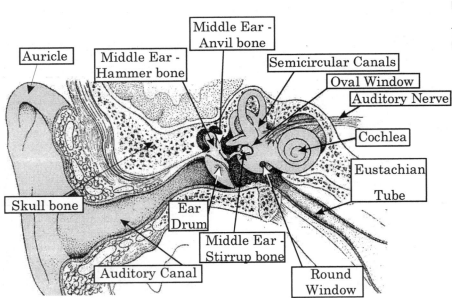

Figure 2-13

The structure of the human ear.

The eardrum is a flexible membrane attached to the wall of the auditory canal. It separates the external ear from the middle ear.[10] The eardrum vibrates when sound waves are present in the auditory canal. Its vibrations set the three small bones[11] of the middle ear in motion, so that the sound waves are transferred through these bones to a flexible membrane *(oval window)* in the skull. The oval window allows the motions of the middle ear bones to set up vibrations in the fluid of fluid-filled chambers in the skull. The oval window transmits the motions of the middle ear bones to the fluid filling the inner ear. These motions induce sound waves in the *cochlea*,[12] the part of the inner ear where sound is detected (see Box 2-2). The function of the middle ear is to transform sound waves in air to sound waves in the fluid of the inner ear (see Box 2-3).

10.   The middle ear is filled with air, whose pressure is equalized with that of the atmosphere through the eustachian tube, which connects the middle ear cavity with the mouth cavity. The importance of this equalization has been impressed upon those who have dived deep into water or taken a trip by air while suffering from a cold.

11.   The bones *(ossicles)* are the "hammer" *(maleus)*, the "anvil" *(incus)*, and the "stirrup" *(stapes)*.

12.   This part of the inner ear is coiled like a snail's shell. "Cochlea" is from the Greek *kochlias*, meaning snail.

Box 2-2   The Cochlea

Figure 2-14

The cochlea is a pea-sized, snail-shell-shaped cavity in the skull which forms part of the inner ear. A magnified view of this part, taken from Fig. 2-13, is shown in Fig. 2-14.

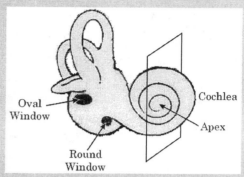

Inner ear.

A cut through the cochlea from its apex to its base (along the plane shown in Fig. 2-17) reveals that the interior of the cochlea is divided into two main fluid-filled chambers connected at the apex of the cochlea (Fig. 2-15).

One of these chambers ends at the oval window, while the other ends at the round window. The flexible round window prevents damagingly high pressures from occurring in the practically incompressible fluid of the inner ear. For example, when the bones of the middle ear push the oval window into the cochlear fluid, the round window bulges outward into the air-filled space of the middle ear (see Box 2-3).

Figure 2-15

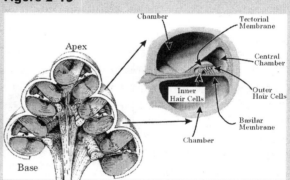

Structure of the cochlea.

The two main chambers are separated by a third, central chamber which is roughly triangular in cross-section. *Reissner's membrane* forms one side. One of the other two sides of the triangle is formed by the *basilar membrane* (base membrane), which vibrates when sound waves are present in the cochlear fluid and which plays a central role in the detection of sound. The *tectorial membrane* (roof membrane) projects over the basilar membrane. *Outer* and *inner hair cells* are anchored to the basilar membrane with the tips of their hairs in close proximity to the tectorial membrane (see Fig. 2-15).

(cont'd.)

The basilar membrane has interesting mechanical properties. It is stiff near the windows, where it is narrow, but thick. The membrane's width gradually increases, and its stiffness (along with its thickness) gradually decreases as the end of the membrane near the apex of the cochlea is approached. Because of this gradation in stiffness, the peak amplitude of vibration of the basilar membrane occurs at different positions along the membrane as the frequency of sound changes. The stiff regions of the membrane close to the windows vibrate most at high frequencies; the slack regions near the apex vibrate most at low frequencies. Thus, the frequency of sound is associated with a particular position of maximum vibration on the basilar membrane (Fig. 2-16).

Figure 2-16

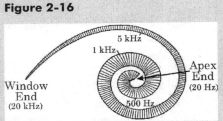

Positions of maximum vibration of the basilar membrane for various frequencies of sounds.

As the basilar membrane vibrates, the outer hair cells, whose hairs are attached to the tectorial membrane, cause the tectorial membrane to vibrate. Because the basilar and tectorial membranes are attached to the rest of the cochlea at different positions, the vibrations of the two membranes result in a transverse shearing motion of the tectorial membrane relative to the basilar membrane. The hairs of the inner hair cells, which are loosely attached to the tectorial membrane, are pushed back and forth as a result of the relative motion of the two membranes. The motions of the hairs stimulate the inner hair cells to generate neural signals, which are carried by the auditory nerve fibers to cells in the auditory centers of the brain. For sound of a given frequency, the hair cells at the position of maximum vibration of the basilar membrane are stimulated the most, resulting in the detection of sound of that frequency.

Box 2-3   Function of the Middle Ear

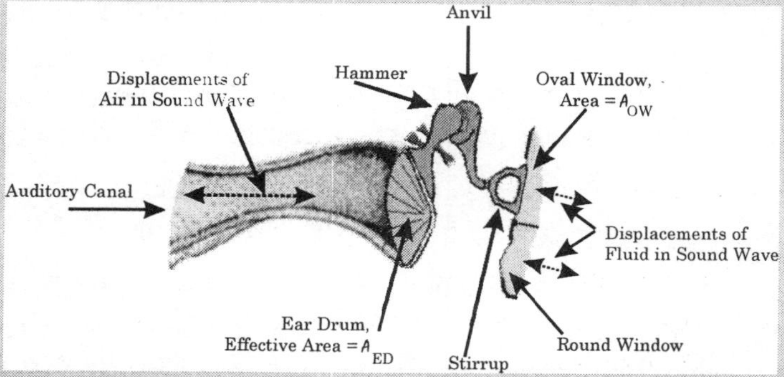

The structures of the middle ear have developed to transmit sound waves in air entering the ear into sound waves in the fluid of the inner ear where the auditory sense cells are located. This development was essential to the transmission of sound because of the great differences in the physical properties (listed in Table 2-1) between air and the fluid of the inner ear,

Figure 2-17

The middle ear.

which is essentially water. One way in which the acoustic difference between substances can be characterized is through the acoustic resistance $R_A = \rho v$. From Table 2-1, $R_{A, Air} \approx 410 \ \text{kg·m}^{-2}\text{·s}^{-1}$ and $R_{A, Fluid} \approx 1.5 \times 10^6 \ \text{kg·m}^{-2}\text{·s}^{-1}$, that is, $R_{A, Fluid} \approx 3.7 \times 10^3 \ R_{A, Air}$. This large difference indicates that the fluid of the inner ear would simply act as a rigid wall and reflect incident sound waves in air, without some intervening mechanism to allow transmission of the waves into the fluid.

The middle ear facilitates the transmission of sound waves into the inner ear by acting as a *pressure amplifier* (Fig. 2-17).

Pressure variations due to sound waves in the external auditory canal exert a force F_{ED} on the eardrum. This force is equal to the actual air pressure relative to atmospheric pressure ΔP_{Air} multiplied by the effective area $\mathcal{A}_{ED}$ of the eardrum (about 44 mm² of the total 65 mm² area of the eardrum):

$$F_{ED} = \Delta P_{Air} \ \mathcal{A}_{ED}$$

The bones of the middle ear transmit F_{ED} to the oval window of the inner ear, where the force is F_{OW}. The three bones act as a lever with arms of unequal lengths, and thus the force F_{OW} is greater than F_{ED} by a factor M, the *mechanical*

(cont'd.)

advantage of the lever system: $F_{OW} = MF_{ED}$. In the case of the adult human ear, $M = 1.3$, so that

$$F_{OW} = 1.3 \, \Delta P_{Air} \, \mathcal{A}_{ED}$$

The force on the oval window acts on an area $\mathcal{A}_{OW}$ equal to 3.2 mm², producing pressure variations in the inner ear ΔP_{IE}, given by

$$\Delta P_{IE} = F_{OW}/\mathcal{A}_{OW} = (1.3 \, \mathcal{A}_{ED}/\mathcal{A}_{OW})\Delta P_{Air} \approx 18 \, \Delta P_{Air}$$

that is, the middle ear amplifies the pressure by a factor of about 18. The intensity of the sound waves in the inner ear depends on the pressure differences and the acoustic resistance in the inner ear. Equation [2-19], applied to the inner ear, gives for the intensity of the sound in the inner ear I_{IE}:

$$I_{IE} = \Delta P_{IE}^2 / (2R_{A,Fluid})$$

Substitution of the approximate relation between ΔP_{IE} and ΔP_{Air}, and between $R_{A,Fluid}$ and $R_{A,Air}$, gives

$$I_{IE} = (18\Delta P_{Air})^2/(2 \times 3.7 \times 10^3 \times R_{A,Air}) = 0.09 \, I_{Air}$$

In other words, the pressure amplifier of the middle ear partially compensates for the differences in acoustic resistance, so that the intensity of the sound in the inner ear is about 10% of that in the auditory canal.

The sound waves in the fluid of the inner ear cause the basilar membrane, which divides the cochlea into two major parts, to vibrate. The physical properties of the basilar membrane are such that high-frequency sounds cause it to vibrate only near the oval window, where the fluid sound waves originate, while low-frequency sounds cause maximal vibration at the opposite end of the basilar membrane. In other words, each position along the basilar membrane corresponds to a sound wave of a particular frequency, which causes the membrane to vibrate with its greatest displacement at that position. Motion of the basilar membrane excites the sensory cells which are situated on it. The cells at the position of maximum displacement are excited more than are the other cells on the basilar membrane. Their neural signal is the largest, and the brain interprets this as sound of a certain pitch. These cells also send information which allows the brain to judge the loudness of the sound.

2.7 Bat Physics: Echolocation and the Doppler Effect

Echolocation

Some bats and marine mammals use echoes of their cries to locate prey and other objects in the space surrounding them. Bats have developed an ability to use these echoes not only for echolocation but also to provide them with information, for example, about the texture of the object reflecting their cries and its size (the so-called *frequency-modulated* or FM bats; Fig. 2-18). Some bats emit cries which allow them to estimate their speed relative to their prey using the Doppler effect (the *constant-frequency* or CF bats; Fig. 2-18).

The advantage of being able to locate prey and then to intercept it using sound is that hunting can take place at night. Bats, which would have to compete with birds for insects during the day, thus have developed a sense which allows them to hunt during the night when most birds cannot.[13]

Echolocation is based on the fact that sound requires time to travel to an object, and that the echo requires time to return to the source of the sound (Fig. 2-19). For the convenience of picturing these situations, it will be assumed that the bat emits spherical waves and that the object reflects spherical waves.[14] The time required for the sound to reach an object a distance d away is just d/v, with v the speed

Figure 2-18

Different types of bats use hunting cries appropriate to the areas in which they hunt. Frequency-modulated (FM) cries used in open areas start at a high frequency and quickly sweep down in a chirp to a low frequency, which is nevertheless above the range of human hearing. Bats hunting in and around trees and shrubs often use a constant-frequency (CF) cry followed by a quick drop in frequency (FM), making CF/FM cries.

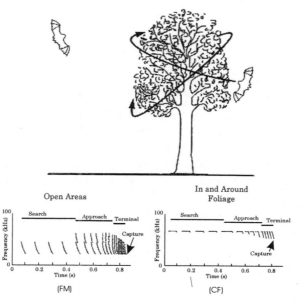

---

13. Owls that hunt on moonless nights also use sound to find their prey, but they use the sound that the prey makes.

14. In reality, the voice apparatus of a bat is constructed so that sound is directed in a narrow cone away from the bat's face, so that echoes return from objects only within that narrow cone. The bat's voice is more like a searchlight beam than like a bare lightbulb.

of propagation of sound in the medium. Assuming that the source and the object are at rest,[15] the time for the echo to return to the source is the same, so that the time elapsed Δt between a pulse of sound from the source and the arrival of the echo back at the source is given by

$$\Delta t = \frac{2d}{v}$$ [2-24]

Since the time elapsed and the distance between the source and the object are proportional, measurement of Δt is equivalent to measurement of the distance d.

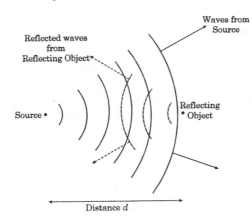

Figure 2-19

Incident and reflected waves obey the inverse square law.

EXAMPLE 2-12

A bat produces 2-ms-long echolocating pulses in the form of spherical waves with a total acoustic power of 50 000 W. An insect reflects the pulses from the bat, and the echo of a pulse arrives back at the bat 0.010 s after the beginning of the pulse. The insect's cross-sectional area is 0.25 cm². Assume that all of the power incident on the insect is reflected.

(a) What is the distance d to the insect?

(b) What is the intensity of the pulse at the insect?

(c) Assuming that the echo produced by the insect is a spherical wave, what is the intensity of the echo at the bat's ear?

---

15. This is reasonable given that, for example, the speed of sound in air is about 340 m/s, or 1.22×10^3 km/h.

SOLUTION

(a) Equation [2-24] relates Δt (= 0.010 s here) to the distance d between the source of the sound and the object producing the echo. Solve for the distance to find $d = \Delta t\, v/2 = 0.01\text{ s} \times 340\text{ m·s}^{-1}/2 = 1.7\text{ m}$.

(b) The bat is assumed to produce spherical sound waves so that the intensity of the bat's voice at the insect is, according to Eq. [2-21], $I_{insect} = P_{source}/(4\pi d^2) = 5.0\times10^4\text{ W}/(4\pi(1.7\text{ m})^2) = 1.4\times10^3\text{ W/m}^2$.

(c) The total power incident on the insect $P_{insect} = I_{insect}\mathcal{A}_{insect} = 1.4\times10^3\text{ W/m}^2 \times 0.25\text{ cm}^2 \times (1\text{ m}/100\text{ cm})^2 = 3.5\times10^{-2}\text{ W}$. It is assumed that all of this power is reflected in the form of the echo, and so it becomes the power of the source of the echo, i.e., $P_{echo} = P_{insect}$. The distance from the insect to the bat is again 1.7 m and so the intensity of the echo I_{echo} at the bat's ear is given by $I_{echo} = P_{echo}/(4\pi d^2) = 3.5\times10^{-2}\text{ W}/(4\pi(1.7\text{ m})^2) = 9.6\times10^{-4}\text{ W/m}^2$.

Doppler Effect

The *Doppler effect*[16] allows certain species of bats to use sound to estimate the velocity of their prey and, in some cases, to detect the beat of their prey's wings. The bat (source of sound) and the prey (reflecting object) are usually both in motion. This complicated situation can be understood by considering three simpler cases: (1) the effect of the bat's flight on the sound it produces; (2) the effect of the reflecting object's motion on the echoes which the object produces; and (3) the effect of the bat's flight on the frequency of the echoes which it hears.

---

16. Christian Johann Doppler (1803–1853), Austrian physicist.

1. *For a Moving Source and Stationary Object*

In the first situation, in which the bat (source) flies with speed v_S while emitting cries (Fig. 2-20), a spherical wave is emitted from the bat during the bat's cry. Since the bat is flying, the source of the spherical waves moves. Suppose that the bat produces a compression at time $t = 0$. The next compression will be emitted one period later, at time $t = T$. During that time, however, the bat has moved a distance $d = v_S T$. If the bat were at rest, the distance between successive compressions would have been λ. But now the compressions ahead of the bat are closer together by the distance d, while behind the bat they are farther apart by that distance. The effective wavelength λ_{front} of the bat's cry in front of the bat is thus just given by

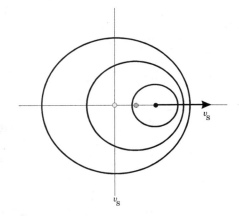

Figure 2-20

Source moving at speed v_S.

$$\lambda_{\text{front}} = \lambda - v_S T$$

Using the relation $v = \lambda/T$ (Eq. [1-23]), where v is the speed of sound, and substituting for T,

$$\lambda_{\text{front}} = \lambda\left(v - \frac{v_S}{v}\right)$$

The frequency of the sound in front of the bat f_{front} is thus higher than the frequency emitted by the bat, and is given by

$$f_{\text{front}} = \frac{v}{\lambda_{\text{front}}} \qquad\qquad \textbf{[2-25]}$$

Using the relation $v = \lambda f$ (Eq. [1-23]), and substituting for λ,

$$f_{\text{front}} = \frac{f}{\left(1 - \dfrac{v_S}{v}\right)}$$

Because the speed of the bat's flight is considerably less than the speed of sound in air, it is possible to approximate Eq. [2-25] with the expression[17]

$$f_{\text{front}} = f\left(1 + \frac{v_S}{v}\right) \qquad \textbf{[2-26]}$$

Behind the bat, the distance between successive compressions is greater than the wavelength of the bat's cry when the bat is at rest, the wavelength being $\lambda_{\text{behind}} = \lambda + v_S T = \lambda(1 + v_S/v)$ and so the frequency of the sound will be lower:

$$f_{\text{behind}} = \frac{f}{\left(1 + \dfrac{v_S}{v}\right)} \qquad \textbf{[2-27]}$$

Equation [2-27] may similarly be approximated by

$$f_{\text{behind}} = f\left(1 - \frac{v_S}{v}\right) \qquad \textbf{[2-28]}$$

The increase in frequency in front of the source and the decrease behind the source is familiar to everyone who has heard the horn of a car sounded continuously as the car drives past. As the car approaches, the sound of the horn has a higher frequency than when the car has passed and is moving away. Just as the car passes, the frequency is exactly what it would have been had the car been stopped with its horn sounding. These shifts in frequency are due to the Doppler effect.

EXAMPLE 2-13

A diesel locomotive with horn blaring (frequency = 400 Hz) sounds a warning to motorists at a level crossing. The speed of the locomotive is 80.0 km/h. What frequency do the motorists hear as the locomotive (a) approaches? (b) departs?

---

17. The quantity $1/(1-x)$ can be represented by the infinite series $1 + x + x^2 + x^3 + \ldots$. If x is small enough, the series can be approximated by $1 + x$, so that $1/(1-x) \cong 1 + x$. For example, for $x = 0.10$, $1/(1 - 0.10) = 1.11$, while $1 + 0.10 = 1.10$.

SOLUTION

(a) First convert 80.0 km/h to m/s:
(80.0 km/hr)(1000 m/km)(1 h/3600 s) = 22.2 m/s.

The frequency while approaching is that in front of the source, and so, according to the approximate Eq. [2-26], f_{front} = 400 Hz × (1 + (22.2/340)) = 426 Hz, which is 26 Hz <u>higher</u> in frequency than is the frequency of the horn.

(b) The frequency behind the source is given by the approximate Eq. [2-28], in which the sign of the ratio of speeds is just the negative of that in Eq. [2-26]. Thus the frequency behind is 26 Hz <u>lower</u> than is the frequency of the horn, or 374 Hz.

---

2.   *For a Stationary Source and Moving Object*

In the second situation, in which sound is reflected by a moving object, there are also frequency shifts. The explanations are slightly different. First, consider the case when the reflecting object (RO) moves with speed v_{RO} toward the source of the sound (Fig. 2-21). The object experiences the successive compressions of the sound wave at a higher rate than it would if it were at rest. This situation is analogous to a boat sailing into the wind and encountering waves driven by the wind: the frequency with which the boat bobs up and down is greater than if it were anchored. The result is that the frequency with which the compressions are encountered f_{toward} is greater than the frequency f' of the sound by the factor $(1 + v_{\text{RO}}/v)$, or

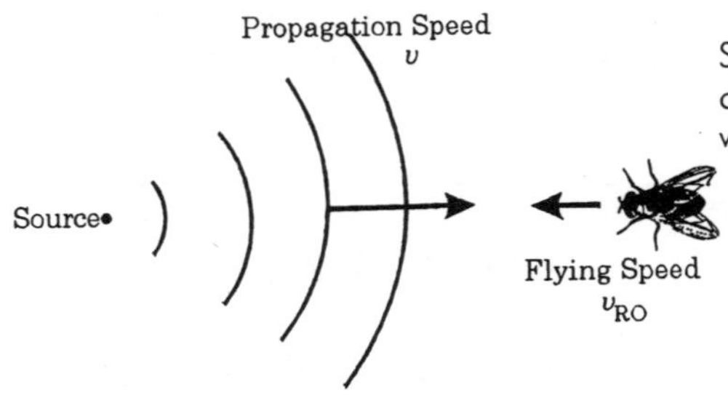

Figure 2-21

Source at rest and reflecting object moving with speed v_{RO}.

$$f_{\text{toward}} = f'\left(1 + \frac{v_{\text{RO}}}{v}\right) \qquad [2\text{-}29]$$

When the object moves away from the source of sound, the frequency f_{away} with which it encounters compressions is less than f':

$$f_{\text{away}} = f'\left(1 - \frac{v_{\text{RO}}}{v}\right) \qquad \textbf{[2-30]}$$

EXAMPLE 2-14

An automobile alarm system in a stationary vehicle emits pulses of sound of frequency 1.0 kHz. You depart the vicinity on your bicycle at a speed of 20 km/h (5.56 m/s). What frequency do you hear?

SOLUTION

The frequency equals
$f_{\text{away}} = 1.0$ kHz $\times$ (1 – (5.56 m·s^{-1}/340 m·s^{-1})) = 0.98 kHz.

3. *For a Moving Source and Moving Object*

In the third situation, in which the bat listens to returning echoes, there are again frequency shifts. Consider the case which is of interest to the bat, in which the bat flies toward the source of the echoes, which have frequency f''. This is identical to the second case, in which the object moves toward the source of sound (Eq. [2-29]), with v_{RO} replaced by the bat's speed v_S and f_{toward} by the frequency of sound detected by the bat's ear f_{ear}:

$$f_{\text{ear}} = f''\left(1 + \frac{v_S}{v}\right) \qquad \textbf{[2-31]}$$

Before dealing with the complicated situation in which both bat and prey are moving, consider the case of a bat flying toward a stationary object, for example, a plant. What is the frequency of the returning echo as it enters the bat's ear? Since the object is in front of the bat and since the bat flies toward the source of the echoes, use Eqs. [2-26] and [2-29] to find that

$$f_{\text{ear}} = f\left(1 + \frac{v_S}{v}\right)\left(1 + \frac{v_S}{v}\right) = f\left(1 + \frac{2v_S}{v}\right) \qquad \textbf{[2-32]}$$

to the same degree of approximation. Thus, the returning echo of the bat's cry has a higher frequency than does the bat's cry. The fractional increase in frequency $(f_{\text{ear}} - f)/f$ is given by

$$\frac{(f_{\text{ear}} - f)}{f} = \frac{2v_S}{v} \qquad \textbf{[2-33]}$$

EXAMPLE 2-15

The horseshoe bat *Rhinolophys ferrumequinuum* produces a cry with a steady frequency of 80.0 kHz in air at 17°C at standard pressure. It normally flies at a speed of 5.00 m/s. A plant toward which the bat flies produces an echo of the bat's cry.

(a) What frequency does the bat hear?

(b) What is the fractional increase of the echo's frequency above the bat's voice?

SOLUTION

(a) $f_{ear} = f(1 + 2v_S/v)$
   $= 80.0 \text{ kHz} \times (1 + 2 \times 5.00 \text{ m·s}^{-1}/340 \text{ m·s}^{-1}) = 82.4 \text{ kHz}$

(b) $(f_{ear} - f)/f = 2v_S/v = 0.0294$, or 2.94% higher in frequency than the bat's cry.

Now it is possible to deal with the case in which the bat listens to echoes produced by flying prey. First suppose that the prey is in front of the moving bat, flying toward the bat. The frequency of the bat's voice in front of the bat f_{front} is given by Eq. [2-26]. This sound strikes the prey, which reflects echoes back toward the bat with frequency f'' given by Eq. [2-29]. As the bat flies into these approaching echoes, it hears them with frequency f_{ear} given by Eq. [2-31]. Altogether then,

$$f_{ear} = f\left(1 + \frac{v_S}{v}\right)\left(1 + \frac{v_{RO}}{v}\right)\left(1 + \frac{v_S}{v}\right) \approx f\left(1 + \frac{(2v_S + v_{RO})}{v}\right) \qquad \textbf{[2-34]}$$

using the fact again that both the bat and the prey fly much more slowly than the speed of sound.

If the prey flies away from the bat, the bat hears sound with frequency

$$f_{ear} = f\left(1 + \frac{(2v_S - v_{RO})}{v}\right) \qquad \textbf{[2-35]}$$

EXAMPLE 2-16

The bat in Example 2-15 detects the echo of a flying insect at 83.3 kHz. What is the insect's flying speed?

SOLUTION

Since the frequency heard by the bat is higher than that produced by a stationary object, which was found to be 82.4 kHz, the insect is flying toward the bat. Equation [2-34] applies in this case:

83.3 kHz = 80.0 kHz × (1 + (2 × 5.00 m/s + v_{RO})/340 m/s).

This gives v_{RO} = 4.03 m/s for the insect's speed towards the bat.

EXAMPLE 2-17

The insect in Example 2-16 hears the bat's cries and begins an evasive manoeuvre by flying directly away from the bat. What frequency does the resulting echo have at the bat's ear?

SOLUTION

Equation [2-34] gives

f_{ear} = 80.0 kHz × [1 + (2 × 5.00 m/s – 4.03 m/s)/340 m/s]
     = 81.4 kHz

Note that this frequency is lower than that produced by a stationary object.

It turns out that CF bats have an inner ear which is especially sensitive to frequencies just above the natural frequency of their voices. For example, in the case of the bat of Example 2-15, the inner ear is most sensitive to frequencies in the range 81–83 kHz. If the echo frequency falls outside this range, the bat changes its flying speed so that the Doppler-shifted frequencies fall into the most sensitive range of the ear. The bat must integrate knowledge of its flying speed with the frequency and intensity characteristics of the echo to gain information about its target's distance and speed.

Exercises

2-1   Atmospheric conditions are such that a sound wave propagates in air with a speed of 350 m/s. The frequency of the sound is 10 Hz. What are the wavelength, the period, and the angular frequency of this wave?

2-2   The length of the auditory canal in a young human is 0.013 m. What are the wavelength and frequency of the longest wavelength standing wave which can be set up in the canal?

2-3   Among the interesting artifacts that archaeologists uncover at a site in Egypt is a tube resembling a flute. The end close to what is presumed to be the mouth hole is stopped up, as in a modern flute, and there are finger holes. With all of the finger holes covered, the distance from the mouth hole to the open end of the tube is 0.500 m. What is the fundamental frequency of this instrument?

2-4   A bat produces a sonar pulse with intensity of 0.020 W/m². Express this as an intensity level in db.

2-5   An FM bat emits a sequence of chirps. It detects an echo 5 ms after the chirps began. How far away is the reflecting object?

2-6   A car approaches at 36 km/hr with its horn blaring at a frequency of 500 Hz. What frequency do you hear?

Problems

2-7   An impoverished university student is about to eat a bright red strawberry in a supermarket when the store manager 10 m away yells at him. The word she uses has a duration of 0.20 s and a frequency of 500 Hz. The acoustic power produced during the word is 0.00020 W.

(a)   How much sound energy actually impinged on the student's ears (total area = 0.020 m²) and made him jump?

(b)   What was the maximum displacement of the air molecules at the student's ear?

2-8   A concert hall is under construction. At the present stage of construction, the walls are bare and the hall is empty. It is 10 m high, 30 m wide, and 40 m long. A loudspeaker connected to an audio oscillator and amplifier is set up in the hall. The acoustic engineer knows that with the apparatus turned on, acoustic resonances (standing waves) will occur in the room at certain frequencies. Calculate the three fundamental frequencies of the resonances that the engineer should expect.

2-9   In moderate conversation the human voice radiates 1.0×10^{-5} W of acoustic power. If there were no other sounds in the universe and if there were no losses of acoustic power in the air, at what distance would a person have to be from another person speaking at a moderate conversational level so that the intensity of sound at the listener's ear would equal 1.0×10^{-12} W/m^2 (i.e., close to the threshold of human hearing)?

2-10 The intensity of a sound is quadrupled. By how much does the intensity level change?

2-11 A group of students camping during the summer are caught in an afternoon thunderstorm. One of the students has brought along, by a stroke of good luck, a sound level meter. A thunder clap registers an intensity level of 100 db on the meter. The sound arrived 4 s after the lightning flash was seen. Assuming that the source of the thunder clap produced spherical waves, calculate the acoustic power produced.

2-12 In the terminology of musicians, very loud and very soft sounds are called "triple forte" *(fff)* and "triple piano" *(ppp)*. Suppose that *fff* corresponds to an intensity of 1.0×10^{-2} W/m^2 and *ppp* corresponds to 1.0×10^{-8} W/m^2. Calculate the intensity levels of these sounds.

2-13 In its search for flying insects, a bat uses an echolocating system based on pulses of high frequency sound. These pulses are 2.0 ms in duration, have a frequency of 50 kHz, and have an intensity level of 100 db at 1.0 m from the bat's mouth. Assume that the bat produces spherical acoustic waves.

(a) What is the acoustic power of each pulse?

(b) How much acoustic energy is there in each pulse?

(c) How much acoustic power propagates through a spherical surface centred on the bat with radius equal to 1.0 m? With radius equal to 5.0 m?

(d) What is the intensity of a pulse at a distance of 1.0 m from the bat? At 5.0 m from the bat?

(e) A June bug is located 5.0 m from the bat. The effective cross-sectional area of the insect is 10 mm^2. How much of the acoustic power emitted by the bat is intercepted by the insect?

(f) What is the amplitude of oscillation of the air molecules at the June bug's position?

2-14 A porpoise sends an echolocating pulse of frequency 60 kHz as it tracks the location of a shark. The power of the pulse is 3.0×10^{-2} W. The intensity of the pulse at the position of the shark is 1.5×10^{-5} W/m^2.

(a) What is the distance between the shark and the porpoise?

    (b)   What is the amplitude of oscillation of the water molecules at the shark's position?

2-15 A CF bat emitting cries of 60.0 kHz flies directly at a wall. The bat hears an echo of frequency 61.9 kHz. How fast is the bat flying?

Answers

2-1   35 m, 0.10 s, 63 radians/s

2-2 5.2×10^{-2} m, 6.5×10^3 Hz

2-3   340 Hz

2-4   103 db

2-5   0.85 m

2-6   515 Hz

2-7   (a) 6.4×10^{-10} J (b) 8.9 nm

2-8   4.3 Hz, 5.7 Hz, 17 Hz

2-9 8.9×10^2 m

2-10 +6 db

2-11 2.3×10^5 W

2-12 100 db, 40 db

2-13 (a) 0.12 W (b) 2.4×10^{-4} J
(c) 0.12 W through both spheres
(d) 1.0×10^{-2} W/m^2 at 1 m; 4.0×10^{-4} W/m^2 at 5 m
(e) 4.0×10^{-9} W (f) 4.5 nm

2-14 (a) 12.6 m (b) 1.2×10^{-11} m

2-15 5.4 m/s

3 Light and the Optics of Vision

3.1 Introduction

The life scientist is interested in light and its related phenomena for many reasons: light interacts with atoms and molecules, and many of these interactions with biomolecules are at the heart of their activity; examples include photosynthesis and the visual process. Optical instruments of interest and importance in the life sciences require us to have a knowledge of the laws of reflection and refraction to understand them. Any of a number of systems could form the basis of this study: the microscope, telescope, or even a vertebrate eye. In the following discussion, the human eye will be used as the example to derive and apply the elementary laws of refractive optics.

Light is studied and understood in the context of two related classical models:[1]

1. **The ray model**: Light is represented as something which travels from a source outward in straight lines in all directions. Its direction of propagation is altered only by reflection or refraction at surfaces according to simple laws.

2. **The wave model**: The propagation of light is represented as a classical wave similar to that of sound as described in previous chapters, with a well-defined speed, wavelength and frequency. The *rays* of the ray model are lines which are always at right angles to the *wave fronts* in this model.

The second model will be particularly relevant to the topics of visual acuity and resolution studied in Section 3.11; at first, the emphasis will be on the ray model.

----

1.   In addition there is the "modern" or quantum description which will not be discussed until Chapters 4 and 5.

3.2 The Nature of Light

Light is an *electromagnetic wave* phenomenon; visible light waves constitute only a small fraction of the electromagnetic (EM) spectrum of waves (see Section 3.3.) An EM wave differs from an acoustical wave in that no particle or material motion is involved. In fact, an EM wave can propagate perfectly well in a vacuum. The wave arises from closely coupled oscillations of electric ($\vec{E}$) and magnetic ($\vec{B}$) fields as illustrated in Fig. 3-1. Note that the wave is a *transverse wave* with $\vec{E}$ and $\vec{B}$ oscillating in directions which are mutually perpendicular, and perpendicular to the direction of propagation (see Box 3-1).

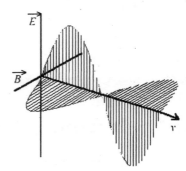

Figure 3-1

Electric ($\vec{E}$) and magnetic ($\vec{B}$) fields in EM waves traveling with speed v.

Box 3-1 Polarization

The fact that light is a transverse wave and sound is a longitudinal wave means that light has an extra property that is absent for sound. This property, which is designated by the direction in space in which the electric field vector ($\vec{E}$) points, is called *polarization*. Light from ordinary sources (e.g., light bulbs, the sun, etc.) normally consists of many waves whose $\vec{E}$ oscillations are in all possible directions perpendicular to the direction of propagation of the wave; such waves are said to be *unpolarized*. Under certain circumstances (e.g., in a laser), light can be made to oscillate in one direction only; such waves are said to be *plane polarized*. There can also be "circularly" and "elliptically" polarized light. The polarization of light is almost irrelevant in human vision. A few species (insects, birds) are able to detect the polarization of light caused by the scattering of sunlight by air molecules, and use it principally for purposes of navigation.

The wave travels with a speed of 2.998×10^8 m/s in a vacuum. This value, which seems to be one of the few genuine constants of our universe, is usually given the symbol c. The EM waves, regardless of their source, travel at the speed c in a vacuum. For the purpose of this discussion it is not necessary to know, in detail, how EM waves are produced; it is sufficient to realize that EM waves are created when electric charges are accelerated. For example, a radio transmitter accelerates electric charge into, and out of, an antenna. The accelerated charges emit energy in the form of EM waves with the same frequency as the action of the transmitter. Since the orbiting electrons in an atom or molecule are also accelerating, then atoms and molecules can also be emitters of EM radiation.[2]

3.3   The Electromagnetic Spectrum

Since light is a wave, it must have the properties we associate with all waves: *wavelength (λ), period (T), frequency (f), amplitude (A)*, and speed *(v)*. Since $T = 1/f$, one of these is redundant. The period is almost never used in discussions involving electromagnetic (EM) waves. The speed, wavelength and frequency are related by the familiar wave relation:

$$v = f\lambda \qquad\qquad \text{[3-1]}$$

For propagation of light in a vacuum, $v = c$.

The most common means of characterizing a wave is probably in terms of its wavelength. The large extent of the EM spectrum in terms of wavelength is shown in Fig. 3-2 (where it should be noted that the horizontal scale of wavelength is logarithmic). The internationally preferred unit of wavelength for UV and visible light is the nanometre (1 nm = 10^{-9} m).[3] Units of frequency, hertz (Hz; cycles per second), are sometimes used, but mostly in the microwave and radio-frequency part of the spectrum. The unit favoured by most molecular spectroscopists is the *wave number*, determined by first finding the wavelength in centimetres and then taking the inverse of that quantity (that is, $1/\lambda$). The unit of the wave number is, therefore, centimetre^{-1} (cm^{-1}). As will be seen in Chapter 4, this unit is directly proportional to the photon energy.

The EM spectrum is conventionally divided into six (or more) regions:

Gamma rays: These are the EM waves of shortest wavelength and, as shown in Chapter 4, the highest energy. With wavelengths shorter than 10^{-11} m (10^{-2} nm), these EM radiations can produce significant

---

2.   In this case there are other "quantum" considerations which must be taken into account as well.

3.   The nm was at one time called the "millimicron," but the use of this term is now severely discouraged. An older unit of wavelength which refuses to go away is the Angstrom (Å) which represents 10^{-10} m; therefore 1 nm = 10 Å.

damage in living systems at the molecular level. Such damage can either be to the detriment of, or of benefit to (as in radiation therapy), the biological system.

X-rays: With wavelengths between 10^{-8} m and 10^{-11} m (10 nm and 10^{-2} nm), these waves are less energetic than gamma rays but still are quite damaging. Their considerable penetrating power in biological tissues makes them important in medical diagnosis.

Ultraviolet (UV) rays: These wavelengths lie between 3×10^{-7} m and 10^{-8} m (300 nm and 10 nm). Some of these waves, while still invisible to the normal human eye, are part of the solar spectrum and can be quite damaging to the external cells of the body, e.g., the eye and the skin.

Visible light: This narrow region of the EM spectrum contains the correct energy waves to stimulate the photochemical reactions in human vision. The wavelength range is from 4×10^{-7} m to 6.9×10^{-7} m (400 nm to 690 nm).

Infrared: In the wavelength region 10^{-7} m (1000 nm) to 1 mm (10^{-3} m), these are "heat" radiations and will denature (cook) proteins.

Radio waves: The shorter radio waves with wavelengths near 10^{-3} m (1 mm) will heat water as in the microwave oven. All wavelengths out to about 10^3 m (1000 m) are commonly used in communications.[4] Whether or not these longer wavelengths have detrimental biological effects is a matter of great controversy.

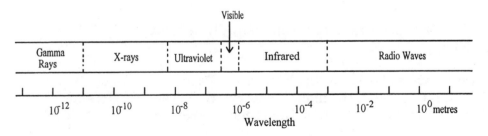

Figure 3-2

The electromagnetic spectrum.

---

4. Communication with submarines is sometimes effected with wavelengths of the order of 10^3 km.

3.4   **Energy, Power and Intensity of Light**

EM waves carry energy, and so the same concepts of wave energy and power that applied to acoustic waves in Chapter 2 are relevant here as well. The *intensity*[5] of light (*I*) is the amount of energy per unit time (or power *P*) passing through unit area, $\mathcal{A}$, perpendicular to the direction of flow, and thus is

$$I = \frac{P}{\mathcal{A}}$$ [3-2]

As with sound, the intensity of light varies inversely as the square of the distance, *r*, from a point source to the receiver (the *inverse square law*), that is

$$I \propto \frac{1}{r^2}$$ [3-3]

3.5   **Reflection and Refraction**

Figure 3-3

Reflection and refraction at an interface between two media of differing refractive indices (*n*).

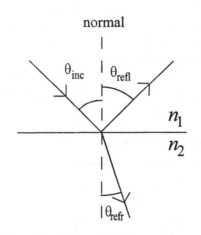

When a ray of light is incident on a transparent interface between two media as shown in Fig. 3-3, some of the light is *reflected* from the surface, and some enters the second medium and, in general, has its direction changed. The ray in the second medium is the *refracted* ray. Refraction is the change in direction of light rays as they pass from one medium to another. A common example of refraction is the effect one observes when looking downward into a calm pond. Objects below the surface appear nearer to the surface than they actually are. A canoe paddle, for example, will seem to have a sharp outward bend at the point where it is submerged. The directions of the incident, reflected and refracted rays are described with respect to the *normal* (or perpendicular) to the surface.

---

5.   In this book the word "intensity" is used for this quantity in keeping with other situations (e.g., sound) of wave power per unit area. However, in more advanced works, when dealing with EM radiation, the correct term is "irradiance."

The laws of reflection are simple and can be stated as two rules:

1. The angle that the reflected ray makes with the normal (θ_{refl}) is equal to the angle that the incident ray makes with the normal (θ_{inc}), i.e., $\theta_{\text{refl}} = \theta_{\text{inc}}$.

2. The incident ray, reflected ray and normal all lie in a common plane.

The refracted ray which enters the second medium has its direction changed so that, in general, $\theta_{\text{refr}} \neq \theta_{\text{inc}}$ as seen in Fig. 3-3. Refraction occurs because light travels more slowly through transparent materials than it does in a vacuum (each material is characterized by a different velocity). The ratio of the speed (c) of light in a vacuum to the speed (v) in a medium is termed the *refractive index* (n) of the medium, so $n = c/v$. Since the frequency of the light does not change in passing from one medium to another and since $v = f\lambda$ as in Eq. [3-1] then, if the speed changes, the wavelength must also change. Thus

$$n = \frac{c}{v} = \frac{\lambda}{\lambda_m} \qquad [3\text{-}4]$$

where λ_m is the wavelength of the light in the medium, and λ is the wavelength in a vacuum. Clearly the refractive index of a vacuum is 1. Since speed v is always less than c then refractive indices of all transparent material media are numbers greater than 1.

The refractive indices for air, water, glass and the various parts of the human eye are given in Table 3-1.

Table 3-1   Refractive Indices

Material	n
Air	1.0003
Water	1.33
Ophthalmic glass	1.50
Human eye	
Cornea	1.38
Aqueous humour	1.34
Lens	1.41
Vitreous humour	1.34

For the moment, consider a monochromatic or single-frequency light ray. Because of the different speeds of the light in different media, there is a discontinuity in the light ray at the interface between two media: the light ray bends, i.e., is refracted (Fig. 3-4). The change in direction that occurs is related to the refractive indices by Snell's law,[6] where the parameters are as defined in Fig. 3-4.[7]

$$n_1 \sin(\theta_1) = n_2 \sin(\theta_2) \qquad\qquad \textbf{[3-5]}$$

Figure 3-4

Effect of refractive index (*n*) on refraction angle (θ_2).

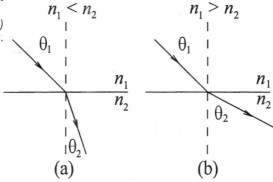

Ordinary white light is made up of many colours (frequencies). Each colour is refracted by a different amount because the speed of light in the medium, and therefore the refractive index, is slightly frequency-dependent.[8] This leads to different image positions in a lens (for example) for different colours, an effect termed *chromatic aberration* which will be discussed in Section 3.10.

EXAMPLE 3-1

A light ray passes from air into water making an angle (in air) with the normal of 30° (θ_1). (Refer to Fig. 3-4a).

(a) What is the angle the ray makes with the normal in water (θ_2)?

(b) What is the speed of light in water?

---

6. Willebrord Snell (1591–1626), Dutch scientist. The law was actually first discovered much earlier by Alhazen (965–1028).

7. If $\theta_2 = 90°$ for the case where $n_1 > n_2$, no light penetrates into medium 2. Then the angle of incidence is termed the critical angle θ_c and is defined by $\sin(\theta_c) = n_2/n_1$. For any angle of incidence greater than this critical angle, *total internal reflection* takes place, i.e., there is no refracted ray. This is what occurs in a light pipe or optical fibre.

8. This dependence of speed of a wave on frequency is called *dispersion*. Sound waves also exhibit dispersion.

(c) A ray of light passes from glass to water making an angle, in the glass, of 20° with the normal. (Refer to Fig. 3-4b.) What is the angle of refraction?

SOLUTION

(a) From Snell's law (Eq. [3-5]):

$(1.00) \sin(30°) = (1.33) \sin(\theta_2)$
$\sin(\theta_2) = (1.00/1.33) \sin(30°) = [(1.00/1.33)](0.5) = 0.3759$
$\theta_2 = \sin^{-1}(0.3759) = 22°$

(b) $v_{water} = c/n_{water} = (2.998 \times 10^8 \text{ m/s})/1.33 = 2.25 \times 10^8 \text{ m/s}$

(c) From Snell's law (Eq. [3-5]):

$(1.50) \sin(20°) = (1.33) \sin(\theta_2)$
$\sin(\theta_2) = (1.50/1.33) \sin(20°) = [(1.50/1.33)](0.342)$
$= 0.3857$
$\theta_2 = \sin^{-1}(0.3857) = 23°$

3.6 The Human Eye as an Optical Instrument

Often the eye is likened to a camera, for its purpose is to form an image on the retina in the same way that a camera forms an image on a photographic plate. Figure 3-5 shows a simple lateral (horizontal) cross-sectional view of the right human eye. The eye is almost spherical (as it must be if it is to have freedom to rotate about any axis)[9] with a slight hemispherical bulge called the *cornea* where light enters the eye. The space between the cornea and the *crystalline lens* is filled with a transparent fluid—the *aqueous humour*. The *iris*, which adjusts the amount of light entering the eye, lies in front of the lens. The *ciliary muscle* is attached to

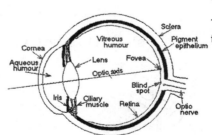

Figure 3-5

The lateral cross section of the human right eye.

---

9. Some mammalian eyes are not spherical and therefore are not able to rotate, e.g., the cat and horse.

the lens and, with it, divides the eye into *anterior* and *posterior chambers*. The posterior chamber is filled with a transparent fluid called the *vitreous humour*.

The *visual* or *optic axis* of the eye extends through the centre of the refracting surfaces (cornea and lens), through the geometrical centre of the sphere, and intersects the back surface in a region of specialized cells—the *fovea*. Extending for about 120° (measured from the centre of the sphere) in all directions about the fovea is the layer of photoreceptor cells, nerve cells and nerve fibres—the *retina*.

The nerve fibres, which actually overlay the photoreceptors, converge in a nerve bundle (*optic nerve*) which leaves the eye about 15° to the inside of the fovea. This region of the retina contains no photoreceptor cells and constitutes a *blind spot*. You can see the blind spot for yourself in the activity in Box 3-2.

Box 3-2   Blind Spot Activity

Cover your right eye with your hand and look at the circle with your left eye only. Move your head slowly toward and away from the figure. At some position (about 10–20 cm), the cross in your peripheral vision will vanish; its image is now on your blind spot. Cover your left eye and, looking at the cross, you can similarly make the circle disappear on the blind spot of your right eye. You cannot make the circle disappear by looking at the cross with your left eye, nor the cross disappear by looking at the circle with your right eye. Since (as is shown later) the eye inverts images, this tells you that the blind spot is toward the centre of the head from the fovea.

Figure 3-6

Sheba looks into the light.

Behind the retina is an optically absorbing layer—the *pigment epithelium*.[10] This layer serves many purposes but one is similar to that of the black paint inside a camera: it suppresses stray reflections of light in order not to blur the sharpness of the image. In cats, visual acuity is not as important as the ability to see in dim light; they have a reflecting epithelium (Fig. 3-6), as anyone

10. The dark pigment in this layer is *melanin*, the same pigment that is responsible for the tanning of the skin and darkening of hair.

knows who has seen a cat's eyes at night, reflecting back the light of a car's headlights.

The Cornea

The cornea is a thin transparent film which is kept in its hemispherical shape by the slightly higher pressure of the aqueous humour in the anterior chamber. The cornea is a transparent section of the white *sclera* which encloses the rest of the eye. The transparency is a result of a regular arrangement of the transparent cells as opposed to their random distribution in the rest of the sclera.[11] This regularity also applies to the lens discussed in the next paragraph. The radius of curvature of a typical adult male cornea is about 8.0 mm.

The Crystalline Lens

The lens is formed of regular layers of specialized cells much like the layers of an onion. It is bounded by two spherical surfaces. In its relaxed (least curved) state, the first (anterior) surface has a radius of curvature $r_1 = 10$ mm, and for the posterior surface $r_2 = 6$ mm. The lens is flexible, and constriction at its edge by the ciliary muscle squeezes the edge of the lens and increases the curvature of the anterior surface, thus increasing the refracting power and giving the eye *accommodation,* or the ability to focus on objects at different distances. As a person ages, more layers are added at the surface of the lens, and the inner layers compress and harden. This makes the lens increasingly stiff as a person ages, and a loss of accommodation results. This normal loss with aging is called *presbyopia* and leads to longsightedness in the elderly (see also Fig. 3-13).

Clearly the refracting properties of the eye are those of four separate spherical surfaces: two in the cornea and two in the lens. It is therefore necessary to derive expressions for the refraction of a ray of light at a spherical surface.

3.7 Image Formation by Spherical Surfaces

Imagine a luminous object of height y at a distance p from a curved surface of radius r as shown in Fig. 3-7. The centre of curvature of the surface is at C. A ray OB originating at O (for "Object") is refracted at the

---

11. For the same reason snow, which is made of transparent ice crystals, looks white when scattering sunlight.

surface at B and crosses the optic axis at I (for "Image"), at a distance q from the surface. The refractive indices in the *object* and *image spaces*[12] are n_1 and n_2 respectively. The incident ray OA along the optic axis is normal to the refracting surface and is undeviated. A ray from O' through the centre of curvature C to I' is also undeviated because it is also normal to the refracting surface.

Figure 3-7

Refraction at a spherical surface.

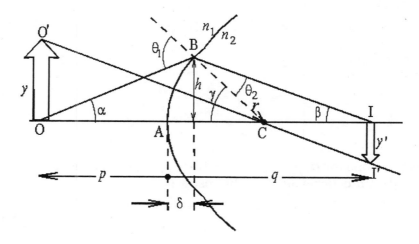

From Fig. 3-7, it is evident in triangle OBC that[13]

$$\theta_1 = \alpha + \gamma \qquad \text{[3-6]}$$

and in triangle BCI that

$$\gamma = \theta_2 + \beta \qquad \text{[3-7]}$$

In addition,

$$\tan(\alpha) = \frac{h}{p+\delta}, \tan(\beta) = \frac{h}{q-\delta}, \tan(\gamma) = \frac{h}{r-\delta} \qquad \text{[3-8]}$$

In order to analyze this situation, use is made of the fact that the rays of light from the object make very small angles (~1°) with respect to the optic axis OACI (the axis of symmetry) of the curved surface; the rays are said to be *paraxial* (close to the optic axis). For small angles,

---

12.   The object and image spaces are the physical regions where respectively the object exists and the image is formed. For the eye, the object space is air and the image space is the interior of the eye.

13.   In each case this follows from the geometrical theorem that "the external angle to a triangle is equal to the sum of the two opposite internal angles."

both the sine and tangent are approximately the angles themselves in radians $[\sin(\theta_1) \approx \theta_1, \sin(\theta_2) \approx \theta_2]$ and the small distance δ may be neglected. Hence Snell's law, Eq. [3-5], for the ray OBI becomes

$$n_1\theta_1 = n_2\theta_2$$

and using Eqs. [3-6] and [3-7],

$$n_1(\alpha + \gamma) = n_2(\gamma - \beta) \qquad \textbf{[3-9]}$$

and Eq. [3-8] becomes

$$\alpha = \frac{h}{p}, \quad \beta = \frac{h}{q}, \quad \gamma = \frac{h}{r} \qquad \textbf{[3-10]}$$

If the expressions for α, β and γ in Eq. [3-10] are substituted in Eq. [3-9], then h cancels out and the result is

$$\frac{n_2 - n_1}{r} = \frac{n_1}{p} + \frac{n_2}{q} \qquad \textbf{[3-11]}$$

The quantity on the left-hand side of Eq. [3-11] is called the *power (P)* of the surface and is said to have units of *diopters* (D) when r is measured in metres; a diopter has units of inverse metres (m^{-1}). Therefore for a spherical surface

$$P = \frac{n_2 - n_1}{r} \text{ (diopters, for } r \text{ in metres)} \qquad \textbf{[3-12]}$$

Powers can be positive or negative; positive powers converge light rays and negative ones diverge them. The usefulness of the concept of power is that for surfaces that are not far apart (compared with other distances in the system), the powers of the surfaces add to give the total power of the system. This fact will be used in the analysis of the cornea in Example 3-2 and in the analysis of a thin lens in Section 3.9.

EXAMPLE 3-2   *The Optical Power of the Cornea*

The cornea consists of two spherical surfaces very close together with the same radius of curvature, $r = 8.0$ mm $= 0.0080$ m. Its power is the sum of the powers of the surfaces taken individually: $P = P_1 + P_2$. Using subscripts "cor" for the cornea and "aqh" for the aqueous humour, and refractive indices as in Table 3-1, Eq. [3-12] gives:

$$P = (n_{cor} - n_{air})/r + (n_{aqh} - n_{cor})/r$$
$$= (n_{aqh} - n_{air})/r = (1.34 - 1.00)/0.0080 = 43 \text{ diopters}$$

Using Eq. [3-11], the shortest distance at which the cornea alone can form an image is obtained when the object distance p is equal to infinity, therefore n_{air}/∞ is negligible:

$$n_{air}/\infty + n_{aqh}/q = 43$$

$$q = 1.34/43 = 0.0312 \text{ m} = 3.1 \text{ cm}$$

The distance 3.1 cm in Example 3-2 is larger than the diameter of any human eye, so it is clear that the cornea alone cannot form an image on the retina; further refractive power supplied by the lens is required.

The symbols p, q and r in Eqs. [3-10] and [3-11] can be either positive or negative, and *sign conventions* are required for consistency. Since the light from an object can be considered to be a flow of energy, we use the direction of this flow as a basis for each rule. These sign conventions are given in Box 3-3. The magnification referred to in the box is defined after Example 3-3. Notice that, according to the sign convention, everything as illustrated in Fig. 3-7 is positive. The sign convention is applied further in Example 3-3.

Box 3-3 Sign Conventions for *p, q, r* and *m*

(See also Fig. 3-8a,b)

1. The object distance, *p*, is positive when the direction **from** the optical surface **to** the object is **against** the flow of light.

2. The image distance, *q*, is positive when the direction **from** the optical surface **to** the image is **with** the flow of light.

3. The radius of curvature, *r*, is positive when the direction **from** the optical surface **to** the centre of curvature of the surface is **with** the flow of light.

4. The magnification, *m*, is positive when the image has the same orientation as the object, e.g., both "up." The magnification is negative when the orientation of object and image are opposite, e.g., object "up," image "down."

EXAMPLE 3-3

Fig. 3-8a shows *p, q,* and *r* are all positive and *m* negative, and Fig. 3-8b shows *p* and *m* are positive and *q* and *r* negative.

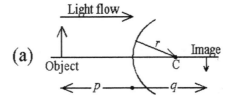

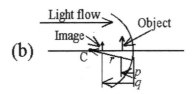

Figure 3-8

(a) *p, q* and *r* positive, *m* negative; (b) *p* and *m* positive, *q* and *r* negative.

Magnification

For an object of_finite height y, the image height y' can also be determined; the ratio $y'/y = m$ is called the *magnification*. In Fig. 3-7, the ray from the object tip through the centre of curvature C to the tip of the image is undeviated because it intersects the refracting surface at right angles. If distances above the optic axis are positive and those below are negative, then from similar triangles

$$\frac{y}{p+r} = \frac{-y'}{q-r}$$

and by cross multiplication

$$m = \frac{y'}{y} = -\frac{q-r}{p+r}$$

Using the expressions for α, β and γ given in Eq. [3-10] gives

$$m = \frac{y'}{y} = -\frac{n_1 q}{n_2 p} \qquad\qquad \textbf{[3-13]}$$

From Eq. [3-13] it can be seen that m is positive for an erect image and negative for an inverted image (see Box 3-4).

EXAMPLE 3-4

Estimate the size of the image on the retina if the object is a man (2.00 m tall) standing 5.0 m from the eye. The image distance (cornea to retina) is about 2.5 cm = 0.025 m. Notice that the refractive index n_2 in the image space is mostly that of the humour, 1.34 (see Table 3-1).

SOLUTION

$$y' = -\frac{n_1 q}{n_2 p}\, y = -\frac{1.00 \times 0.025 \text{ m} \times 2.00 \text{ m}}{1.34 \times 5.0 \text{ m}} = -0.0075 \text{ m}$$

Box 3-4   Image Properties

Images are characterized as:

Figure 3-9

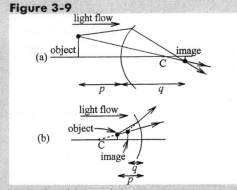

- real or virtual
- erect or inverted

A **real** image results when the rays from the object actually converge at the image. Such an image not only can be seen but can be caught on a screen. The situation is shown in Fig. 3-9a and is the situation with a camera, motion-picture projector, or the eye. Real images have positive q.

(a) Real and inverted image;
(b) virtual and erect image.

One situation for a **virtual** image is shown in Fig 3-9b. In this case all light rays diverge in the image space, but they seem to originate at the image. This image can be seen but cannot be caught on a screen. Your image in the bathroom mirror is such a virtual image. Virtual images have negative q.

An **inverted** image is shown in Fig. 3-7 and Fig. 3-9a; the image is inverted with respect to the object.

An **erect** image is shown in Fig. 3-9b; the image has the same orientation ("up") as the object.

The Helmholtz Relaxed Eye

It is now possible to calculate the imaging properties and the total refracting power of the relaxed eye, i.e., the eye looking at a very distant object. Since the four refracting surfaces are not all close together, it is not sufficient just to add the powers of the surfaces. Instead, the position of the image formed by the cornea must first be calculated. This image becomes the object for the anterior surface of the lens. Finally the image formed by the anterior surface becomes the object for the posterior surface to form the final image on the retina. The geometrical arrangements of the surfaces are given in Table 3-2 and in Fig. 3-10a,b,c.

Figure 3-10

The image formed (a) by the cornea; (b) by the first surface of the lens; (c) by the back surface of the lens.

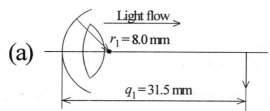

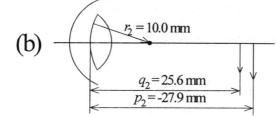

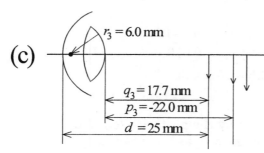

Table 3-2   Properties of the Helmholtz Relaxed Eye

Distance from front surface of cornea to front surface of lens	3.6 mm
Thickness of lens	3.6 mm
Cornea (r_1)	8.0 mm
Anterior surface of lens (r_2)	10.0 mm
Posterior surface of lens (r_3)	6.0 mm
Refractive indices:	
Aqueous and vitreous humour	1.34
Lens	1.41

The relaxed eye is looking at infinity, so the first object distance is infinity. Therefore the first two surfaces (the cornea) are dealt with by using Eq. [3-11] or by repeating the calculation of Example 3-2, as shown in Fig. 3-10a.

$$\frac{n_{air}}{p_1} + \frac{n_{aqh}}{q_1} = \frac{n_{aqh} - n_{air}}{r_1}$$

$$\frac{1}{\infty} + \frac{1.34}{q_1} = \frac{1.34 - 1}{0.0080\,m}$$

$$q_1 = 0.0315 = 31.5 \text{ mm}$$

Next find the image formed by the anterior lens surface as shown in Fig. 3-10b. The image formed by the cornea is the object for this surface, therefore $p_2 = (q_1 - 0.0036 \text{ m}) = (0.0315 \text{ m} - 0.0036 \text{ m}) = 0.0279$ m. However, the direction from the anterior lens surface to this object is with the flow of the light, so the object distance is negative.

$$\frac{n_{aqh}}{p_2} + \frac{n_{lens}}{q_2} = \frac{n_{lens} - n_{aqh}}{r_2}$$

$$\frac{1.34}{-0.0279} + \frac{1.41}{q_2} = \frac{1.41 - 1.34}{0.010\,m}$$

$$q_2 = 0.0256 = 25.6 \text{ mm}$$

Finally the last image is formed by the posterior lens surface as shown in Fig. 3-10c, using the image at q_2 as its object, i.e., $p_3 = 0.0256$ m - 0.0036 m = 0.0220 m, which is also measured with the flow of the light, i.e., negative. In this case the radius of curvature is also negative as can be seen by applying the conventions of Box 3-3 to Fig. 3.10c:

$$\frac{n_{lens}}{p_3} + \frac{n_{vh}}{q_3} = \frac{n_{vh} - n_{lens}}{r_3}$$

$$\frac{1.41}{-0.0220\,m} + \frac{1.34}{q_3} = \frac{1.34 - 1.41}{-0.0060\,m}$$

$$q_3 = 0.0177 = 17.7 \text{ mm}$$

The final image is 17.7 mm beyond the posterior surface of the lens, or (17.7 + 3.6 + 3.6) mm = 25 mm behind the cornea (d in Fig. 3-10c), which is the position of the retina in the normal adult eye.

The above calculation shows that, with the combined power of the cornea and relaxed lens, the eye is able to bring to focus on the retina the image of an object placed at a great distance. Further, the lens is flexible, and constriction at its edge by the ciliary muscle increases the curvature of the anterior surface, thus increasing its power and permitting focus to be maintained on objects down to a distance of about 25 cm in front of the cornea. Not all animals achieve accommodation by this method of distorting the shape of the lens. Another type of vertebrate eye is discussed in Box 3-5.

Box 3-5   The Eye of the Teleost (Bony) Fish

Animals that live in water are presented with a severe visual problem in that the refracting power of the cornea is reduced almost to zero since the index of refraction of water and of the aqueous humour are almost equal. All of the refracting power must be provided by the lens. Thus the curvature of the lens surfaces must be large even before the problem of accommodation is faced.

Figure 3-11

The teleost fish's eye.

Many animals accomplish this by using a focusing sphere (providing all the refraction) in place of a lens, and moving the sphere back and forth for accommodation rather than inducing curvature distortions. Such an eye is shown in Fig. 3-11. The muscle that moves the sphere back and forth to achieve accommodation is the structure to the right of the sphere. The cornea has very little curvature since it can contribute little to the refracting power; it is simply a window.

Animals that must see both in and out of the water have a particularly severe situation as any power of the cornea in air will vanish in water.

3.8   The Reduced Eye Model

For purposes of simple calculation, the eye can be represented as a single optical surface of variable power with air on one side (the object space) and a uniform medium of $n = 1.34$ in the image space. All object and image distances are measured from a position 1.7 mm behind the surface (i.e., halfway between the cornea and the anterior surface of the lens) as shown in Fig. 3-12.

In this model the object distance $= d + 0.0017$ m and the image distance is $(0.0250 - 0.0017)$ m $= 0.0233$ m. The reduced eye equation is

$$P = \frac{1}{d + 0.0017\,\text{m}} + \frac{1.34}{0.0233\,\text{m}}$$

$$= \frac{1}{d + 0.0017\,\text{m}} + 57.5\,\text{diopters}$$

[3-14]

Clearly if $d = \infty$ then $P = 57.5$ diopters; this is the power of the relaxed eye in this model. It is interesting to see how much power the lens accommodation must add as the object is brought from infinity. This is summarized in Table 3-3.

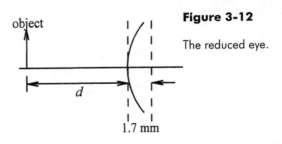

Figure 3-12

The reduced eye.

Table 3-3   Change in Power with Object Distance for the Human Eye

d (m)	P (diopters)	ΔP* (diopters)
∞	57.5	0
10	56.6	0.1
1	58.5	1
0.5	59.5	2
0.25	61.5	4
0.1	67.3	9.8
0.05	76.8	19

*ΔP represents accommodation required at a certain d compared to that required by the relaxed eye (at $d = \infty$).

Table 3-3 shows that for object distances from infinity down to 10 m, the amount of accommodation required is very small: even down to 1 m it amounts to only 1 diopter. The greatest accommodation is required in the working distance from 1 m to 10 cm. At $d = 5$ cm, more accommodation is required ($\sim$20 diopters) than the lens can provide, and the 10 diopters required at $d = 10$ cm is possible only for the very young and is a strain even for them. This can be seen by looking at Fig. 3-13. This presbyopia as a function of age is a natural process.

Figure 3-13

Accommodation vs. age (presbyopia).

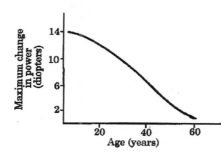

EXAMPLE 3-5

Check the calculation for ΔP for one of the lines in Table 3-3, say the line for $d = 0.05$ m.

SOLUTION

Using Eq. [3-14]:

$P = 1/(0.05 \text{ m} + 0.0017 \text{ m}) + 57.5 \text{ m}^{-1} = 76.8 \text{ m}^{-1}$

$\Delta P = 76.8 - 57.5 = 19$ diopters

3.9   Image Formation by Thin Lenses

There are many situations where two refracting surfaces are sufficiently close together that the total refracting power is just the simple sum of the two powers. This is the case for a thin lens where the thickness of the lens is negligible with respect to all other relevant distances. Spectacle lenses are an example, and the results of this section will be applied to them in Section 3.10.

Consider a biconvex lens as shown in Fig. 3-14. For the sake of generality the two radii of curvature are taken to be different, and the indices of refraction of the media on either side of the lens are also different. In the following, the same method is used as for the four refracting surfaces in the eye (Section 3.7). The object–image relation is derived for each of the two surfaces in turn; the image distance for the first surface becomes the object distance (with a minus sign) for the second surface. Applying Eqs. [3-11] and [3-12] to the first surface (Fig. 3-15a):

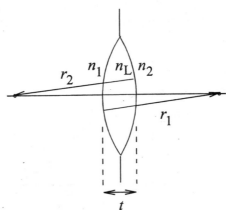

Figure 3-14

The thin biconvex lens (where *t* is the thickness of the lens).

$$P_1 = \frac{n_L - n_1}{r_1} \text{ and } P_1 = \frac{n_1}{p} + \frac{n_L}{q_1}$$

Note that no sign is assigned to r so that if the surface is convex, r is positive whereas if it is concave, r is negative.

For the second surface (Fig. 3-15b), $p_2 = -(q_1 - \delta)$, and thus

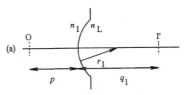

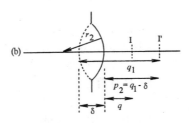

Figure 3-15

Object–image relationships for a thin lens: (a) first surface; (b) second surface.

$$P_2 = \frac{n_2 - n_L}{r_2} \text{ and } P_2 = \frac{n_L}{-(q_1 - \delta)} + \frac{n_2}{q}$$

where q is the final image distance. Again no sign is assigned to r.

For a thin lens, $\delta << q_1$ and so δ is negligible, so that

$$P_2 = \frac{n_L}{-q_1} + \frac{n_2}{q}$$

Adding P_1 and P_2 gives

$$P_1 + P_2 = \left(\frac{n_1}{p} + \frac{n_L}{q_1}\right) + \left(\frac{n_L}{-q_1} + \frac{n_2}{q}\right)$$

$$P = \frac{n_1}{p} + \frac{n_2}{q}$$

The power can also be written

$$P = P_1 + P_2 = \frac{n_L - n_1}{r_1} - \frac{n_L - n_2}{r_2}$$

These two expressions for the total power can be equated and, if the same medium is on either side of the lens, as is usually the case, then $n_1 = n_2 = n$ and the *lensmaker's equation* is obtained:

$$\frac{n}{p} + \frac{n}{q} = \left(n_L - n\right)\left(\frac{1}{r_1} - \frac{1}{r_2}\right) = P \qquad \textbf{[3-15]}$$

If, further, the medium on each side of the lens is air ($n = 1$) then Eq. [3-15] becomes

$$\frac{1}{p} + \frac{1}{q} = \left(n_L - 1\right)\left(\frac{1}{r_1} - \frac{1}{r_2}\right) = P \qquad \textbf{[3-16]}$$

The quantity in the middle of Eq. [3-16] is an expression for the power of the lens that contains only physical properties of the lens (curvatures and material). It can be replaced with a simpler concept. Suppose the object for the lens is at infinity ($p \rightarrow \infty$), then $1/p \rightarrow 0$, and the image distance f is given by

$$\frac{1}{f} = \left(n_L - 1\right)\left(\frac{1}{r_1} - \frac{1}{r_2}\right)$$

This special distance f is called *the focal length* and the power equals $1/f$ and is a constant for a given lens. The focal length is the position where the image is formed of a very distant object ($p \rightarrow \infty$) and so, like the power P, is a unique property of a given lens. Therefore Eq. [3-16] can be written for a lens in air

$$P = \frac{1}{p} + \frac{1}{q} = \frac{1}{f} \qquad \textbf{[3-17]}$$

Equation [3-17] is called the *thin lens equation*. The power and therefore the focal length can be positive or negative; a lens is often referred to as "positive" or "negative," indicating the sign of its power. Positive lenses converge light rays and negative lenses diverge them. The physical form of a positive lens does not have to consist of two convex surfaces as shown in Fig. 3-14. Different forms of positive and negative lenses are shown in Fig. 3-16. Using Eq. [3-17] it is easy to verify the sign of f in each case. Notice that positive lenses are thicker at the centre than at the edge, and vice versa for negative lenses.

The magnification, M, of a lens is simply a product of the individual magnifications corresponding to each optical surface; therefore $M = m_1 m_2$. Using Eq. [3-13], and remembering that $p_2 = -q_1$, one can show that the magnification of a lens having two different media on front and back optical surfaces is

$$M = -\frac{n_1 q}{n_2 p} \qquad \textbf{[3-18a]}$$

In the case where the media on each side of the lens are the same,

$$M = -\frac{q}{p} \qquad \text{[3-18b]}$$

converging or positive lens   diverging or negative lens

Figure 3-16

Types of lenses.

EXAMPLE 3-6

An object 1.5 cm high is placed successively at (a) 10 cm, (b) 5.0 cm and (c) 2.0 cm from a lens in air with a focal length of 5.0 cm. Where is the image in each case? Is it real or virtual; erect or inverted? What can be said in each case about its size?

SOLUTION

(a) Using Eq. [3-17] with $p = 10$ cm,

$$1/p + 1/q = 1/f$$

$$1/(10 \text{ cm}) + 1/q = 1/(5.0 \text{ cm})$$

$$1/q = 1/(5.0 \text{ cm}) - 1/(10 \text{ cm}) = 1/(10 \text{ cm})$$

$$q = 10 \text{ cm}$$

The image is located 10 cm on the far side of the lens from the object; since $q > 0$, the image is real.

Using Eq. [3-18b],

$$M = -q/p = -10 \text{ cm}/10.0 \text{ cm} = -1.0$$

Since magnification is the ratio of image height to object height $M = y'/y$, then the image height

$y' = -1.0 \times 1.5$ cm $= -1.5$ cm; the minus sign indicates that it is inverted.

(b) If $p = 5.0$ cm, then, from Eq. [3-17], $q \to \infty$; since $q > 0$, the image is real.

$$M = -q/p = -\infty$$

The image is infinitely far away (i.e., the rays emerging from the lens are parallel) and its size is not defined. The minus sign indicates that it is inverted. This is very nearly the case of a movie projector where q is much greater than p, and the image is inverted and quite large.

(c) If $p = 2.0$ cm, then $q = -(10/3)$ cm, i.e., the image is 3.3 cm from the lens on the object side and is virtual ($q < 0$).

$$M = -q/p = -(-10/3)/2 = 1.67.$$
The image is 1.67×1.5 cm $= 2.5$ cm high and is erect.

A two-lens system solved by these analytic methods is in Example 3-7.

EXAMPLE 3-7

Find the magnification and image position if an object is placed 0.15 m from the first lens of the two-lens system shown in Fig. 3-17. The focal lengths are 0.10 m for the first lens and 0.11 m for the second lens. The two lenses are separated by a distance of 0.16 m.

SOLUTION

For the first lens:

Figure 3-17

A two-lens example.

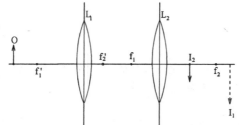

$$\frac{1}{p_1} + \frac{1}{q_1} = \frac{1}{f_1}$$

$$\frac{1}{0.15\,\text{m}} + \frac{1}{q_1} = \frac{1}{0.10\,\text{m}}$$

$$q_1 = 0.30\,\text{m}$$

The position of the image I_1 at q_1 yields the object distance p_2 for the second lens. Since q_1 is "downstream" from the second lens, p_2 is negative (refer to Box 3-3). Thus,

$$p_2 = -(0.30 - 0.16)\,\text{m} = -0.14\,\text{m}$$

$$\frac{1}{p_2} + \frac{1}{q_2} = \frac{1}{f_2}$$

$$\frac{1}{-0.14\,\text{m}} + \frac{1}{q_2} = \frac{1}{0.11\,\text{m}}$$

$$q_2 = 0.062\,\text{m}$$

Thus the final image I_2 is 0.062 m to the right of the second lens.

The corresponding magnification of this system of lenses is

$$M = M_1 M_2 = \left(-\frac{q_1}{p_1}\right)\left(-\frac{q_2}{p_2}\right)$$

$$M = \left(-\frac{0.030}{0.15}\right)\left(-\frac{0.062}{(-0.14)}\right) = -0.89$$

The real image is inverted and 0.89 times the size of the object.

Graphical methods for solving the lens equation are given in Box 3-6.

Box 3-6   Graphical Ray Methods

Graphical ray diagrams, drawn to scale on graph paper, can also be used to determine image positions as shown in Fig. 3-18 for the case of a positive and a negative thin lens. Some of the important rules to be considered in making such diagrams are as follows.

Figure 3-18

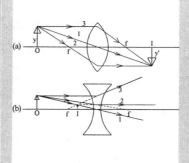

Graphical ray methods for (a) a positive lens and (b) a negative lens.

For the positive lens:

(a) A ray through the centre of the (thin) lens is undeviated, labeled 1 in Fig. 3-18a.

(b) A ray passing through the focal point in the object space is rendered parallel to the optic axis in the image space, labeled 2 in Fig. 3-18a.

(c) A ray parallel to the optic axis from the object passes through the focal point of the lens in the image space, labeled 3 in Fig. 3-18a.

(d) Where two or more real rays cross, a real image is formed; if a real ray intersects with a ray projection then a virtual image is formed (see negative lens, this box).

For the negative lens:

(e) A ray through the centre of the (thin) lens is undeviated, labeled 1 in Fig. 3-18b.

(f) A ray aimed at the focal point in the image space is made parallel to the optic axis in the image space, labeled 2 in Fig. 3-18b.

(g) A ray parallel to the optic axis from the object is refracted in the image space so that its backward projection goes through the focal point in the object space, labeled 3 in Fig. 3-18b.

(h) The image is in front of the lens, is erect, and is virtual. All object positions for negative lenses give rise to virtual images.

EXAMPLE 3-8

Solve the two-lens case of Example 3-7 by graphical methods.

SOLUTION

The solution is shown in Fig. 3-19.

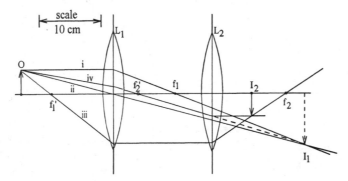

Figure 3-19

Graphical solution to Example 3-7.

1. Lens L_1 with its focal points f_1 and f_1' is shown.

2. Ray i is parallel to the axis and goes through f_1 (Rule (c), Box 3-6); Ray ii goes through the centre of the lens and is undeviated (Rule (a), Box 3-6). These determine the position of the image I_1 which would be formed if L_2 were not present.

3. Lens L_2 with its focal points f_2 and f_2' is shown 16 cm to the right of L_1.

4. Ray iii passed through f_1' and emerged from L_1 parallel to the axis (Rule (b), Box 3-6). It is intercepted by L_2 and, since it is parallel, goes through f_2 (Rule (c), Box 3-6).

5. Of all the rays from O there is one (Ray iv) which went through f_2' headed for I_1 but was intercepted by L_2 and emerged parallel.

6. Rays iii and iv intersect at I_2, the final image which is:
   • 6.2 cm to the right of L_2
   • 0.89 times the size of the object
   • real
   • inverted

The Compound Microscope

The method of analysis for the two-lens system in Example 3-7 can be applied to an understanding of the function of a *compound microscope*. The microscope has assisted the development of knowledge of cellular structure and function more than any other single device.[14] In recent years microscopes have become more and more complex, with the use of phase contrast, dark-field illumination, and polarizing optics. The principal components of all microscopes are the same, however, consisting of two lenses: the objective and the eyepiece (Fig. 3-20). The object distance is slightly larger than the focal length of the objective lens. This results in the production of a real, inverted image located inside the focal point of the eyepiece. This image becomes the object of the eyepiece. Because this new object distance is less than the focal length of the eyepiece, the resulting image is virtual, still inverted and considerably magnified. The placement of the image is about 25 cm from the eyepiece, which locates it just outside the closest distance at which an average viewer can focus.

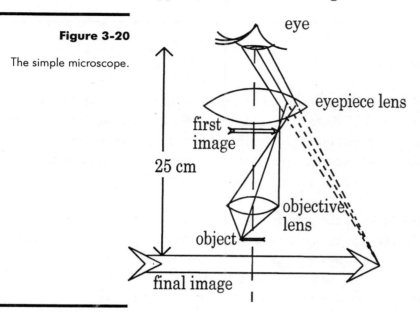

Figure 3-20

The simple microscope.

3.10   Optical Defects of the Eye and Their Corrections

The ability to perceive a clear, undistorted image of an object varies considerably from one individual to another. Persons who see well at all distances are the exception rather than the rule and, after middle age, even this group's visual ability deteriorates. The genetic, environmental or behavioural reasons for this variability are not

---

14.   The first such instrument was developed by Galileo in 1610.

understood. The invention of spectacles (Box 3-7) led to the art and science of optometry, and in the present era humans in all modern societies lead more active lives because of artificially improved vision.

Box 3-7   Eyeglasses

The history of the invention of spectacles is very obscure. Pliny records in 77 A.D. that the emperor Nero watched gladiatorial games through an "emerald." Since the Romans knew how to grind and polish lenticular shapes in glass and gems, it is thought by some that this might have been used to correct myopia. It also could have been used just in the sense of sunglasses to reduce glare.

The problem of longsightedness in the elderly was recognized in the scriptoria of monasteries. Here the elderly monks, whose life work was copying manuscripts, were unable to do their work because of presbyopia. By about 1350 there are paintings showing monks with double lenses suspended on the nose or hanging from caps. The invention is sometimes credited to the monk Roger Bacon (1214–1292), who wrote extensively on optics, but he was probably only offering explanations for what was already known practically. Subsequently it was found (by whom and when is not known) that concave lenses corrected myopia, and by 1600 the trade of "spectacle maker" was well established.

The use of a cylindrical lens to correct astigmatism is credited to both William Wollaston (1766–1828) and Thomas Young (1773–1829). Bifocal spectacles (small reading lenses mounted in the bottom of a spectacle lens) were invented by Benjamin Franklin (1706–1790).

Contact lenses were first used by the Zurich physician A.E. Fick in 1887, but did not become practical until K.M. Touhy in the U.S.A. patented hard plastic contact lenses in 1947. Soft contact lenses were invented by O. Wichterle in Czechoslovakia in 1960.

The three most common visual anomalies are (i) *hypermetropia* or farsightedness,[15] in which the sharp image of a distant object falls beyond the retina; (ii) *myopia* or nearsightedness, in which the sharp image of the distant object falls in front of the retina; and (iii) *astigmatism*, in which the eye cannot focus clearly on horizontal and vertical lines simultaneously, because the surface of the cornea is not spherical but is more sharply curved in one plane than in the other.

Figure 3-21 illustrates image formation for the normal, hypermetropic and the myopic eye. Note that in the hypermetropic eye, the curvature of the cornea is insufficient (i.e., power too small) to set the image on the back surface of the eye. In myopia, the cornea has too much curvature (i.e., power too large) with respect to the shape of the eye. Thin spectacle lenses can be used to correct for hypermetropia, myopia and astigmatism. Figure 3-21 also shows the effect of the lenses required to correct hypermetropia and myopia.

Figure 3-21

Hypermetropia, myopia, and their corrections.

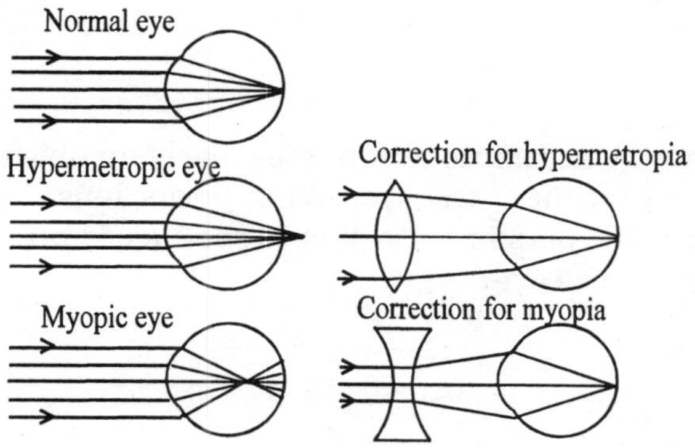

For hypermetropia and myopia, the defect can be defined in terms of two points measured from the front surface of the eye:

1. The *near-point* is the object position closest to the eye for which the image can be clearly brought to focus on the retina. In the normal eye this is about 25 cm.

2. The *far-point* is the object position furthest from the eye for which the image can be clearly brought to focus on the retina. In the normal eye this is at infinity.

---

15. Sometimes called *hyperopia*.

The Hypermetropic Eye

A person with this defect cannot focus a sharp image on the retina for an object closer than some near-point greater than 25 cm.[16] The use of the proper lens with positive power in front of the eye brings the rays to a sharp focus on the retina as shown in Fig. 3-21. Sometimes lenses of different powers are required for distant, intermediate and close-up vision. These are often compounded into bifocal and trifocal spectacles.

EXAMPLE 3-9

Suppose a person can, by accommodation, see clearly an object only as close as 1.0 m in front of the eye. What power lens would be required for this person to be able to see clearly an object 0.25 m in front of the eye?

SOLUTION

Since this unaided eye cannot focus anything closer than 1.0 m, a lens must be used which will create an erect image 1.0 m from the eye when the object is only 0.25 m away. The eye then sees clearly the image (but does not directly see the object). A converging lens is necessary. Figure 3-22 shows the effect.

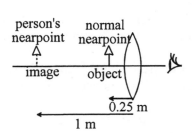

Figure 3-22

Correction for hypermetropia.

The object at a distance $p = 0.25$ m forms a virtual image at a distance $q = -1.0$ m. Using Eq. [3-17], the power of the lens is

$$P = \frac{1}{p} + \frac{1}{q}$$

$$P = 1/(0.25 \text{ m}) + 1/(-1.0 \text{ m}) = 3 \text{ diopters}$$

16. An extreme case results when the lens is removed (the *aphasic eye*) in cataract surgery. In this case there is insufficient power in the cornea to bring an object to sharp focus at any distance unless the cornea is reshaped, for example with laser surgery.

The Myopic Eye

The myopic eye focuses parallel light in front of the retina and hence cannot see well at a distance. The corrective lens must render distant light less convergent (i.e., less power) and produce a virtual image at the person's far-point.

EXAMPLE 3-10

The far-point of a myopic eye is 0.50 m. What power is required in order for the eye to see clearly an object at infinity?

SOLUTION

This unaided eye cannot focus on anything more than 0.50 m away. To see the object at a considerable distance, a lens must be used that will create an image 0.50 m in front of the eye, where it can be seen. A diverging lens is required (Fig. 3-23).

Figure 3-23

Correction for myopia.

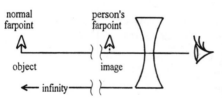

The object at infinity forms a virtual image at a distance $q = -0.50$ m. Using Eq. [3-17], the power of the lens is

$$P = \frac{1}{p} + \frac{1}{q}$$

$$P = 1/\infty + 1/(-0.50\,\text{m}) = -2.0\,\text{diopters}$$

The Astigmatic Eye

People with this defect have distorted vision because of the nonuniform curvature of the cornea or lens surfaces. In the simplest case the refracting elements have two focal lengths. One image axis will be in focus while the other perpendicular axis is out of focus. The correction in this simple case is a cylindrical lens, that is, one with no curvature in one plane and sufficient curvature in the other to bring the two image axes into coincidence.

Chromatic Aberration

Before leaving the topic of visual aberrations, something should be said about chromatic aberration of the lens of the eye. The refractive index of a material varies with the wavelength of the light so that all colours will not be focused at the same point by an optical system. To eliminate chromatic aberrations in optical systems, double lenses, which compensate for various wavelengths, can be used. But the eye contains one lens made of one material, so the eye has a high degree of chromatic aberration. However, the effect doesn't bother us very much for the following reasons. The effect is moderate in the red and green, but becomes very pronounced in the ultraviolet. However, the lens of the eye is relatively opaque in the ultraviolet.[17] Also, in dim light, we see by vision which is most sensitive in the blue-green (see Chapter 4). This is a coarse-grained type of vision. In bright light, the eye shifts to high acuity vision, where sensitivity is better in the red. Here the chromatic aberration is least. Nature is very clever.

Binocular Vision

Humans have binocular vision; that is, they have two eyes, both of which look in the forward direction, the fields of view overlapping for about 170°. This arrangement allows for very accurate determination of distance and relative position. On the other hand, some animals, such as chameleons, have panoramic vision; that is, their eyes look in different directions with no overlap of the fields of view. Because the human eyes are separated, a slightly different image is viewed by each eye—a phenomenon called *binocular parallax*. The two images so formed must be meshed by the visual system; this gives rise to depth perception. For short distances, we use binocular parallax to judge distance. For very distant objects, the two eyes see practically the same view, and binocular parallax effects no longer play a role. Then distance is judged by the size of familiar objects, the amount of light scattered by intervening air, and the curvature of the horizon.

3.11   Visual Acuity and Resolution

Up to this point it has been assumed that light can be treated by the ray concept. But the laws of geometrical optics are valid only if the dimensions of the physical objects that the light encounters are much

---

17. Hence the ability of people with *aphasia* (see Footnote 16) to see in the ultraviolet region of the spectrum.

larger than the wavelength of the light. However, when light passes through small apertures, some *diffraction* effects (bending of the light ray by the obstacle) will be seen.

Consider the image, formed by a lens on a screen, of a small pinhole which is uniformly illuminated with parallel light. The image should be a geometrical point and it will be very nearly so if the aperture is large. However, if the aperture is small, diffraction will occur, leading to a spreading of the image. Light through a circular aperture is focused not as a disc of finite radius but as a "fuzzy" disc surrounded by concentric bright and dark rings.

The pattern is shown in Fig. 3-24. A graph of the intensity measured along a line through the centre of the pattern is shown at the top of the figure.

If this pattern is formed by light of wavelength λ falling on a circular aperture of diameter *a*, then the angular separation of the central maximum and the first minimum is given by:

$$\alpha = 1.22 \frac{\lambda}{a} \qquad \text{[3-19]}$$

Two point sources can be resolved (i.e, imaged clearly as two images) by an optical system if the corresponding diffraction patterns are sufficiently small or sufficiently separated to be distinguishable. The usual criterion for resolvability is that stated by Lord Rayleigh: "Two equally bright point sources can just be resolved by an optical system if the central maximum of the diffraction pattern of one source coincides with the first minimum of the other." This is illustrated in Fig. 3-25 where the intensity graph of two diffraction patterns and their sum are shown well-resolved, just resolved at the Rayleigh limit, and unresolved. At the limit of resolution, then, the angular separation between two just-resolved sources α is just that given by Eq. [3-19].

Figure 3-24

The diffraction pattern of a circular aperture (bottom) with its intensity profile across a diameter (top).

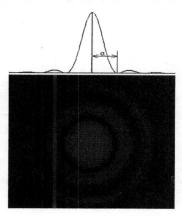

In the case of the human eye, the aperture creating the diffraction patterns of two sources is the eye's pupil with diameter a. Note the inverse dependence on pupil diameter; wider pupils give better resolution.[18]

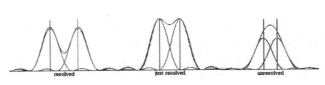

Figure 3-25

Two diffraction patterns near the Rayleigh limit of resolution.

resolved   just resolved   unresolved

The minimum diameter of the pupil is about 2 mm, and the eye is most sensitive to light of a wavelength in air of about 555 nm. If we assume that all refraction takes place at the cornea (Fig. 3-26) then the resolution limit of the eye is

$$\alpha_2 = 1.22 \times (555 \times 10^{-9} \text{ m}/1.34)/0.002 \text{ m} = 2.5 \times 10^{-4} \text{ rad},$$

where 555×10^{-9} m/1.34 is the wavelength of the light inside the eye.

At an image distance of 25 mm (the diameter of the eye) two such Rayleigh discs would be separated by d as shown in Fig. 3-26. Since for small angles the arc length (d) is equal to the angle in radians (α_2) times the radius of the sector (ℓ) then $d = 2.5 \times 10^{-4} \times 0.025$ m $= 6.3 \times 10^{-6}$ m or 6.3 μm. Since the diameter of the cone cells is about 3 μm, two such images would just fall on separate cells and so be capable of being distinguished as separate images. The cell size in the eye has developed to just utilize all the resolution that the Rayleigh criterion allows.

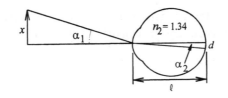

Figure 3-26

Object and image size at the limit of resolution.

The short calculation above also demonstrates that the angular separation inside the eye of objects that can just be resolved must be $\alpha_2 = 2.5 \times 10^{-4}$ radians (Fig. 3-26). The required separation (x) of two objects placed at the near-point of the normal eye as shown in Fig. 3-26 can now be found.

Recall that the small-angle version of Snell's law is $n_1\alpha_1 = n_2\alpha_2$. With $n_1 = 1$ (for air) then $\alpha_1 = (1.33/1)2.5 \times 10^{-4} = 3.3 \times 10^{-4}$ radians. Again for small angles,

---

18. This is the same reason that astronomers use bigger telescopes: larger telescopes can resolve stars that are separated by smaller angles.

$$\alpha_1 = \frac{x}{\text{near-point distance}}$$

[3-20]

$$x = (0.25)(3.3 \times 10^{-4}) = 8.3 \times 10^{-5}\,\text{m} = 83\,\mu\text{m}$$

This is approximately, then, the nearest separation at which we can still see distinct objects (Box 3-8).

Box 3-8   Visual Acuity

The ability of the eye to resolve small objects is called *acuity*. It is measured by ophthalmologists by means of the *Snellen Chart** familiar to everyone who has had an eye examination. A person with normal vision can just clearly resolve details which subtend an angle at the eye of 1 minute of arc (5.4 µm on the retina). So that the eye will be in a relaxed state while testing, the chart is placed 6 m (20 ft.) from the eye. The test objects are letters (mostly with straight-line segments) that are made up of elements of 1 arc-minute and are 5 elements high. Therefore the letters in the "6-metre letters" are $(5/60)(\pi/180)6000 = 8.7$ mm high. A person with normal vision can read such a line of letters at 6 m (20 ft.) and is said to have "6/6 vision" (20/20 in North America). Letters with a height 13.1 mm subtend an arc of 5 arc-minutes at 9 m (30 ft.). A person who can read only down to the "9-metre letters" at a distance of 6 m is said to have 6/9 (20/30) vision.

* Invented in 1862 by Herman Snellen (1834–1908), professor of ophthalmology at Utrecht.

Obviously, with the aid of the appropriate lens system, such as a microscope, we can do much better. The resolution of a microscope is diffraction-limited, just as is the eye, but gains a significant advantage because an object can be placed very close to its objective lens. The smallest resolvable separation x_m for an ordinary microscope is given by

$$x_m = \frac{\lambda}{2n\,\sin(\theta)}$$

[3-21]

where λ is the wavelength of the light in air, θ is half the viewing angle of the objective lens, and n is the index of refraction of the medium in which the object is immersed (usually air). The quantity $n\,\sin(\theta)$ is called the *numerical aperture* of the objective. The better microscope objectives have numerical aperture values of about 0.5. Thus with the aid of a microscope and 500 nm light, we can see objects distinctly at separations in air of only

$$x_m = \frac{500 \times 10^{-9}}{2 \times 0.5} = 0.5\,\mu m$$

This value can be slightly improved by placing oil between the object and the objective as is done in the so-called "oil-immersion" microscope. In this type of microscope the resolution is increased because of the relatively large value of the refractive index of oil.

Exercises

3-1 If red light has a wavelength of 620 nm in a vacuum, what are the wavelength, frequency and speed of this light in the vitreous humour of the eye? ($n = 1.34$)

3-2 What is the speed of light in medium 2 in the diagram below?

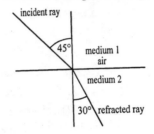

3-3 The minimum diameter of the pupil of the human eye is about 2 mm and the maximum diameter is about 8 mm. How much more light energy per unit time enters the eye with the pupil fully dilated than with the pupil at its minimum size?

Problems

3-4 A thin lens has radii of curvature $r_1 = +2.0$ cm and $r_2 = +1.0$ cm. The glass of which it is made has refractive index $n = 1.5$.

    (a) What is its power in air?

    (b) What is its power in water? ($n = 1.33$)

    (c) Now suppose that it is placed in a wall on one side of which is air and on the other water. (The surface having radius r_2 is in contact with water.) What is the power in this case?

    (d) In questions (a), (b), and (c) find out where the image of an object placed 10 cm from the lens is formed. What power would the lens in question (b) have to have in order that the image distance be the same as in question (c)?

3-5 Show that a uniform glass sphere of radius r and refractive index n_2, immersed in a medium of index n_1, forms an image of a very distant object at a distance $rn_2/2(n_2 - n_1)$ from its centre. (See Box 3-5.)

3-6　The physical properties of the relaxed eye are given in Table 3-2. With these properties, the distance from the cornea to the retina is 25 mm and an object infinitely far away is in focus on the retina. If an object is placed a distance of 1.0 m from the cornea, the eye accommodates by changing the radius of curvature of the posterior surface of the lens. Find the new radius of curvature. Assume that the changes are small enough that the distances between the surfaces remain unchanged.

3-7　The reduced eye (i.e., the simplified model of the human eye) is useful in calculating where images are formed and what sorts of corrections must be made to improve a faulty lens system in the eye. Suppose the only thing that happens when the eye accommodates is that the power of the lens changes. What does it have to change to in order that an object 20 cm from the front surface of the eye is in focus on the retina? What increase in power above the power of the relaxed eye must accommodation therefore provide?

3-8　Two lenses having equal focal lengths of 0.50 m are separated by 0.80 m. If a 0.30-m high object is placed 4.0 m in front of this lens system, find the position and height of the image.

3-9　Wally Whoop has been a bird watcher for many years. However, he has always been frustrated in his desire to observe the nesting habits of his favourite birds, cranes. When he came too close, they flew away. Too poor to buy binoculars or a telescope, Wally had just enough money to buy two lenses and a jar of rubber cement. The focal length of one lens, the objective, was 0.30 m. This lens was placed at one end of the cardboard tube from a roll of wax paper. Wally didn't know the focal length of the other lens (the eye lens) but he kept repositioning it until he obtained a system which would magnify objects 50 m away by a factor of ten.

(a)　What is the separation of the lenses?

(b)　What is the focal length of the eye lens?

(c)　Redo parts (a) and (b) if the magnification is –10; i.e., if the final image is inverted as in an astronomical telescope.

Hint: The image of the lens system should be roughly at infinity (assume 50 m from the objective). Note also that for distant objects, the image falls on the focal point.

3-10　A bacterial cell of length 3.0 µm is placed on the stage of a microscope. When the cell is in focus, the distance between the cell and the objective lens is 1.45×10^{-3} m. The focal length of the objective lens is 1.40×10^{-3} m. If the magnification factor, M, of the eyepiece is 10, find the overall magnification of the microscope and the size of the image.

3-11　The far-point of a myopic eye is 2.00 m. What is the power of the spectacle lens which will correct the myopia? What is the focal length of this lens?

3-12 A person with hypermetropia has a near-point of 1.00 m. A lens of power +2.00 D is placed directly in front of the eye. What is the near-point of the eye-plus-lens system?

3-13 The fovea of the human eye is approximately 0.30 mm in diameter. At what distance would a circle of diameter 10 cm have to be so that the image covers the fovea?

3-14 Two stars make an angle of 10^{-3} radians at the eye of an observer. If a person who can just distinguish two spots, 0.5 cm apart, at a distance of 10 m views the stars, can he distinguish them as two distinct objects?

3-15 Two small lights are placed 10 m from your head in a darkened room. Calculate the minimum separation of these lights such that you still recognize them as being two. (Choose a wavelength of 550 nm and a pupil diameter of 3.0 mm.)

Answers

3-1 463 nm, 4.84×10^{14} Hz, 2.24×10^{8} m/s

3-2 2.12×10^{8} m/s

3-3 16

3-4 (a) –25 D (b) –8.5 D (c) 8.0 D
(d) –0.029 m, –0.061 m, –0.665 m, 11.3 D

3-6 –5.4 mm

3-7 5 D

3-8 0.38 m downstream from 1st lens,
–0.08 m high

3-9 (a) 0.27 m (b) –0.030 m (c) $d = 0.33$ m, $f = 0.030$ m

3-10 –280X, 0.84 mm

3-11 –0.5 D, –2 m

3-12 1/3 m

3-13 600 cm (approx.)

3-14 $\alpha < 10^{-3}$ rad, yes

3-15 2.2 mm

4 Absorption and Emission of Light by Molecules

4.1   Introduction

Many biological molecules are particularly important because of their ability to convert the energy of an electromagnetic wave into chemical energy. Well-known examples of these molecules are the various pigments involved in vision, which are linear molecules, and the cyclic molecule chlorophyll found in green plants. Much scientific effort has gone into studies of light absorption by molecules, resulting in the development of many new disciplines such as molecular spectroscopy, photochemistry, photobiology, and photophysiology. The theories that have evolved are certainly not simple ones; however, light absorption is so vital for life itself that it is necessary to understand the process.

To begin, it is important to remember that ultraviolet (UV) and visible light waves constitute only a small fraction of the electromagnetic spectrum (see Fig. 3-2). A *photon*[1] of light at the gamma ray end of the spectrum has an energy of about 10^{-13} J ($\sim$1 MeV), while a photon near the radio end of the spectrum has an energy of only 10^{-21} J ($\sim$0.01 eV, see Box 4-1). Thus, there is an enormous difference between the energies of photons from different parts of the spectrum. It must also be remembered that the energies involved in most chemical reactions are of the order of 1 to 4 eV or 2×10^{-19} to 7×10^{-19} J per molecule (30–100 kcal/mol). Photons in the UV and visible part of the spectrum have energies which are in this range. It should, therefore, not be very surprising to learn that these photons are involved in photochemical and photobiological reactions. Since chemical bonds are primarily interactions between electrons, it is probable that the initial effect of the absorption of a photon is the redistribution of the bonding electrons in the molecule. Thus, photons in the UV–visible region have a special ability to initiate chemical reactions when they are absorbed.

---

1.   A photon and its energy will be defined in the section on the photoelectric effect.

Box 4-1   Units of Energy

The fundamental unit of energy is the joule (J) which is a newton·metre (N·m). This unit is, however, not the most appropriate for describing energy processes at the atomic level. Chemists measure such energies in kcal/mol which specifies the energy in heat units distributed over one Avogadro's number of molecules. To describe the energy possessed by one electron in one atom requires a smaller unit. The common one is the "electron volt" (eV). This is the kinetic energy acquired by a single electron starting from rest and accelerating through a potential difference of one volt. This energy is the product of the charge and the potential difference, i.e., $1.0 \text{ eV} = 1.602 \times 10^{-19} \text{ C} \times 1 \text{ V} = 1.602 \times 10^{-19} \text{ J}$. Therefore $1 \text{ eV} = 1.602 \times 10^{-19} \text{ J}$. The energy of chemical bonds is of the order of a few electron volts. The energy of the inner-shell electrons of metal atoms is of the order of kilo-electron volts ($1 \text{ keV} = 10^3 \text{ eV}$), and the energy of particles in the nuclei of atoms is of the order of mega-electron volts ($1 \text{ MeV} = 10^6 \text{ eV}$).

By contrast, photons in the X-ray or γ-ray region are so energetic that they break molecular bonds; hence, such photons are said to be *ionizing radiation*. Studies of the interaction of ionizing radiation and matter have led to development of such disciplines as radiation physics, radio chemistry and radiation biology, subject areas that will be discussed in Chapter 6. At the other end of the spectrum, photons of infrared and microwave radiation have very little energy and so have a much more subtle effect on the molecule. These photons induce changes in the ways in which molecules vibrate and rotate. Detailed investigations of infrared absorption have led to an extensive knowledge of the structure of molecules. Thus, while the basic physical characteristics of all photons are the same, the particular effect a photon has on a molecule depends solely on the photon energy.

For light to be absorbed by a molecule, there must be some kind of interaction between the light wave and the particles which constitute the molecule. Considerable space has already been devoted to a discussion of the properties of electromagnetic waves in Chapter 3. It remains to discuss the properties of the particles which make up molecules (in particular, the electrons), and to discuss their role in the light absorption process. These topics lead to a detailed discussion of the visual process in Chapter 5.

4.2 **Electromagnetic Waves and Photons**

The electromagnetic (EM) wave was introduced in Chapter 3, where it was seen that it consists of an oscillating (and propagating) electric field.[2] Chapter 15 on the topic of electricity will explain that electric fields exert forces on electric charges. It is not surprising, then, that EM waves exert forces on the electrons in molecules. As these forces are exerted, work is done and energy can be transferred from the EM wave to the electrons and thus to the molecule. The laws that apply to this transfer, however, are not those that derive from Newton's mechanics and the classical electricity discussed in Chapter 15; they are a new set of laws of *quantum physics* that apply to the interaction of EM waves with atoms and molecules. There are many ways, both experimental and theoretical, to introduce the quantum laws. We will adopt here one of the most straightforward experimental phenomena for this purpose: the *photoelectric effect*.

The Photoelectric Effect

Knowing that EM waves exert forces on (and can transfer energy to) electrons and that all matter contains electrons, it is not surprising to learn that EM waves (light) can cause electrons to be ejected from certain materials. This is especially true for metals in which the conduction electrons are already very loosely bound to the metallic atoms. The emission of electrons from metals by the absorption of light is called the photoelectric effect.[3] What is surprising is that measurements on the energy of the emitted electrons cannot be explained by classical physics, but require new rules.

Figure 4-1 shows an experimental apparatus for making these measurements. Light from a monochromatic source (S) at frequency f (and wavelength λ) illuminates a metal surface (M) where electrons are emitted, some of which are traveling in the direction of the negatively charged electrode (C). The electric potential V of the electrode can be varied and, since like charges repel, the electrons will be slowed down on their way. Those that do make it will contribute to a current that can be detected by the ammeter (A). If the potential of C is large (and negative) then the electron will be brought to rest well before reaching C and will be turned back as shown in trajectory 1. Trajectory 2 is for

---

2. The simultaneous presence of the magnetic field will not enter into this simplified discussion.

3. Discovered in 1888 by the German physicist Heinrich Rudolf Hertz (1857–1894).

the case of a lower
potential of C.
Trajectory 3 indicates
the case for which the
electron has just the
right amount of energy
to make it to the
electrode with no kinetic
energy ($K = \frac{1}{2}mv^2$) left
(i.e., $v = 0$).

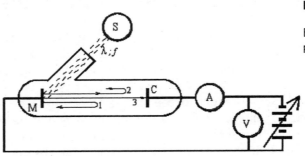

Figure 4-1

Experiment to investigate the
photoelectric effect.

Naively one might
suppose that random
processes would result in electrons being emitted with a wide range of
energies from small to large values, and that a current would always
show in the ammeter for all reasonable values of V. This is not the case;
it is found that there is a particular (absolute value) of V above which
no electrons reach the electrode. This $|V|$ is known as the "stopping
potential" V_s, and is related to the maximum K (K_{max}) of the emitted
electrons. As shown in Chapter 15, the work done on a charge q as it
passes through a potential difference V is qV. An electron (charge $-e$)
emitted with K_{max} will just be stopped when it reaches electrode C if $|V|$
$= V_s$. The work done ($-eV_s$) equals the loss in kinetic energy ($-K_{max}$),
and thus $eV_s = K_{max}$.

Increasing the intensity of the light, which certainly increases the
total available energy, has no effect on the value of V_s. The current is
increased (there are more electrons so energy is conserved) but the
maximum energy of the electrons (as measured by V_s) does not change.
If the frequency of the incident light is changed, V_s also changes, and it
is further found that ΔV_s is proportional to Δf. If a different metal is
used with the same frequency, V_s will change. Such experiments show
that the maximum energy of the emitted electrons depends only on the
type of metal and the frequency of the incident light; the intensity of
the light affects only the number of electrons. In addition there is a
threshold frequency f_0 such that, for frequencies below f_0, no matter how
intense the light, no electrons are emitted,[4] i.e., $V_s = 0$. A plot of V_s vs. f
for a metal is shown in Fig. 4-2 where the linear increase of V_s with
frequency is evident and the threshold frequency f_0 can also be seen.

---

4. Most of this experimental work was by the Hungarian physicist Philipp
    Lenard (1862–1947) around 1902.

Since classical physics will not explain these simple experimental results, they were for many years a major puzzle in physics. The explanation was supplied by Einstein[5] in 1905. In order to explain the emission of radiation from hot glowing objects (the so-called *blackbody radiation* discussed in Chapter 14), Planck,[6] in 1900, had introduced the *quantum hypothesis*. In this model, radiation is assumed to have the properties of a wave during propagation, but has some particle-like properties when emitted from atoms. It behaves as though it were emitted in indivisible "lumps" called *quanta* or *photons*. Planck further assumed that the energy *E* of a quantum of radiation was proportional to the frequency of the radiation, and described it in Planck's relation:

$$E = \mathrm{h}f = \frac{\mathrm{hc}}{\lambda} \qquad \qquad \text{[4-1]}$$

The constant of proportionality h is called Planck's constant and has the value 6.626×10^{-34} J·s; c is the speed of light in a vacuum and has the value 2.998×10^8 m/s.[7]

EXAMPLE 4-1

The peak of the human eye's sensitivity is near a wavelength of 555 nm. What is the energy of a photon of this light in joules and in electron volts?

SOLUTION

$E = \mathrm{hc}/\lambda \ = (6.626 \times 10^{-34} \text{ J·s} \times 2.998 \times 10^8 \text{ m/s})/555 \times 10^{-9} \text{ m}$
$\qquad = 3.58 \times 10^{-19}$ J

$E = 3.58 \times 10^{-19}$ J$/(1.602 \times 10^{-19}$ J/eV$) = 2.24$ eV

---

5. Albert Einstein (1879–1955). For this work he was awarded the Nobel Prize in 1921.

6. Max Karl Ernst Ludwig Planck (1858–1947), a German physicist, was awarded the Nobel Prize in 1918.

7. A particularly useful form of Planck's relation (Eq. [4-1]) is known as the *Duane–Hunt relation* which gives the energy of a photon in electron volts when the wavelength is specified in nanometres: $E(\mathrm{eV}) = 1240/\lambda(\mathrm{nm})$. See Problem 4-6.

ALTERNATE SOLUTION
(Using the Duane–Hunt relation, see Footnote 7)

$E = 1240/555 = 2.23$ eV

Einstein reasoned that if radiation was emitted in quanta by atoms then it probably was absorbed in quanta as well. This means that when an electron in the metal absorbs a quantum of radiation it must take all of it; this photon cannot be subdivided leaving some of its energy unabsorbed. What happens to this absorbed energy? Some of it becomes the K of the free electron, but some is required to unbind the electrons from the metal surface.

The strength with which electrons are bound to the surface of a given metal varies, and hence the energy required to unbind them is not constant. However, the minimum energy required to unbind an electron from a particular metal is constant and is known as the *work function* of the metal, usually represented by the symbol Φ.[8] If an electron can be unbound with energy Φ, then it will leave the surface with the maximum possible amount of kinetic energy K_{max}.

Applying the principle of conservation of energy

$$\mathrm{h}f = K_{max} + \Phi \qquad\qquad \textbf{[4-2]}$$

which is *Einstein's photoelectric equation*. If K_{max} is replaced by eV_s in Eq. [4-2] and the resulting equation solved for V_s then

$$V_s = \frac{\mathrm{h}}{e}f - \frac{\Phi}{e}, \text{ for } f \geq f_0 \qquad\qquad \textbf{[4-3]}$$

which is the equation of a straight line for V_s vs. f and is exactly depicted in Fig. 4-2. The existence of the threshold frequency f_0 arises for electrons that are given just enough energy to unbind from the metal (Φ) with nothing left over to supply K, i.e., $V_s = 0$. For frequencies below f_0 there is not enough energy to unbind the electrons. If any one of the three quantities Φ, e or h are known,

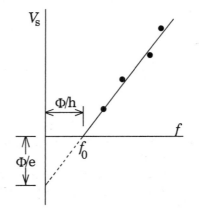

Figure 4-2

Graph of stopping potential vs. frequency of light.

---

8. Φ is the uppercase Greek letter *phi*.

then photoelectric experiments allow for the determination of the other two. Notice that if energies are measured in electron volts then Eq. [4-3] is particularly simple since the division by e converts the terms on the right side to electron volts; it becomes

$$V_s = E_{photon} - \Phi \text{ (for energies measured in electron volts)} \quad \textbf{[4-4]}$$

EXAMPLE 4-2

A metal is illuminated in a vacuum with light of wavelength 414 nm. The stopping potential of the photoelectrons is 2.0 V. What is the work function of the metal in electron volts?

SOLUTION

The energy of the photon $= hf = hc/\lambda$
$= (6.626 \times 10^{-34} \text{ J·s} \times 2.998 \times 10^8 \text{ m·s}^{-1})/414 \times 10^{-9} \text{ m}$
$= 4.80 \times 10^{-19} \text{ J} = 4.80 \times 10^{-19} \text{ J}/1.602 \times 10^{-19} \text{ J·eV}^{-1}$
$= 3.0 \text{ eV}$

Using Eq. [4-4]:

$\Phi = E_{photon} - V_s = 3.0 - 2.0 = 1.0 \text{ eV}$

ALTERNATE SOLUTION
(Using the Duane–Hunt relation, see Footnote 7)

The energy of the photon $= 1240/414 = 3.0 \text{ eV}$ (then proceed as above)

4.3   The Wave Nature of Electrons

The revolution in physics brought about by the ideas of Planck and Einstein showed that phenomena previously thought to be purely waves had, under the right conditions, some of the properties of particles. An obvious question that can be raised, then, is whether things which had heretofore been considered particles could exhibit wavelike properties. By 1927 there was already experimental evidence for this from the reflection of electron beams from the surface of nickel (Ni) crystals,[9] but it required the application of the earlier theoretical speculations of de Broglie[10] to provide the explanation.

A purely wavelike phenomenon is that of *interference* and *diffraction* discussed in Chapter 3 for light. If particles have wavelike properties, can they exhibit diffraction when passing through apertures or reflecting off fine gratings? It was this phenomenon which Davisson and Germer had observed. In Fig. 4-3 are shown photographs of the fluorescent screen of an electron diffraction tube. A beam of electrons, similar to the beam in a television picture tube, is made to pass through a very thin foil of polycrystalline carbon and then fall on the screen. If the electrons continued to behave as particles one might expect the beam to be spread out a bit because of interactions with the carbon atoms, but in addition there is a series of *diffraction rings*. If the carbon were a single crystal, the screen would show a series of diffraction spots. However, the beam is wide enough to include many small carbon crystallites with random orientations, and the randomly oriented diffraction patterns add up to a series of rings. In the lower photograph of Fig. 4-3, the voltage accelerating the electrons is greater than that in the upper photograph. This means that the electrons in the second case have a higher K and therefore a higher velocity and momentum. As can be clearly seen, the rings have a smaller diameter; de Broglie's hypothesis explains all these facts.

De Broglie's hypothesis was that moving particles would have a wavelength given by

$$\lambda = \frac{h}{p} = \frac{h}{mv} \qquad \textbf{[4-5]}$$

where $p = mv$ is the linear momentum of the particle of mass m and speed v. (The units of momentum are kg·m/s.) When this simple relation is applied to the experiments of Davisson and Germer or to the electron diffraction rings of Fig. 4-3, the results are explained perfectly (see Box 4-2).

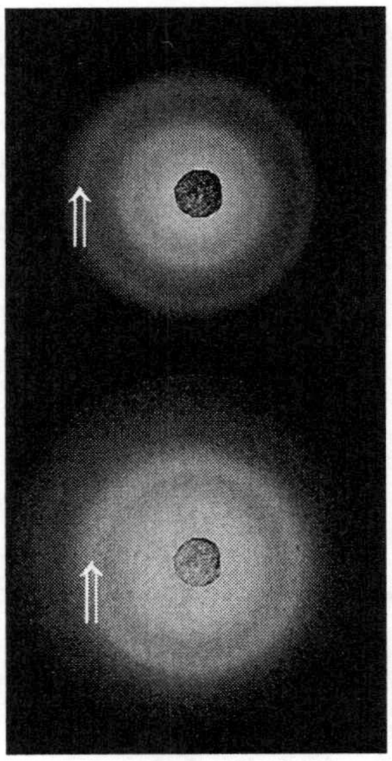

Figure 4-3

Electron diffraction rings from polycrystalline carbon. Upper: 3000 V; Lower: 4000 V. The arrows indicate the same diffraction ring.

9.   Obtained by American physicists Clinton J. Davisson (1881–1958) and L. Germer (1896–1971) at the Bell Telephone Laboratories. Davisson was awarded the Nobel Prize in 1932.

10. Prince Louis de Broglie (1892–1987), French philosopher and physicist, awarded the Nobel Prize in 1929.

Box 4-2 Analysis of Electron Diffraction Rings

By direct measurement on Fig. 4-3, the ratio of the diameters of the most prominent diffraction rings (upper over lower) is 1.12.

When electrons of charge –e are accelerated through a potential difference V (see Chapter 15), their energy is equal to eV which appears as $K = \frac{1}{2}mv^2$ where m is the mass of an electron and is a constant (m_e). Thus

$$2m_e eV = (m_e v)^2 = p^2$$

The diameter d of the diffraction ring is proportional to the wavelength λ of the diffracting radiation (see Chapter 3) and so using

$$p = m_e v = \sqrt{2m_e eV}$$

the wavelength is

$$\lambda = \frac{h}{p} = \frac{h}{\sqrt{2m_e eV}} = \frac{constant}{\sqrt{V}}$$

The accelerating voltages in the two pictures are 3000 V (above) and 4000 V (below), therefore

$$\frac{d_1}{d_2} = \frac{\lambda_1}{\lambda_2} = \sqrt{\frac{V_2}{V_1}} = \sqrt{\frac{4000}{3000}} = 1.15$$

The agreement with the measured ratio, 1.12, is within 3%.

EXAMPLE 4-3

What is the de Broglie wavelength of the electrons producing the lower diffraction pattern in Fig. 4-3 where the accelerating potential was 4000 V? (The electron mass is 9.109×10^{-31} kg.)

SOLUTION

Using the analysis of Box 4-2, $p = (2m_e eV)^{1/2}$
$= (2 \times 9.109 \times 10^{-31} \text{ kg} \times 1.602 \times 10^{-19} \text{ C} \times 4000 \text{ V})^{1/2}$
$= 3.42 \times 10^{-23}$ kg·m·s^{-1}

The wavelength is $\lambda = h/p$
$= 6.626 \times 10^{-34}$ J·s $/3.42 \times 10^{-23}$ kg·m·s$^{-1} = 1.94 \times 10^{-11}$ m, which is the size of an atom!

If particles have wavelike properties, what is it that is "waving"? The answer to this question is difficult to visualize; we say that the wave is represented by a *wave function* with the symbol ψ.[11] This wave function is related to a more concrete quantity called the *probability density (P)* through the relation $P = \psi^2$. In situations where electrons are confined to moving in one dimension only (say along the length of a long molecule) then the probability density is the probability per unit length of locating the electron in some small region of the molecule. If the probability density at some point is large, then the likelihood of finding the electron near that point is large. If the probability density at some point is low, then the probability of finding the electron near that point is low.

Since particle waves are used to describe electrons in atoms and molecules before and after light absorption, it is fairly easy to confuse the properties of particles and electromagnetic waves. Table 4-1 lists the properties of both waves and particles and will help remove some of the confusion.

---

11. ψ is the lowercase Greek letter *psi*.

Table 4-1   Comparison of Electromagnetic and Particle Waves

Electromagnetic Waves	Particle Waves
magnetic and electric fields oscillate	ψ and P oscillate
travel in packets called quanta or photons	particles travel singly or in beams
produced by accelerating a charge	particle beams can be produced by accelerators or reactors, for example
energy of a photon $= hf$	energy of particle $= \frac{1}{2}mv^2$
travel with the speed of light c	can travel at any speed $(< c)$
wavelength $\lambda = c/f = hc/E$	wavelength $\lambda = h/p = h/mv$

4.4 **Electrons in Molecules**

In the introduction to this chapter, it was stated that the initial event following the absorption of a UV or visible photon was a restructuring or redistribution of the bonding electrons in the molecule. This implies that if the process is to be understood, some knowledge is required of the distribution of electrons before and after the absorption event. This involves a consideration of the wave properties of the electron. Thus, in this section use is made of standing particle waves to obtain probability densities and hence the distribution of electrons. This approach yields further benefits in that the energies of the electrons in the molecule and the approximate wavelengths of light the molecule will absorb can be calculated. This *quantum mechanical* approach may seem complicated and difficult, but a large number of chemists, molecular biologists and biophysicists use the theory to explain various biological phenomena such as mutations, macromolecular structure, and enzyme reactions in addition to the photobiological processes considered here.

The molecules under consideration are mostly those formed of carbon–carbon (C–C) bonds, either as a linear chain or a ring array, such as the retinal and cytosine molecules shown in Fig. 4-4. In all such molecules the bonding electrons can be divided into two distinct classes. The first group is one in which the electrons are strictly localized on the

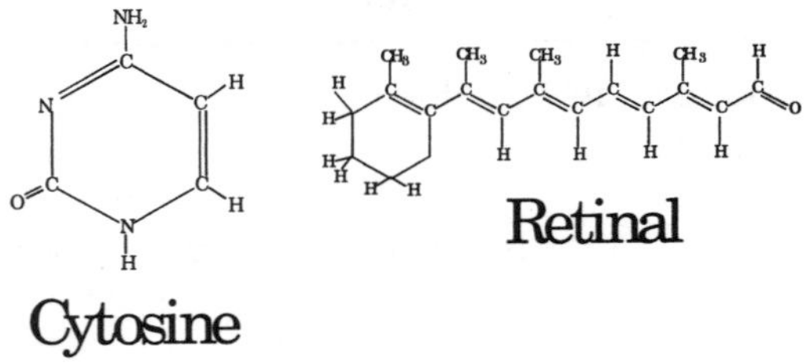

constituent atoms.[12] These
are so tightly bound that they
do not enter into the
light-absorption processes
considered here. The second
group, because of the
presence of the first group,
find themselves at a greater
distance from the atoms and
become almost totally
de-localized from the atoms
which "donated" them to the
molecule.[13] These π-electrons
are members of the molecule as a whole, much as the conduction
electrons in metals are members of the metal as a whole, and are not
localized on a particular atom. The number of π-electrons present in
these molecules equals the number of atoms in the conjugated chain
(alternating single and double bonds) of the molecule.[14] Thus retinal in
Fig. 4-4 has 12 π-electrons and cytosine has 6.

Figure 4-4

Linear and cyclic conjugated
π-bonded molecules.

Cytosine

Retinal

In these molecules, the π-electrons are free to roam anywhere in the
conjugated double-bond chain. Since the π-electrons are not as tightly
bound as σ-electrons, they undergo excitation comparatively easily. The
approximation is made that the π-electrons alone are involved in the
absorption of UV or visible light. Thus, it is assumed that absorption of
UV or visible light brings about a restructuring or redistribution of the
π-electrons only. While the σ-electrons are ignored in the following
treatment of light absorption, it could be mentioned that these
electrons do play a role in the absorption of very short wavelength UV
and X-ray radiation.

All of the 12 π-electrons in the retinal molecule are free to move
throughout the long chain of conjugated double bonds. Since the
particle wave associated with the electron is related to the probability
density of the electron, it too must exist along this conjugated
double-bond sequence. It is necessary to find a particle wave function
which corresponds to each of the 12 π-electrons in the conjugated
double-bond sequence. The double bonds are first imagined to be
strung out in a straight line to simplify the calculation. This string of

12. These are referred to as the σ-electrons (*sigma* electrons) in molecular
     orbital theory.
13. These are referred to as the π-electrons (*pi* electrons) in molecular orbital
     theory.
14. This means that, of the four outer-shell electrons brought into the molecule
     by each carbon atom, three are involved in σ-bonding and one in
     π-bonding.

Figure 4-5

The linear π-bonded chain of retinal.

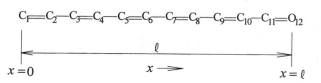

$$C_1\!\!=\!\!C_2\!\!-\!\!C_3\!\!=\!\!C_4\!\!-\!\!C_5\!\!=\!\!C_6\!\!-\!\!C_7\!\!=\!\!C_8\!\!-\!\!C_9\!\!=\!\!C_{10}\!\!-\!\!C_{11}\!\!=\!\!O_{12}$$

$x = 0$ ℓ $x \longrightarrow$ $x = \ell$

carbons corresponds to a one-dimensional "box" in which the x (horizontal) axis runs along the length ℓ of the molecule as shown in Fig. 4-5. The length of the box is roughly the length of eleven C–C bonds which is $11 \times 0.15 = 1.65$ nm.[15]

4.5   Linear Molecules

The Wave Function and Probability Density for a Particle in a Box

Imagine adding the 12 π-electrons one at a time to the box. Once the first one is added, there must be a finite probability of finding the particle at some point inside the box, but the probability of finding the electron outside the box must be zero. We need a ψ, therefore, which has a finite value inside the box, but goes to zero at the ends and everywhere else. Many functions could be chosen to satisfy these requirements, but the simplest is

$$\psi(x) = b \sin\left(\frac{\pi x}{\ell}\right) \qquad\qquad \textbf{[4-6]}$$

Figure 4-6

Simplest π-electron wave function.

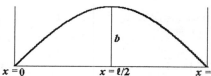

$x = 0$ b $x = \ell/2$ $x = \ell$

where x can have values between 0 to ℓ only (i.e., $0 \leq x \leq \ell$) and b is a constant. If $x = 0$ or ℓ, then the sine is zero as must be the case if the probability of finding the electron at, or beyond, the ends of the molecule is to be zero.

The function, when superimposed on the box, looks like the one shown in Fig. 4-6. Note that a cosine function will not satisfy the boundary condition that ψ = 0 at the ends, and it cannot be considered as a possible wave function.

---

15.   In all molecules of this type, the C–C bond length is very nearly constant at 0.15 nm.

According to the ideas presented earlier, this wave function describes the shape and density of the electron distribution. At the centre of the box (which corresponds to the midpoint of the bond between C_6 and C_7), the

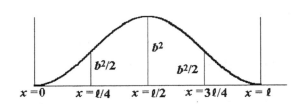

Figure 4-7

Simplest π-electron probability density.

cloud is most dense. The probability density $[P(x)]$ of finding the electron is greatest in this region. To calculate this probability density at any position x, the equation $P(x) = \psi^2(x)$ is used, where $\psi(x)$ is the value of ψ at the point x. There are many rigorous reasons why $P(x)$ is equated to $\psi^2(x)$, rather than to $\psi(x)$, but one obvious justification is that one cannot have a negative probability, and $\psi^2(x)$ is always positive (even if $\psi(x)$ is negative). The graph of $P(x)$ corresponding to Fig. 4-6 is shown in Fig. 4-7, which is just the graph of the square of the sine function.

What of the constant b which appears in front of the sine in Eq. [4-6]? At each point along x, there is a particular probability density $P(x)$; for example at $x = \ell/4$ the probability density $P(x)$ of finding the electron is $0.5\, b^2$. However, since one electron was put in the box, if the sum is made of all the probability densities over the whole box, the total probability of finding the electron in the box will be found. Since the electron cannot be anywhere else, that probability is 1.

Mathematically, this summation corresponds to an integration of the probability density over the box. Thus,

$$b^2 \int_{x=0}^{x=\ell} \left(\sin \frac{\pi x}{\ell} \right)^2 dx = 1 \qquad \text{[4-7]}$$

On performing this integration (see Box 4-3) and solving for b, the result is $b = (2/\ell)^{1/2}$ and the wave function is

$$\psi = \sqrt{\frac{2}{\ell}} \sin\left(\frac{\pi x}{\ell} \right) \qquad \text{[4-8]}$$

Box 4-3   Evaluation of the Constant "b"; "Normalization" of the Wave Function

Using Eq. [4-7],

$$b^2 \int_{x=0}^{x=\ell} \left(\sin \frac{\pi x}{\ell} \right)^2 dx = 1$$

(cont'd.)

Let $\pi x/\ell = q$, therefore $(\pi/\ell)dx = dq$; and

$$1 = \frac{\ell}{\pi} b^2 \int_0^\pi \sin^2 q \, dq$$

The integral of $\sin^2$ is given by (see any table of integrals):

$$\int \sin^2 q \, dq = \frac{1}{2} q - \frac{1}{4} \sin(2q)$$

Therefore

$$1 = \frac{\ell}{\pi} b^2 \left[\frac{1}{2} q - \frac{1}{4} \sin(2q) \right]_0^\pi$$

$$1 = \frac{\ell}{\pi} b^2 \left[\frac{\pi}{2} - \frac{1}{4} \sin(2\pi) - 0 + \frac{1}{4} \sin(0) \right]$$

$$1 = \frac{\ell}{\pi} b^2 \left[\frac{\pi}{2} \right]$$

$$b^2 = \frac{2}{\ell}$$

Equation [4-8] is one of a whole family of functions of the form

$$\psi_n(x) = \sqrt{\frac{2}{\ell}} \sin\left(\frac{n\pi x}{\ell}\right) \quad n = 1, 2, 3, \ldots \qquad \textbf{[4-9]}$$

each of which is a wave function which satisfies the boundary requirements. Wave functions ψ_2 and ψ_3 are shown in Fig. 4-8a.[16]

Figure 4-8

π-electron (a) wave function and (b) probability density for $n = 2, 3$.

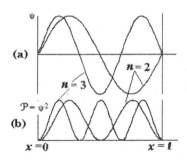

Inspection of this figure indicates that the wave functions for different values of n are similar to the various possible standing waves on a string fixed at the ends (see Chapter 1). Indeed, the wave functions described by Eq. [4-9] are standing particle waves of the π-electrons in the molecule. The time-dependent part of the standing wave (the $\cos(\omega t)$ part) has been

16. The quantum number n is <u>not</u> the same "principal" quantum number n describing the electronic energy in atoms. This n describes the energy and momentum of the π-electrons in molecules only.

ignored because our interest has been centered on the spatial location of the electron. The number $n = 1, 2, 3, \ldots$ is called a *quantum number* and specifies the number of electron half-wavelengths $(\lambda/2)$ which must appear in the box. The probability density curves which correspond to the wave functions ψ_2 and ψ_3 are shown in Fig. 4-8b.

The dimensions of the probability density $P(x)$ are apparent by squaring Eq. [4-9] to obtain $P(x)$, the probability per unit length, i.e., the dimensions of P are length^{-1}. To obtain a probability $\mathcal{P}$ (which is a pure number with no dimensions), the probability density must be multiplied by a length. The probability of finding the electron in a region of length Δx is

$$\mathcal{P} = P(x)(\Delta x) \qquad \text{[4-10]}$$

If $P(x)$ is constant (or very nearly so) over the region Δx then this equation is satisfactory. However, if $P(x)$ is changing across the region then an integration is necessary:

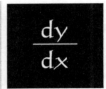

$$\mathcal{P} = \int_{x}^{x+\Delta x} P(x) \, dx \qquad \text{[4-11]}$$

In most of the problems considered in this text, the length, Δx, is small. In this situation Eq. [4-10] yields satisfactory values for $\mathcal{P}$.

EXAMPLE 4-4

(a) Find the probability density of an $n = 3$ electron at the point indicated by the arrow in the wave function in Fig. 4-9.

(b) Find the probability of finding the electron in the region of Fig. 4-9 marked as Δx.

(c) What would be the probability of finding the electron between $x = 0$ and $x = \ell/3$?

Figure 4-9

Probability density with $n = 3$.

SOLUTION

(a) Using Eq. [4-9] for ψ

$$P(x) = \psi_3^2(x) = \frac{2}{1.65}\sin^2\left(3\pi\left(\frac{x}{\ell}\right)\right)$$

where $\dfrac{x}{\ell} = \dfrac{1}{6}$

$$P(x) = \frac{2}{1.65}\sin^2\left(\frac{\pi}{2}\right) = 1.21 \text{ nm}^{-1}$$

(b) Eq. [4-10] can be used if $P(x)$ is assumed to be constant along Δx. (From Fig. 4-9 this is not too bad an approximation since Δx is small.)

Thus,

$$\mathcal{P} = 1.21 \text{ nm}^{-1} \times 0.050 \text{ nm} = 0.061$$

Therefore the probability of finding the $n = 3$ electron in the short region is 0.061 or 6.1 percent of the time.

(c) If the function in Fig. 4-9 were squared to give the probability density, one-third of the area under the resulting curve would be between $x = 0$ and $x = \ell/3$. The probability of finding the electron between $x = 0$ and $x = \ell$ is 1, which is the total area under the curve. Therefore the probability of finding the electron in the $x = 0$ to $x = \ell/3$ region is 1/3, or 33.3 %.

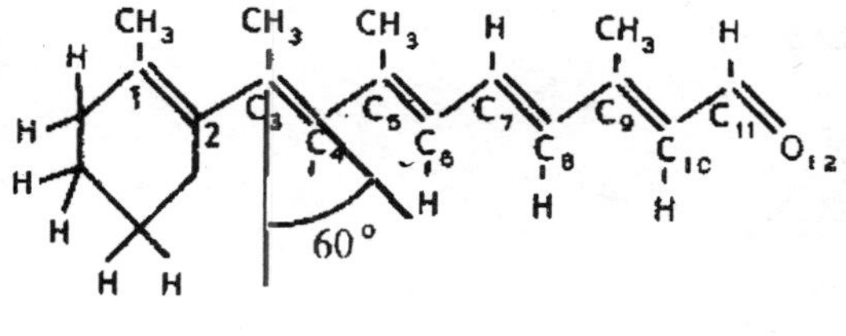

Figure 4-10

The geometry of the bonds in the π-bonded chain of retinal.

In all of the preceding it has been assumed that the length of the molecule in question, and therefore the length of the one-dimensional box that confines the electron, is the sum of the individual bond lengths. This is not exactly the case because the adjacent bonds are at an angle to each other. In all of the linear molecules considered here, the bond angle is about 120°. The situation is shown for retinal in Fig. 4-10 (compare with its depiction in Fig. 4-5). Clearly the effective bond distance, which determines all lengths along the molecule, is given by $(0.15 \text{ nm}) \sin(60°) = 0.15 (0.866) = 0.13$ nm, i.e., all distances as previously calculated should be multiplied by 0.866.

EXAMPLE 4-5

Solve Example 4-4 using the length corrected for the bond angle of 120°.

SOLUTION

Only the answers to parts (a) and (b) are affected.

(a) $2/\ell$ becomes $2/(1.65 \text{ nm} \times 0.866) = 1.4 \text{ nm}^{-1}$; the argument of the sine is unaffected.

Therefore,

$P(x) = 1.4 \text{ nm}^{-1}$

(b) $\mathcal{P} = 1.4 \text{ nm}^{-1} \times 0.05 \text{ nm} = 0.07$

Convince yourself that the answer to part (c) is unaffected.

The Energy of a Particle in a Box

What is the energy of a π-electron having quantum number n? Since n is the number of half-wavelengths of the electron in the box, then $n\lambda/2 = \ell$. Using $\lambda = h/m_e v$ (Eq. [4-5]) as λ, the wavelength of the electron, where m_e is the electron's mass, gives,

$$\frac{n\text{h}}{2\text{m}_e v} = \ell$$

Squaring the above gives

$$\frac{n^2\text{h}^2}{4\text{m}_e^2 v^2} = \ell^2$$

The $m_e^2v^2$ in this equation can be replaced as follows. The kinetic energy of a particle such as the electron is $K = \frac{1}{2}m_ev^2 = m_e^2v^2/2m_e$. This kinetic energy is very nearly the total energy of the π-electron, E_n.[17] Thus

$$m_e^2v^2 = 2m_eE_n$$

which gives

$$\frac{n^2h^2}{8m_eE_n} = \ell^2$$

Thus the energy of the π-electron with quantum number n in a linear π-bonded molecule is

$$E_n = \frac{n^2h^2}{8m_e\ell^2} \quad n = 1, 2, 3, \ldots \qquad \textbf{[4-12]}$$

EXAMPLE 4-6

What is the energy (in electron volts and joules) of an electron in the state $n = 6$ in retinal? Evaluate it both with and without the bond-angle correction. (The mass of an electron $m_e = 9.109 \times 10^{-31}$ kg; h = 6.626×10^{-34} J·s)

SOLUTION

The length of the molecule without the bond-angle correction is 11×0.15 nm = 1.65 nm = 1.65×10^{-9} m.

Using Eq. [4-12] with $n = 6$
$E_6 = [6^2 (6.626 \times 10^{-34} \text{ J·s})^2]/[8(9.109 \times 10^{-31} \text{ kg})(1.65 \times 10^{-9} \text{ m})^2]$
$\qquad = 7.97 \times 10^{-19}$ J
$\qquad = 7.97 \times 10^{-19}$ J $/(1.602 \times 10^{-19}$ J/eV) = 4.97 eV

If the correct bond angle is used, then ℓ must be multiplied by $\sin(60°) = 0.866$ and therefore the energy is increased by $1/(0.866)^2 = 1.33$

$E_6 = 7.97 \times 10^{-19}$ J $\times 1.33 = 10.6 \times 10^{-19}$ J = 6.6 eV

---

17. This approximation is reasonable here since the π-electrons are relatively far from the positively charged nuclei, and so the electric potential energy is small. It is therefore assumed to be zero.

4.6   Electronic States of Linear Molecules

As can be seen in Eq. [4-12], the energy of a π-electron increases as n increases; in fact it increases with n^2. The absorption of light by the molecule involves promoting an electron from one value of n to one of higher value, the energy of the photon supplying the energy difference. But what is the initial value of n, and by how much does it change?

Before the first question can be addressed, it must be noted that another quantum state (and therefore another quantum number) must be introduced to describe a π-electron. This other state is called *electron spin* because, in some naive models of the electron, it can be thought of as a spinning top. Tops can spin clockwise or counterclockwise and so the electron can have two spin states. These states are described by a *spin quantum number (s)* which can have only one of two values, viz., $s = \frac{1}{2}$ or $s = -\frac{1}{2}$.[18] These are sometimes referred to as "spin up" and "spin down," or represented by the two arrow symbols $\uparrow$, $\downarrow$, which are more useful here. Finally it is necessary to invoke the general principle known as the *Pauli exclusion principle* which dictates that:

> **In a given system (molecule), no two electrons may occupy the same quantum state (have the same set of quantum numbers).**

Any electron, then, with a given value of n can have one of two spin states, $\uparrow$ or $\downarrow$. If one imagines adding the π-electrons to the molecule one after another starting in the lowest energy state ($n = 1$), then there can be two electrons in that state and in each of the other n states until all the electrons are accounted for. For example, Fig. 4-11 shows the electron configuration of lowest energy (*ground state*) for retinal. Recall from the discussion above that each atom in the π-bonded chain contributes one π-electron for a total of 12 in this case. Thus for retinal, in its ground state, the n levels are filled to $n = 6$, and the first empty state is $n = 7$. Such a diagram is called an *electronic state diagram*.

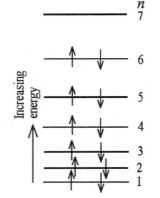

Figure 4-11

The electronic state diagram for the ground state of retinal.

The second question that was asked above was: "By how much does n change when a photon is absorbed?" The answer to a question of this

---

18.   The numerical values are of little consequence here. The value ½ arises in more detailed quantum theories.

kind, in quantum theory, is in the form of a selection rule. The answer in this case is that n can change by any integer amount, but by far the most probable is for it to change by 1. Thus we have

Selection Rule: $\Delta n = \pm 1$

When the ground state of the molecule is considered, then only the $+1$ change is possible.[19] In retinal, then, the lowest possible energy transition is from $n = 6$ to $n = 7$. The original and final state diagrams are shown in Fig. 4-12. In order for this transition to take place, the energy of a photon of frequency f and wavelength λ must be absorbed such that

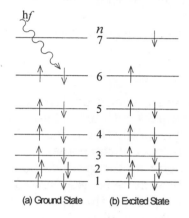

Figure 4-12

The (a) ground and (b) first excited state of retinal.

$$E_{n+1} - E_n = hf = \frac{hc}{\lambda} \quad \textbf{[4-13]}$$

EXAMPLE 4-7

What is the wavelength of the photon which is absorbed when an electron in the molecule retinal makes the lowest possible energy transition? Use the proper bond-angle configuration.

SOLUTION

From Fig. 4-12 it is seen that the lowest energy transition is that for $n = 6 \rightarrow n = 7$. From Example 4-6 it was seen that the energy of the electron in retinal for $n = 6$ was

$$E_6 = 10.6 \times 10^{-19} \text{ J} = 6.6 \text{ eV}$$

Since the energy simply increases as n^2 then the energy in the $n = 7$ state will be

$$E_7 = (7^2/6^2)E_6 = 14.4 \times 10^{-19} \text{ J} = 9.0 \text{ eV}$$

---

19. In the ground state, if n were to change by -1 for any electron, it would have to go to a state which is already fully occupied, and so is prevented by the Pauli exclusion principle.

The energy difference $E_7 - E_6 = (14.4 - 10.6) \times 10^{-19}\,\text{J} = hc/\lambda$ (using Eq. [4-1]), from which

$$\lambda = 5.2 \times 10^{-7}\,\text{m} \approx 520\,\text{nm}$$

ALTERNATE SOLUTION

Using the Duane–Hunt relation (see Footnote 7):

$$E_7 - E_6 = 9.0 - 6.6 = 2.4\,\text{eV} = 1240/\lambda, \text{ from which } \lambda \approx 520\,\text{nm}$$

In Example 4-7 it is seen that the simple theory presented here predicts an absorption maximum at 520 nm for retinal. In fact the absorption peak in retinal is at 510 nm. The prediction is faulty because of the approximations, in particular the neglect of the potential energy *(U)*. The *U* is not zero nor is it constant between the ground and excited state. In almost all cases the *U* in the excited state is greater, and so some of the energy of the photon must also provide this increase. Thus the energy of the photon must be greater than predicted by the simple theory and have a shorter wavelength as required. The discrepancy between theory and measurement is worse for shorter molecules.

4.7   Energy Level Diagrams

The electronic state diagrams of Figs. 4-11 and 4-12 serve only to determine the appropriate values of *n* for the ground state, and for the *excited states* which are states in which one or more of the electrons do not have their lowest *n* value. Fig. 4-12a shows the first excited state of retinal. Having fulfilled this function, such diagrams offer little further insight. Of more use is the *energy level diagram*. This is a "ladder" diagram in which the electron's (and therefore the molecule's) energies are represented to scale. Since energies below the ground state are inaccessible, the energies are represented relative to the ground state, i.e., $E_{\text{ground state}} = 0$. The energies of the five levels above $n = 5$ of retinal are given in Table 4-2, of which the first two can be found in Example 4-7. It is here assumed that only one electron is excited to the 7, 8, 9, 10 levels; the simultaneous excitation of more than one electron in the same molecule is a very improbable event. Figure 4-13 shows the energy of four excited states of retinal relative to the energy of the ground state. The lowest energy transition which was calculated in Example 4-7 is indicated by the upward pointing arrow; this transition would be accompanied by the absorption of a photon of $\lambda = 520$ nm.

Table 4-2   Electron Energies in Retinal

n	E_n (eV)	E (eV) relative to n = 6
6	6.6	0
7	9.0	2.4
8	11.8	5.2
9	14.9	8.3
10	18.4	11.8

Figure 4-13

The energy level diagram of retinal.

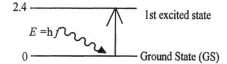

Energy in eV above GS

8.3 ——————— 3rd excited state

5.2 ——————— 2nd excited state

2.4 ——————— 1st excited state

$E = hf$

0 ——————— Ground State (GS)

4.8   Cyclic π-Bonded Molecules

As a second and slightly more complicated example, consider light absorption by porphyrin rings which are the chromophores (light-absorbing groups) of chlorophyll, hemoglobin, myoglobin and many other important molecules. Figure 4-14 illustrates the porphyrin systems found in several different biological molecules.

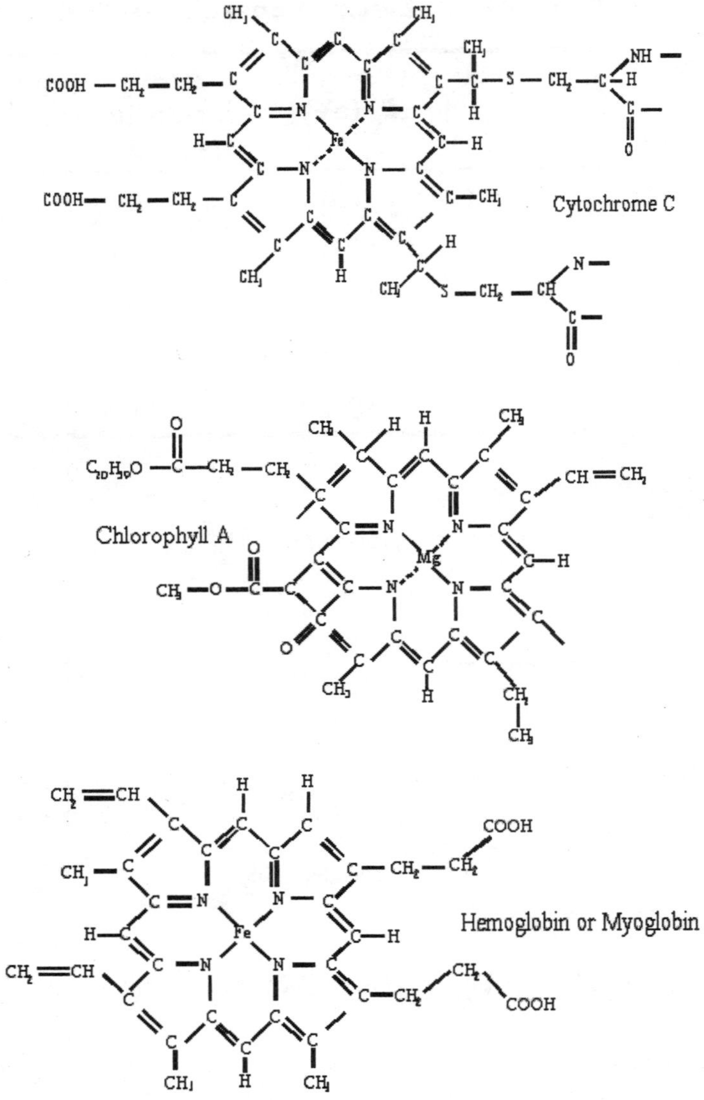

Figure 4-14

Porphyrin ring molecules:
(a) cytochrome C,
(b) chlorophyll A,
(c) hemoglobin or myoglobin.

(a) Cytochrome C

(b) Chlorophyll A

(c) Hemoglobin or Myoglobin

To reduce excessive computation, a porphyrin ring can be considered simply as a circle. In this problem we have to fit electron wavelengths around the ring. The electron wave function $\psi(x)$, where x is a position on the ring, must have the following property: if the electron starts off at some point x, having a wave function $\psi(x)$, then when it has gone once around the ring it has traveled a distance ℓ so that its wave function is now $\psi(x + \ell)$. But it has to come back to its original starting point where its wave function is $\psi(x)$. Hence,

$$\psi(x + \ell) = \psi(x) \qquad \text{[4-14]}$$

This property of periodicity is possessed by both the sine and cosine functions so that $\psi(x)$ can be either $\sin(kx)$ or $\cos(kx)$. Note that $\cos(kx)$ cannot be discarded as in the case of a linear chain of conjugated bonds because in this case there is no boundary condition on the wave function to exclude it. Therefore, applying Eq. [4-14]

$$\sin k(x + \ell) = \sin(kx)$$

and

$$\cos k(x + \ell) = \cos(kx)$$

which means that $k\ell = 2\pi n$ where $n = 1, 2, 3, \ldots$. Thus, it is necessary to fit whole wavelengths rather than half-wavelengths on the ring, as was discussed in Chapter 1 for standing elastic waves on a ring. The properly normalized wave functions are

$$\psi_{s,n} = \sqrt{\frac{2}{\ell}}\sin\left(\frac{2\pi n x}{\ell}\right) \quad n = 1, 2, 3, \ldots$$

$$\psi_{c,n} = \sqrt{\frac{2}{\ell}}\cos\left(\frac{2\pi n x}{\ell}\right) \quad n = 1, 2, 3, \ldots \qquad \textbf{[4-15]}$$

Notice that there are two possible wave functions for each value of $n = 1, 2, 3, \ldots$ whereas there was only one for the linear molecules.[20] This means that for cyclic molecules there can be four electrons in each $n > 0$ state: two with spins ↑, ↓ with a sine wave function, and two (↑, ↓) with a cosine.

However, there is still one other state. Although the sine function is zero for $n = 0$, the cosine is not, so there is an additional state $\psi_{c,0}$ which has quantum number 0 and can accommodate two electrons. For the quantum number to be 0, the momentum of the electron must be zero. From $\lambda = h/p$ (Eq. [4-5]), then λ must be infinitely long and the electronic wave function must be a constant over the length ℓ. Since the total probability of finding the electron on the molecule must be $\mathcal{P} = 1$, then the probability density P must be given by $P \times \ell = 1$ or $P = 1/\ell$, and

$$\psi_{c,0} = \sqrt{\frac{1}{\ell}} \qquad \textbf{[4-16]}$$

Figure 4-15 shows the electronic state diagram for a cyclic molecule; the electrons for chlorophyll, which has 20 π-electrons, are entered in the lowest energy (ground) state. The two electrons in the $n = 5$ state

20. This is called a "degeneracy" in quantum theory; these electron wave functions are "doubly degenerate."

could be both sine or both cosine rather than one of each as indicated. The important thing is that the $n = 5$ state is not filled, and the lowest energy transition in chlorophyll would be one in which an $n = 4$ electron (with proper spin) went into the $n = 5$ state as is also indicated in Fig. 4-15.

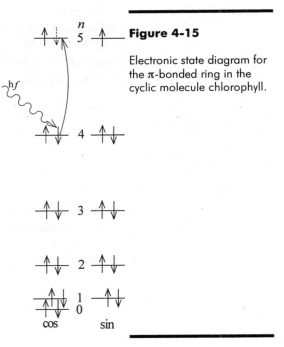

Figure 4-15

Electronic state diagram for the π-bonded ring in the cyclic molecule chlorophyll.

The length ℓ depends on the number of bonds in the ring. For porphyrin rings the effective bond length is 0.12 nm; this value takes into account the bond angles. For a 20-atom ring, then, $\ell = 20 \times 0.12$ nm $= 2.4$ nm. Notice, an n-bond ring has a length $n \times$ (effective bond length), whereas a linear molecule has a length $(n - 1) \times$ (effective bond length).

4.9 π-Electronic Energy in a Ring Molecule

The electron's energy is evaluated in a manner similar to the derivation of Eq. [4-12], with the same approximations. The only difference is that the de Broglie wavelength of the electron is now an integer multiple of ℓ rather than a half-integer, i.e., $n\lambda_e = \ell$ and, using Eq. [4-5], ($\lambda = h/m_e v$) for λ_e

$$\frac{nh}{m_e v} = \ell$$

Squaring the above gives

$$\frac{n^2 h^2}{m_e^2 v^2} = \ell^2$$

But the kinetic energy of a particle such as the electron is $K = \frac{1}{2}m_e v^2 = m_e^2 v^2/2m_e$. Again this kinetic energy is very nearly the total energy of the π-electron, E_n. Thus,

$$m_e^2 v^2 = 2m_e E_n$$

$$\frac{n^2 h^2}{2m_e E_n} = \ell^2$$

The energy of the electron with quantum number n in a cyclic π-bonded molecule is

$$E_n = \frac{n^2 h^2}{2m_e \ell^2} \quad n = 0, 1, 2, 3, \ldots \qquad \textbf{[4-17]}$$

EXAMPLE 4-8

What is the energy of the ground state and of the first excited state of chlorophyll?

SOLUTION

As can be seen from the electronic state diagram in Fig. 4-15, the energy of the lowest completely filled state of chlorophyll is that of an $n = 4$ electron.

Using Eq. [4-17],

$$E_4 = \frac{4^2 h^2}{2m_e \ell^2} = \frac{4^2 (6.626 \times 10^{-34})^2}{2(9.109 \times 10^{-31})(2.4 \times 10^{-9})^2}$$

$$= 6.7 \times 10^{-19}\, J = 4.2\ eV$$

The energy of an $n = 5$ electron is just $5^2/4^2$ of this value, i.e., 6.6 eV.

The rest of the electronic energies of chlorophyll from $n = 4$ to $n = 8$ are given in Table 4-3.

Table 4-3   Electron Energies in Chlorophyll

n	E_n (eV)	E (eV) relative to $n = 4$
4	4.2	0
5	6.6	2.4
6	9.5	5.3
7	12.9	8.7
8	16.8	12.6

Since the states below $n = 4$ in chlorophyll are inaccessible, then the ground state energy is taken as 0 and all excited states are relative to that. Using the values of Example 4-8, the energy of the first excited state is $(6.6 - 4.2)$ eV $= 2.4$ eV. The first four levels of the energy level diagram are shown in Fig. 4-16. The lowest energy transition in the molecule is indicated which corresponds to the electron transition from $n = 4$ to $n = 5$ shown in Fig. 4-15.

Energy in eV above GS

8.7 ——————————— 3rd excited state

5.3 ——————————— 2nd excited state

2.4 ——————————— 1st excited state

$E = hf$

0 ——————————— Ground State

Figure 4-16

The energy level diagram of chlorophyll.

EXAMPLE 4-9

What is the wavelength of the photon which is absorbed when an electron in a molecule of chlorophyll makes the lowest possible energy transition?

SOLUTION

From Fig. 4-15 it is seen that the lowest energy transition is that for $n = 4 \rightarrow n = 5$. From Example 4-8 it was seen that the energy of the electron in chlorophyll for $n = 4$ was

$E_4 = 6.7 \times 10^{-19}$ J (4.2 eV) and in the 5th level
$E_5 = 10.5 \times 10^{-19}$ J (6.6 eV)

The energy difference $E_5 - E_4 = (10.5 - 6.7) \times 10^{-19}$ J $= hc/\lambda$, from which $\lambda = 520 \times 10^{-9}$ m $= 520$ nm.

ALTERNATE SOLUTION

Using the Duane–Hunt relation:

$E_5 - E_4 = 6.6 - 4.2 = 2.4$ eV $= 1240/\lambda$ nm, from which $\lambda = 520$ nm.

Again, as for the linear molecules, the correct wavelength is shorter than the prediction given in Example 4-9. For chlorophyll *a* the peak of the absorption is at 428 nm and for chlorophyll *b* at 457 nm. Again, the major omission is the neglect of the potential energy, and as for the linear molecules, including it would correct the prediction in the right direction, i.e., more photon energy and toward a shorter wavelength.

4.10   The Beer–Lambert Law

The previous sections have discussed the absorption of light at the individual molecular level. Experiments, however, are constrained to measuring the absorption of light in samples containing large numbers of molecules, and so the observed absorption will be an average of some kind. The instrument which makes these measurements is called a *spectrophotometer*.

The Spectrophotometer

A spectrophotometer consists of a source of light covering the wavelength range of interest, followed (as described below) by a *monochromator*, a sample chamber, and a *detector*.

The light source used depends on the wavelength region of interest. Hydrogen lamps are usually used for the ultraviolet spectrum and tungsten lamps are used for the visible spectrum.

The monochromator is a form of wavelength filter that accepts the polychromatic light (multi-coloured light containing many wavelengths) and passes only a very narrow band of wavelengths. Various techniques are available to accomplish this (prism, adjustable interference filter, etc.), but the most common is the *diffraction grating*. This is usually a plane mirror on which a large number of fine grooves (~500 per mm) have been cut. The interference of light when reflected from these grooves spreads the light out according to its wavelength. The spread-out light is allowed to fall on an exit slit and, by rotation of the grating, a monochromatic beam of adjustable wavelength emerges to fall on the sample in the sample chamber.[21]

---

21.   Both the light source and monochromator can be replaced in some specialized devices with a "tunable" laser. A laser is a highly monochromatic source of light, and some may be tuned over narrow wavelength regions.

The detector is a light-sensitive device such as a photocell, a photodiode or a photomultiplier. Typically, such a device works by producing an electrical current which is proportional to the power of the light falling on it. This electrical current, after amplification, is displayed on a meter or recorded in some other way. The meter usually is designed so that either *absorbance* or *percent transmittance* can be read. In the more sophisticated devices, the wavelength setting of the monochromator can be increased or decreased automatically, and an on-line computer is programmed to control the whole operation and to monitor absorbance or percent transmittance automatically. A graph of absorbance vs. wavelength is what is referred to as an *absorption spectrum*.

The absorption spectrum of metmyoglobin obtained with such a spectrophotometer is shown in Fig. 4-17.

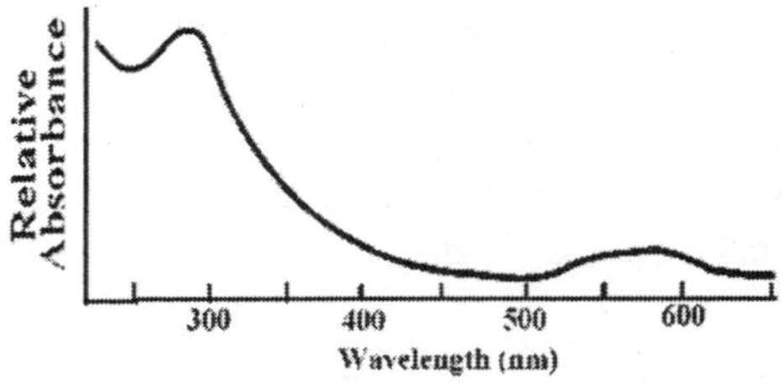

Figure 4-17

The absorption spectrum of metmyoglobin.

Transmittance

The amount of light absorbed by a solution, a solid, or even tree leaves can be expressed in two ways. The first is called *transmittance* with the symbol T if it is expressed as a fraction $(0 \le T \le 1)$, or $\%T$ if expressed as a percent $(0 \le \%T \le 100)$.

If a beam of light at wavelength λ of intensity $I_0(\lambda)$ is incident on a sample where absorption takes place, then the emerging beam will be of a reduced intensity $I(\lambda)$.[22] The transmittance is defined as

$$T = \frac{I(\lambda)}{I_0(\lambda)}$$
and
$$\%T = \frac{I(\lambda)}{I_0(\lambda)} \times 100\%$$ **[4-18]**

Suppose, as in Fig. 4-18, a beam of monochromatic radiation is traveling in the x-direction through an absorbing medium. At a particular point, consider a thin layer of the medium of thickness dx and of area $\mathcal{A}$, perpendicular to the beam. The intensity incident on the face of this layer is I and, as it passes through the layer, I changes by a small amount dI due to absorption. The fraction of the radiation

22.   Other processes may also contribute to reducing the beam intensity: in particular, scattering. This will be ignored in the following discussion.

absorbed in the layer is $-dI/I$.[23]

If the concentration of the absorbing molecules is C (molecules/volume), then the number of molecules "seen" by the incident radiation is $C\mathcal{A}(dx)$. (It is assumed that C and dx are small enough that no molecule is hidden behind another.) If the effective absorbing area of each molecule is σ, then the total effective area for absorption is $\sigma C\mathcal{A}(dx)$. "Effective" means that if a photon is incident on this area, it will be absorbed. The fraction of the radiation absorbed must equal the fraction of area $\mathcal{A}$ available for absorption, i.e.,

Figure 4-18

Derivation of the Beer–Lambert law ($\mathcal{A}$ is the cross-sectional area).

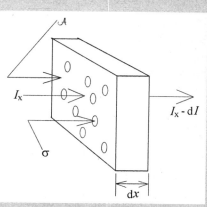

$$-\frac{dI}{I} = \frac{\sigma C\mathcal{A}(dx)}{\mathcal{A}} = \sigma C(dx)$$

This equation defines the quantity σ; it is the fraction of radiation absorbed per unit concentration and per unit distance, and has units of m^2/molecule, or more usually litre/mol·cm; notice that it has the dimensions of an area. The quantity σ is often called the *absorption cross-section*.

Moving to a more macroscopic level, instead of considering the thin slice dx, consider all of the absorbing material from $x = 0$ where $I = I_0(\lambda)$ to any general point x where $I = I(\lambda)$. Assuming the medium is homogeneous, integration of the above equation describes how $I(\lambda)$ decreases with x. The result is

$$I(\lambda) = I_0(\lambda)e^{-\sigma Cx} = I_0(\lambda)e^{-\mu x} \qquad \textbf{[4-19]}$$

where $\mu = \sigma C$. The quantity μ is called the *linear absorption coefficient* and it has the dimensions length^{-1}; it is usually measured in cm^{-1}. Since μ depends on the concentration it is not a molecular property like σ, but is a bulk property of the medium. Like σ it is a function of the wavelength.

Equation [4-19] is known as the *Beer–Lambert law* (or sometimes just "Beer's law").[24] A statistical derivation of this equation can be found in Appendix 1 as an application of Poisson statistics.

23. dI is negative since I decreases, therefore $-dI/I$ is a positive fraction.
24. August Beer (1829–1863), German chemist; and Johann Heinrich Lambert (1728–1777), German mathematician.

EXAMPLE 4-10

Chlorophyll a (molar mass 902.5 g/mol) absorbs strongly at $\lambda = 660$ nm. At this wavelength a chemist found that a chlorophyll a solution in acetone of concentration 11 µg/cm³, in a glass cell of length 1.00 cm, transmitted 12.5% of the incident light.

a) What is the linear absorption coefficient of this solution?

b) What is the absorption cross-section of the chlorophyll a molecule in solution for this wavelength?

SOLUTION

a) $x = 1.00$ cm and, since 12.5% is transmitted, then $I(\lambda)/I_0(\lambda) = 0.125$

Using Beer's law, Eq. [4-19], $I(\lambda) = I_0(\lambda)e^{-\mu x}$, or

$$\ln[I(\lambda)/I_0(\lambda)] = -\mu x$$

$$\mu = [\ln(I(\lambda)/I_0(\lambda))]/(-x) = [\ln(0.125)]/(-1.00 \text{ cm})$$
$$= 2.08 \text{ cm}^{-1}$$

b)   The concentration is

$$C = 11 \times 10^{-6} \frac{\text{g}}{\text{cm}^3} \times \frac{10^6 \text{cm}^3}{1\text{m}^3} \times \frac{1\text{mol}}{902.5\text{g}} \times \frac{6.022 \times 10^{23} \text{molecules}}{1\text{mol}}$$

$$= 7.34 \times 10^{21} \text{ molecules/m}^3$$

The absorption cross-section is

$$\sigma = \frac{\mu}{C} = \frac{208 \text{ m}^{-1}}{7.34 \times 10^{21} \text{molecules/m}^3}$$

$$= 2.83 \times 10^{-20} \text{ m}^2/\text{molecule} \times \left(\frac{10^9 \text{nm}}{1\text{ m}}\right)^2$$

$$= 0.028 \text{ nm}^2/\text{molecule}$$

The chlorophyll *a* molecule is approximately planar with a physical size of about 2 nm×2 nm or an area of 4 nm². Clearly the absorption cross-section in solution for $\lambda = 660$ nm is much less than the geometric area.

Absorbance

The second way in which absorption of light can be described follows directly from Eq. [4-19]. Taking the logarithm of the equation gives

$$\ln \frac{I(\lambda)}{I_0(\lambda)} = -\mu x \quad \text{or} \quad \ln \frac{I_0(\lambda)}{I(\lambda)} = \mu x$$

From this it can be seen that the quantity $\ln[I_0(\lambda)/I(\lambda)]$, which depends linearly on path length and concentration, is a measure of the ability of a solution to absorb light of wavelength λ. Thus absorbance *(A)* is defined as

$$A = \log \frac{I_0(\lambda)}{I(\lambda)} \qquad \textbf{[4-20a]}$$

Note that absorbance is defined with the logarithm to the base 10, and $\log N = 0.4343 \ln N$. Thus

$$A = 0.4343 \ln \frac{I_0(\lambda)}{I(\lambda)} = 0.4343 \mu x$$

The quantity $0.4343\mu = 0.4343\sigma C$ is redefined as εC, where ε is called the *extinction coefficient.* Thus $\varepsilon = 0.4343\sigma$ and

$$A = \log \frac{I_0(\lambda)}{I(\lambda)} = 0.4343 \mu x = \varepsilon C x \qquad \textbf{[4-20b]}$$

The usefulness of absorbance is apparent when the transmission of several absorbing species in series, each with its own cross-section, path length and concentration, is wanted. In the case of absorbers in series, or a number of different absorbers in the same sample, it is the absorbances which add, i.e.,

$$\begin{aligned} A_{\text{total}} &= 0.4343 \left(\mu_1 x_1 + \mu_2 x_2 + \mu_3 x_3 + \ldots \right) \\ &= \varepsilon_1 C_1 x_1 + \varepsilon_2 C_2 x_2 + \varepsilon_3 C_3 x_3 + \ldots \end{aligned} \qquad \textbf{[4-21]}$$

EXAMPLE 4-11

Measurements in a spectrophotometer on three leaves from a maple tree give the following information (for $\lambda = 450$ nm):

Leaf Number	Linear Absorption Coefficient (μ)	Leaf Thickness
1	1.39×10^2 cm^{-1}	0.100 mm
2	1.43×10^2 cm^{-1}	0.120 mm
3	1.48×10^2 cm^{-1}	0.091 mm

If the three leaves are arranged one behind the other, what percentage of the incident 450 nm light is transmitted by the combination?

SOLUTION

For Leaf 1 its absorbance is $A_1 = 0.4343\mu_1 x_1$
$= 0.4343(1.39 \times 10^2$ cm$^{-1})(0.100$ mm$)(1$ cm$/10$ mm$) = 0.604$

Similarly for   Leaf 2, $A_2 = 0.745$
Leaf 3, $A_3 = 0.585$

With all three leaves in series, the absorbances add
$A = A_1 + A_2 + A_3 = 1.934 = \log(I_0/I)$ from Eq. [4-20a]
Therefore, using Eq. [4-18],
$I_0/I =$ antilog $1.934 = 85.9$

$T = I/I_0 = (85.9)^{-1} = 0.012$

$\%T = 1.2\%$ of the incident light is transmitted.

4.11   Absorption of Infrared and Microwave Radiation by Molecules

Whereas UV and visible light cause transitions from one electron distribution to another in molecules, the absorption of infrared (IR) light leads to transitions between different vibrational energy levels. These vibrational energy levels arise because the atoms and bonds in a molecule act just like a combination of masses connected by springs and have, therefore, various natural frequencies of vibration. In Chapter 1 it was shown that the frequency of vibration of a mass m connected to a spring of force constant k is $f = [1/(2\pi)](k/m)^{1/2}$ (Eq. [1-13]). At the molecular level, the natural frequencies are of this form as well, but they are quantized, that is, the allowed frequencies may have only certain values. This means that the allowed energies of vibration are also quantized. The quantized energy of vibration is given by

$$E_v = hf(v + \tfrac{1}{2}) = \frac{h}{2\pi}\sqrt{\frac{k}{m_{eff}}}(v + \tfrac{1}{2}) \quad v = 0, 1, 2, 3, \ldots \qquad \textbf{[4-22]}$$

where h is Planck's constant, k is the effective spring constant of the chemical bond involved, m_{eff} is the effective mass attached to that spring,[25] and v is the *vibrational quantum number*.[26] There are two things to note about this expression.

1. E_v can never be zero, i.e., the molecule can never stop vibrating entirely. Even for $v = 0$, the lowest energy level, $E_0 = \tfrac{1}{2}hf$, which is called the *zero point energy*.

2. The allowed energy levels are equally spaced (since they depend on v to the 1st power), with the separation between adjacent levels being $\Delta E = 2E_0 = hf$.

Figure 4-19

Vibrational energy levels and fluorescence.

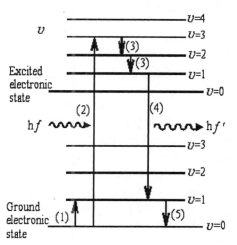

The energy level diagram of a quantized vibrator is a "ladder" of equally spaced levels, as shown in both the ground and first excited electronic states in Fig. 4-19.

The selection rule for changes of v is similar to that for n in the case of electronic transitions.

---

25. In a complex molecule, the calculation of this effective mass may be very complicated.

26. The addition of the number ½ to v in Eq. [4-22] is a complexity of the quantum theory beyond the scope of this discussion.

Vibrational Selection Rule: $\Delta v = \pm 1$

The energies involved in vibration are much less than those involved in electronic transitions, so that the photons absorbed when a molecule changes only its vibrational state have a much smaller energy and lie in the infrared (IR) region of the spectrum. Because the energy levels are equally spaced, transitions between any two adjacent vibrational levels can be induced by photons of the same energy. This contrasts strongly with the electronic energy levels considered earlier, which are not equally spaced. The fact that the vibrational levels are equally spaced plus the restriction that $\Delta v = \pm 1$ means that each bond type will absorb IR light at one wavelength only. At moderate temperatures, almost all molecules are in the $v = 0$ vibrational state of the ground electronic state, so that the transition labeled 1 in Fig. 4-19 is the usual one that produces absorption in the IR spectral region; it is called the *fundamental vibrational transition*. This absorption has been an invaluable asset to chemists, since molecules can be fingerprinted by their characteristic IR spectra. A brief list of bonds and the corresponding IR wavelength absorbed is given in Table 4-4.

Table 4-4 Infrared Absorption Bands for Some Chemical Bonds

Bond	Wavelength (μm)
C–H (aliphatic)	3.3 to 3.7
C–H (aromatic)	3.2 to 3.3
C–O	9.55 to 10.0
C=O (aldehydes)	5.75 to 5.80
S–H	3.85 to 3.90

This manifold of vibrational states exists for each electronic state, although the spacing of the vibrational levels will be different in each electronic state as shown in Fig 4-19. When an electronic transition $\Delta n = +1$ occurs, the vibrational quantum number can change by any amount.[27] Such a transition is labeled 2 in Fig. 4-19. The energy of this transition is greater than for that for the "ground state, $v = 0$" to "excited state, $v = 0$" transition that has been considered in the previous sections. Thus, the electronic spectrum recorded in the visible or UV region is not a sharp line at a well-defined frequency, but is a broad band of frequencies in the region of the electronic transition; this is clearly evident in the spectrum of metmyoglobin in Fig. 4-17.

---

27.   The selection rule $\Delta v = \pm 1$ applies only if there is no simultaneous electronic transition.

If only the vibrational transitions complicated the spectrum, it would show a series of equally spaced lines associated with the electronic transition. That it usually does not (see Fig. 4-17) is because many other effects contribute to the spectrum and broaden the individual lines.

Among these other effects is the fact that, as well as having electronic and vibrational energy, the molecule can rotate and has rotational energy. It will now come as no surprise to find that this motion is also quantized. The rotational energy states involve even less energy than do the vibrational states, and so each vibrational level in the diagram of Fig. 4-19 has an accompanying manifold of rotational levels attached to it. The transition labeled 2 could therefore end in the upper state on any one of several closely spaced rotational levels associated with that vibrational level, further altering the absorption wavelength and smearing out the absorption spectrum.

When rotational transitions occur alone, they absorb energy in the microwave region at energies much less than for the IR. Box 4-4 shows the complete absorption spectrum of water including all electronic, vibrational and rotational transitions.

Box 4-4   The Complete Absorption Spectrum of Water

Figure 4-20 plots the linear attenuation coefficient for water across the whole of the electromagnetic spectrum. Because water does not contain π-bonds, it does not absorb in the visible region of the spectrum. Our atmosphere, which contains a great deal of water vapour, is optically clear in the visible region, allowing most of the visible light from the sun to reach Earth's surface. If this were not the case, it would be impossible for Earth to support life. As the graph shows, water is a strong absorber of UV and the shorter wavelengths that are especially damaging to life forms (see Section 4.13). Hence, the water in Earth's atmosphere helps to protect us

Figure 4-20

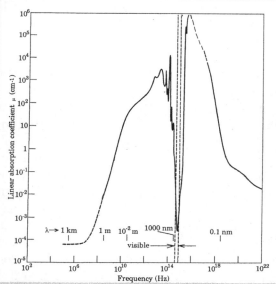

The complete absorption spectrum of water.

(cont'd.)

from these short wavelength rays. The spiked region in the IR corresponds to the vibrational absorption bands of the water molecule. These bands help absorb the "blackbody" radiation from Earth (see Section 14.5) which otherwise would be lost to space, and help to moderate Earth's temperature via the "Greenhouse Effect." In the microwave region, the rotation absorption bands of water are so close together that one loses the "spikes"; however, this absorption, like the IR, also helps to moderate atmospheric temperature. As well, the microwave oven (which operates at $f = 2.45$ GHz [gigahertz]) relies on this absorption to cook food.

4.12   **Fluorescence and Phosphorescence**

Fluorescence

As has been stated, at room temperature most molecules in the ground electronic state will be in the lowest vibrational energy level. If a molecule does find itself in a high vibrational level it will, after several collisions, drop to a lower level and give up its excess energy to other molecules in its vicinity. We have, therefore, to distinguish between "hot" and "excited" molecules. "Hot" molecules are those which find themselves in higher <u>vibrational</u> energy levels than their neighbours. "Excited" molecules are those which are in an <u>electronic</u> state other than the ground state. A molecule can be in an "excited" state and yet be "cold" in the vibrational sense, i.e., it will be in the $v = 0$ state.

These situations arise often when one considers the light absorption cycle shown in Fig. 4-19. Initially, a photon of energy hf is absorbed, leading to transition (2). The molecule is now both "hot" and "excited." (The reason for this heating of the excited molecule is described in Box 4-5.) It then collides with its neighbours, giving up its excess vibrational energy by increasing the velocity (i.e., K) of the collision partners; the temperature of the system rises correspondingly. Once the molecule has dropped through its vibrational manifold to some lower level of the excited electronic state (e.g., the transitions (3) in Fig. 4-19), a process called *vibrational relaxation*, it can return to the ground electronic state by radiating a photon of energy hf'(4); this process (4) is called *fluorescence*. After further vibrational relaxation down to the ground state vibrational manifold (5), the molecule is found at its original energy. The complete cycle takes about 10^{-7} s.

Box 4-5   The Franck–Condon Principle

When a molecule makes a transition from the ground to an excited electronic state it usually starts in the $v = 0$ vibrational state in the ground electronic state. After the transition to the excited state the molecule will probably not be in the $v = 0$ state but will, with a high probability, be in a higher vibrational state. (It is "hot.") This is known as the *Franck–Condon principle*. The physical principle can be seen from a simple mechanical model.

Take as a simple model of a vibrating molecule a single mass m suspended on a spring of stiffness constant k as shown in Figure 4-21a. This represents the molecule in its ground electronic state with $v = 0$, and so is at rest at level 1. (We are ignoring the zero-point energy.) In the real molecule the stiffness k is determined by the electronic bonding which, in the ground state, usually has the maximum value.

Figure 4-21

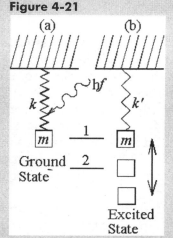

The Franck–Condon principle:
(a) before photon absorption;
(b) after photon absorption

The absorption of a photon by a π-electron has the effect of suddenly weakening the spring (reducing k to k') before the mass has a chance to move. The equilibrium position with the weaker spring is at level 2 as shown in Figure 4-21b. The immediate result is that the mass m falls and the system is set into oscillation.

The reverse process also results in oscillation. The hot, excited molecule loses its vibrational energy by collision with other molecules and so is in equilibrium at level 2. The emission of a photon (fluorescence) makes a stiff spring of constant k before the mass has time to move. The stiffer spring jerks m upward, setting it into oscillation. Therefore de-excitation will seldom result in a molecule returning to the $v = 0$ vibrational state.

Several other important observations can be made here:

1. Transitions can be from the ground electronic state $v = 0$ level to any of several vibrational and rotational levels of the excited electronic state.

2. The energy of the fluorescent photon must be equal to or (usually) less than the energy of the absorbed photon (see Fig. 4-19). Therefore the wavelength of the fluorescent light is longer than that of the absorbed light.[28]

3. The difference in energy appears as thermal energy in the system due to the vibrational relaxation process.

Phosphorescence

So far, only those situations have been described in which the spin of the excited electron is the same as it was in the ground state because the electron spin does not change during an electronic transition. Once the molecule is in the excited state, however, the spin of this electron can "flip" under the appropriate conditions.[29] The result of this spin-flip is shown in Fig. 4-22. The new state is called a "triplet" excited state, whereas the unflipped excited state is called a "singlet" state.[30] The excited triplet state is slightly lower in energy than is the corresponding singlet state shown in Fig. 4-22.

Consider an example in which the triplet state is populated as shown in Fig. 4-23. After the absorption of the incoming photon (1), the molecule begins to descend the vibrational manifold of the first excited singlet state (2). If, during this time, a spin-flip occurs, the molecule suddenly finds itself in a triplet state (3). It can then descend through the vibrational manifold of the triplet state until it reaches the $v = 0$ level (4).

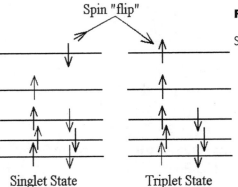

Spin "flip"

Singlet State　　Triplet State

Figure 4-22

Singlet and triplet states.

---

28. For example, when irradiated with ultraviolet light, starch fluoresces blue.

29. For example, the excited molecule finds itself in an inhomogeneous (not constant in space) electric or magnetic field.

30. The origins of the words "singlet" and "triplet" are a result of the detailed quantum theory of spin states and will not be pursued here.

At this point the molecule is trapped. It cannot immediately go to the ground state as in the case of fluorescence because the electron has the wrong spin and is prevented by the Pauli exclusion principle. It cannot return to the first excited singlet state because energy would be required. The molecule remains in the triplet state until some event occurs in its environment and induces another spin-flip. The molecule is then free to drop to the ground state and does so, simultaneously emitting light (5). This emission, which can at times be up to several seconds after the original photon was absorbed, is called a *phosphorescence*. Note that the phosphorescent photon, h*f*″, is of less energy than either the absorbed photon or the fluorescent photon.

Figure 4-23

Phosphorescence.

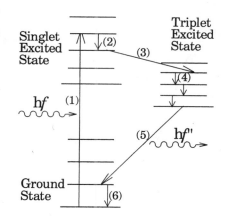

While the excited singlet and triplet states are the sources of fluorescent and phosphorescent light, they also are the intermediates in photobiological reactions. First of all, they have considerably more energy than the ground state, and this excess energy is in just the range necessary to drive most chemical reactions. Secondly, molecules have completely different physical properties in excited states than they do in the ground state because the electrons have been redistributed. They can, therefore, take part in chemical reactions even though the ground state is often unreactive. Of the two excited state species, the triplet state is involved in more photobiological reactions than is the singlet. This, in most cases, is simply a matter of time; the singlet state lifetime is only about 10^{-7} s and this state has little chance to react. The triplet state, on the other hand, is relatively long-lived and has plenty of opportunity to react.

4.13   Biological Effects of Ultraviolet and Visible Radiation

The ultraviolet (UV) region of the solar spectrum (200–350 nm) is especially harmful to living systems. Cellular injury, mutations and lethality have all been demonstrated to be biological effects of intense doses of this light. In some cases, such as in the use of germicidal lamps, ultraviolet light can be used to sterilize rooms and equipment. However, great care must be taken in many clinical and industrial situations to ensure that people are not exposed unwittingly to significant UV light levels. Fortunately for systems living on earth, very little of the intense solar UV light can penetrate the atmosphere (see Box 4-4).

The molecular aspects of UV-induced damage in living systems have been an object of intensive study by photobiologists for many years. Central to their interest was the observation that deoxyribonucleic acid (DNA), the genetic carrier molecule, was involved. Some of the initial evidence for this was obtained by comparison of the *action spectrum*[31] of bacterial cell death with the absorption spectrum of DNA as shown in Fig. 4-24. It was immediately apparent that the wavelength of UV that was most efficient at bringing about cell death was precisely the wavelength that DNA absorbed most efficiently, and damage to protein was not involved.

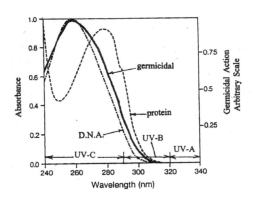

Figure 4-24

UV absorbance of DNA and protein and the germicidal action spectrum.

An offshoot of the intense study of this type of damage in bacterial DNA was the observation that under appropriate conditions, the damage could be repaired by the cell. This new knowledge made a great deal of sense to any researchers who believed that significantly more UV light reached Earth's surface millions of years ago. The ability of the cell to combat some of the UV damage gave it a definite evolutionary advantage. One process of dimer repair is an enzymatic one involving recognition and excision of a length of DNA containing the damage.

Unless the intensity is very high, such as one finds in a laser beam, *visible light* is essentially harmless to most living systems. There are some conditions and some diseases in which living systems can be sensitized to visible light. The result of this sensitization can, just as for ultraviolet light, be cellular injury, mutations and lethality. One characteristic of these processes is that a sensitizer molecule, such as some dye molecule, absorbs the light to form an excited state, which is usually a triplet state, followed by one of several possible mechanisms which lead to damage of a cellular molecule such as DNA. In one class of the reactions known as *photodynamic reactions* it is known that oxygen is involved. Damage in these cases can be especially severe, and

31. To obtain an action spectrum, living cells are exposed to a constant number of photons in a series of experiments in which only the wavelength is changed. The fraction of killed cells vs. wavelength gives the action spectrum.

in humans the result can be painful burns and sometimes death. In recent years, certain tranquilizers, drugs, perfumes, and soaps have been found to contain powerful photosensitizing agents. Many unfortunate experiences have been the result.

Figure 4-25

The action spectrum of erythema and cancer of the skin.

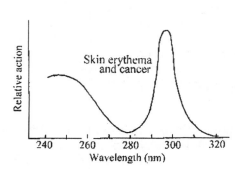

Most people have experienced sunburn. This condition occurs if one is exposed to a substantial amount of sunlight without the skin being previously pigmented (tanned). While the result is a burn similar to that obtained from a photodynamic reaction, the mechanism is entirely different. The action spectrum for sunburn is shown in Fig. 4-25. The action spectrum indicates that the maximum erythemic (skin reddening) effects occur at approximately 295 nm, a wavelength that is present in the sun's rays, even at Earth's surface. The skin reddening that we notice is a result of substances that are vasoactive (affecting the vascular system) being released in the epidermis by the radiation. Following the initial reddening, skin pigmentation or tanning occurs as melanin (the brown tanning substance) migrates to the skin surface. The action spectrum of non-melanoma skin cancer is identical to that for sunburn, indicating an intimate link between the two.

Exercises

4-1 Show that an energy of 1.00 kcal/mol is equal to 0.0435 eV per molecule (1 kcal = 4186 J).

4-2 Ultraviolet light in the 260-nm region is especially damaging to the DNA of cells. How much energy is involved when a DNA molecule absorbs a photon of this light?

4-3 An electron at a certain energy level in a molecule can jump to an adjacent level 6.0×10^{-19} J higher. Will the molecule containing this electron absorb light in the wavelength region 400–800 nm?

4-4 Light of wavelength 496 nm falls on a metal surface and produces electrons of maximum $K = 1.5$ eV. What is the work function (Φ) of the metal?

4-5 What is the probability density for an $n=2$ electron at the midpoint of a linear π-bonded molecule?

Problems

4-6 Verify the value of the constant, 1240, in the Duane–Hunt relation (see Footnote 7).

4-7 The light-sensitive molecule in photographic film is silver bromide (AgBr). The activation of the film requires the dissociation of this molecule, which has a dissociation energy of 24 kcal/mol. What is the maximum wavelength of a photon which will just dissociate AgBr? Use this result to explain why the light from a distant star will expose photographic film, but the constant exposure to the signals from a 20 000-W FM radio station broadcasting at a frequency of 91.1 MHz has no effect.

4-8 The diameter of the image of the sun on the retina is only about 160 μm. Because of the sun's brightness, we will normally blink so that the image lasts only about 0.20 s. The intensity of the light at the retina for such an image is about 10^4 W/m^2. If we imagine that the light of the sun is of one wavelength only, say 500 nm, how many photons does this short glimpse of the sun involve?

4-9 The following data were obtained from photoelectric measurements on clean potassium surfaces in a vacuum.

λ (nm)	200	300	400	500
V_s (volts)	4.11	2.05	1.03	0.41

By plotting an appropriate graph, determine the work function of potassium and the value of Planck's constant, assuming that the value of e = 1.602×10^{-19} C is known.

4-10 A biophysicist is using a beam of neutrons to study structural changes in connective tissue brought about by muscular dystrophy. In these experiments it is very important that the wavelength of the neutrons is known. By a technique called time-of-flight, she determines the speed of the neutrons to be 1.3×10^3 m/s. What is the wavelength of the neutrons? (The mass of a neutron is 1.674×10^{-27} kg.)

4-11 Consider a π-electron in the $n = 2$ molecular energy level of butadiene. The wave function (ψ) for this situation is as shown:

Butadiene may be sketched as

and the wave function looks like

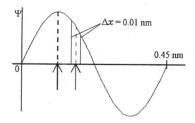

What is the probability density at the points indicated by the arrows? What is the probability of finding an $n = 2$ electron in the indicated region of width 0.01 nm? Sketch the probability density in the $n = 2$ and $n = 3$ states.

4-12 Below is the molecule 1,3-hexadiene, and those carbons which comprise a conjugated system are numbered:

$$CH_2 \!=\! CH \!-\! CH \!=\! CH \!-\! CH_2 \!-\! CH_3$$
$$1 2 3 4$$

What is the expression for the wave function of the second electronic state? For an electron in this level, what is the probability density at the position of the second carbon atom?

4-13 Consider an electron in the $n = 1$ electronic state of retinal. What is the probability density of the electron at the position of carbon four? Sketch the wave function in the $n = 2$ energy state.

Retinal

4-14 The molecule hexatriene is similar to the six-carbon atom structure illustrated in Problem 4-12, except that it has a fully conjugated π-bond system throughout. What is the probability of finding an $n = 3$ π-electron between the positions $x = 0$ and $x = \ell/3$ in this molecule?

4-15 What is the energy of an electron in the $n = 2$ electronic molecular state of butadiene? (See Problem 4-11.) Include the effect of the bond angles.

4-16 You are given two distinct electronic states to be occupied by two (indistinguishable) electrons. Sketch all those diagrams which show ways of doing this that are consistent with the Pauli exclusion principle. (There should be four diagrams.)

4-17 What is the energy of the lowest unfilled electronic molecular state in the ground state of butadiene? Include the effect of the bond angles.

4-18 In a certain linear conjugated π-system, the $n = 5 \rightarrow n = 6$ transition is induced by light of wavelength 548 nm. What is the length of the system?

4-19 A linear molecule in which there is a sequence of conjugated double bonds 12 atoms long is observed to have a maximum absorption at 500 nm. For a similar molecule, in which the conjugated sequence is only 8 atoms long, at what wavelength would you expect to find the maximum absorption?

4-20 In order to obtain a standing wave on a ring, how many half-wavelengths must there be around the circumference?

4-21 Calculate the wavelength of light which would induce the lowest energy transition of benzene (see diagram). The C–C bond length is 0.12 nm.

4-22 The absorbance of a solution is 0.60. What is the percent transmittance?

4-23 About 95 percent of the photons striking a sample of tree leaves are absorbed. What is the absorbance and the percent transmittance of the leaves?

4-24 Suppose that the percent transmittance of a given solution of length 1.0 cm is 50 percent. If the length of the solution in the light path were three times larger, what would the percent transmittance be?

4-25 A solution of biological molecules absorbs 380 nm light most strongly and fluoresces most strongly at 610 nm. What is the amount of energy per photon lost (i.e., converted into thermal energy) in the process?

4-26 A sample is irradiated at its absorption maximum ($\lambda_{max} = 500$ nm) with 1.00×10^{18} photons. The absorbance of the sample was found to be 0.680. Calorimetric analysis indicated that 0.100 J of the incident light energy were converted into heat in the sample. What is the approximate value of λ'_{max} of the fluorescent light, assuming a quantum yield for fluorescence of 1.00? In this context, a quantum yield of 1.00 means that, for every photon absorbed, one photon is fluoresced.

4-27 An atomic absorption spectrum consists of individual lines. Why then for a complex molecule does the absorption spectrum consist of broad bands?

4-28 Let E_r represent the minimum energy required to bring about a transition between adjacent rotational energy levels in a molecule. E_v will represent the minimum energy required to bring about a change between vibrational energy levels in a molecule and E_e will represent the minimum energy to bring about a change in electronic configuration. What is the correct ordering of these energies from the largest to the smallest?

4-29 Molecules may become excited by absorbing electromagnetic radiation. The absorbed energy may be used to (i) increase the rotational energy of the molecule, (ii) increase the vibrational energy, (iii) place the electrons in a different distribution. If a molecule absorbs a photon of (a) ultraviolet radiation, and (b) microwave radiation, which of (i), (ii), or (iii) could occur?

Answers

4-2   4.8 eV

4-3   No

4-4   1.0 eV

4-5   0

4-7   1188 nm

4-8 1×10^{14}

4-9 $\Phi = 2.1$ eV

4-10 0.3 nm

4-11 4.4 nm^{-1}, 2.2 nm^{-1}, 0.02

4-12 3.3 nm^{-1}

4-13 0.7

4-14 ⅓

4-15 1.6×10^{-18} J

4-17 3.6×10^{-18} J

4-18 1.35 nm

4-19 290 nm

4-20 Even number

4-21 1.4×10^{-7} m

4-22 25%

4-23 1.3, 5%

4-24 12.5%

4-25 2×10^{-19} J

4-26 733 nm

4-28 E_e, E_v, E_r

4-29 (a) (i), (ii), (iii) (b) (i)

5 Quantum Nature of Vision

5.1 Introduction

The human eye focuses light on the retina and forms real optical images of the world there, as described in Chapter 3. Some of the light in the image on the retina is absorbed by specialized retinal cells, called *photoreceptor cells*, which contain the chromophore *retinal*.[1] A photoreceptor cell responds to the absorption of light by producing changes in the electric potential difference (voltage)[2] across its cell membrane, which results in chemical communication with neighbouring cells. The focus of this chapter is on the structure of the retina, in particular of the photoreceptor cells in the retina, and on the exquisite sensitivity of the photoreceptor cells to the presence of light.

5.2 Structure

The human eye is a simple eye whose gross structure has been described in Chapter 3 (see Fig. 3-5). The inside surface of the eyeball behind the crystalline lens is covered with the retina, a thin layer of nerve tissue. Figure 5-1 shows the structure of the human retina in two regions: at the centre of the fovea, where the retina is about 0.1 mm thick, and in the retina well away from the fovea (the *peripheral retina*), where the thickness is about 0.2 mm. The human visual system is constructed so that the image of an object that is under close inspection is formed at the centre of the fovea, which is responsible for high acuity vision. The region of the retina outside the fovea is responsible for peripheral vision in humans.

---

1. The process of absorption of light by molecules is discussed in Chapter 4.
2. Electric potential differences are discussed in Chapter 15.

In both regions, the surface of the retina facing the vitreous humour, i.e., facing the incoming light, is covered with the axons of the output cells, called *ganglion cells*, whose cell bodies are located just below the surface of the retina. These axons converge toward the blind spot of the eye (see Box 3-2) where they form the optic nerve. The optic nerve exits from the eye at the blind spot and ends in a centre in the midbrain, the *lateral geniculate nucleus*, where the nerve fibres make contact with other cells. The blind spot is also the region through which blood vessels enter and leave the eye; the blood vessels spread out, forming a treelike network over the surface of the retina, which shades the cells in the retina below the network.

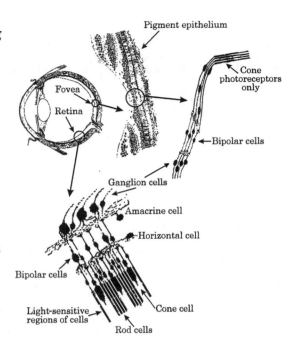

Figure 5-1

Structure of the human fovea and the retina outside the fovea.

Just below the surface of the retina in regions away from the fovea, there is a layer of cells making contact with the ganglion cells. These are the *amacrine cells* which carry visual information laterally over short distances. Elongated cells, called *bipolar cells,* also make contact with the ganglion cells, and with the amacrine cells. At their opposite ends, the bipolar cells make contact with photoreceptor cells and with *horizontal cells* which, like the amacrine cells, carry visual information laterally across the retina. The light-sensitive parts of the photoreceptor cells, the so-called *outer segments*, face the *pigment epithelium*, the darkly pigmented layer that lines the inside of the outer white surface of the eye (the sclera). This is one of the curious features of the human eye and, more generally, the vertebrate eye: the light-sensitive regions of the photoreceptor cells face towards the pigmented layer, away from the light coming to them through the corneal bulge, the humours, the crystalline lens, and finally past the blood vessels and the transparent layer of cells lying between the vitreous humour and the photoreceptor cell bodies. Two types of photoreceptor cells lie outside the fovea, whose shapes suggest the names *rod cells* and *cone cells*. At the centre of the fovea, the only photoreceptor cells are cone cells, which turn out to be different from the cone cells found outside the fovea. Furthermore, there are no amacrine or horizontal cells in the central part of the fovea.

The distribution of photoreceptor cells in the human retina is shown in Fig. 5-2. Some of the cone cells in the fovea respond best to red light (wavelength of maximum sensitivity, λ_{max} ~575 nm), while the rest of the cone cells there respond best to green light (λ_{max} ~545 nm; Fig. 5-3). The cone cells of the fovea are responsible for human day vision (*photopic vision*). This organization of the retina into two distinct functional regions is called the *duplex retina*. The photoreceptor cells in the rest of the retina surrounding the fovea are a mixture of rod cells that are responsible for our night vision (*scotopic vision*) and respond best to yellow light, and a type of cone cell that responds best to blue light (λ_{max} ~450 nm; Fig. 5-3).

Figure 5-2

Distribution of photoreceptor cells in the human retina. The origin of the retinal eccentricity axis is taken to be the point at which the optic axis of the eye intersects the retina (i.e., the fovea).

Figure 5-3

Responses of cone photoreceptor cells to light of different wavelengths (— red cone; − − green cone;−·−· blue cone). The rod cell response (not shown) resembles the green cone response except that rod cell absorption is greatest at about 510 nm.

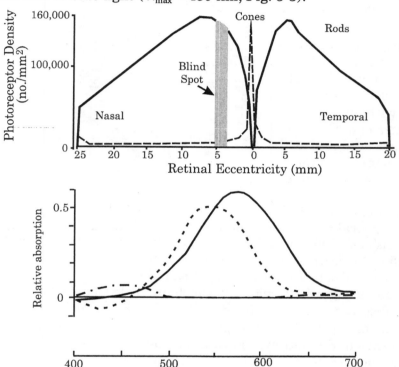

The photopic visual system is sensitive to the color of an object, unlike the scotopic system which is colour blind. The mechanisms of colour vision have been the subject of research for many decades, and that research is still active. Whatever the exact mechanisms, examination of Fig. 5-3 showing the broad absorption spectra of the three cone pigments suggests that the detection of colour is based on the differences in the responses of the different types of photoreceptor cells to light of a given wavelength. One possibility may be that the red and green cones of the fovea provide information to cells which produce a red/green summed response and a red/green difference response. Similarly, the blue cones and the rod cells may produce in subsequent cells a summed response (yellow) and a difference response (blue). These four responses can provide information about the colour composition and the intensity of light coming from an object.

In the human eye, there are about 100 million rod cells and 6 million cone cells of all types. The number of cone cells in the rod-cell-free part of the fovea (~0.2 mm in diameter) is only about 5000. This part of the fovea is responsible for the high acuity of human photopic vision. Because the rod cells are so numerous and so much more is known about them than about cone cells, discussion of the structure and function of photoreceptor cells which follows will be confined to the rod cells.

The structure of the rod cell is shown in Fig. 5-4. The light-sensitive part of the cell that faces the pigment epithelium is cylindrical in shape with a radius of about 1 μm (see Box 5-1); it is filled with membrane in the form of *discs*. The discs are produced at the junction of the outer segment with the rest of the cell. The outer segment of a rod cell in

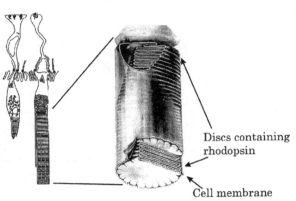

Discs containing rhodopsin

Cell membrane

Figure 5-4

Cone and rod cell photoreceptors in the peripheral retina in the same orientation as in Fig. 5-1. The outer segment of the rod cell has been magnified to show the structure of the discs.

the peripheral retina surrounding the fovea is about 25 μm long and contains about 900 discs. Each disc is a continuous *phospholipid bilayer* (Fig. 5-5) made up of phospholipid molecules which enclose a fluid-filled space; each disc is similar in shape to a red blood cell. The phospholipid molecule has two fatty acid tails which are hydrophobic[3] and a head group (represented by the filled circle in Fig. 5-5) which is hydrophilic.[4] This makes it possible for the phospholipid molecules to arrange themselves in the pattern shown in Fig. 5-5 and thus form a thin membrane.

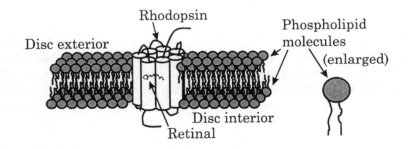

Rhodopsin

Disc exterior

Phospholipid molecules (enlarged)

Disc interior

Retinal

Figure 5-5

A rhodopsin molecule in the phospholipid bilayer membrane of a disc.

---

3.   Literally: "water hating."
4.   Literally: "water loving."

The disc membrane also contains proteins involved in vision, including the *visual pigment* molecule *rhodopsin*, which is composed of the glycoprotein *opsin* and the chromophore retinal attached to the

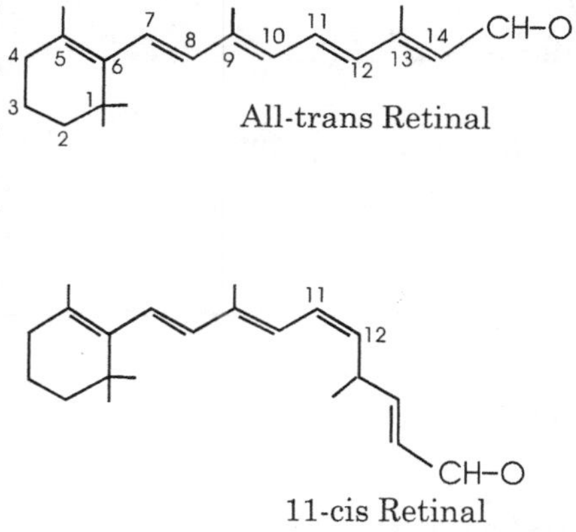

Figure 5-6

The two forms (isomers) of retinal found in rhodopsin in rod cells. The 11-cis isomer is the form of retinal before exposure to light.

All-trans Retinal

11-cis Retinal

seventh of opsin's seven membrane-spanning segments (represented by cylinders in the figure). Before the chromophore retinal has absorbed a photon of light, it is in its 11-cis isomeric form, as shown in Fig. 5-6.

The rod cells dominate the light-absorbing characteristics of the retina. The retina of an eye, when it is dark-adapted, has a purplish hue. This results from the fact that rod cells contain rhodopsin which absorbs yellow light preferentially. The light which comes back to the observer from the dark-adapted retina is thus deficient in yellow light, and appears purple. In fact, the original name for the visual pigment was *visual purple*, the present name "rhodopsin" having been made from two Greek words meaning "purple" and "vision." If the dark-adapted retina is exposed to enough light, it turns to a yellowish-white color. This phenomenon is called *bleaching* of the visual pigment.

Box 5-1   Light Diffraction and Cell Size

The effects of diffraction on the image formed by the eye's optical system on the retina were discussed in Chapter 3. Consider a thoroughly dark-adapted human eye, in which the pupil diameter is at its maximum of about 6.0 mm. Light with wavelength equal to 500 nm, at which the visual pigment in the rod cells absorbs best, has a Rayleigh criterion angle α given by Eq. [3-19]:

$$\alpha = 1.22\lambda/a$$

where λ is the wavelength of light in the eye and a is the diameter of the aperture.

$$\alpha = 1.22 \times 500 \times 10^{-6}\ \text{mm}/6.0\ \text{mm} = 1.0 \times 10^{-4}\ \text{radians}$$

Within the vitreous humour of the eye (refractive index = 1.34), this angle equals

$$\alpha_{eye} = 1.0 \times 10^{-4}/1.34 = 7.5 \times 10^{-5}\ \text{radians}$$

(cont'd.)

The distance from the optical centre of the eye to the retina is 23.3 mm (see Section 3.8) and so this angle corresponds to a distance d on the retina given by

$$d = 7.5 \times 10^{-5} \text{ radians} \times 23.3 \times 10^{-3} \text{ m} = 1.7 \text{ } \mu m.$$

The diameter of a human rod cell is about 2 μm, very close to the size of the diffraction effect on the retina. In other words, the cells in the eye are no smaller in diameter than they have to be.

5.3   Mechanism of Vision

The Fate of the Photons Incident on the Eye

Light must reach the rod cell outer segments in order to be absorbed and start the process of vision. Of the light falling onto the cornea, about 2.5% is reflected from the corneal surface. A further 9% is scattered or absorbed by the material of the cornea. The front and back surfaces of the crystalline lens reflect light.[5] Light is also scattered and absorbed by the humours and material of the lens. About 57% of the light incident on the eye reaches the surface of the retina. Further scattering and absorption in the retina reduces the amount of light reaching the rod outer segments to about 50% of the incident light.

The rod outer segment has a refractive index (Eq. [3-4]) which is greater than that of the extracellular fluid bathing the rod cell. This difference in refractive index can cause light inside the outer segment to reflect from the sides of the outer segment, making the outer segment behave like a light guide similar to the optical fibers used, for example, in telecommunications. About one-half of the light reaching a rod outer segment comes from a direction which makes an angle with the cylindrical axis of the outer segment too large for the outer segment to guide the light along its length. This light passes across the outer segment with little chance of being absorbed by the rhodopsin in the outer segment. The rest of the light (about 25% of the light incident on the eye) is aligned well enough with the outer segment's axis that the light is guided along the outer segment's axis, thus maximizing the chance that this light is absorbed.

---

5.   Johannes Purkinje (1787–1869), Czech physiologist and founder of the first department of physiology at the University of Breslau, observed these reflections. They have been used to study the changes in the shape of the lens during accommodation.

For light with wavelength in air of about 500 nm, the peak absorbance (Chapter 4) per unit length of the rod outer segment turns out to be 0.015 μm⁻¹, so that a 25 μm long outer segment has an absorbance equal to 0.38 (see Box 5-2). In other words, about 58 (= 100 × (1 – 10⁻⁰·³⁸)) of every hundred photons guided along the rod outer segment are absorbed in the discs of the outer segment. The quantum efficiency of rhodopsin is 0.67, which means that 67% of the absorbed photons produce the isomerization of retinal necessary to begin the process of signaling the presence of light in the eye. Taken altogether then, only about 0.67 × 0.58 × 25% ≅ 10% of the light incident on the eye results in isomerizations of the retinal in rhodopsin molecules, thus initiating a neural signal in the rod photoreceptor cell.

Box 5-2   Extinction Coefficient of Rhodopsin

There are about 140 million rhodopsin molecules in a rod outer segment. For light traveling along the cylindrical axis of an outer segment of length 25 μm and radius 1.0 μm, the absorbance A is 0.38. From this information, it is possible to calculate the extinction coefficient ε (Chapter 4) of rhodopsin in the disc membranes within the outer segment. First, the number of moles of rhodopsin in the outer segment equals

$$1.4 \times 10^8 \text{ molecules}/(6.022 \times 10^{23} \text{ molecules/mole}) = 2.3 \times 10^{-16} \text{ mol}$$

These are contained in a cylinder (the outer segment) with volume equal to the area of the base (πr^2) times the height (x) of the cylinder:

$$\pi r^2 x = \pi (1.0 \times 10^{-6} \text{ m})^2 \times 25 \times 10^{-6} \text{ m} = 7.9 \times 10^{-17} \text{ m}^3$$

The rhodopsin is contained in this volume, so that rhodopsin concentration

$$C_{\text{rhodopsin}} = \#\text{moles/vol}$$

$$= 2.3 \times 10^{-16} \text{ mol}/7.9 \times 10^{-17} \text{ m}^3$$

$$= 2.9 \text{ mol/m}^3 = 2.9 \times 10^{-3} \text{ mol/L}$$

(cont'd.)

Equation [4-20b] relates absorbance A, concentration C, and length x:

$$A = \varepsilon C x$$

so that

$$\varepsilon = A/Cx$$
$$= 0.38/(2.9 \times 10^{-3} \text{ mol/L} \times 25 \times 10^{-4} \text{ cm})$$
$$= 5.2 \times 10^{-4} \text{ L/mol} \cdot \text{cm}$$

In solution, rhodopsin has an extinction coefficient of only about 80% of this. Rhodopsin molecules in solution are randomly oriented in three dimensions, and thus so is the retinal in them. In the disc membranes, however, retinal lies approximately parallel to the disc membrane. Retinal is thus in an ideal orientation to absorb light traveling along the outer segment axis, because the electric field of the light is perpendicular to the direction of propagation of the light and consequently also parallel to the disc membrane. In solution, on the other hand, some of the retinal is oriented perpendicular to the electric field of the light and cannot absorb the light.

EXAMPLE 5-1

The length of the outer segments of rod cells close to the fovea is about 31 μm. What fraction of light with wavelength 500 nm, traveling along the axis of such an outer segment, is absorbed by it? (The absorbance per unit length of the outer segment is 0.015 μm⁻¹.)

SOLUTION

The total absorbance of the cell is $A = 0.015$ μm⁻¹ $\times 31$ μm $= 0.47$. The fraction $T = I(\lambda)/I_0(\lambda)$ (Eq.[4-18]) transmitted through the outer segment $= 10^{-0.47} = 0.34$ (see Eq. [4-20a]). Thus, the fraction absorbed $= 1 - T = 0.66$.

An Absorbed Photon Triggers a Neural Signal

After the absorption of a photon (within 1 ns), retinal changes its shape to that of the all-trans isomer (see Fig. 5-6). This transition signals the glycoprotein to which retinal is attached to begin a series of configurational changes. When rhodopsin reaches one particular configuration[6] within a millisecond, it becomes a biologically active enzyme. In its active form, rhodopsin initiates a cascade of enzymatic reactions, ending with changes in the electrical properties of the membrane of the rod outer segment. The electrical changes in the rod outer segment trigger changes at the other end of the rod cell, producing a chemical communication with the bipolar and horizontal cells in contact with the rod cell there.

The biologically active form of rhodopsin is inactivated by other enzymes in the disc membrane after a time of the order of minutes. Following inactivation, perhaps as much as 30 minutes later, the inactivated rhodopsin splits into opsin and all-trans retinal. The chromophore in its all-trans configuration is transported out of the cell to the pigment epithelium where it is transformed back into the 11-cis isomer by an enzyme there. After the chromophore is re-isomerized, it is returned to a rod outer segment, where it binds to an opsin to form a rhodopsin molecule, ready to absorb a photon.[7] This is a major component of the process of *regeneration of rhodopsin*.

Regeneration of the Visual Pigment Rhodopsin

Humans suddenly exposed to bright light, such as from a photographic flash, have difficulty seeing for some time after the exposure. Even for eyes exposed to ordinary room illumination, the time to adapt to total darkness is considerable. Adaptation to darkness involves several processes, including the regeneration of rhodopsin molecules and changes in the state of physiological adaptation of the rod and cone cells to light. Studies of dark adaptation are complicated by the human duplex retina which contains both cone and rod photoreceptor cells. Cones dominate vision except under conditions of very low illumination. They adapt to darkness at a very different rate than rods do.

Some humans with functioning rod cells lack any cone vision, and these people (called *rod monochromats*[8]) make good subjects for

6.   One of the metastable configurations called *metarhodopsin*.
7.   Details of the processes described can be found in books devoted to vision.
8.   In their eyes only the rod visual pigment is functional.

studying the relationship between regeneration of rhodopsin and psychophysical tests of dark adaptation. It turns out that the time dependence of dark adaptation in rod monochromats matches the time dependence of the recovery of rhodopsin from exposure to light (Fig. 5-7). The bleached visual pigment disappears with an exponential time dependence;[9] the half-life of the bleached pigment is about 5.2 min, giving a time constant[10] $\tau = 5.2$ min/ln(2) = 7.5 min. This means that the concentration C of bleached rhodopsin following a flash of light obeys the equation

$$C = C_0 e^{-t/\tau} \qquad\qquad \textbf{[5-1]}$$

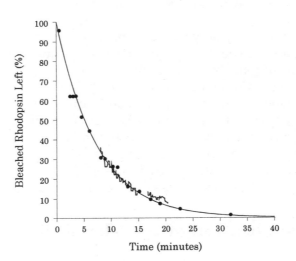

Figure 5-7

Recovery of rhodopsin from exposure to light. The dots are measurements of the regeneration of rhodopsin made in a rod monochromat; measurements in normal human eyes follow the same time dependence. The wavy lines are a plot of the rod monochromat's dark adaptation in time, the time dependence of which matches that of the regeneration of rhodopsin.

EXAMPLE 5-2

How long does it take for 99% of the rod visual pigment to regenerate in the normal human eye?

SOLUTION

99% of the bleached pigment must disappear, that is, $C = 0.01C_0$. Insert this value into Eq. [5-1] and solve for the time t which it takes for the concentration of bleached rhodopsin to reach this level:

9. Exponential decreases and half-life are discussed in Chapter 6.
10. The time constant τ (Greek lowercase *tau*) is defined as the reciprocal of the decay constant λ.

$$C/C_0 = 0.01 = e^{-t/7.5 \text{ min}}$$

Hence, $\ln(0.01) = -t/7.5$ min

$-4.6 = -t/7.5$ min

$t = 4.6 \times 7.5$ min $= 35$ min

Limit of Sensitivity of Rod Cells

How many photons must be absorbed by a rod cell's outer segment in order that the cell respond to light? The implications of the discussion above are that a single photon generates a neural signal in the retina. Because the absorption of photons by molecules is a random process, it should be true then that the human response to very dim light close to the threshold of vision is random. In the 1940s, decades before it became possible to observe directly the response of photoreceptor cells to very dim light, Hecht, Schlaer, and Pirenne[11] studied the absolute threshold of vision in human subjects, using traditional techniques of psychophysics,[12] and found that their subjects' responses to dim flashes of light could be explained if they assumed that a rod cell which absorbed just one photon produced a neural response.

Hecht, Schlaer, and Pirenne used light with a wavelength of about 500 nm for their experiments, because that is the kind of light to which

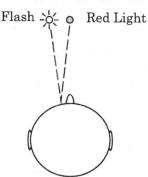

Flash ☀ ○ Red Light

the rod cells, which are the only photoreceptor cells capable of responding to very dim light, respond best. The subjects' eyes were fixed on a dim red spot of light (Fig. 5-8), visible to the cone cells in the fovea but not to the rod cells, which do not absorb red light well.[13] The test light of about 500 nm wavelength was positioned so that its image fell on the region of the

Figure 5-8

Arrangement of fixation light (red) visible to cone cells and test flash in the experiments of Hecht, Schlaer, and Pirenne.

11. S. Hecht, S. Schlaer, and M.H. Pirenne, at Columbia University. See *Journal of General Physiology*, Vol. 25, pp 819–840 (1942) and *Journal of the Optical Society of America*, Vol. 38, pp 196–208 (1942).

12. These involve presenting subjects with controlled stimuli and asking them to report what they perceive.

13. Because of this, dark adaptation of the human eye can be preserved in the presence of red light.

retina where the rod cells are densest (Fig. 5-2). The test light was flashed on for a short period of time (0.1 s) and the subjects were asked whether they had seen the flash or not. The total power of the flash of light was varied until the subjects' responses were correct 60% of the time. The power at the threshold of vision depended on the subject and varied between about 2×10^{-16} and 6×10^{-16} W. In order to produce flashes with such low power, an ordinary light source was used with a colour filter to produce yellow light and *neutral density filters*[14] to attenuate the power of the light source.

For these 0.1-s flashes, the range in power at threshold corresponds to a range in energy ($E = Pt$) of between 2×10^{-17} and 6×10^{-17} J. Light of wavelength 500 nm consists of photons each with energy given by Eq. [4-1]:

$$E_{500 \text{ nm}} = hc/\lambda = 4.0 \times 10^{-19} \text{ J}$$

Thus, between 50 and 150 photons fell on a subject's cornea at the threshold of vision. Since, as stated earlier, only about 10% of these were absorbed by rod cell outer segments, between 5 and 15 photons absorbed by outer segments during the 0.1-s flash resulted in the perception of light in 60% of the flashes.

Hecht, Schlaer, and Pirenne estimated that the image of the flash on the retina where the rod cells are densest covered about 500 rod cells. As a result, there was a probability of only 5/500 to 15/500, or 1 to 3%, that a given rod cell absorbed a photon at the threshold of seeing. Because the absorption of a photon is independent of the absorption of another photon, the probability for a rod cell to absorb 2 photons in a given flash is just the square of the probability to absorb one photon. Thus, the probability to absorb 2 photons in a given flash is between about 0.01 and 0.1%, which is negligible. In most flashes, then, any given rod cell on which the image of the flash falls absorbs at most 1 photon. The conclusion to be drawn from these results is that the absorption of a single photon in a rod cell outer segment causes a neural response in the rod cell.

Even though a rod cell responds to the absorption of a single photon, no subject reported that a flash with power corresponding to 1 photon absorption per flash was visible. Thus, even though a rod cell responds to the absorption of a single photon, the visual system requires a larger number of rod cells in a region of the retina to respond within a short time interval for the subject to perceive a flash of light.

---

14. The term *neutral density* indicates that a filter absorbs light of all wavelengths equally. It is characterized by its absorbance A as defined in Eq. [4-20b].

In fact, as the experiments of Hecht, Schlaer, and Pirenne showed, that number is on average between 5 and 15 photons absorbed within 0.1 s in an area in which there are of the order of 500 rod cells. The circuitry of the neurons carrying the visual information from the photoreceptor cells out of the retina toward the brain is designed to filter out noise from photoreceptor cells excited, not by light, but by chance. In addition to measuring the absolute sensitivity of rod cells to light, the experiments of Hecht, Schlaer, and Pirenne thus also provided information about this filter.

Because the absorption of photons by the materials of the eye and finally by rhodopsin molecules in rod outer segments is a random process, the actual number of photons absorbed by rod outer segments in a given flash is random and, as the discussion above indicates, small. Random occurrences of this type are described by the *Poisson distribution* described in Appendix 1. Figure 5-9 shows the Poisson probabilities for flashes in which the average number a of photons absorbed is 1, 5, and 10. For the case of $a = 5$, for example, flashes in which exactly 4 or exactly 5 photons are absorbed have the highest probability, which is about 0.18. In about 2% of these flashes, 10 photons are absorbed (see Example 5-3).

Figure 5-9

Poisson probability $\mathcal{P}(n;a)$ for various n and average $a = 1$, 5 and 10.

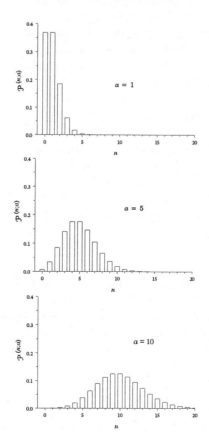

EXAMPLE 5-3

The average number of photons absorbed in a series of flashes is 5. Calculate the probability that exactly 10 photons are absorbed in a particular flash.

SOLUTION

Using Eq. [A1-1] from Appendix 1,

$$\mathcal{P}(n;a) = \mathcal{P}(10;5) = e^{-5} \times 5^{10}/10! = 0.018,$$

which is approximately 2%.

The random nature of the absorption of photons predicts that the responses of the subjects in the experiments of Hecht, Schlaer, and Pirenne are also random. For example, if the retinal circuitry requires that at least 7 rod cells respond within 0.1 s in a region containing 500 rod cells, the subject will respond positively to flashes in which 7, 8, 9, . . . photons are absorbed. The probability that in a given flash at least 7 photons are absorbed is equal to the probability that in the flash 7 or 8 or 9 or . . . photons are absorbed:

$$\mathcal{P}(\geq 7;a) = \mathcal{P}(7;a) + \mathcal{P}(8;a) + \dots$$

This infinite sum turns out to be easy to evaluate because of a property of probabilities: the total probability that 0 or more photons are absorbed must equal 1. In other words,

$$\mathcal{P}(0;a) + \mathcal{P}(1;a) + \dots + \mathcal{P}(6;a) + \mathcal{P}(7;a) + \mathcal{P}(8;a) + \dots = 1,$$

which implies that

$$\mathcal{P}(\geq 7;a) = 1 - [\mathcal{P}(0;a) + \mathcal{P}(1;a) + \dots + \mathcal{P}(6;a)].$$

More generally, if the retinal circuitry of a subject requires at least n rod cells to respond in order that a given flash be visible to the subject, the probability that a given flash will be seen by the subject is $\mathcal{P}(\geq n;a)$, with

$$\mathcal{P}(\geq n;a) = 1 - [\mathcal{P}(0;a) + \mathcal{P}(1;a) + \dots + \mathcal{P}(n-1;a)] \qquad \textbf{[5-2]}$$

Figure 5-10

Poisson probability $\mathcal{P}(\geq n;a)$ that at least n rod cells respond to a flash in which a respond on average. The threshold for seeing a flash is taken to be the value of a at which this probability $= 0.6$ (dashed line).

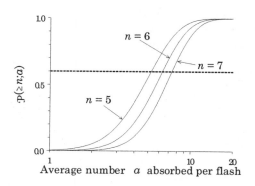

Figure 5-11

Responses to flashes of three subjects plotted on the curves from Fig. 5-10.

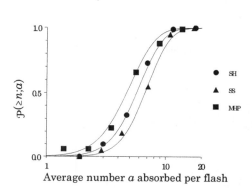

Figure 5-10 shows how this probability varies with the average number a, in cases in which at least 5, 6, or 7 photons must be absorbed for the subject to respond positively. For example, if the average number, a, of photons absorbed per flash is 5, the probability of at least 6 rod cells responding to a flash is approximately 0.4.

The responses of subjects in this type of experiment depend on the average number of photons absorbed by rod outer segments in the way predicted. Figure 5-11 shows the responses of 3 subjects[15] who participated in the original experiments of Hecht, Schlaer, and Pirenne. These responses are plotted on the curves predicting the probability of seeing a flash, as shown in Fig. 5-10.

EXAMPLE 5-4

For subject NNN, at least 7 photons must be absorbed during the 0.1-s flash of light in order for NNN to see it. In the flashes to which NNN is exposed, 5.4 photons are absorbed on average. What fraction of flashes will NNN see?

SOLUTION

The fraction desired is given by Eq. [5-2]:

$$\mathcal{P}(\geq 7;5.4) = 1 - [\mathcal{P}(0;5.4) + \mathcal{P}(1;5.4) + \ldots + \mathcal{P}(6;5.4)]$$

---

15. The initials of the subjects in Fig. 5-11 are those of the authors of the experiments.

Using Eq. [A1-2] it is easy to calculate these probabilities (results shown are to 3 significant digits, but for the purpose of calculation, numbers were not rounded until the final step):

$\mathcal{P}(0;5.4) = e^{-5.4} = 0.00452$

$\mathcal{P}(1;5.4) = 5.4 \times \mathcal{P}(0;5.4) = 0.0244$

$\mathcal{P}(2;5.4) = (5.4/2) \times \mathcal{P}(1;5.4) = 0.0659$

$\mathcal{P}(3;5.4) = (5.4/3) \times \mathcal{P}(2;5.4) = 0.119$

$\mathcal{P}(4;5.4) = (5.4/4) \times \mathcal{P}(3;5.4) = 0.160$

$\mathcal{P}(5;5.4) = (5.4/5) \times \mathcal{P}(4;5.4) = 0.173$

$\mathcal{P}(6;5.4) = (5.4/6) \times \mathcal{P}(5;5.4) = 0.156$

Thus,

$\mathcal{P}(0;5.4) + \mathcal{P}(1;5.4) + \ldots + \mathcal{P}(6;5.4) = 0.704$

and

$\mathcal{P}(\geq 7;5.4) = 1 - 0.704 = 0.30.$

In other words, about 30% of these flashes are visible to subject NNN.

Problems

5-1   The outer segments of rods in the toad eye are 3.0 μm in radius and 60 μm in length. There are 3.0×10^9 rhodopsin molecules in the outer segment. Assuming that the extinction coefficient of rhodopsin for light with wavelength equal to 500 nm is 5.0×10^4 L·mol^{-1}·cm^{-1}, what fraction of photons traveling along the outer segment is absorbed?

5-2   The visual pigment in cone cells regenerates more quickly than in rod cells: the time for one-half of the pigment to regenerate is 1.4 min. How long does it take for 99% of the visual pigment in cone cells bleached in a bright flash of light to regenerate?

5-3   In an experiment using very dim flashes of light produced by using a very dense neutral density filter, it is found that the average number of photons per flash is 1.6.

    (a)   What is the probability that a particular flash contains exactly 0 photons?

    (b)   What is the probability that a particular flash contains at least 1 photon?

5-4   For a particular subject, 4.0% of the photons directed at the pupil of the subject's eye are absorbed by rod cells. An experiment is performed in which 50 flashes of light (duration of each equal to 0.1 ms) containing on average 80 photons are presented to the subject. In how many of these 50 flashes will the subject's rod cells absorb 1 or 2 photons?

5-5   An experiment is devised to determine the minimum number of photons which must be absorbed in the photoreceptor cells contained in one unit of the compound eye of the horseshoe crab in order that a nerve impulse be generated in the optic nerve fibre coming from that unit. Using neutral density filters, it is found that short flashes of light contain on average 100 photons at the threshold for generating a nerve impulse. Estimates of reflection from the compound eye's surface and absorption in the materials of the eye indicate that only 5.3% of the photons are absorbed by the rhodopsin of the photoreceptor cells. To calculate the minimum number of photons which must be absorbed, it turns out to be necessary to calculate the fraction of these flashes in which fewer than 3 photons are absorbed. What is that fraction?

5-6   Suppose that for the subject described in Problem 5-4, flashes of light are visible to the subject if the rod cells in the region of the retina where the image of each flash falls absorb at least 4 photons within the 0.1-ms duration of each flash. How many of the 50 flashes are expected to be visible to the subject?

Answers

5-1   87%

5-2   9.3 min

5-3   (a) 0.2 (b) 0.8

5-4   17

5-5   0.10

5-6   20

6 Radiation Biophysics

6.1 Introduction

Studies of the effects of high-energy radiation on biological molecules, cells and whole systems has been one of the most popular areas of biophysics for many years. The use of high-energy radiation in the diagnosis and treatment of many human illnesses has been a boon to medicine. Also, in order to deal fairly and intelligently with the technological and political issues related to nuclear energy, citizens should become knowledgeable about nuclear processes.

6.2 Structure of Nuclei

The Rutherford[1] model views the atom as having a very small dense atomic nucleus that contains nearly all the mass of the atom and all of its positive electrical charge. Orbiting around this nucleus in a volume about 10^{24} times greater is a number of negatively charged electrons, equal in number to the number of positive charges in the nucleus, so that the atom as a whole is electrically neutral. These electrons determine the chemical nature of the atom (e.g., a neutral atom with six electrons is always carbon), and in what follows there is very little further interest in them; the interest is with the nucleus.

The nucleus is composed of two types of particles of almost equal mass: the proton and the neutron. The total number of protons and neutrons in the nucleus is the *baryon number*[2] (A) of the nucleus. The proton has a single positive electrical charge and the neutron has no charge. Some basic properties of these particles are given in Table 6-1.

---

1.   Ernest Rutherford (1871–1937), British physicist.
2.   This number has also been called the "atomic mass number" although it is not exactly the atomic mass. To two or three significant figures the baryon number, A, is the same as the molar mass. For example, the baryon number of ^{16}O is 16; its molar mass is 15.99491 g/mol.

A chemical element can exist with different nuclear masses; for this to be the case the number of neutrons must vary, since the number of protons is fixed. For example, the element carbon exists with baryon numbers 10, 11, 12, 13 and 14. Since all carbons have six protons, then these forms of carbon must have four, five, six, seven and eight neutrons in their nuclei. These are called the *isotopes* of carbon.[3]

Table 6-1   Properties of Proton, Neutron, Electron

Particle	Mass	Charge*
Proton	1.673×10^{-27} kg	e
Neutron	1.675×10^{-27} kg	0
Electron	9.109×10^{-31} kg	$-e$

*$e = 1.602 \times 10^{-19}$ C

The notation used to specify a particular isotope of element X is

$$_{Z}^{A}X_{N} \qquad \textbf{[6-1]}$$

where Z is the *atomic number*, that is, the number of protons in the nucleus (also equal to the number of orbiting electrons in the neutral atom), N is the number of neutrons in the nucleus, and A, as defined previously, is the baryon number. Clearly A = Z + N and so one of the numbers is redundant and can be omitted; usually N is omitted and the symbol becomes,

$$_{Z}^{A}X \qquad \textbf{[6-2]}$$

The isotopes of carbon mentioned above are $_{6}^{10}$C, $_{6}^{11}$C, $_{6}^{12}$C, $_{6}^{13}$C and $_{6}^{14}$C.

EXAMPLE 6-1

How many protons and neutrons are in the nucleus of uranium $_{92}^{238}$U?

SOLUTION

The atomic number is 92, so there are 92 protons in the nucleus.

---

3.   A common jargon for these is "carbon-10, carbon-11," etc.

The baryon number is 238, so there are 238 – 92 = 146 neutrons in the nucleus.

There are about 100 different elements, but many elements have more than one <u>stable</u> isotope; there are about 300 stable isotopes. For example, beryllium has only one stable isotope ($^9_4$Be) whereas tin (Sn) has 10. Not every imaginable mixture of protons and neutrons will form a stable nucleus. Almost all stable nuclei have a number of protons (Z) which is less than the number of neutrons (N = A – Z). This stability can be understood on the basis of simple electrostatics. If there are too many protons, the mutual Coulomb electrical repulsion (see Chapter 15) of the positive charges overcomes the attractive nuclear forces holding the nucleus together, which come from both the protons and the neutrons. For almost all nuclei, N is somewhat greater than Z, but not by a large amount.

Fortunately for the present discussion, a detailed theory of the energetics of nuclei is not required. It is sufficient to know that any nucleus having too few or too many neutrons relative to protons will be unstable, and that such a nucleus will change in some way to redress the imbalance. It does this by emitting various particles in a process called *radioactivity*.

6.3   Radioactivity

It might be expected that a nucleus with an excess of either protons or neutrons might simply emit the requisite number of those particles and so produce a new stable nucleus called the *daughter* nucleus. Because of the internal structure of the nucleus this never happens: neutron-emitting or proton-emitting nuclei are unknown. The unstable nucleus achieves stability by emitting two other types of particles, sometimes in a series of transformations. In the early days of nuclear physics these two particles were unidentified and were simply labeled "alpha" (α) and "beta" (β). It was also recognized that another radiation often accompanied α and β and it was labeled "gamma" (γ). Very quickly these radiations were identified; their properties are given in Table 6-2. Note that the β-radiation has two different forms, electrons and positrons (positive electrons), depending on the charge of the emitted particle. Electrons and positrons are not baryons, and on the nuclear scale their mass is negligible so their baryon number is zero.

Table 6-2   Radioactive Emissions

Particle	Identity
Alpha (α)	Nucleus of helium atom $_2^4\text{He}$
Beta (β)	β- Ordinary electron $_{-1}^{0}e$
	β+ Positive electron or positron $_1^0\bar{e}$
Gamma (γ)	Electromagnetic wave of very short wavelength and thus high energy

Alpha Emission

Some nuclei at the high-mass end of the periodic table achieve stability by the emission of a tightly bound cluster of neutrons and protons that constitute the nucleus of normal helium $_2^4\text{He}$. An example of practical importance is $_{92}^{238}\text{U}$ whose decay scheme is

$$_{92}^{238}\text{U} \rightarrow {}_{90}^{234}\text{Th} + \alpha$$

(where Th is thorium) or, using the knowledge that an α-particle is a helium nucleus,

$$_{92}^{238}\text{U} \rightarrow {}_{90}^{234}\text{Th} + {}_2^4\text{He}$$

These transformations are subject to certain conservation laws that determine the balance of the two sides of the equation:

1. Electrical charge must be conserved; this is the same as the atomic number (subscript). It therefore follows that the atomic numbers on both sides of the equation must add up to the same thing (92 = 90 + 2).

2. The baryon number is also conserved; the superscripts on each side of the equation must add up to the same thing (238 = 234 + 4).

The α-particles are emitted with well-defined energies, typically a few MeV,[4] and because of their large mass and charge, they interact strongly with matter. As a result they are easily shielded, being effectively stopped by a sheet of paper. Alpha-emitters tend to have long lifetimes.

---

4.   1 MeV = 10^6 eV = 1.602×10^{-13} J

Because of their large mass, α-particles are not easily deviated from a straight-line path in matter. Their well-defined energy (from a given type of radioactive nucleus) means that their penetration distances in matter are almost all the same. The penetration distance of 5.3 MeV α-particles in air is about 30 mm, and in water or human tissue is about 30 μm. It can be shown that the loss of energy per unit path length of a charged particle in matter is inversely proportional to the energy, i.e.,

$$\frac{\mathrm{d}E}{\mathrm{d}x} \propto \frac{1}{E} \qquad\qquad \textbf{[6-3]}$$

Thus, the energy deposited in matter per unit length of path is greatest at the end of the path where the energy is the lowest. Charged particles deposit most of their energy near the end of their path. This fact can be very useful, for example, in the treatment of some deep-seated tumours.

The quantity $\mathrm{d}E/\mathrm{d}x$ is called the *linear energy transfer* (LET). A large LET means a short path; α-particles have a large LET.

Beta Emission

By far the predominant method of radioactive adjustment for unstable nuclei is by β-emission. For example, $^{3}_{1}$H (tritium) undergoes radioactive decay by emitting a β-particle. Using the knowledge that a β-particle is an electron, the transformation equation describing this process is written:

$$^{3}_{1}\text{H} \rightarrow {}^{3}_{2}\text{He} + {}^{0}_{-1}\text{e} + \bar{\nu}$$

Note that the rules for the conservation of atomic number (charge) and baryon number still hold. The symbol $\bar{\nu}$ (antineutrino) is discussed further in the next paragraph. Note that in the process a nucleus of hydrogen has been transformed into one of helium; no further transformations will take place in this case as this isotope of helium is stable.

Unlike the case of alphas, all the betas from a given radioisotope do not have the same energy. What is observed is a continuous spectrum of energies from zero to some maximum. Since the initial and final nuclear masses are fixed, then from Einstein's mass–energy relation ($E = \Delta mc^2$), it might be expected that any difference in mass should appear as the energy of the β-particle, and hence the β-energy should also be fixed. This paradox can be resolved if another particle is released along with the β to share the energy. Such a particle was postulated and was subsequently found. It is a particle without charge, and if it has a mass it is very small. The particle, called the *antineutrino*[5] ($\bar{\nu}$), has a speed essentially equal to that of light and is very difficult to stop or detect.[6] Neutrinos will not be considered further as they are irrelevant to practical terrestrial problems. The average energy of β-particles is about one-third of their maximum energy.

An example of a nucleus that emits a *positron,* or *positive electron*, is $^{11}_{6}$C; its decay is given by

$$^{11}_{6}\text{C} \rightarrow {}^{11}_{5}\text{B} + {}^{0}_{1}\text{e} + \nu$$

Again the daughter boron nucleus is stable. Note that when a positron is emitted it is accompanied by a *neutrino* (ν).

Beta-particles from radioactive nuclei have speeds close to that of light, and kinetic energies of the order of 1 MeV. They travel for about 3 m in air or a few millimetres in water or human tissue before coming to rest. In the process of coming to rest in tissue they can do much damage. It is rather easy to shield a β-emitter; a plastic sheet 1 cm thick affords complete protection. If, however, β-emitting materials are ingested via food, air, or water, the betas can cause considerable internal damage.

Because their mass and charge are much smaller than those of α-particles, β-particles do not interact as strongly with matter and so have a smaller LET. They are also more easily scattered and deviated from their path. The result is that their penetration depth in matter is longer and more diffuse than for alphas. A quantity called *range* is defined such that

$$\text{Range} = (\text{Penetration Depth}) \times \text{density } (\rho) \qquad \textbf{[6-4]}$$

Note that the dimension of range is length $\times$ (mass)(length)$^{-3}$ = mass(length)$^{-2}$. For β-particles of energy greater than 0.6 MeV, the range is given approximately by the empirical equation

$$\text{Range} \, (\text{kg/m}^2) = 5.42 \, E \, (\text{MeV}) - 1.33 \qquad \textbf{[6-5]}$$

---

5. ν is the lowercase Greek letter *nu*.

6.   Nuclear particles exist in two classes: particles and antiparticles. These have the property of mutual annihilation. For example, a proton annihilates with an antiproton to produce two (or more) γ-rays; similarly for electrons and anti-electrons (positrons).

EXAMPLE 6-2

What is the penetration depth of a 0.80-MeV β-particle in air and water?

SOLUTION

Using Eq. [6-5],

Range = 5.42 × 0.80 – 1.33 = 3.0 kg/m²

In air, ρ = 1.3 kg/m³

Penetration depth = (Range)/ρ = (3.0 kg/m²)/(1.3 kg/m³)
$$= 2.3 \text{ m}$$

In water, ρ = 1000 kg/m³

Penetration depth = (3.0 kg/m²)/(1000 kg/m³) = 3.0×10⁻³ m
$$= 3.0 \text{ mm}$$

In both cases the distance the particle travels is about twice the penetration depth because of the scattering and the resultant zig-zag path.

Gamma Emission

The γ-rays are very short-wavelength electromagnetic waves. After the emission of an α- or β-particle, the daughter nucleus, in most cases, is left with excess energy in an excited state. This excess energy is emitted as a γ-ray, very soon (10⁻¹⁴ s) after the primary event, and lets the nuclear particles readjust into their lowest energy (ground) state. This is similar to the readjustment of electrons in excited atoms or molecules, where low-energy electromagnetic waves are emitted as X-rays or light.

Gamma-rays are very penetrating, having energies around 1 MeV. Typically several centimetres of lead are required to attenuate them to an acceptable level and form an effective shield. The concept of "penetration depth" and "range" does not apply to γ-rays; the absorption of these rays is discussed in Section 6.8 on absorption of radiation.

Producing Isotopes

With nuclear reactors and high-energy particle accelerators, stable nuclei can be transmuted into radioactive ones by adding or removing neutrons or protons. For example, the cobalt isotope ^{60}Co is produced by bombarding ^{59}Co with neutrons (n) (charge = 0, baryon number = 1) in a nuclear reactor,

$$^{59}_{27}\text{Co} + ^{1}_{0}\text{n} \rightarrow ^{60}_{27}\text{Co}$$

Cobalt as ^{60}Co is long-lived; each radioactive nucleus decays by emitting a β-particle followed by two γ-rays:

$$^{60}_{27}\text{Co} \rightarrow ^{60}_{28}\text{Ni} + ^{0}_{-1}\text{e} + 2\gamma$$

The highly penetrating nature of these gammas enables them to reach and destroy deep-seated tumours.

Box 6-1   Natural Radioactive Isotopes

Few radioactive isotopes occur naturally in substantial quantities because their lifetime must be very long to have survived since their formation in whatever cosmological event was involved, e.g., the formation of the universe itself (20×10^9 yr) or the formation of the solar system (5×10^9 yr). Examples are $^{238}_{92}$U (α-emitter, $T_{\frac{1}{2}} = 4.5 \times 10^9$ yr), $^{235}_{92}$U (α-emitter, $T_{\frac{1}{2}} = 7.1 \times 10^8$ yr) and $^{40}_{19}$K (β-emitter, $T_{\frac{1}{2}} = 1.3 \times 10^9$ yr). A few unstable nuclei are produced continuously by the action of cosmic rays in the atmosphere, but the quantities are minuscule. Examples are the production of $^{14}_{6}$C, so important in *carbon dating* in archaeology, and $^{3}_{1}$H or *tritium*, the radioactive form of hydrogen found in trace quantities in terrestrial water. Most of the exposure to radiation we experience comes from a small number of natural radioactive isotopes (see Box 6-2).

6.4   Radioactive Series

Some very heavy nuclei are so very far from nuclear stability that they require many radioactive events to occur before they achieve stability (see Box 6-1). This results in a *radioactive series*. Such a series usually begins with a long-lived parent whose slow rate of decay determines how many of each of the subsequent species are found downstream in the various daughter nuclei.

An example of such a series is that which begins with $^{238}_{92}$U and ends with lead $^{206}_{82}$Pb. The series with its emissions and lifetimes are given in Fig. 6-1. Several other series are known as well: one begins with $^{235}_{92}$U and ends with $^{207}_{82}$Pb, and another begins with $^{232}_{90}$Th and ends with $^{208}_{82}$Pb.

Figure 6-1

The ^{238}U radioactive series.

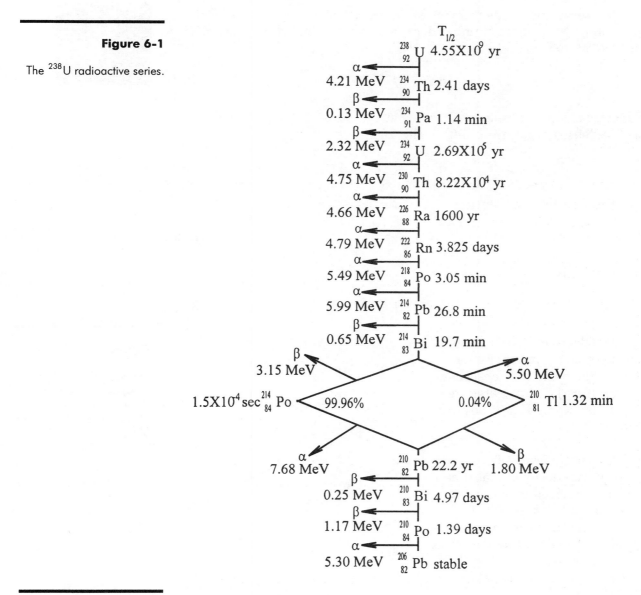

6.5   Radioactive Decay and Half-Life

Suppose that a sample contains a number N_0 of radioactive nuclei at time $t = 0$. The time at which a given nucleus decays is entirely random, so only the average behaviour of a large number of nuclei can be considered. Let λ be the probability that in unit time a given nucleus will decay; this is called the *decay constant*.

If after a time t, the number of nuclei remaining is N, then in the next short time dt the number decaying will be proportional to both N and dt; therefore,

$$dN = -\lambda N\, dt \qquad\qquad \textbf{[6-6]}$$

The minus sign expresses the fact that the number N can only decrease as t increases. Equation [6-6] in the form,

$$dN/N = -\lambda\, dt \qquad\qquad \textbf{[6-7]}$$

has the well-known solution

$$N = N_0 e^{-\lambda t} \qquad\qquad \textbf{[6-8a]}$$

or, alternatively,

$$\ln(N/N_0) = -\lambda t \qquad\qquad \textbf{[6-8b]}$$

The decrease takes place in an exponential manner over time. The behaviour described by Eq. [6-8a] is illustrated in Fig. 6-2a and by Eq. [6-8b] in Fig. 6-2b.

A characteristic of exponential decay is that it can be characterized by a unique time called the *half-life* $(T_{1/2})$, that is, the time for any given starting number N_0 to decrease to $\tfrac{1}{2}N_0$. Substituting $N = \tfrac{1}{2}N_0$ at $t = T_{1/2}$ into Eq. [6-8a] or [6-8b] gives

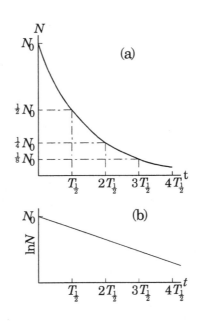

Figure 6-2

Radioactive decay:
a) exponential decay showing half-life;
b) the semilogarithmic plot of (a).

$$T_{\frac{1}{2}} = 0.693/\lambda \qquad \textbf{[6-9]}$$

The half-life is clearly illustrated in Fig. 6-2a. The half-life is usually specified for radioisotopes in preference to decay constants since it immediately conveys the important information about how long the isotope will survive. For example, plutonium ^{239}Pu, which causes great concern because of its cancer-inducing properties, has a half-life of 24 000 years.

EXAMPLE 6-3

How long will it take for stored radioactive ^{239}Pu waste to decay to 1.0% of its present level?

SOLUTION

$\lambda = 0.693/24\ 000\ \text{yr} = 2.89 \times 10^{-5}\ \text{yr}^{-1}$

$N = 0.010\ N_0$

Using Eq. [6-8a],

$0.010 = e^{-\lambda t}$

$t = -[\ln (0.010)]/\lambda = -[\ln (0.010)]/2.89 \times 10^{-5}\ \text{yr}^{-1}$

$\quad = 1.6 \times 10^5\ \text{yr}$

6.6 Biological and Effective Half-Life

If an amount of radioactive material is ingested by a living organism, the effective half-life of the radioactivity within the organism can be significantly altered by the biological activities of the organism. Although the isotope is decaying with a physical half-life of $_pT_{\frac{1}{2}}$ (decay constant = λ_p), the organism may be eliminating the isotope in some manner: excretion, perspiration, exhalation, etc. The rate of elimination is often proportional to the amount present, so the amount present in the organism also decays exponentially with a biological half-life $_bT_{\frac{1}{2}}$ (decay constant = λ_b).

Using the reasoning for the Poisson distribution given in Appendix 1, the probability that there are no physical decays

in time t is

$$\mathcal{P}_p(0) = e^{-\lambda_p t}$$

Similarly, the probability that there is no biological elimination in the same time t is

$$\mathcal{P}_b(0) = e^{-\lambda_b t}$$

The probability that there is no physical decay <u>and</u> no biological elimination is the product of the two probabilities

$$\left[\mathcal{P}_p(0)\right]\left[\mathcal{P}_b(0)\right] = \left[e^{-\lambda_p t}\right]\left[e^{-\lambda_b t}\right] = e^{-\left(\lambda_p + \lambda_b\right)t} = e^{-\lambda_e t}$$

where λ_e is the *effective decay constant* and

$$N = N_0 e^{-\left(\lambda_p + \lambda_b\right)t} = N_0 e^{-\lambda_e t}$$

Therefore,

$$\lambda_p + \lambda_b = \lambda_e \qquad\qquad \textbf{[6-10]}$$

From Eqs. [6-9] and [6-10] it follows that

$$1/_p T_{\frac{1}{2}} + 1/_b T_{\frac{1}{2}} = 1/_e T_{\frac{1}{2}} \qquad\qquad \textbf{[6-11]}$$

EXAMPLE 6-4

When iodine (I) is ingested by humans, they eliminate it such that one-half the body's iodine content is excreted every 4.0 days. Radioactive ^{131}I with a physical half-life of 8.1 days is administered to a patient. When will only 1.0% of the isotope be in the patient's body?

SOLUTION

From Eq. [6-11],

$1/_e T_{\frac{1}{2}} = 1/_p T_{\frac{1}{2}} + 1/_b T_{\frac{1}{2}} = 1/8.1 \text{ d} + 1/4 \text{ d} = 0.37 \text{ d}^{-1}$

Therefore

$_e T_{\frac{1}{2}} = 1/0.37 = 2.7 \text{ d}$

and

$$\lambda_e = 0.693/{_e}T_{\frac{1}{2}} = 0.693/2.7 \text{ d} = 0.26 \text{ d}^{-1}$$

$$N/N_0 = e^{-0.26t} = 0.010$$

$$\ln(0.010) = -0.26\, t$$

$$t = 18 \text{ d}$$

6.7   Activity

Activity is a term that refers to the number of radioactive nuclei that disintegrate per second and can be considered a measure of the *strength* of the sample. It is clear that the activity, A,[7] will depend on both the number of nuclei present and the half-life; the shorter the half-life, the faster the nuclei decay and the greater the strength. Using Eq. [6-7],

$$A = |dN/dt| = \lambda N \qquad \qquad \textbf{[6-12]}$$

Since N decays exponentially, then so also will A and at the same rate. Using Eq. [6-8a], it follows that

$$A = A_0 e^{-\lambda t} = \lambda N_0 e^{-\lambda t} \qquad \qquad \textbf{[6-13]}$$

Thus the amount of radiation—α, β or γ—emitted per second falls off exponentially.

The current unit of activity is called the *becquerel*[8] and is defined as one disintegration per second; the abbreviation is Bq. An older unit that is gradually losing favour is the *curie*[9] (Ci) for which

$$1 \text{ Ci} = 3.7 \times 10^{10} \text{ Bq} \qquad \qquad \textbf{[6-14]}$$

---

7.   Not to be confused with the baryon number A defined in Section 6.2.
8.   Antoine Henri Becquerel (1852–1908), French physicist and discoverer of radioactivity.
9.   Marie Curie (née Sklodowska) (1867–1934), Polish-French chemist, and Pierre Curie (1859–1906), French physicist.

EXAMPLE 6-5

A 3.7×10^{14} Bq (10 kCi) source of ^{60}Co is used for cancer treatment. Each disintegrating nucleus emits two γ-rays, one of energy 1.17 MeV and one of 1.33 MeV. What is the mass of ^{60}Co present in the source, and how much energy is emitted per second in the form of γ-rays? The half-life of ^{60}Co is 5.3 years.

SOLUTION

$\lambda = 0.693/(5.3 \text{ yr} \times 365 \text{ day/yr} \times 24 \text{ h/day} \times 3600 \text{ s/h})$
$= 4.14 \times 10^{-9} \text{ s}^{-1}$

Since $A = \lambda N$ (Eq. [6-12]) we can write the number of radioactive nuclei as

$N = A/\lambda = 3.7 \times 10^{14} \text{ s}^{-1}/4.14 \times 10^{-9} \text{ s}^{-1} = 8.94 \times 10^{22} \text{ atoms}$

Since 6.022×10^{23} atoms of ^{60}Co have a mass of 60 g or 0.060 kg, 8.94×10^{22} atoms have a mass of $8.94(0.06/60.22) = 0.0089$ kg.

The energy per second $= 3.7 \times 10^{14} \text{ s}^{-1}(1.17 + 1.33) \text{ MeV}$
$= 9.25 \times 10^{14} \text{ MeV/s}$
$= 9.25 \times 10^{14} \text{ MeV/s} \times 1.602 \times 10^{-13} \text{ J/MeV}$
$= 1.5 \times 10^{2} \text{ J/s} = 1.5 \times 10^{2} \text{ W}$

6.8   Absorption of Radiation

When alpha, beta or gamma radiation enters matter, it interacts with the atoms and molecules of the matter via many processes. In these interactions, the energy of the radiation is transferred to the atoms and molecules which may become altered in important ways, such as the ionization of atoms or the production of free radicals from molecules. The mechanisms of energy loss are important to understand from the point of view of *shielding* from radiation, and the effect on the absorbing matter has important consequences for radiation damage.

It was mentioned earlier in this chapter that α–particles can be stopped by a sheet of paper, and β-particles by 1 cm or so of plastic. This makes shielding an almost trivial matter in radioisotope or research laboratories or in factories where source activities of 10^{4} Bq ($\sim\mu$Ci) up to 10^{18} Bq ($\sim$100s of MCi) are normally encountered. When activities are of the order of 10^{19} Bq ($\sim$1000s of MCi) or greater, two problems arise. First, the material is weakened by the constant radiation damage to its structure; and second, the kinetic energy deposited by the particles as they slow down in the shielding ends up as heat, and the shielding may warp or even melt. Some high-level fission wastes have to be stored in metal tanks that are constantly cooled.

With γ-rays, there is a totally different situation. Being EM radiation, these are much more penetrating than are α- or β-particles. If a beam of high-energy photons enters a material, several processes remove photons from the beam. As the beam goes through the material, the number of photons in it steadily decreases both by losing energy in various absorption mechanisms and by scattering out of the beam. In contrast with the situation described in Chapter 4 for the Beer–Lambert law, where scattering was taken to be negligible in the materials of interest, scattering with high-energy photons cannot be neglected. If, however, the scattering is a random process then it will also follow an exponential relation as does absorption. The *linear attenuation coefficient* (μ) is a constant which will now include the effects of both scattering and absorption. The number decreases exponentially with distance so that if N_0 photons strike a unit area per second, then the number surviving at a distance x downstream is

$$N = N_0 e^{-\mu x} \qquad \textbf{[6-15]}$$

where μ is the linear attenuation coefficient (unit length^{-1}) which is a characteristic of the material and the photon energy. Material of high atomic number Z has a larger μ than does material of low Z. It is for this reason that lead (with atomic number 82) is an effective and convenient shield against γ-rays.

From this discussion it is important to realize that heavy shielding such as lead in radiation structures and equipment is there not for the α- and β-particles but for the γ-rays. If the gammas are adequately shielded, the alphas and betas will be taken care of automatically.

EXAMPLE 6-6

What thickness of lead is needed to reduce a flux of 1.5-MeV γ-rays by a factor of 100? (For 1.5-MeV γ-rays, μ = 57 m^{-1} in lead.)

SOLUTION

Since $N/N_0 = 0.0100$, we have

$0.0100 = e^{-\mu x}$, or

$-\mu x = \ln(0.01) = -4.6$

Therefore, $x = 4.6/\mu$

For lead, $x = 4.6/(57 \text{ m}^{-1}) = 8.1 \times 10^{-2} \text{ m} = 8.1 \text{ cm}$

In the energy range of 1 to 2 MeV, the attenuation of γ-rays is almost all due to interaction with the orbital electrons of the atoms, and so the linear attenuation coefficient is roughly proportional to the material's density ρ. If the linear attenuation coefficient μ is replaced by a product $\mu_m \rho$, then the quantity μ_m, called the *mass attenuation coefficient,* will be nearly a constant for all materials. Eq. [6-15] then becomes

$$N = N_0 e^{-\mu_m \rho x} \qquad \textbf{[6-16]}$$

The mass attenuation coefficient for γ-rays at an energy of 1.5 MeV is well approximated by

$$\mu_m = 5.0 \times 10^{-3} \text{ m}^2/\text{kg} \qquad \textbf{[6-17]}$$

EXAMPLE 6-7

Repeat Example 6-6 for steel that has a density of 7800 kg/m³.

SOLUTION

Using Eq. [6-16] and [6-17],

$N = N_0 e^{-\mu_m \rho x}$

or $0.0100 = e^{-0.0050 \times 7800 x}$

$$\ln(0.0100) = -0.0050 \times 7800x$$

$$x = 0.12 \text{ m} = 12 \text{ cm}$$

At energies around 10 keV, the mass attenuation coefficient of X-rays is much higher than 0.0050 m²/kg and it falls rapidly with increasing energy, leveling out at higher energies. As a result, an exposure to X-rays at 10 keV will produce a severe radiation burn which is slow to heal. Diagnostic X-rays, however, are produced at energies near 100 keV at which the tissue is much more transparent, because of the decreased coefficient, and the contrast with bone, for example, can be recorded.

6.9   Mechanisms of γ-Ray Absorption

It is important to distinguish between attenuation, scattering and absorption. *Attenuation* is the description of how a beam of radiation diminishes in intensity with distance. All of the attenuated energy need not, however, be deposited in the sample; some of it may be merely *scattered*, i.e., have its propagation direction changed. Only the amount that has actually transferred energy to the sample is *absorbed*. The energy absorbed by a material will always be less than that inferred from the attenuation.

When γ-rays enter matter, their interaction is almost exclusively with the orbital electrons of the atoms. This is easy to see on the grounds of geometry alone; the electrons are spread out over a volume of the order of an atomic size (diameter $\sim 10^{-9}$ cm) whereas the nuclear volume is much smaller (diameter $\sim 10^{-13}$ cm), and so presents a much smaller target. The interaction of the γ-rays with the electrons takes place via three mechanisms whose relative importance depends on the energy of the γ-rays. These mechanisms are the photoelectric effect, Compton scattering,[10] and electron–positron pair production.

The Photoelectric Effect

In this process, which was discussed in detail in Chapter 4 for visible light, a γ-ray of energy E interacts with an orbital electron, ejecting it with an energy equal to the difference between the γ-ray

---

10.   Arthur Holly Compton (1892–1962), American physicist.

energy and the electron's binding energy (called the "work function" in Chapter 4). That is,

$$E_{\text{electron}} = E - E_{\text{binding}} \qquad \textbf{[6-18]}$$

In this process the probability of interaction is greatest when the γ-ray energy and the binding energy are nearly equal. For visible light (see Chapter 4) this means that the outer loosely bound electrons will be ejected preferentially since their binding energy is of the order of a few eV which is also the energy of visible-light photons. Clearly, however, for high-energy photons, the inner-shell electrons are most likely to be ejected.

The cross-section (probability, σ_{pe}) for the photoelectric detachment of an inner-shell electron cannot be written in simple form, but an approximate relation for photon energies around 100 keV is

$$\sigma_{\text{pe}} \sim \mu_{m(\text{pe})} \sim Z^4/E^3 \qquad \textbf{[6-19]}$$

where μ_m is the mass attenuation coefficient, E is the energy of the photon and Z is the atomic number of the absorbing material. Of course $\sigma_{\text{pe}} = 0$ until the photon energy reaches the binding energy and E_{electron} in Eq. [6-18] becomes positive; at this point the probability of absorption becomes very large. With further increases in the energy, the cross-section decreases as E^{-3}. The cross-section in water (tissue) falls off so rapidly that at energies above 100 keV the photoelectric effect becomes negligible compared with the Compton scattering which is described in the next section.

For the atoms in tissue (mostly H and O in water), the binding energy of the inner electrons is only of the order of a few hundred eV at most, so from Eq. [6-18], the energy of the ejected electron constitutes nearly all of the energy of the incoming γ-ray. The result of the photoelectric effect, then, is to produce a high-speed electron in the medium.

Compton Scattering

At higher energies, as the cross-section for photoelectric scattering becomes less important, another photon–electron process becomes important. This interaction is with the weakly bound outer-shell electrons in the atoms, and so the electron binding energy can be neglected and the electrons can be considered essentially free. This process can be viewed as simply a "billiard-ball" type of collision between a photon of initial energy E and an electron of initial energy equal to zero. After the interaction, the photon and electron move off at angles θ and Φ with respect to the original direction, with final

Figure 6-3

Compton scattering.

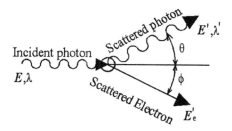

energies E' and E_e' as shown in Fig. 6-3. The derivation of the final results for this Compton scattering can be found in any textbook on modern or atomic physics, and only the final results will be presented in this text.

Since the incoming photon gives up energy to the electron, the outgoing photon has less energy and therefore, from Planck's relation (Eq. [4-1]), a reduced frequency or a longer wavelength. The wavelength difference is

$$\lambda' - \lambda = \frac{h}{m_e c}\left[1 - \cos(\theta)\right] \qquad \textbf{[6-20]}$$

In Eq. [6-20], m_e is the rest mass of the electron.

The kinetic energy of the recoiling electron is given by

$$E_e' = \frac{E\left[1 - \cos(\theta)\right]}{1 - \cos(\theta) + \dfrac{m_e c^2}{E}} \qquad \textbf{[6-21]}$$

Again the result of the interaction of γ-rays with matter is the production of a high-energy electron. The scattered γ-ray may undergo further interactions or it may leave the sample. The fraction of the incident energy transferred to the electron increases, with the incident energy being 2% at 5 keV and 95% at 5 MeV.

Electron–Positron Pair Production

The third process of photon absorption is the production of an electron–positron pair from a γ-ray of sufficient energy according to Einstein's mass–energy equivalence

$$E = 2m_e c^2 \qquad \textbf{[6-22]}$$

where m_e is the rest mass of the electron (or positron). The energy equivalence of an electron (or a positron) is 0.51 MeV and so an electron–positron pair can be produced by γ-rays of at least 1.02 MeV. When such γ-rays interact with a heavy nucleus, this conversion may take place. The threshold for the process is 1.02 MeV, and any excess energy appears as kinetic energy of the pair of particles. Again the result is the production of high-energy electrons. The cross-section for pair production increases with increasing energy and is not significant below about 2 MeV. It is therefore not very important in the context of radiological effects.

Total Attenuation

Total attenuation and absorption from all mechanisms for water is shown in Fig. 6-4. It can be clearly seen how the Compton effect dominates in the energy region 0.1 to 10 MeV, and how the total mass absorption coefficient $\mu_{m,a}$ is almost constant over this range. The total mass absorption coefficient is well approximated by (compare to Eq. [6-17]),

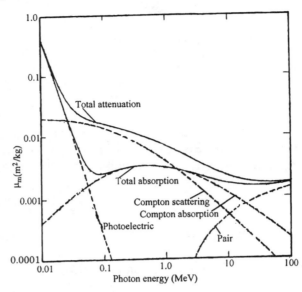

Figure 6-4

Mass absorption, attenuation, and scattering coefficients for water:
Total absorption = Photoelectric + Pair + Compton absorption;
Total attenuation = Total absorption + Compton scattering.

$$\mu_{m,a} = 3.0 \times 10^{-3} \text{ m}^2/\text{kg} \qquad \textbf{[6-23]}$$

6.10 Radiation Exposure and Absorbed Dose

As seen in Section 6.9, the final result of all types of radiation entering matter is to produce high-speed electrons. These particles lose energy to the molecules of the material in a sequence of ionizations which, on average, transfer about 35 eV of energy on each interaction. Therefore, each electron results in a large number of affected molecules. A 1-MeV electron will disrupt $10^6/35 = 3 \times 10^4$ molecules. These high-speed electrons rarely collide with the nucleus. Nuclear reactions are rare in irradiated matter; the flux of the radiation must be very large such as is found in the core of a nuclear reactor. Ionization damage can be classified in two ways in living matter:

1. The ionization process may break a valence bond in a macromolecule such as DNA. The resulting rearrangement of bonds in the molecule or subsequent chemical reactions with the disturbed site on the molecule may then disrupt the proper functioning of the molecule and the cell in which it resides.

2. Much of the material in a cell is water. Incoming particles may disrupt the water molecule leaving molecular fragments called *free radicals* such as H and OH; these are chemically very reactive and

attack biological molecules, doing great damage. In fact, most radiation damage to living material is of this type (see Section 6.14).

It is important to distinguish between *exposure* and *dose*. An object may be exposed to a radiation field but may absorb more or less energy from the field depending on the nature of the radiation in the field and the properties of the absorber. The energy absorbed is a measure of the dose and will be discussed later. It is important first to be able to measure the exposure.

Since radiation ionizes matter, the strength of a radiation field (or exposure) is measured by the ionization it produces in some standard mass of material. It follows that the units of exposure are coulombs/kg. Since the early days of X-ray research, the standard material has been dry air at standard conditions. The unit of exposure is the *roentgen*[11] and is defined as:

One roentgen of radiation is that radiation that will produce 2.58×10^{-4} C/kg of charge of either sign in dry air at NTP.[12]

The amount of biological damage is determined by the amount of energy deposited by the radiation. *Dose (D)* is defined as the amount of energy deposited per unit mass. Absorbed doses are now usually expressed in *grays* (Gy) and absorbed dose rates in grays per second, where

$$1 \text{ gray} = 1 \text{ J/kg} \qquad \textbf{[6-24a]}$$

An older unit of dose is the *rad* (not to be confused with the radian measure of angles) which is defined as

$$1 \text{ rad} = 0.01 \text{ J/kg} \qquad \textbf{[6-24b]}$$

Clearly from Eq. [6-24a] and [6-24b], 1 rad = 10^{-2} Gy.

EXAMPLE 6-8

An individual ingests 3.7×10^4 Bq (1 µCi) of a 2.0 MeV β-emitter and it distributes uniformly throughout his body.

---

11. Wilhelm Konrad Röntgen (1845–1923), German physicist and discoverer of X-rays.
12. NTP means normal temperature and pressure, or 20°C and 1 Atm.

The effective half-life is 28 yr. Calculate the initial absorbed dose rate in Gy/s and rad/hr and the total absorbed dose received in the first 28 years. Take the person's mass as 70 kg.

SOLUTION

The energy per second = 3.7×10^4 decay/s $\times$ 2.0 MeV/decay
= 7.4×10^4 MeV/s $\times$ 1.602×10^{-13} J/MeV
= 1.2×10^{-8} J/s

The initial absorbed dose rate = 1.2×10^{-8} J·s^{-1}/70 kg
= 1.7×10^{-10} J·kg^{-1}·s^{-1}
= 1.7×10^{-10} Gy·s^{-1} $\times$ 3600 s·h^{-1}/0.01 Gy·rad^{-1}
= 6.1×10^{-5} rad/h

Using Eq. [6-9],

$\lambda = 0.693/T_{1/2} = 0.693/(28 \text{ yr} \times 365 \text{ d/yr} \times 24 \text{ h/day} \times 3600 \text{ s/h})$
    = 7.85×10^{-10} s^{-1}

Using Eq. [6-12], at the start the number of atoms is
$N = A/\lambda = (3.7 \times 10^4 \text{ s}^{-1})/(7.85 \times 10^{-10} \text{ s}^{-1}) = 4.7 \times 10^{13}$

In 28 years, one-half of these emit 2-MeV betas for a total absorbed dose of $(\frac{1}{2}) \times 4.7 \times 10^{13}$ decays $\times$ 2 MeV·decay^{-1} $\times$ 1.602×10^{-13} J·MeV^{-1}/70 kg = 0.11 Gy = 11 rads

---

Dose calculations for γ-rays are not so straightforward but, as discussed previously (see Eq. [6-23]), the total mass absorption coefficient for γ-rays in the energy range 100 keV to 5 MeV is almost constant at 3.0×10^{-3} m^2/kg as can be seen in Fig. 6-4. Therefore the absorbed dose due to a beam of N γ-rays per second per m^2 of tissue delivered in time t is

$$D = 3.0 \times 10^{-3} NEt \quad \text{J·kg}^{-1} \text{ (Gy)} \qquad \text{[6-25]}$$

where E is the energy in joules of one γ-ray photon.

EXAMPLE 6-9

An individual stands 5.0 m from a 3.7×10^7 Bq (1 mCi) source of 0.50 MeV γ-rays. What absorbed dose does the person receive by remaining there for one hour? The person's frontal area is 0.75 m^2.

SOLUTION

First the intensity (number, N per m² per s) of the γ-rays at a distance of 5 m must be evaluated. If a sphere of 5 m radius is drawn, clearly all the γ-rays must go through the surface of that sphere and the number per unit area per second is given by $3.7 \times 10^7/4\pi r^2$ where $r = 5.0$ m.

$N = 3.7 \times 10^7/4\pi 25 = 1.18 \times 10^5$ m$^{-2 \cdot}$s^{-1}

$E = 0.50$ MeV $= 0.50$ MeV $\times 1.602 \times 10^{-13}$ J/MeV
  $= 0.80 \times 10^{-13}$ J

Using Eq. [6-25],

Dose rate (D/t)/m² $= 3.0 \times 10^{-3} NE$
$= 3.0 \times 10^{-3} \times 1.18 \times 10^5$ m$^{-2 \cdot}$s$^{-1} \times 0.80 \times 10^{-13}$ J
$= 2.8 \times 10^{-11}$ Gy/s/m² $= 2.8 \times 10^{-9}$ rad/s/m²

In one hour the absorbed dose D is
2.8×10^{-11} Gy/s/m² $\times 3600$ s $\times 0.75$ m² $= 7.6 \times 10^{-8}$ Gy

---

6.11 **Equivalent Dose**

When radiation enters living tissue, the damage it produces, whether directly or as a result of making chemically active species, has various consequences for the organism. Some damage is repairable; it would be remarkable if this were not so, since living organisms have had to evolve in an environment that includes natural sources of ionizing radiation. Some damage is not repairable, however, and that is the topic of this section.

A simple calculation of the absorbed dose in grays (or rads), as was done in the previous section, does not tell the whole story as far as biological effects are concerned. The biological effects also depend on the spatial distribution of the energy deposited. As mentioned in Section 6.10, one particle with an energy of 1 MeV produces about 30 000 ionizing events. This is true for both the α- and the β-particle. The difference is that the easily stopped α does all this damage in a region a few micrometres deep, whereas the damage caused by the β is more sparsely distributed over a depth of a few millimetres. A single cell can be damaged by α-particles at several places so that it is unable to recover, but the less concentrated damage caused by the β may be repairable.

The quantity that accounts for this difference in the relative effect of radiation has had several names in the past; among these are the *quality factor* (QF) and *relative biological effectiveness* (RBE). The most recent name is the *radiation weighting factor* (w_R). Although these quantities have slightly different definitions, for the purpose of this discussion they are the same and simply give, as a dimensionless multiplication factor, the relative effectiveness of a given radiation at producing biological damage. The quantity w_R has been determined by comparing the dose needed to produce a specific effect with some arbitrary standard. This standard is taken to be the dose of X-rays and γ-rays that are among the least damaging of radiations; the w_R of this standard is taken to be 1. It is expected, then, that values of w_R will be numbers equal to or greater than 1. With this standard it is found that alphas require only one-twentieth the dose to produce the same damage as do X-rays, and therefore, for alphas $w_R = 20$. The values of w_R are summarized in Table 6-3 for all the radiations of biological interest and concern.

Table 6-3   Values of w_R

Radiation	w_R
Photons, all energies	1
Electrons, all energies	1
Neutrons, energy:	
10 keV	5
10 keV to 100 keV	10
100 keV to 2 MeV	20
2 MeV to 20 MeV	10
> 20 MeV	5
Protons, energy > 2 MeV	5
α-particles, fission fragments	20

Adapted from 1990 Recommendations of the International Commission on Radiological Protection (ICRP) Standard.

Using w_R, the *equivalent dose* (*H*) is defined, which is a more accurate measure of the biological damage than is absorbed dose alone; the equivalent dose must be summed over all the radiations incident on the tissue of interest. Thus

$$H = \sum w_R D_R \qquad \textbf{[6-26]}$$

The unit of equivalent dose, for dose in grays, is the *sievert* (Sv) (if the dose is in rads, the equivalent dose is in rems).[13]

EXAMPLE 6-10

A small point-like radioactive source of 3.7×10^4 Bq (1 μCi) is spilled on a researcher's hand. The source is polonium ^{210}Po which is an α-emitter with an energy of 5.4 MeV. The range of 5.4 MeV α-particles in skin (essentially water) is 0.0020 cm. What is the absorbed dose and the equivalent dose, received in 1 hour?

SOLUTION

On the surface of the skin only one-half of the radiation enters the skin; the volume affected is a hemisphere of volume
$V = (½)(4/3)\pi r^3$
$= (2\pi/3)(2.0 \times 10^{-5} \text{ m})^3 = 1.7 \times 10^{-14} \text{ m}^3$

The mass affected $= 1.7 \times 10^{-14} \text{ m}^3 \times 1000 \text{ kg·m}^{-3}$
$= 1.7 \times 10^{-11} \text{ kg}$

Only half of the radiation enters the skin, so the activity
$= ½(3.7 \times 10^4) \text{ Bq} = 1.85 \times 10^4 \text{ Bq}$

The energy per second
$= 1.85 \times 10^4 \text{ s}^{-1} \times 5.4 \text{ MeV} \times 1.602 \times 10^{-13} \text{ J·MeV}^{-1}$
$= 1.6 \times 10^{-8} \text{ J·s}^{-1}$

Dose rate $= 1.6 \times 10^{-8} \text{ J·s}^{-1}/1.7 \times 10^{-11} \text{ kg} = 940 \text{ Gy·s}^{-1}$

In 1 hour, the absorbed dose
$D = 940 \text{ Gy·s}^{-1} \times 3600 \text{ s} = 3.4 \times 10^6 \text{ Gy}$

Using Table 6-3, the equivalent dose
$H = w_R \times D = 20 \times 3.4 \times 10^6 \text{ Gy} \cong 7.0 \times 10^7 \text{ Sv}$

The sources and the effective annual dose of environmental radiation (both natural and artificial) are given in Box 6-2. A summary of the new and old units of dose is given in Box 6-3.

13.   "Rem" stands for "roentgen equivalent man"; 1 rem = 10⁻² Sv.

Box 6-2   Sources of Environmental Radiation

The annual exposure to high-energy radiation from natural sources is far higher than from any artificial source. The natural and artificial average annual doses are summarized in the table.

Source of Radiation	Annual Effective Dose in mSv	%
Natural		
Radon	2.0	55
Cosmic rays	0.27	8
Terrestrial	0.28	8
Internal	0.39	11
Total natural	3.0	82
Artificial		
Medical	0.53	15
Consumer products	0.10	3
Total artificial	0.63	18
Total	**3.6**	**100**

Adapted from 'Health Effects of Exposure to Low Levels of Ionizing Radiation,' BEIR V, 1990.

By far the largest contribution, on average, is from radon, which is a natural decay product of uranium. This contribution, however, is also the most variable depending on location. The "terrestrial" contribution comes from uranium, thorium and potassium in the rocks and thus in building materials. The "internal" contribution is largely from potassium in our cells. Medical uses (mostly X-rays) provide the bulk of the average artificial exposure; the nuclear fuel-cycle contribution is unmeasurable.

Box 6-3   Summary of Radiation Units

Since the older units (rad, rem, etc.) are not passing from use as rapidly as might be wished, and since many regulations are still specified in these units, the two systems are summarized in the table below.

Comparison of Traditional and SI Units for Radioactivity-Related Quantities

Quantity	Traditional Name (Symbol)	SI Name (Symbol)
Activity	curie (Ci)	becquerel (Bq)
Energy	electron volt (eV)	joule (J)
Exposure	roentgen (R)	coulomb/kilogram (C/kg)
Absorbed dose (D)	rad	gray (Gy)
Equivalent dose (H)	rem	sievert (Sv)

1 Ci $= 3.7 \times 10^{10}$ Bq
1 MeV $= 1.602 \times 10^{-13}$ J
1 R $= 2.58 \times 10^{-4}$ C/kg
1 rad $= 10^{-2}$ Gy
1 rem $= 10^{-2}$ Sv

6.12   **Target Theory**

As shown in the previous section, the result of all processes which extract energy from radiation in matter is the production of a high-energy electron. These primary events will be spaced far apart for radiations with a low LET (e.g., γ-rays), or close together (perhaps all in a single cell) for radiations with a high LET (e.g., α-particles). As the primary charged particle travels through the matter, it loses energy by detaching secondary electrons from molecules as it passes by, with an average loss of 35 eV per interaction. Since the binding energy of an electron in a molecule is only of the order of 3 eV, then even the secondary electrons will be traveling at considerable speed and will cause further ionizations within a small volume (diameter ~ 1 nm) of the secondary event.

If all of this molecular disruption occurs near certain critical locations in a living cell, the cell will be *inactivated* and will die. For this reason these are called *inactivating events*. It is important to realize that an inactivating event has only the <u>potential</u> to kill a cell; most will not, as they occur in non-critical locations which, in fact, constitute most of the volume of the cell. There is a good likelihood that an α-particle will kill a living cell if it enters it, since many inactivating events are produced in each cell and the probability of one occurring near a critical target is high. Alpha-emitters are use to treat skin cancers for this reason. By contrast, the probability that γ-rays will produce more than one inactivating event in a cell is small unless the intensity of the radiation is very high.

If I represents the number of inactivating events per m³, then experiments show that

$$I = 6.1 \times 10^{19}\, D \qquad\qquad \textbf{[6-27]}$$

where D is the absorbed dose in grays.[14]

Since the inactivating events are distributed randomly through the tissue, the Poisson distribution (see Appendix 1) can be used to calculate the probability that a cell will receive 0, 1, 2, or more events. Notice this does <u>not</u> give the number of cells that will be killed, only those that receive an inactivating event with the potential for killing. The average number of inactivating events per cell is $I\mathcal{V}$ where I is given in Eq. [6-27] and $\mathcal{V}$ is the volume of the cell. Using $a = I\mathcal{V}$ in the Poisson distribution (Eq. [A1-1], Appendix 1), the probability $\mathcal{P}(n)$ of n inactivating events occurring in the same cell is

$$\mathcal{P}(n) = \frac{e^{-I\mathcal{V}}(I\mathcal{V})^n}{n!} \qquad\qquad \textbf{[6-28]}$$

EXAMPLE 6-11

A tissue containing spherical cells of diameter 8.00 μm is irradiated to a dose of 0.10 mGy. What is the probability that 0, 1, 2 and more than 2 inactivating events will occur in any cell?

---

14. For doses measured in rads, the formula is $I = 6.1 \times 10^{17}\, D$.

SOLUTION

The average number of events per cell is $I\mathcal{V}$
$$= (6.1\times10^{19}\,\text{events/m}^3/\text{Gy})(0.10\times10^{-3}\,\text{Gy})(4/3)\pi(4.00\times10^{-6}\,\text{m})^3$$
$$= 1.64$$

Fraction receiving no events
$$= \mathcal{P}(0) = e^{-1.64} = 0.194 \text{ or } 19\%$$

Fraction receiving 1 event
$$= \mathcal{P}(1) = \mathcal{P}(0)(a/n) \text{ where } a = I\mathcal{V} \text{ and } n = 1:$$
$$\mathcal{P}(1) = (0.194)(1.64/1) = 0.318 \text{ or } 32\%$$

Fraction receiving 2 events
$$= \mathcal{P}(2) = \mathcal{P}(1)(a/n) = (0.318)(1.64/2) = 0.260 \text{ or } 26\%$$

Fraction receiving more than 2 events
$$= 1 - [\mathcal{P}(0) + \mathcal{P}(1) + \mathcal{P}(2)]$$
$$= 1 - (0.194 + 0.318 + 0.260) = 0.228 \text{ or } 23\%$$

The probabilities found by using Eq. [6-28] (and shown in Example 6-11) allow one to find the fraction of cells receiving a certain number of inactivating events, but provide no information regarding the fraction of cells which would be killed by a certain dose. It is often useful to describe another volume, called the *sensitive volume*, $\mathcal{V}_s$, which, when hit by an inactivating event, will indeed lead to cell death. This volume, which is much smaller than that of the whole cell, is a *target* which must be hit to kill the cell; otherwise the cell will survive. It has been shown by experiment that $\mathcal{V}_s$ is the volume of the cell's DNA.[15] If each cell contains one target then the probability that a cell will survive is the probability that the sensitive volumes are not hit. If N_0 is the original number of cells and N the number which survive, then the probability of survival is N/N_0 and

$$\mathcal{P}(0) = \frac{N}{N_0} = e^{-I\mathcal{V}_s}$$

or using Eq. [6-27],

$$\ln\frac{N}{N_0} = -I\mathcal{V}_s = -6.1\times10^{19}\,D\mathcal{V}_s \qquad \textbf{[6-29]}$$

---

15.   This work was performed on fruit flies by Russian geneticist N.N. Timofeev-Resovsky (1900–1981) with Zimmer and Delbrück in Germany before WWII.

Equation [6-29] provides an experimental method to determine a cell's sensitive volume. Experiments are performed in which living cells are exposed to increasing doses of radiation. A semilogarithmic graph of N/N_0 vs. D will give a straight line whose slope will be $-6.1 \times 10^{19}\, \mathcal{V}_s$.

If the system being inactivated is an enzyme molecule of density ρ, then the molar mass M of this protein will be

$$M = \mathcal{V}_s\, \rho N_A \qquad\qquad \textbf{[6-30]}$$

where N_A is Avogadro's number. Typical values of ρ for protein lie around 1300 kg/m³. Using this method, molar masses of 30 000 and 230 000 have been found for ribonuclease and the enzymatic portion of myosin, respectively. These values are in agreement with those obtained by other methods. This method is not always successful, and quite poor estimates of molar masses are found if the sensitive volume for the enzyme is much different from its true volume.

6.13 **Multiple Targets**

In some cases the system being irradiated may have more than one sensitive target, all of which must be hit to inactivate the cell. If there are p targets in the cell then there must be p hits in order for the cell to die. The probability of hitting one of the targets is one minus the probability of a miss; that is

$$1 - \mathcal{P}(0) = 1 - e^{-I\mathcal{V}_s}$$

If there are p targets, the probability that all of them will be hit is the product of their individual probabilities, or

$$\left(1 - e^{-I\mathcal{V}_s}\right)^p$$

Since this is the probability of hitting all p targets, the probability of their <u>not</u> all being hit is the probability of survival N/N_0, thus

$$\frac{N}{N_0} = 1 - \left(1 - e^{-I\mathcal{V}_s}\right)^p \qquad\qquad \textbf{[6-31]}$$

Equation [6-31] is simplified in two cases:

1. For low doses I is very small and so $I\mho_s \to 0$ and $e^{-0} \to 1$; Eq. [6-31] becomes

$$N/N_0 = 1 - (1 - 1)^p = 1$$

In other words, all the cells survive as expected at vanishingly small doses.

2. For large doses

$$\frac{N}{N_0} = 1 - \left(1 - e^{-I\mho_s}\right)^p$$
$$= 1 - \left[1 - pe^{-I\mho_s} + \tfrac{1}{2}p(p-1)e^{-2I\mho_s} + \text{higher order terms}\right]$$

Since $e^{-2I\mho}$ and all higher order terms are small with respect to $e^{-I\mho}$ they can be neglected, and

$$\frac{N}{N_0} \approx pe^{-I\mho_s} \quad \text{or} \quad \ln\frac{N}{N_0} \approx \ln(p) - I\mho_s \qquad \textbf{[6-32]}$$

Figure 6-5 shows the result of a semilogarithmic plot of survival fraction vs. I (or D) for the case of $p = 2$. At large doses the straight-line portion has a slope $-I\mho_s$ as before. At low doses the curve bends over and ends up at 1 as it must. (The probability of survival cannot be greater than 1.) The straight-line portion extrapolates to intersect the vertical axis at p, the number of targets.

Target theory has been used for many years to evaluate the size of DNA in cells, the number of targets, the size of enzymes and virus genetic materials, and other quantities. Often it is successful, but sometimes it gives poor results and must be applied cautiously along with other experimental checks.

Figure 6-5

Survival fraction vs. dose for multi-targets.

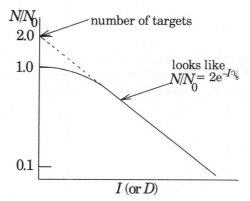

6.14 Action of Radiation at the Molecular Level

Radiation damage to biological systems occurs because crucial molecules are altered directly or indirectly as radiation passes through the cell. A bond in a DNA molecule may be broken due to direct ionization by the radiation or, more likely, free radicals produced by the ionization of water may attack the DNA and cause indirect damage. In some species, if only one strand of the DNA is altered and the cell is in a resting stage, a special mechanism can cleave out and replace the damaged section. Many species do not have this repair mechanism, and even if they do there might not be time to effect the repair if the cell is in rapid growth. Both the indirect effect and the repair mechanisms can produce uncertainty in experiments based on target theory.

The nucleic acids are not the only molecular types subject to damage. Damage to protein molecules such as enzymes can also seriously impair cell function. Scission (breaking of the primary chain) or cross-linking between two parts of the primary chain is often sufficient to impair the action of an enzyme. Even the removal of a group as small as an NH_3 group may be serious, especially if it is located in the active site of an enzyme.

Some of the indirect (solvent) type of effects are brought about by processes initiated by the ionizing radiation such as

(a) $\qquad H_2O \rightarrow HO\cdot + H\cdot$

(b) $\qquad H_2O \rightarrow e^- + H_2O^+$

and $H_2O^+ \rightarrow HO^+ + H\cdot$ or $H^+ + HO\cdot$

(c) $H_2O + e^- \rightarrow H_2O^-$

and $H_2O^- \rightarrow OH\cdot + H^-$ or $OH^- + H\cdot$

The dots in the above reactions signify free radicals which are extremely reactive. They may alter proteins and nucleic acids directly, or recombine to form compounds such as H_3O which are chemically very reactive.

Exercises

6-1 Complete the following reaction equations and identify W, X, Y and Z. A periodic table is required.

(a) $^{226}_{88}\text{Ra} \rightarrow {}^{222}_{86}\text{Rn} + \text{W}$

(b) $^{210}_{82}\text{Pb} \rightarrow {}^{210}_{83}\text{X} + \text{Y}$

(c) $^{4}_{2}\text{He} + {}^{9}_{4}\text{Be} \rightarrow {}^{12}_{6}\text{C} + \text{Z}$

6-2 What is the half-life of a radioactive substance for which the decay constant is $2.0 \times 10^{-6}\,\text{s}^{-1}$?

6-3 The mean lethal radiation dose in humans is about 5 Gy (500 rads). How much energy is deposited per kg of tissue by a dose of this magnitude?

6-4 A dose of 5 Gy of X-rays is required to destroy 63% of a particular bacterial population. It requires 1 Gy of β-particles of mean energy 1.0 MeV to accomplish the same result. What is w_R for the β-particles in this population of bacteria?

Problems

6-5 (a) What is the minimum thickness of lead (density 11 400 kg/m³) required to completely shield 1.0 MeV β-particles?

(b) In human tissue, through how many typical cells of thickness 20 μm would they penetrate?

6-6 Ingested ^{131}I (the so-called "radioactive cocktail") is regularly used in clinics to destroy overactive or diseased thyroid tissue. One day a counter positioned near a patient's thyroid records 1.0×10^4 counts per second. Forty-eight hours later the count rate is 6.0×10^3 counts per second. What is the biological half-life of iodine in the thyroid if the physical half-life of ^{131}I is 8.1 days?

6-7 The physical half-life of a radioactive isotope is 15 h. The biological half-life is 30 h. If, after the ingestion of this isotope, the count-rate of a blood sample is 4.0×10^4 counts per second, how long will it take for the count-rate to fall to 4.0×10^3 counts per second?

6-8   Two mice each eat one-third of a piece of radioactive cheese, leaving one-third behind. Forty minutes later the two mice are caught and eaten by an owl. Six hours after this meal the owl is caught and has a total body count of 200 counts per minute which, by coincidence, is the same as the remaining piece of cheese at the same time. The effective half-life for the radioactive isotope in an owl is 4.00 h and in a mouse 1 h and 20 min.

   (a)   What is the biological half-life for this isotope in a mouse?

   (b)   What was the initial count-rate for the whole piece of cheese? The accompanying diagram might help you to organize your thoughts.

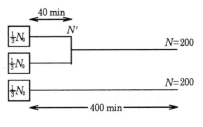

6-9   The half-life of ^{28}Al is 2.3 min and of ^{24}Na is 15 h. If a source initially contains 1000 times as many aluminum as sodium atoms, what will be the ratio of the number of aluminum atoms to sodium atoms remaining after (a) 10 min and (b) 30 min?

6-10   A piece of bone of thickness 3.0 cm and density 3000 kg/m^3 is placed between a γ-source and a detector. The count-rate drops from 1000 counts per minute to 200 counts per minute. What is the mass attenuation coefficient for the bone?

6-11   The count-rate from a γ-source is measured to be 1000 counts per minute. When 1.0 cm of lead is placed between the source and the detector, the count-rate is reduced to 100 counts per minute. What additional thickness of lead would have reduced the count-rate to 50 counts per minute?

6-12   By direct measurement with a millimetre scale on Fig. 6-4 (a magnifying glass will help), show that, for water at an energy of 100 keV, the mass absorption coefficient is 3×10^{-3} m^2/kg and the mass attenuation coefficient is 2×10^{-2} m^2/kg.

6-13   An X-ray unit that operates at 100 000 volts will produce X-rays of maximum energy of 100 000 eV. What will be the wavelength of these X-rays? Through what thickness of human tissue will a beam of these X-rays pass before the intensity is decreased to one-half? (Use the data from Problem 6-12.)

6-14 Mammalian cells each having a volume of 3.6×10^{-17} m³ are irradiated by γ-rays. It is found that 1.0×10^{17} inactivating events are produced per m³ of the cells. What fraction of the cells experience no inactivating events?

6-15 There are about 6.1×10^{19} inactivating events per m³ per Gy. How many inactivating events per m³ are produced by a typical medical X-ray of dose 0.10 mGy (10 mrad)?

6-16 If the volume of a cell is about 2.0×10^{-17} m³, what fraction of the cells of the person being X-rayed in Problem 6-15 would be hit?

6-17 After a group of cells has been irradiated by a dose of 2.0 Gy (200 rad), the fraction surviving is 0.57. If there is only one target per cell, what is the sensitive volume?

6-18 HeLa cells are a particular strain of malignant human cells which can be cultured quite easily. The following table shows the surviving fraction of HeLa cells as a function of X-ray dose.

(a)   Find the number of sensitive targets in the HeLa cell.

(b)   Find the sensitive volume (volume of the DNA) of the HeLa cell. (The solution will require plotting a graph on semilogarithmic paper.)

Surviving fraction	Dose (Gy)
0.82	0.6
0.72	0.85
0.5	1
0.45	1.5
0.31	2
0.18	2.5
0.11	3
0.06	4
0.01	5

Answers

6-1 W = $_2^4$He, X = $_{83}^{210}$Bi, Y = $_{-1}^0$e, Z = $_0^1 n$

6-2 3.5×10^5 s

6-3 5 J/kg

6-4 5

6-5 (a) 0.36 mm (b) 200

6-6 4.1 d

6-7 33 h

6-8 (a) 100 min (b) 1200 counts per minute

6-9 (a) 49 (b) 0.12

6-10 0.018 m^2/kg

6-11 3 mm

6-13 0.0124 nm, 3.5 cm

6-14 0.027

6-15 6.1×10^{15}

6-16 0.115

6-17 4.6×10^{-21} m^3

6-18 (a) 2 to 2.5 targets (b) 1.5×10^{-20} m^3

7 Mechanics of Biological Systems: Kinematics

7.1   Introduction

Mechanics is that branch of physics concerned with understanding the relationship between the motion of an object and the forces acting on it. The object can range in size from the tiniest subatomic particle to a galaxy. Thus it is not difficult to view the laws of mechanics as being among the most fundamental laws of nature. In this and the next two chapters, we shall describe the principles of classical (or Newtonian[1]) mechanics in the context of how they can be used to analyze common animal motions such as walking, running, and jumping. Occasionally we shall apply the term *biomechanics* to this material, although like its parent term "mechanics," this usually encompasses a broader range of topics including the biomechanics of deformable solids (Chapter 10) and the biomechanics of fluids (Chapters 11 and 12).

The form and lifestyle of any animal are determined by the physical forces which dominate its environment. For mammals, including humans, the most important of these forces is gravity. This clearly places significant restrictions on our movements (such as how high we can jump) and, as will be seen in Chapter 10, plays a major role in determining animal size and strength. The human body is an exceedingly complex entity. It moves with a smoothness and coordination that no machine can emulate, yet the motions that it can perform are still subject to the laws of classical mechanics.

While the emphasis in these chapters will be on animal mechanics, as these can undergo self-propelled motion, several aspects of biomechanics are also relevant to plant life. For example, a number of primitive unicellular organisms such as bacteria propel themselves by a flagellar motion of their tails or their *cilia*, analogous to the motion of a tadpole (see Chapter 12). The principles of static equilibrium, which are

---

1.   Sir Isaac Newton (1642–1727), English scientist and mathematician.

a subset of Newton's laws of motion, also play a significant role in the construction of plants such as grasses and trees—for example, in determining the ability of a stalk of grass to resist bending in the presence of wind. To deal with the latter topic, we shall also have to take into account the elastic properties of biological tissues, describing their ability to withstand stresses due to external forces. The subject of elasticity of materials will be discussed separately in Chapter 10.

7.2   One-Dimensional Motion

Almost all animals are capable of motion. It is necessary for survival, e.g., to search for food or to escape from danger. Some animals are capable of great speed (e.g., the cheetah can sprint at 110 km/h). Others, while slower moving, possess greater endurance (e.g., the horse can maintain a speed of 24 km/h for a distance of 55 km). In this section, we will consider kinematics, which is the study of motion without regard to its physical causes. General motion consists of *linear* (or *translational*) contributions and *rotational* contributions. For example, during athletic activities, the human body executes linear and rotational motion simultaneously, and the two must be coordinated for the best result; the swivelling shoulder blades of the horse and cheetah add many centimetres to their strides. We shall see in this section how to analyze linear motion. All aspects of rotational motion will be covered in Chapter 9.

Translational kinematics begins by introducing the *displacement* of a moving object, which is the change in position of the object during its motion. We shall initially consider motion in one dimension only. This is not a serious constraint, as any general three-dimensional motion can be analyzed in terms of its components along three independent directions. Taking motion to be along the x-direction, if the initial and final positions of an object are denoted x_1 and x_2, respectively, then the displacement of the object is $x_2 - x_1$. This is written as

$$\Delta x = x_2 - x_1 \qquad \text{[7-1]}$$

If the time taken to travel from x_1 to x_2 is $t_2 - t_1 = \Delta t$, then the *average velocity* is the rate of change of position, i.e.,

$$v_{av} = \frac{\Delta x}{\Delta t} = \frac{x_2 - x_1}{t_2 - t_1} = \frac{\text{displacement}}{\text{elapsed time}} \qquad \text{[7-2]}$$

Note that Δx and v_{av} could be either positive or negative, depending on whether the motion is in the direction of increasing or decreasing position x. We have to distinguish between *displacement* and *distance* traveled, as well as between *average speed* and magnitude of the average velocity. The average speed over some time interval is equal to the average velocity only if the object does not reverse direction. The more general definition of average speed is (distance traveled)/(elapsed time). For example, if an object travels 10.0 m in the +x-direction for 5.0 s, and then travels 10.0 m in the –x-direction for 5.0 s, the displacement of the object and hence its average velocity are zero, but the average speed is 20.0 m/10.0 s = 2.0 m/s.

Figure 7-1a illustrates the definition of v_{av} between two times t_1 and t_2 on a plot of position x as a function of time. If the elapsed time from t_1 to t_2 is divided into several smaller subintervals, then it may turn out that the average velocity differs from one subinterval to the next.

Figure 7-1

(a) Graph of the position x of some object as a function of time t, illustrating that the average velocity between times t_1 and t_2 is the slope of the line P$_1$P$_2$; (b) The instantaneous velocity at time t_1 is the slope of the tangent line to the x–t curve at point P$_1$.

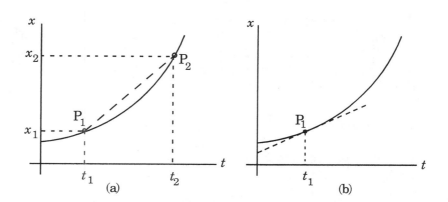

For example, while driving your car along an expressway, your speed may vary from larger to smaller values depending on (among other things) the degree of traffic. The velocity at any specific instant of time is called the "instantaneous" velocity v: the magnitude of this is the *instantaneous speed*, which is the value registered at any time by the car's speedometer.

The instantaneous velocity is defined as the ratio of the displacement to elapsed time when the latter approaches zero, i.e.,

$$v = \lim_{\Delta t \to 0} \frac{\Delta x}{\Delta t} = \frac{dx}{dt}$$ [7-3]

where dx/dt is the derivative of x with respect to t. It is easy to imagine letting t_2 approach t_1 (i.e., $\Delta t \to 0$) in Fig. 7-1a: in this limit, the slope of the line P$_1$P$_2$ connecting positions x_1 and x_2 approaches the slope of the *tangent* line to the curve of $x(t)$ at time t_1 (Fig. 7-1b). Henceforth, the term "velocity" will refer to instantaneous velocity.

Uniform Motion

One distinguishes two types of motion: *uniform* and *nonuniform*. Uniform motion is that occurring at constant velocity (i.e., constant speed in a straight line). In such a case, a plot of position x vs. time t

will be a straight line rather than a curved line as in Fig. 7-1. The instantaneous velocity is the same at all times, and equals the average velocity over any time interval. In this case, v_{av} can be replaced by v in Eq. [7-2] to give:

$$\Delta x = v\Delta t$$

$$\text{or } x_2 = x_1 + v(t_2 - t_1)$$

[7-4]

EXAMPLE 7-1

A cheetah can sprint at a speed of 110 km/h. The best a human is capable of is a speed of 35.0 km/h. A man and a cheetah are initially 0.400 km apart. Assuming that both man and cheetah are running at their top speed, how long does it take the cheetah to overtake the man?

SOLUTION

It is always a good idea in problems of this nature to sketch a diagram showing the initial and final conditions (Fig 7-2).

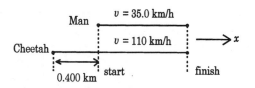

Figure 7-2

Example 7-1: Comparison of the speed of a man and a cheetah.

We apply Eq. [7-4] twice:
once for the man and once for the cheetah. We'll take the initial time $t_1 = 0$, and at this time choose the initial position of the cheetah to be x_1 (cheetah) = 0, so that x_1 (man) = 0.400 km. We then solve for the time t_2 when x_2 (cheetah) = x_2 (man), i.e.,

$$(110\,\text{km/h})t_2 = 0.400\,\text{km} + (35.0\,\text{km/h})t_2$$

Therefore:

$$t_2 = \frac{0.400\,\text{km}}{(110\,\text{km/h} - 35.0\,\text{km/h})} = 5.33 \times 10^{-3}\,\text{h}$$

$$= 19.2\,\text{s}$$

Nonuniform Motion

In nonuniform motion, the velocity generally changes with time. This is the case in Fig. 7-1 where the velocity in the x-direction is continually increasing with time. Analogous to the relation between displacement and velocity, the *acceleration* of an object is defined as the rate of change of its velocity. If the velocities at two times t_1 and t_2 are v_1 and v_2, then the average acceleration between these times is

$$a_{av} = \frac{v_2 - v_1}{t_2 - t_1} = \frac{\Delta v}{\Delta t}$$ **[7-5]**

The instantaneous acceleration at any instant is the limiting value of a_{av} as Δt approaches zero,

$$a = \lim_{\Delta t \to 0} \frac{\Delta v}{\Delta t} = \frac{dv}{dt}$$ **[7-6]**

Just as v at any point equals the slope of the tangent line on a graph of x vs. t at that point, the acceleration a is equal to the slope of the tangent line to a curve of v vs. t at the particular point.

If we express velocity in units of metres per second and time in seconds, then the unit of acceleration is (m/s)/s, more simply written as m/s^2.

In the remainder of this chapter, we shall consider only the simplest type of nonuniform or accelerated motion, for which the acceleration a is constant. In this case, the instantaneous and average accelerations are equal, and we can replace a_{av} in Eq. [7-5] by simply a. It is usually convenient to set the initial time t_1 to be zero and to write the later time t_2 simply as t. Also, the velocity at the initial time is usually written as v_0 and the velocity at later time t is simply denoted v. Then Eq. [7-5] becomes

$$a = \frac{v - v_0}{t} \quad \text{or} \quad v = v_0 + at$$ **[7-7]**

For this case of constant acceleration, we will also wish to know how the position x varies with time t, generalizing the relation in Eq. [7-4] which holds for constant velocity, that is, for $a = 0$. This can be derived in several ways, for example, by graphical methods or analytically by re-expressing Eq. [7-3] in integral form. Here we shall simply state the result, which is

$$x = x_0 + v_0 t + \tfrac{1}{2} a t^2$$ **[7-8]**

where x is the position at time t, while x_0 is the initial position at time $t = 0$.

Those with a calculus background can verify, by taking two successive time derivatives of Eq. [7-8], that this equation satisfies the earlier relations in Eqs. [7-3] and [7-6], i.e., $(dx/dt) = v$ (with v given by Eq. [7-7]) and $(dv/dt) = a$.

Two other relations linking the variables x, v_0, v, a and t for the case of constant acceleration can be derived. The first is obtained by solving Eq. [7-7] for t in terms of v, v_0, and a [$t = (v - v_0)/a$], and inserting the result into Eq. [7-8]. On rearrangement, this becomes

$$v^2 = v_0^2 + 2a(x - x_0) \qquad \textbf{[7-9]}$$

Alternatively, by solving Eq. [7-7] for a in terms of v, v_0, and t, and then substituting the result into Eq. [7-8], we find

$$x - x_0 = \left(\frac{v_0 + v}{2}\right)t \qquad \textbf{[7-10]}$$

This last equation is equivalent to Eq. [7-2] together with the identification $v_{av} = (v_0 + v)/2$, which follows straightforwardly from the fact that the velocity changes linearly with t when a is constant, so that v_{av} is the mean of the initial and final velocities.

Equations [7-7] to [7-10] are the equations of motion for an object with constant acceleration in one dimension: sometimes these are known as the Galilean[2] equations. As seen from their derivation, the last two equations are not independent of the first two, but are just different ways of expressing the same information. Note that none of these equations contains all of the variables x, x_0, v_0, v, a, and t. Which of these equations to use in problem solving depends on which variables are given and which ones are unknowns to be determined. In particular, one should choose an equation from [7-7] to [7-10] which contains only the single unknown.

It cannot be overemphasized that Eqs. [7-7] to [7-10] are valid <u>only</u> if the acceleration is constant. If the acceleration changes (either in magnitude or direction), these equations cannot be used!

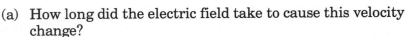

EXAMPLE 7-2

An ion traveling east through a biological membrane at 150 m/s enters an electric field which reduces its velocity to 100 m/s toward the east. While this acceleration (assumed constant) is occurring, the ion undergoes a displacement of 50.0 Å in the same direction (1 Å = 10^{-10} m).

(a) How long did the electric field take to cause this velocity change?

2.   Galileo Galilei (1564–1642), Italian astronomer and physicist.

(b) What acceleration did the ion experience?

SOLUTION

(a) We are given the initial and final velocities of the ion while it is in the electric field, namely $v_0 = 150$ m/s and $v = 100$ m/s (both in the same easterly direction, which we identify with the positive x-direction), and the displacement of the electron, $x - x_0 = 50.0$ Å. Perusal of the kinematic equations above shows that Eq. [7-10] is the one needed to find the time t. Solving that equation for t,

$$t = 2\frac{(x - x_0)}{v_0 + v}$$

$$= \frac{2 \times 50.0 \times 10^{-10}\,\text{m}}{(150 + 100)\,\text{m/s}} = 4.00 \times 10^{-11}\,\text{s}$$

(b) From the information given in the question, the acceleration a could be determined from Eq. [7-9]. Alternatively, since we have determined t from part (a), either of Eqs. [7-7] or [7-8] could also be used. Each method will give the same result. Here we will use Eq. [7-7],

$$a = \frac{v - v_0}{t} = \frac{(100 - 150)\,\text{m/s}}{4.00 \times 10^{-11}\,\text{s}}$$

$$= -1.25 \times 10^{12}\,\text{m/s}^2$$

The negative sign for a means that the ion has slowed down, i.e., its velocity toward the east has decreased. In such a case, we say that the ion has "decelerated." Alternatively, we could say that the acceleration has a magnitude of 1.25×10^{12} m/s^2 toward the west.

EXAMPLE 7-3

A cheetah (again!) hiding in the bushes spots a gazelle sprinting along at 17.0 m/s. When the gazelle passes him, the cheetah takes off after the gazelle with a constant acceleration of 3.00 m/s². How long will it take the cheetah to overtake the gazelle and how far has it traveled in that time?

SOLUTION

We take the origin to be where the cheetah is initially hiding, so $x_0 = 0$. The motion of the gazelle from this point is at constant velocity, so its position vs. time is described by Eq. [7-8] ($x = x_0 + v_0 t + \frac{1}{2} a t^2$) with $a = 0$ and $v_0 = 17.0$ m/s,

$$x \text{ (gazelle)} = (17.0 \text{ m/s}) \, t$$

while the position of the cheetah is given by Eq. [7-8] with $x_0 = 0$, $v_0 = 0$, and $a = 3.00$ m/s^2,

$$x(\text{cheetah}) = \frac{1}{2} (3.00 \text{ m/s}^2) \, t^2$$

As in Example 7-1, we solve for the time t when the positions of the two animals are equal,

$$(17.0 \text{ m/s}) \, t = (1.50 \text{ m/s}^2) \, t^2$$

Dividing both sides by t and rearranging, this gives

$$t = \frac{17.0 \text{ m/s}}{1.50 \text{ m/s}^2} = 11.3 \text{ s}$$

Having found t, we can now go back and calculate the distance traveled by the cheetah,

$$x \text{ (cheetah)} = (1.50 \text{ m/s}^2) \, t^2$$
$$= (1.50 \text{ m/s}^2)(11.3 \text{ s})^2$$
$$= 192 \text{ m}$$

7.3   Acceleration Due to Gravity

The most familiar example of motion with (nearly) constant acceleration is that of a body falling under the influence of Earth's gravitational attraction. When air resistance can be neglected, it was first demonstrated by Galileo that all bodies undergo the same acceleration when they are in "free fall," regardless of their size or weight. The magnitude of this acceleration is denoted as g. At or near Earth's surface, the value of g is 9.80 m/s^2 to three significant digits. (We shall discuss in Chapter 8 the relation between g and gravitational force.) Since free fall involves constant acceleration, the Galilean equations, Eqs. [7-7] to [7-10], can be applied. Because the motion is along a vertical line, we will make one change when writing those equations, namely to denote position by y rather than x. Later, when we treat two-dimensional motion, x will be used for horizontal motion and y for vertical motion.

EXAMPLE 7-4

A ball is thrown straight up from the edge of a cliff with an initial speed of 15.0 m/s. As it falls back, it just misses the edge and strikes the ground 75.0 m below the point from which it was thrown. What height does the ball reach and how long is it in the air?

SOLUTION

When applying the kinematic equations, it is extremely important to choose an origin of coordinates and to define which axis direction (e.g., up or down) is positive (Fig. 7-3). These choices are arbitrary and are usually a matter of convenience.

For example, if we choose the positive direction to be upward, then the downward acceleration due to gravity is negative, i.e., $a = -g = -9.80$ m/s^2, and the initial upward velocity is positive ($v_0 = 15.0$ m/s). Let us make this choice for the current problem, and also take the origin at the initial position of the ball at the top of the cliff, so that $y_0 = 0$. In this problem, the given quantities are, therefore: $v_0 = 15.0$ m/s and $a = -9.80$ m/s^2.

Note that when the ball reaches the top of its flight, it stops going up ($v > 0$) and starts going down ($v < 0$). Hence, $v = 0$ just at the instant it reaches its highest point. The required quantity at the top of the flight is the maximum height y relative to the top of the cliff, and the appropriate equation is Eq. [7-9] (with x_0 and x replaced by y_0 and y):

$$v^2 = v_0^{\,2} + 2a(y - y_0)$$

Setting $v = 0$, $y_0 = 0$, $a = -g$, we find $y = y_{max}$ to be

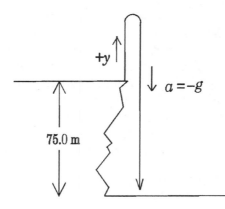

Figure 7-3

Example 7-4: The path of a ball thrown up from the edge of a cliff.

75.0 m

$+y$

$a = -g$

$$y_{max} = \frac{v_0^{\,2}}{2g} = \frac{(15.0 \text{ m/s})^2}{2(9.80 \text{ m/s}^2)} = 11.5 \text{ m}$$

To find the total time the ball is in the air, which we'll denote as t_f ("time of flight"), we know, as one additional given quantity, the final position of the ball, $y = -75.0$ m. Note this is negative since the positive direction was chosen to be upward. The appropriate equation in this case is (from Eq. [7-8]):

$$y = y_0 + v_0 t + \tfrac{1}{2} a t^2 \qquad \text{[7-11]}$$

Substituting $y_0 = 0, y = -75.0$ m, $v_0 = 15.0$ m/s, and $a = -g = -9.80$ m/s², and rearranging yields a quadratic equation for $t = t_f$:

$$(4.90)t_f{}^2 - (15.0)t_f - 75.0 = 0$$

Applying the standard formula[3] for the roots of a quadratic equation, we get

$$t_f = \frac{15.0 \pm \sqrt{(-15.0)^2 - 4(4.90)(-75.0)}}{9.80} = \frac{15.0 \pm 41.2}{9.80} \text{ s}$$

Since only the positive solution (i.e., time after $t = 0$) of this equation makes physical sense, the result is

$$t_f = (15.0 + 41.2)/9.80 = 5.73 \text{ s}$$

Several aspects of the preceding example are important to emphasize. We've seen that the velocity of the ball changes direction at its maximum height y_{max} and is instantaneously zero at that point. Throughout its motion, however, the acceleration remains constant at the value $a = -9.80$ m/s². At the end of its flight, when the ball reaches the bottom of the cliff, the value of the final position y that we use is the displacement relative to the origin, i.e., $y = -75.0$ m. Students often tend to believe that one should include in the final displacement the contribution of the total distance moved up and down (to and from y_{max}) during the flight. The reasons for not doing so are probably best revealed by plotting the position y as a function of time t as predicted by

---

3. For the quadratic equation $ay^2 + by + c = 0$, the solution is

$$y = \frac{-b \pm \sqrt{b^2 - 4ac}}{2a}$$

Eq. [7-11]. This is shown in Fig. 7-4a. The initial slope of the graph, equal to v_0, is positive, but the negative acceleration means that eventually the curve of y vs. t "turns around" at the maximum height y_{max} where the corresponding time is denoted t_{up}. The value of y eventually returns to zero, i.e., when the ball has returned instantaneously to its original position, and thereafter y becomes increasingly negative as the ball drops below the top of the cliff. The time of flight t_f corresponding to the final position $y_f = -75.0$ m is indicated.

Figure 7-4

Position (a) and velocity (b) as functions of time for a ball thrown upward with an initial positive velocity v_0.

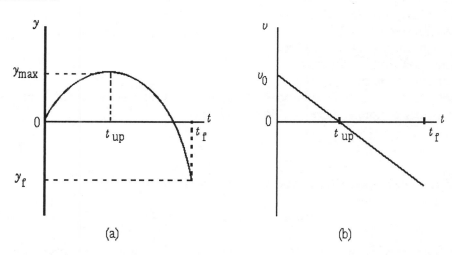

(a)   (b)

Figure 7-4b shows a graph of the corresponding velocity as a function of t, indicating that v decreases from its initial positive value v_0, passing through zero at $t = t_{up}$ and thereafter becoming increasingly negative, i.e., increasing in speed as the ball approaches the ground.

7.4   **Two-Dimensional Kinematics**

Most types of motion we experience occur in either two or three dimensions. In this text, we will limit treatment to two-dimensional motion, which is sufficient to describe a large variety of examples, such as the movement of animals traveling on flat ground, the paths of planets orbiting the sun, and the curved path of a ball thrown in the air. In general, the position, velocity and acceleration must then be considered as *vectors*. Fortunately, as remarked at the beginning of Section 7.2, any general motion can be described using the *components* of these vectors along orthogonal axes; the components along each axis obey the same one-dimensional relations described in Sections 7.2 and 7.3.

We will consider motion in a two-dimensional *x–y* plane. The position of a moving object with respect to the origin will be denoted by the vector $\vec{r}$. The bold notation and the arrow sign point out that $\vec{r}$ is a vector, which refers to a parameter that has both a magnitude and a direction (see Box 1-1). In Fig. 7-5, an object moves from initial position $\vec{r}_1$, having components x_1 and y_1, to a final position $\vec{r}_2$, with components x_2 and y_2. The displacement vector of the object is denoted $\Delta\vec{r}$ and is defined by

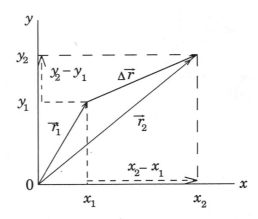

Figure 7-5

The position vector of an object, relative to the origin 0, changes from $\vec{r}_1$ to $\vec{r}_2$. The displacement of the object is $\Delta\vec{r} = \vec{r}_2 - \vec{r}_1$

$$\Delta\vec{r} = \vec{r}_2 - \vec{r}_1 \qquad \text{[7-12]}$$

and the corresponding *x*- and *y*-components are

$$\Delta x = x_2 - x_1 \text{ and } \Delta y = y_2 - y_1 \qquad \text{[7-13]}$$

The generalization of Eq. [7-2] for average velocity is now the vector equation

$$\vec{v}_{av} = \frac{\vec{r}_2 - \vec{r}_1}{t_2 - t_1} = \frac{\Delta\vec{r}}{\Delta t} \qquad \text{[7-14]}$$

As for any vector quantity, the velocity $\vec{v}_{av}$ can be represented either in terms of its components or in terms of its magnitude $|\vec{v}_{av}|$ and direction.

The displacement $\Delta\vec{r}$ may turn out to be the vector sum of several individual displacements, as illustrated in Example 7-5.

EXAMPLE 7-5

A honeybee is visiting a clover patch. Its average speed between flowers is 2.00 m/s. Starting at one flower, it travels north for 0.200 s to another flower, then east for 0.300 s to another flower, and finally southeast for 0.100 s to another flower. What is the bee's final displacement and bearing from the first flower?

SOLUTION

We denote the three individual displacements of the bee as $\Delta\vec{r}_a$, $\Delta\vec{r}_b$, and $\Delta\vec{r}_c$, schematically illustrated in Fig. 7-6, where the origin is at the first flower. From Eq. [7-14], $\Delta\vec{r}_a = \vec{v}_{av,a}\Delta t_a$, etc., where the common magnitude of $\vec{v}_{av}$ is 2.00 m/s. The displacements $\Delta\vec{r}_a$ and $\Delta\vec{r}_b$ are in the north (N) and east (E) directions, respectively, which we take to correspond with the $+y$- and $+x$-directions, hence their components are

Figure 7-6

Example 7-5: Movement of a honeybee.

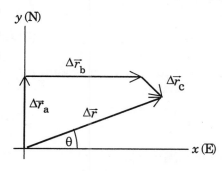

$$\Delta x_a = 0, \Delta y_a = (2.00\text{ m/s})(0.200\text{ s}) = 0.400\text{ m}$$

$$\Delta y_b = 0, \Delta x_b = (2.00\text{ m/s})(0.300\text{ s}) = 0.600\text{ m}$$

The third displacement $\Delta\vec{r}_c$ is toward the southeast, with components

$$\Delta x_c = [2.00\text{ m/s}\cos(45.0°)](0.100\text{ s}) = 0.141\text{ m}$$

$$\Delta y_c = -[2.00\text{ m/s}\sin(45.0°)](0.100\text{ s}) = -0.141\text{ m}$$

The components of the final displacement are then

$$\Delta x = \Delta x_a + \Delta x_b + \Delta x_c = 0.600\text{ m} + 0.141\text{ m} = 0.741\text{ m}$$

$$\Delta y = \Delta y_a + \Delta y_b + \Delta y_c = 0.400\text{ m} - 0.141\text{ m} = 0.259\text{ m}$$

Hence, using the Pythagorean theorem (also known as Pythagoras' theorem) the magnitude of the final displacement is

$$\Delta r = \sqrt{(\Delta x)^2 + (\Delta y)^2} = \sqrt{(0.741\text{ m})^2 + (0.259\text{ m})^2}$$
$$= 0.785\text{ m}$$

The direction of the displacement $\Delta\vec{r}$ can be specified by its angle θ with respect to the x-axis (Fig. 7-6). This can be obtained from

$$\theta = \tan^{-1}\left(\frac{\Delta y}{\Delta x}\right) = \tan^{-1}\left(\frac{0.259\text{ m}}{0.741\text{ m}}\right)$$
$$= 19.3° \text{ north of east}$$

The previous example described a case where the average speed was constant but the direction of the velocity underwent changes. In the most general situation either, or both, the magnitude and direction of the velocity may change. The instantaneous velocity $\vec{v}$ is then defined by the generalization of Eq. [7-3],

$$\vec{v} = \lim_{\Delta t \to 0} \frac{\Delta \vec{r}}{\Delta t} = \frac{d\vec{r}}{dt} \qquad \text{[7-15]}$$

The x- and y-components of this equation are just the corresponding time derivatives of the coordinates x and y, that is

$$v_x = \frac{dx}{dt} \text{ and } v_y = \frac{dy}{dt} \qquad \text{[7-16]}$$

Before taking the limit in Eq. [7-15] (or using Eq. [7-14], $\vec{v}_{av} = \Delta \vec{r}/\Delta t$), note that one can write

$$\Delta \vec{r} = \vec{v}_{av}\, \Delta t$$
$$= \vec{v}\, \Delta t \text{ as } \Delta t \to 0 \qquad \text{[7-17]}$$

This shows that the displacement $\Delta \vec{r}$ is in the direction of the average velocity $\vec{v}_{av}$. As $\Delta t \to 0$, $\vec{v}_{av} \to \vec{v}$, which leads to the conclusion that along any curved path, the instantaneous velocity vector $\vec{v}$ is tangent to the curve at the instant involved. This feature will be used later in Section 9.2.

If the velocity changes during the motion, in magnitude and/or direction, then the object is characterized by a vector acceleration. The generalizations of Eqs. [7-5] and [7-6] should be obvious: in particular, the analog of the latter is

$$\vec{a} = \lim_{\Delta t \to 0} \frac{\Delta \vec{v}}{\Delta t} = \frac{d\vec{v}}{dt} \qquad \text{[7-18]}$$

where $\Delta \vec{v}$ is the change in instantaneous velocity during the time interval Δt.

Projectile Motion

As for one-dimensional motion discussed in Section 7.2, we shall limit consideration to two-dimensional situations where the acceleration $\vec{a}$ is constant. This means that the individual components a_x and a_y are both constant (which includes the possibility that either or both are zero). In these cases, one can immediately write out separate sets of equations for the motion in the x- and y-directions which are completely analogous to the Galilean equations [7-7] to [7-10] with appropriate adjustments in notation. Actually, we shall focus on a particular subclass of two-dimensional motion with constant acceleration, namely *projectile motion*. This describes the motion of any body which has been given some initial velocity $\vec{v}_0$ and then follows a path which is determined solely by gravitational acceleration (i.e., once in motion, the body does not have any self-propulsion system). Examples are a cat leaping from a tree, a swimmer diving off a diving board, or a ball thrown from one person to another. This type of motion is always confined to a vertical plane determined by the direction of the initial velocity $\vec{v}_0$. We will define this to be the x–y plane, with the x-axis horizontal (parallel to the ground) and the y-axis vertical.

Projectile motion can be treated as two separate problems, one involving horizontal motion with zero acceleration, $a_x = 0$, and hence constant velocity v_x, and the other involving vertical motion with constant acceleration a_y fixed by the acceleration of gravity, i.e., $|a_y| = g$.[4] The kinematic equations for both these types of motion are obtained by appropriate adjustment of Eqs. [7-7] to [7-10], and are summarized in Table 7-1. As earlier, we will choose the initial time to be zero and let t denote an arbitrary later time while the object is still in motion. The initial positions in the x- and y-directions are denoted x_0 and y_0, while the components of the initial velocity $\vec{v}_0$ in these directions are denoted v_{0x} and v_{0y}, respectively.

Table 7-1   Kinematic Equations for Projectile Motion

Horizontal (x) Motion:	($a_x = 0$)
[1]	$v_x = v_{0x}$
[2]	$x = x_0 + v_{0x}t$
Vertical (y) Motion:	($|a_y| = g$)
[3]	$v_y = v_{0y} + a_y t$
[4]	$y = y_0 + v_{0y}t + \tfrac{1}{2}a_y t^2$
[5]	$v_y^{\,2} = v_{0y}^{\,2} + 2a_y(y - y_0)$
[6]	$y = y_0 + \tfrac{1}{2}(v_{0y} + v_y)t$

The sign of a_y depends on the (arbitrary) choice of whether the positive y-direction is upward or downward. Since the gravitational acceleration is always toward Earth, $a_y = -g$ if the upward direction is chosen to be positive.

---

4.   Mention should be made that we are neglecting air resistance and several other effects due to the curvature and rotation of Earth.

The final solution to problems of projectile motion usually involves combining the results obtained from consideration of the separate horizontal and vertical motions. A key point to keep in mind is that the time t is a common variable for these two types of motion.

EXAMPLE 7-6

When a baseball leaves a pitcher's hand at an elevation of 1.50 m above the ground, it is moving horizontally with a speed of 40.0 m/s. If the distance to the plate is 20.0 m, how high is the ball above the ground when it passes over the plate?

SOLUTION

Since the initial velocity is horizontal, then $v_{0y} = 0$ and $v_{0x} = 40.0$ m/s. Here we will choose the origin to be at the position of the pitcher, and the positive y-direction to be upward, so that the initial height above the ground is $y_0 = 1.50$ m, and we are asked to find the final height y when the ball passes over the plate. In this example, we are given the horizontal displacement $x - x_0 = 20.0$ m. We can use the latter information to find the time taken by the ball to reach the plate from Eq. [2] in Table 7-1:

$$t = \frac{x - x_0}{v_{0x}} = \frac{20.0 \text{ m}}{40.0 \text{ m/s}} = 0.500 \text{ s}$$

Then we can use Eq. [4] in Table 7-1 to find y (noting that $a_y = -g$):

$$y = 1.50 \text{ m} + \tfrac{1}{2}(-9.80 \text{ m/s}^2)(0.500 \text{ s})^2$$
$$= 0.275 \text{ m}$$

EXAMPLE 7-7

A cat leaps horizontally with a velocity of 4.00 m/s from a tree branch 2.00 m above the ground. What is its velocity when it strikes the ground?

SOLUTION

As in the previous example, since the cat leaps horizontally, its initial vertical velocity $v_{0y} = 0$. In this situation, the time in the air will be exactly the same as if it had been dropped from rest, i.e., the horizontal motion does not prolong the time it is in the air. Since the choice of positive y-direction is arbitrary, it is convenient in this example to choose it as <u>downward</u>.

To find the final y-component of the velocity, we use Eq. [5] in Table 7-1, noting that the final vertical displacement is $y - y_0 = 2.00$ m and $a_y = g$:

$$v_y{}^2 = 2(9.80 \text{ m/s}^2)(2.00 \text{ m}) = 39.2 \text{ m}^2/\text{s}^2$$

Hence

$$v_y = \sqrt{39.2 \text{ m}^2/\text{s}^2} = 6.26 \text{ m/s}$$

Note that v_y is taken as the positive square root of $v_y{}^2$, since we know that the final velocity must have a downward vertical component. The x-component of the velocity remains constant at its initial value, $v_x = v_{0x} = 4.00$ m/s. Hence the resultant final velocity has magnitude

$$v = \sqrt{(6.26 \text{ m/s})^2 + (4.00 \text{ m/s})^2} = 7.43 \text{ m/s}$$

and is in the direction (see Fig. 7-7)

$$\theta = \tan^{-1}(v_y/v_x) = \tan^{-1}(6.26 \text{ m·s}^{-1}/4.00 \text{ m·s}^{-1})$$
$$= 57.4°$$

Thus, the final velocity is 7.43 m/s at 57.4° below the horizontal.

Figure 7-7

Example 7-7: Velocity of a cat leaping from a tree.

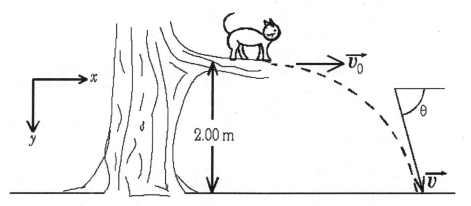

EXAMPLE 7-8

A ball is thrown upward at speed v_0 and angle θ_0 relative to the ground. What are (a) the range R and (b) the maximum height attained by the ball, in terms of the given parameters v_0, θ_0, and g?

SOLUTION

(a) The range R refers to the horizontal distance traveled by the ball from the initial point to the final point at which it returns to the ground (for simplicity, we'll assume the ball initially starts at ground level). From common experience, one expects that the path (or *trajectory*) in the x–y plane, followed by the ball, is that illustrated in Fig. 7-8, where we have chosen the initial position (at $t = 0$) to be at the origin, so that $x_0 = y_0 = 0$. The horizontal and vertical components of the initial velocity $\overrightarrow{v_0}$ are (taking the positive y-direction to be upward)

$$v_{0x} = v_0 \cos(\theta_0)$$

$$v_{0y} = v_0 \sin(\theta_0) \qquad\qquad [7\text{-}19]$$

In Fig. 7-8, the position of the ball at a sequence of times is indicated by dots, and the corresponding velocity vector $\overrightarrow{v}$ at these times is also indicated. The x-component v_x remains constant, according to Eq. [1] in Table 7-1, while the y-component v_y continuously decreases with time according to Eq. [3] in the table. At the highest point in the trajectory, $v_y = 0$, and thereafter $\overrightarrow{v}$ starts pointing downward as the ball falls back to the ground.

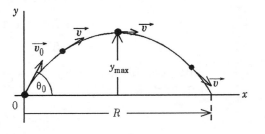

Figure 7-8

The trajectory of a projectile having initial velocity $\overrightarrow{\mathbf{v}}_0$, illustrating the definition of the horizontal range R and maximum height y_{max}.

Let us denote the total time the ball is in the air, i.e., the time it takes for the ball to return to the ground, as t_f or "the time of flight," as in Example 7-4. The horizontal range R is then just equal to the displacement $x - x_0$ at this time. From Eq. [2] in Table 7-1,

$$R = v_{0x} t_f \qquad\qquad [7\text{-}20]$$

The time of flight t_f can be found from the equation relating vertical displacement $y - y_0$ and time t, Eq. [4] in

Table 7-1. Taking $a_y = -g$ and setting $y - y_0 = 0$ (at landing), this results in a quadratic equation for t_f:

$$(g/2)t_f^2 - v_{0y}t_f = 0 \qquad \textbf{[7-21]}$$

One solution for this equation is simply $t_f = 0$, which really corresponds to the time at which the object initially leaves the ground. The other solution is the one we desire, namely,

$$t_f = \frac{2v_{0y}}{g} \qquad \textbf{[7-22]}$$

Substituting this into Eq. [7-20] and using Eq. [7-19] gives

$$R = \frac{2v_{0x}v_{0y}}{g} = \frac{2v_0^2 \sin(\theta_0)\cos(\theta_0)}{g}$$
$$= \frac{v_0^2 \sin(2\theta_0)}{g} \qquad \textbf{[7-23]}$$

where the second line follows from a trigonometric identity.[5] We do not recommend that you memorize the formula of Eq. [7-23]; rather, it is the method of analysis which is the important aspect to grasp.

(b) The maximum height y_{max} attained by the projectile can be determined by a method entirely analogous to that described in Example 7-4. At this height, $v_y = 0$, so that Eq. [5] in Table 7-1 can be used (setting $y_0 = 0$ and $y = y_{max}$):

$$y_{max} = \frac{v_{0y}^2}{2g} = \frac{v_0^2 \sin^2(\theta_0)}{2g} \qquad \textbf{[7-24]}$$

Similarly, the time t_{up} to reach this height can be found using Eq. [3] in Table 7-1,

$$t_{up} = \frac{v_{0y}}{g} = \frac{v_0 \sin(\theta_0)}{g} \qquad \textbf{[7-25]}$$

Notice that t_{up} is exactly one-half of the total time of flight t_f, which should be obvious from the symmetry of this problem (see Fig. 7-8).

The path or trajectory of the object illustrated in Fig. 7-8 (that is, the variation of vertical position y as a function of

---

5. $2 \sin(\theta) \cos(\theta) = \sin(2\theta)$

horizontal position x) is *parabolic*. This is generally true of all projectile motions. The general relation between y and x follows by expressing time t in terms of x using Eq. [2] in Table 7-1 and substituting this into Eq. [4] of the table, again setting $x_0 = y_0 = 0$,

$$y = \left(\frac{v_{0y}}{v_{0x}}\right)x + \left(\frac{a_y}{2v_{0x}^2}\right)x^2 \qquad \textbf{[7-26]}$$

which is of the form $y = bx + cx^2$, the mathematical equation for a parabola.

7.5   Relative Velocity

We conclude this chapter with a further discussion of velocity and its vector character. The main idea can be illustrated by the one-dimensional motion of a car, A, on a highway as seen by two observers, one standing by the highway and the other riding in a second car, B, which is being passed by car A. The velocities of both cars in the direction of the highway, as seen by the stationary observer C, are denoted v_{AC} and v_{BC}.[6] However, as we all know from experience, the velocity of car A as seen by the rider in car B is the <u>difference</u> $v_{AC} - v_{BC}$. For example, if $v_{AC} = 120$ km/h while $v_{BC} = 100$ km/h, the observer in car B sees car A passing by with a velocity of only 20 km/h. This shows that the velocity of an object depends on its motion relative to that of the observer. We say that the velocity of an object relative to a particular observer is the *relative velocity*. Each observer may be pictured as carrying a coordinate system relative to which velocity can be measured, forming what is called a *frame of reference*. It is clear that <u>all</u> velocities are relative to some frame of reference.

Now we extend the above concept to motion in more than one dimension. We denote the velocity of object A relative to a frame of reference C as $\vec{v}_{AC}$. The velocity of a second object B relative to the same frame of reference is $\vec{v}_{BC}$. We can introduce a third relative velocity, namely that of object A relative to the frame of reference of object B, $\vec{v}_{AB}$. Generalizing the example of the two cars on a highway, these three relative velocities are related by

$$\vec{v}_{AB} = \vec{v}_{AC} - \vec{v}_{BC} \qquad \textbf{[7-27a]}$$

---

6.   In this and other one-dimensional situations, boldface type for vector notation will be dropped.

which can be rewritten as

$$\vec{v}_{AC} = \vec{v}_{AB} - \vec{v}_{BC} \qquad \textbf{[7-27b]}$$

Note the ordering of the subscripts in Eq. [7-27]. It may sometimes also be useful to note that the velocity of A relative to B and the velocity of B relative to A are related by

$$\vec{v}_{BA} = -\vec{v}_{AB} \qquad \textbf{[7-28]}$$

Before going on to two-dimensional examples, let's give another one-dimensional illustration, in this case of Eq. [7-27b]. Suppose A is an airplane which is traveling east, while B is the air or wind affecting the plane's motion, which is also blowing toward the east, which we'll take to be the positive direction. Let the velocity v_{AB} of the plane relative to the air, called the *air velocity*, be 450 km/h, and the air's velocity v_{BC} relative to the ground C be 60 km/h. Then according to Eq. [7-27b], the velocity of the plane relative to the ground, v_{AC}, is

$$v_{AC} = v_{AB} + v_{BC}$$
$$= 450 \text{ km/h} + 60 \text{ km/h} = 510 \text{ km/h, east}$$

Therefore, the wind causes the velocity of the plane relative to the ground to increase, which makes perfect sense. Alternatively, if the wind is blowing west rather than east, then $v_{BC} = -60$ km/h (with respect to the positive east direction), and the plane's velocity relative to the ground is reduced,

$$v_{AC} = v_{AB} + v_{BC}$$
$$= 450 \text{ km/h} - 60 \text{ km/h} = 390 \text{ km/h, east}$$

which is again to be expected.

Similar ideas apply to an object such as a boat moving in a flowing river, as in Examples 7-9 and 7-10.

EXAMPLE 7-9

A small boat sets out to cross a river which flows toward the west at 3.00 m/s. The boat can travel in still water at 5.00 m/s. The boat's pilot aims the boat due north, directly across the river, which is 100 m wide. What is the velocity of the boat relative to the shore, and how far downstream does it move before reaching the other side of the river?

SOLUTION

Figure 7-9 illustrates the situation. Let B denote the frame of reference of the boat, W that of the water, and S that of the shore. The velocity of the boat relative to the shore, $\vec{v}_{BS}$, can be obtained by appropriate change of the subscripts in Eq. [7-27b]:

$$\vec{v}_{BS} = \vec{v}_{BW} - \vec{v}_{WS}$$

The velocity of the boat relative to the water, $\vec{v}_{BW}$, is given as being toward the north with magnitude 5.00 m/s, while the velocity of the river relative to the shore, $\vec{v}_{WS}$, is due west with magnitude 3.00 m/s. From Fig. 7-9, it is clear that

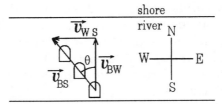

Figure 7-9

Example 7-9: Vector diagram for a boat aimed across a river.

$$\left|\vec{v}_{BS}\right| = \sqrt{\left|\vec{v}_{BW}\right|^2 + \left|\vec{v}_{WS}\right|^2} = \sqrt{(5.00 \text{ m/s})^2 + (3.00 \text{ m/s})^2}$$

$$= 5.83 \text{ m/s}$$

while the direction of $\vec{v}_{BS}$ is indicated by the angle θ given by

$$\theta = \tan^{-1}\left(\frac{\left|\vec{v}_{WS}\right|}{\left|\vec{v}_{BW}\right|}\right) = \tan^{-1}\left(\frac{3.00 \text{ m/s}}{5.00 \text{ m/s}}\right)$$

$$= 31.0° \text{ west of north}$$

Thus, $\vec{v}_{BS}$ is 5.83 m/s at 31.0° west of north.

The figure shows that the component of $\vec{v}_{BS}$ in the northerly direction is just $|\vec{v}_{BW}| = 5.00$ m/s, and therefore it takes the boat a time $t = (100 \text{ m}/5.00 \text{ m·s}^{-1}) = 20.0$ s to cross the river. In this 20.0-s interval, the boat is swept downstream a distance $|\vec{v}_{WS}|\, t = 3.00$ m/s $\times$ 20.0 s $= 60.0$ m, since $|\vec{v}_{WS}|$ is the component of the boat's velocity in the direction of the flowing river.

We have included in Fig. 7-9 pictures of the boat itself at several successive intervals during its travel. These indicate that while the boat is aimed (or, we sometimes say, is headed) across the river, its resultant motion relative to the shore follows the vector $\vec{v}_{BS}$.

EXAMPLE 7-10

In what direction must the boat in the previous example be headed in order to travel directly across the river? How long does it now take to cross the river?

SOLUTION

In this case, it is desired that $\vec{v}_{BS}$, the velocity of the boat relative to the shore, be toward the north. We wish to find the direction that the boat must be aimed (i.e., the heading of the boat) to accomplish this. This is the direction of $\vec{v}_{BW}$, the velocity of the boat relative to the river. (However, the magnitude of $\vec{v}_{BW}$ is the same as in Example 7-9, $|\vec{v}_{BW}| = 5.00$ m/s.) To find $\vec{v}_{BW}$, we rewrite the equation in the previous example as

Figure 7-10

Example 7-10: Vector diagram for a boat traveling across a river.

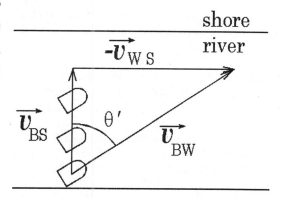

$$\vec{v}_{BW} = \vec{v}_{BS} - \vec{v}_{WS}$$

This vector construction is illustrated in Fig. 7-10. The quantity to be determined is the angle θ'. We can find this from

$$\theta' = \sin^{-1}\left(\frac{|\vec{v}_{WS}|}{|\vec{v}_{BW}|}\right) = \sin^{-1}\left(\frac{3.00 \text{ m/s}}{5.00 \text{ m/s}}\right)$$

$$= 36.9° \text{ east of north}$$

The magnitude of $\vec{v}_{BS}$ is

$$\left|\vec{v}_{BS}\right| = \sqrt{\left|\vec{v}_{BW}\right|^2 - \left|\vec{v}_{WS}\right|^2} = \sqrt{(5.00 \text{ m/s})^2 - (3.00 \text{ m/s})^2}$$

$$= 4.00 \text{ m/s}$$

Hence the time taken by the boat to cross the river is (100 m)/(4.00 m/s) = 25.0 s.

Exercises

7-1 A car driver sees an emergency and reacts by putting on the brakes. During the 0.30-s reaction time, the car maintains its uniform velocity of 28 m/s forward. What is the car's displacement during the time it takes for the driver to react?

7-2 One person can run 100 m in 10.0 s and another person in 13.0 s, both at a constant speed. If two runners pass by a reference position at the same instant, by what distance is the faster runner ahead of the slower runner after the former has run 100 m?

7-3 A rocket is launched vertically upward. Immediately after launch, the rocket engines produce an upward acceleration of 50.0 m/s^2, and this acceleration is maintained until all the fuel is exhausted. If the rocket must have a final upward velocity of 900 m/s, how long must the fuel last?

7-4 An object is initially traveling in the +x-direction with velocity +2.00 m/s. It then undergoes an acceleration of +3.00 m/s^2 for 10.0 s. In this time, how far has it traveled?

7-5 An engine on a straight track passes a station at a speed of 15.0 m/s. It then accelerates at a constant rate of 0.0200 m/s^2 and reaches the next station 10.0 min later. What is the distance between the two stations?

7-6 A car traveling at 80.0 km/h applies its brakes and comes to rest in 4.50 s.

   (a) Convert 80 km/h to units of m/s.

   (b) Assuming a constant acceleration, calculate its value.

   (c) How far did the car travel while the brakes were on?

7-7 In a game of curling, a disc is sent along the ice where it experiences a constant acceleration of 0.0490 m/s^2 in a direction opposite to the initial velocity. The disc travels 25.0 m from release before coming to rest. Determine the disc's (a) initial speed and (b) travel time.

7-8 A high jumper jumps 1.20 m straight up. With what speed did he leave the ground?

7-9 (a) How high does a ball rise if it is thrown vertically upward at 7.00 m/s?

   (b) How long does it take to reach its highest position?

   (c) How long does it take to return to your hand after leaving it?

   (d) What is its velocity when it strikes your hand?

7-10 A butterfly is flying in a garden. First it travels due south for 6.00 m and then 37.0° W of N for 10.0 m. Assume that the positive x-direction is east and the positive y-direction is north.

    (a)   What are the (x,y)-coordinates of the butterfly's final position relative to its initial position?

    (b)   What are the magnitude and direction of the butterfly's displacement (relative to its initial position)?

7-11 A boat, having an initial velocity of 6.60 m/s east, undergoes an average acceleration of 2.20 m/s² south for 2.50 s. Determine the final velocity (magnitude and direction) after this short time interval.

7-12 A man makes a running jump off the high-dive board at the local pool. His velocity as he leaves the board is 3.00 m/s horizontally, and the board is 5.00 m above the water.

    (a)   What is his speed when he hits the water?

    (b)   At what angle does he enter the water?

7-13 Two children are playing catch. One throws the ball with an initial velocity of 20.0 m/s at an angle of 30.0° from the horizontal. How far away should the other child be to catch the ball?

7-14 A cat leaps horizontally from a tree branch 3.00 m above the ground. It lands at a point 4.00 m horizontally from the point below which it leaped. What are the magnitude and direction of the cat's velocity at the instant it touches the ground?

7-15 A quarterback wants to throw a football at 22.0 m/s to a receiver 31.0 m away. At what angle above the horizontal should he throw it? Assume that the receiver catches the ball at the same height that the quarterback released it.

7-16 A bird is flying due north with a speed of 15.0 m/s. A train is traveling due east with the same speed. What are the magnitude and direction of the velocity of the bird as seen by a passenger on the train?

7-17 A bird is migrating non-stop a distance of 2.00×10^3 km. Its destination is due south of its present position. In still air, the bird can fly with a speed of 40.0 km/h. However, there is a wind of 25.0 km/h from the west.

    (a)   In order to make the trip following a straight-line path, in what direction should the bird head?

    (b)   How long will the trip take?

7-18 A boat is going to travel due north, but a tide is running east with a speed of 11.0 m/s. If the boat can move with a maximum speed of 15.0 m/s in still water, what is the maximum speed with which it can travel due north?

Problems

7-19 The human body can survive the trauma due to a sudden stop if the magnitude of the acceleration is less than 245 m/s^2 (= 25 g). If you are in an automobile accident with an initial speed of 85.0 km/h and are stopped by an inflating air bag, over what distance must the air bag stop you if you are to survive the crash?

7-20 A deer starts from rest and accelerates at a rate of 1.20 m/s^2 for 15.0 s. After that, it runs at constant speed for 30.0 s and then slows down at a rate of 2.50 m/s^2 until it stops. Find the total distance covered by the deer.

7-21 A cheetah hiding in the bushes spots a gazelle sprinting along at 20.0 m/s. The cheetah accelerates from rest when the gazelle passes him. If he accelerates to 25.0 m/s in 5.50 s and then maintains this speed, (a) how long will it take and (b) how far will he travel until he catches up with the gazelle?

7-22 A world-class sprinter starts from rest and accelerates to his maximum speed in 3.80 s. If the runner finishes a 100-m race in 9.90 s, what is the runner's acceleration (assumed constant) during the first 3.80 s?

7-23 A baseball pitcher throws a ball vertically straight upward and catches the ball 4.40 s later.

(a)   With what velocity did the ball leave the pitcher's hand?

(b)   What was the maximum height reached by the ball?

7-24 A boy wishes to test how high a building is by dropping a stone from its roof. The boy drops the stone from rest, and 4.80 seconds later hears the sound of the stone striking the ground. How tall is the building? The speed of sound is 350 m/s.

7-25 A skater travels in a straight line for 855 m in a direction 25.0° north of east, then 560 m in a straight line 22.0° east of north.

(a)   What is the skater's resultant displacement?

(b)   If the motion took 4.40 min, determine the skater's average speed and average velocity.

7-26 A bird flying horizontally has initial velocity components $v_{0x} = 4.20$ m/s and $v_{0y} = 3.35$ m/s at time $t = 0$. From $t = 0$ to $t = 10.0$ s, the bird has an average acceleration of magnitude 0.760 m/s^2 and direction 47.0E measured from the $+x$-axis toward the $+y$-axis. At $t = 10.0$ s,

(a)   what are the x- and y-components of the bird's velocity?

(b)   what are the magnitude and direction of the bird's velocity?

7-27 A swimmer dives from a diving board 15.0 m above the water's surface. His initial velocity is 2.10 m/s at an angle of 32.0° above the horizontal.

(a)   How long does it take him to hit the water?

(b)   When he hits the water, what is the magnitude of his displacement from the diving board?

7-28 A football is placed on a line 25.0 m from the goalpost. The placement kicker kicks the ball directly toward the goalpost, giving it a velocity of 22.0 m/s at an angle of 46.0° above the horizontal. The horizontal bar of the goalpost is 2.90 m above the field. How far above or below the horizontal bar of the goalpost will the ball travel?

7-29 (a)   A golfer wants to drive a ball to a distance of 240 m. If he launches the ball with an elevation angle of 14.0°, what is the appropriate initial speed of the ball?

(b)   If the speed is too great by 0.6 m/s, how much farther will the ball travel when launched at the same angle?

(c)   If the elevation angle is 0.5° larger than 14.0°, how much farther will the ball travel if launched with the same speed as calculated in part (a)?

7-30 A baseball is hit and given an initial velocity of 39.0 m/s at an angle of 50.0° above the horizontal. Air resistance may be ignored. An outfielder standing at a distance of 125 m from home plate (in the direction of the hit) starts running to catch the ball at the instant it is hit. How fast must he run in order to catch the ball just before it reaches the ground?

7-31 A bird starts flying with an air velocity of 35.0 km/h, 30.0° south of west. The wind is blowing from the north at 67.0 km/h. What is the location of the bird (relative to its starting point) 1.50 h later?

7-32 An airplane is headed southeast (i.e., 45.0° south of east) with an airspeed of 320 km/h. A wind is blowing at 79.0 km/h, 22.0° east of north relative to the ground. What is the velocity of the airplane relative to the ground?

Answers

7-1   8.4 m forward

7-2   23 m

7-3   18.0 s

7-4   170 m

7-5   12.6 km

7-6   (a) 22.2 m/s (b) –4.93 m/s^2 (c) 50.0 m

7-7   (a) 1.57 m/s (b) 31.9 s

7-8   4.85 m/s

7-9   (a) 2.50 m (b) 0.714 s (c) 1.43 s
(d) 7.00 m/s downward

7-10   (a) (–6.02 m, 1.99 m)
(b) 6.34 m, 18.3° N of W

7-11   8.59 m/s, 39.8° S of E

7-12   (a) 10.3 m/s (b) 73.1° from the horizontal

7-13   35.3 m

7-14   9.22 m/s at 56.3° below horizontal

7-15   19.4°

7-16   21.2 m/s north-west

7-17   (a) 38.7° W of S (b) 64.1 h

7-18   10.2 m/s

7-19   1.14 m

7-20   740 m

7-21   (a) 13.8 s (b) 275 m

7-22   3.29 m/s^2

7-23   (a) 21.6 m/s upward (b) 23.7 m

7-24   100 m

7-25   (a) 1.32×10^3 m, 41.8° N of E
(b) 5.36 m/s; 5.00 m/s, 41.8° N of E

7-26   (a) $v_x = 9.38$ m/s, $v_y = 8.91$ m/s
(b) 12.9 m/s at 43.5° from the x-axis

7-27   (a) 1.87 s (b) 15.4 m

7-28   9.9 m above

7-29   (a) 70.8 m/s (b) 4 m (c) 8 m

7-30   4.6 m/s

7-31   135 km, 70.3° S of W from starting point

7-32   298 km/h, 30.9° S of E

8 Mechanics of Biological Systems: Forces and Motion

8.1 Newton's Laws

In Chapter 7 we considered kinematics—the study of motion without any consideration of the forces causing it. Now we will turn to the subject of *dynamics*, the relationship of motion to the forces that cause it. Again, we will restrict the motion to purely translational motion and leave rotation until Chapter 9.

The basic laws relating motion to forces are Newton's laws. These laws govern the motion of objects as large as planets and stars, and as small as red blood cells. (As the objects undergoing motion approach molecular and submolecular sizes, Newton's laws have to be replaced by the laws of quantum mechanics, discussed in Chapter 4.) Newton's laws are used in analyzing a wide variety of practical problems in engineering and other fields, including biomechanics.

Before introducing Newton's laws, it is worthwhile to comment briefly on the notion of *force*. A common example of a force is that which you must exert to push or pull an object (e.g., a chair) across the floor, or alternatively to lift the object off the floor. In the latter case, the force you are exerting is actually used to overcome the force of gravity[1] $\vec{w}$ exerted by Earth on the object. This gravitational force pulls vertically downward on all objects near the surface of Earth. However, as we all know, an object resting on the floor does not actually move downward, precisely due to the presence of the floor, i.e., the floor exerts an upward supporting force[2] ($\vec{N}$) which exactly counterbalances the gravitational force if the object is at rest (Fig. 8-1). This discussion illustrates several key ideas involved in Newton's laws, such as the

---

1. "$\vec{w}$" is the "weight" of the object.
2. "$\vec{N}$" stands for the "normal" force, i.e., the force perpendicular to the surface.

distinction between a *net* (or *resultant*) force and the separate individual forces that may be acting on an object.

After this preliminary, we now state in Box 8-1 Newton's three laws of motion.

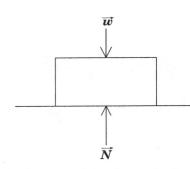

Figure 8-1

An object at rest on the ground is acted on by two forces: its weight **w** and the supporting force of the ground **N⃗**.

Box 8-1   Newton's Laws of Motion

1. A body remains in a state of rest or of uniform motion in a straight line unless it is acted on by a non-zero resultant force. The body is said to be in *equilibrium* when this condition applies.

2. If the resultant force $\vec{F}_R$ acting on an object is not zero, the object experiences an acceleration in the direction of this force. The magnitude of the acceleration is directly proportional to the magnitude of the force and inversely proportional to the mass m of the body.

3. For every force, there is always an equal and opposite reaction force: that is, if object A exerts a force $\vec{F}$ on object B, then B exerts an equal but opposite force $-\vec{F}$ on A.

Mathematically, the second law reads

$$\vec{a} = \frac{1}{m}\vec{F}_R \qquad \text{[8-1a]}$$

or, as more commonly expressed,

$$\vec{F}_R = m\,\vec{a} \qquad \text{[8-1b]}$$

The first form, Eq. [8-1a], emphasizes the fact that the net force $\vec{F}_R$ is the cause of the resulting acceleration $\vec{a}$. One may also note that the first law is really but a special case of the second law since, if the resultant force $\vec{F}_R = 0$, then Eq. [8-1a] gives $\vec{a} = 0$. Hence, the resulting motion is uniform with constant velocity (including the possibility $|\vec{v}| = 0$, namely that the object is at rest). Historically, however, the first law has been presented as a separate law.

The key word used in both the first and second laws is "resultant," i.e., in both these laws, what matters is the net or resultant force, the vector sum of all individual forces that may be acting on an object. If these individual forces are labelled $\vec{F}_1$, $\vec{F}_2$, $\vec{F}_3$, etc., then[3]

$$\vec{F}_R = \vec{F}_1 + \vec{F}_2 + \vec{F}_3 + \ldots = \Sigma \vec{F} \qquad \text{[8-2]}$$

Thus we will often write Newton's second law as

$$\Sigma \vec{F} = m \vec{a} \qquad \text{[8-3]}$$

where we sum over *all* the forces acting on the body. In practice, for a general two-dimensional problem, the summation is performed by resolving each individual force $\vec{F}_n$ into its components along suitably chosen x- and y-axes. By similarly resolving the acceleration $\vec{a}$, Newton's second law, Eq. [8-1b], is most usefully written in component form,

$$\sum F_x = F_{R,x} = ma_x$$
$$\sum F_y = F_{R,y} = ma_y \qquad \text{[8-4]}$$

Newton's second law is intuitively plausible; the law states that, for any given object (of mass m), a larger acceleration is produced by applying a larger net force. For example, the harder you push an object across the floor, the larger is the resulting acceleration of the object. On the other hand, the more massive the object, the greater must be the force applied to accelerate it by any given amount. Acceleration encompasses the acts of starting or stopping the object (if it is initially in motion), or changing the speed or direction of the object's motion. The resistance to changes in the motion of an object is sometimes called the object's *inertia*. Newton's second law states that inertia is directly proportional to the object's mass m.

The SI unit for force, called the *newton* (N), follows from the second law. A newton is the magnitude of the force that produces an acceleration of magnitude 1 m/s² when applied to a body of mass 1 kg. Hence, by substituting these values into both sides of Eq. [8-1b] and taking absolute values, we find that 1 N = 1 kg·m/s².

---

3.   The uppercase Greek letter Σ (*sigma*) denotes the "sum of."

Example 8-1 is a simple illustration of Newton's second law, assuming that motion occurs only in one dimension.

EXAMPLE 8-1

You hit a hockey puck of mass 0.150 kg at rest on ice, giving it an initial velocity of 10.0 m/s. The puck returns to rest after traveling a distance of 25.0 m. What is the frictional force of the ice on the puck, assuming this to be constant?

SOLUTION

In this question, we are concerned with the motion of the puck <u>after</u> it has been set in motion by the action of the hockey stick. Let us take the motion to be in the positive x-direction. We are given the initial velocity, $v_0 = 10.0$ m/s, the final velocity $v = 0$ (when the puck has returned to rest), and the displacement of the puck, $\Delta x = x - x_0 = 25.0$ m. The acceleration can then be found from Eq. [7-9],

$$v^2 = v_0^2 + 2a(x - x_0)$$

Hence,

$$a = \frac{-v_0^2}{2\Delta x} = \frac{-(10.0 \text{ m/s})^2}{2(25.0 \text{ m})} = -2.00 \text{ m/s}^2$$

The acceleration is negative, as in Example 7-2, since the puck's velocity in the initial direction of motion has decreased. The only force[4] F acting on the puck which can cause this deceleration is friction with the ice surface, and hence we find this force to be

$$F = ma = (0.150 \text{ kg})(-2.00 \text{ m/s}^2) = -0.300 \text{ N}$$

Again, the negative sign of this force simply means that it is acting in the negative x-direction, opposite to that of the puck's motion.

---

4. Since the motion is one-dimensional, "F" is not boldface—it actually represents the x-component of the force, F_x.

In Example 8-1, if friction between the puck and ice were completely negligible, then the puck would continue to move indefinitely at a constant speed of 10.0 m/s once it had been set into motion. This is the content of Newton's first law. Instead of a puck sliding across ice, suppose you are pushing, say, a book across a table top. In this case, after you stop pushing the book, it does <u>not</u> continue to move indefinitely but comes to rest almost immediately due to the retarding force of friction with the table. This is not in contradiction with Newton's first law, because after you stop pushing the book, the net force on it is not zero.

Finally, let us comment on Newton's third law. While at first this law may appear to be less significant than the first two laws, this is not really the case, as it always comes into play whenever a force is exerted on an object. For example, suppose you push against a wall with a force of 10 N. According to the third law, the wall will push back on you with a reaction force of 10 N. If the surface you are standing on is frictionless (approximately true of an ice surface), then the reaction force of the wall is the <u>only</u> horizontal force acting on you, and consequently, you will be accelerated away from the wall according to Newton's second law. Another example is the act of walking. When we walk, our feet push backward on the ground, and consequently, the ground pushes forward on our feet, moving us forward. The propellers of a boat push backward on the water, so the water pushes forward on the propellers, moving the boat. Finally, a rocket expels hot gases downward, while the gases push upward to move the rocket.

The third law may appear to contradict the second law, which explains what happens when a resultant force acts on a <u>single</u> object. However, the equal and opposite forces described by the third law always act on <u>different</u> objects. In principle, it does not matter which of the two forces is considered to be the *action* and which one the *reaction*. There is no cause-and-effect relationship implied by the third law, since both action and reaction forces are applied simultaneously.

Before discussing further applications of Newton's laws, let us examine in greater depth the most common force encountered in daily life, namely the gravitational force.

8.2   Gravitational Force and Acceleration

An object near the surface of Earth is attracted downward by the gravitational force exerted on it by Earth. As described earlier in Section 7.3, if the object is "freely falling," it then experiences a downward acceleration of magnitude $g = 9.80$ m/s². This corresponds to a gravitational force of magnitude

$$w = mg \qquad \text{[8-5]}$$

where we have used Newton's second law, noting that gravity is the only force acting on a freely falling object (if air resistance is neglected). This force is identical to what is usually called the "weight" of the object.

One realizes that the gravitational force $\vec{w}$ on an object is always present, given by Eq. [8-5], regardless of whether or not the object is actually undergoing free fall. For example, a person of mass m standing on a rigid floor still experiences the force $\vec{w} = m\vec{g}$ acting downward. As mentioned at the beginning of this section, the fact that the person does not "fall" is because the floor exerts a compensating upward force $\vec{N}$ (see Fig. 8-1), so that the net downward force is zero. The person is then in *translational equilibrium* and does not undergo an acceleration.

Although equal and opposite, $\vec{w}$ and $\vec{N}$ are not an action–reaction pair in the sense of Newton's third law, since they act on the same object. The reaction force paired to $\vec{w}$ is rather an upward gravitational force exerted by mass m on Earth, which will be discussed further shortly.

The gravitational acceleration $\vec{g}$ and hence the weight $m\vec{g}$ do not have precisely the same values at all points on Earth's surface. Furthermore, g decreases slowly with increasing elevation above Earth's surface, and g has quite different values on other planets. These effects follow mainly from Newton's law of universal gravitation, which he used to explain the motion of planets around the Sun. This law states that

Every particle of matter in the universe attracts every other particle with a force $\vec{F}_G$ directed along the line between them. The magnitude of the force is proportional to the product of the masses m_1 and m_2 of the particles and inversely proportional to the square of the distance r between them.

Mathematically, we express this as

$$|\vec{F}_G| = G\frac{m_1 m_2}{r^2} \qquad \text{[8-6]}$$

where G is the *universal gravitational constant*; in SI units, $G = 6.673 \times 10^{-11}$ N·m²/kg². According to this law, $\vec{F}_G$ is a force of mutual attraction between two particles. Conforming with Newton's third law, each of the two interacting particles feels the same magnitude of the attractive force $|\vec{F}_G|$ toward the other particle, that is, the two particles experience equal but oppositely directed forces.

Equation [8-6] strictly applies only to "point" particles, which in practice we can consider to be objects whose size is very small compared with the distance between them. It turns out that Eq. [8-6] also applies to the gravitational force between two extended objects provided they have spherically symmetric mass distributions, in which case r is the distance between the centres of the objects.[5] This enables us to derive an expression for the magnitude of the gravitational acceleration g from Eq. [8-6] by applying the latter to the force of gravity between Earth, of mass m_E, and an object of mass m located at Earth's surface. In this case, the distance r is equal to the radius r_E of Earth (we consider the size of mass m to be much smaller than r_E). Then Eq. [8-6] gives $F_G = (G\, m_E /r_E^2)\, m$. Equating F_G with the weight $w = mg$ yields

$$g = \frac{Gm_E}{r_E^2} \qquad \textbf{[8-7]}$$

Note that g depends on G and the parameters m_E and r_E of the Earth, but not on the mass of the object being considered, as was stated in Section 7.3. Substituting the values for G, m_E (5.98×10^{24} kg) and r_E (6.38×10^6 m) into Eq. [8-7] gives the result $g = 9.80$ m/s^2, in agreement with our previous usage.

A relation analogous to Eq. [8-7] can be used to obtain the magnitude of the gravitational acceleration g on other planets or astronomical bodies, simply by substituting the appropriate mass and radius in place of m_E and r_E. Equation [8-7] can also be extended to evaluate g at an arbitrary height h above Earth's surface, on replacing r_E with $r_E + h$, the distance from Earth's centre. This shows why g decreases with increasing height, although that effect is very small when h itself is small compared with Earth's radius (r_E).

EXAMPLE 8-2

What is the magnitude of the acceleration due to gravity on the surface of Earth's moon? The moon's mass (m_M) is 0.0122 times Earth's mass (m_E) and the moon's radius (r_M) is 0.273 times Earth's radius (r_E).

SOLUTION

Analogous to Eq. [8-7], the magnitude of the acceleration of gravity on the moon is given by

---

5.   It can also apply to extended objects of any shape if the force is taken to act at a special point called the *centre of gravity* of the object.

$$g_M = \frac{Gm_M}{r_M^2}$$

Taking the ratio of this to Eq. [8-7], we get

$$\frac{g_M}{g} = \frac{m_M}{m_E}\left(\frac{r_E}{r_M}\right)^2$$

$$= 0.0122\left(\frac{1}{0.273}\right)^2 = 0.164$$

Hence $g_M = 0.164\,g = 0.164(9.80 \text{ m/s}^2) = 1.60 \text{ m/s}^2$, significantly smaller than the gravitational acceleration on Earth.

8.3   Translational Equilibrium

Newton's three laws of motion are the foundation of classical mechanics, including the mechanics of living organisms. Applications of these laws exist all around us, from the design of an automobile or suspension bridge to the motion of a bicycle or a bird. Here we will start by analyzing equilibrium situations, which involve the first law alone, and then progress to dynamic situations involving the second law. Most of the problems we will discuss concern equilibrium or motion in two dimensions, and hence are vectorial in nature.

In this chapter, the objects being considered will be conceptualized as mathematical points, without extended size or shape. Later, in Chapter 9, we will introduce further considerations needed when a body cannot be adequately represented as a point particle. In the present case, the fundamental physical principle is $\vec{F}_R = \Sigma\vec{F} = 0$, i.e., the vector sum of all forces acting on an object is zero. In two-dimensional situations, this condition can be resolved into separate equations for the x- and y-components of the forces,

$$\sum F_x = 0 \text{ and } \sum F_y = 0 \qquad\qquad \textbf{[8-8]}$$

which are special cases of Eq. [8-4] ($\sum F_x = ma_x, \sum F_y = ma_y$).

As a preliminary example, suppose we observe that a bird is flying at a constant velocity of 3.0 m/s in a direction 30° up from the horizontal. What is the resultant force on the bird? The key phrase to note here is "constant velocity." There is no acceleration, and thus by Newton's first law, the resultant force on the bird is zero. Since we know that the force of gravity $\vec{w}$ is always present and acts vertically downward on the bird, a zero value of the resultant force means another upward force must be acting on the bird which exactly cancels $\vec{w}$. This force is an aerodynamic lift, although our concern in the present problem is just to emphasize that such a force must be present.

Next, we will use a one-dimensional example to illustrate several important concepts.

EXAMPLE 8-3

A gymnast of mass $m = 50.0$ kg suspends herself from the lower end of a rope hanging from a gymnasium ceiling, as shown in Fig. 8-2. Assume that the rope has a negligible mass. What are the magnitudes and directions of the forces acting (a) on the gymnast and (b) on the rope?

SOLUTION

(a) At equilibrium, the gymnast is at rest and hence the resultant force on her is zero. As in the preceding discussion, the gravitational force $\vec{w}$ acting downward on her must be balanced in an upward force exerted by the rope. This force is called the *tension* of the rope, denoted $\vec{T}$. Generally, a tension force occurs whenever a rope (or string or cable) is used to support or pull an object.

To make this more concrete, we draw in Fig. 8-3a a *free-body diagram* (FBD) for the gymnast. This is a diagram which shows the body by itself (in this case, the gymnast) and the forces acting on it due to its surrounding or other bodies. Thus, Fig. 8-3a shows the two forces acting on the gymnast, the gravitational force (of magnitude $w = mg$) acting downward and the upward tension (of magnitude T) exerted by the rope. The gymnast is represented by the dot. Taking the vertical direction to be along y, with positive direction upward, we can find T from the equilibrium condition $\sum F_y = 0$, which becomes

$$\sum F_y = T + (-w) = 0$$

where the negative sign before w follows because the weight acts downward. Hence,

$$T = w$$

Figure 8-2

Example 8-3: A gymnast suspended from a rope.

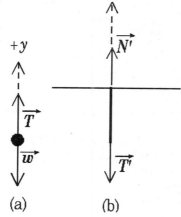

Figure 8-3

Example 8-3: Free-body diagrams for (a) the gymnast and (b) the rope.

(a)   (b)

which shows that T must equal the gymnast's weight (in magnitude), as could be expected. The value of w (and hence T) is obtained from $w = mg$, and thus
$T = 50.0 \text{ kg} \times 9.80 \text{ m/s}^2 = 490 \text{ N}$.

(b) To find the forces acting on the rope, we note that, by Newton's third law, if the rope pulls upward on the gymnast with a force $\vec{T}$, the gymnast must pull downward on the rope with a force $\vec{T}\,'$ of the same magnitude as $\vec{T}$. This downward force on the rope is shown in Fig. 8-3b, which is the FBD for the rope, here represented by a thick line segment.

The force $\vec{T}\,'$ acting downward on the rope is the third law reaction to the upward force $\vec{T}$ exerted by the rope on the gymnast. Notice that the forces $\vec{T}\,'$ and $\vec{T}$ of the action–reaction pair act on different bodies, which distinguishes them from the pair of forces $\vec{T}$ and $\vec{w}$ which, although equal in magnitude, act on the same body. Considering just the forces acting on the rope, it is obvious that the downward pulling force $\vec{T}\,'$ must be balanced by an upward pulling force exerted on it by the ceiling, which we denote $\vec{N}\,'$. Then the equilibrium condition for the rope, $\sum F_y' = 0$, reads explicitly

$$\sum F_y' = N' - T' = 0$$

or

$$N' = T'$$

where N' and T' are the magnitudes of $\vec{N}\,'$ and $\vec{T}\,'$. Since T' is equal to T, which in turn is equal to mg, we have thus determined the magnitudes and directions of all the forces acting on both the gymnast and the rope: in particular, the tension $\vec{T}\,'$ exerted on the rope is 490 N downward and the pulling force of the ceiling $\vec{N}\,'$ is 490 N upward. Note that the forces $\vec{N}\,'$ and $\vec{T}\,'$ acting on the rope are both pulling *away* from the rope, which is generally true for a rope or similar object under tension.

Example 8-3 serves as an illustration of both Newton's first and third laws, as well as introducing the use of free-body diagrams. Now let us go on to several two-dimensional examples of translational equilibrium.

EXAMPLE 8-4

A child on a swing weighs 200 N. Find the size of the horizontal force required to pull the child back so that the swing makes an angle of 30.0° with the vertical direction (Fig. 8-4a).

SOLUTION

The FBD of the child is shown in Fig. 8-4b. The forces acting on the child are her weight $\vec{w}$ acting vertically downward, the horizontal pulling force denoted $\vec{F}_p$ (of magnitude F_p) that we are asked to find, and the tension $\vec{T}$ (of magnitude T) in the cable holding the swing. (In reality, there are <u>two</u> cables supporting the child, each with tension $\vec{T}/2$ acting in the same direction, but we can replace these forces by their vector sum.)

Figure 8-4

Example 8-4: A child on a swing.

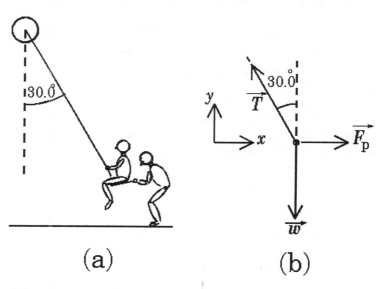

(a)   (b)

Choosing x- and y-directions to be horizontal and vertical as shown, and resolving the tension into its x- and y-components, the equilibrium conditions for the child are:

$$\sum F_x = 0, \text{ therefore } F_p - T\sin(30.0°) = 0$$
$$\sum F_y = 0, \text{ therefore } T\cos(30.0°) - w = 0$$

These are two simultaneous equations in two unknowns, F_p and T. Note that we are given the value of the child's weight, $w = 200$ N, rather than the child's mass m. Since w is given, we can solve the second equation above for T,

$$T = w/\cos(30.0°) = \frac{200\text{N}}{\cos(30.0°)} = 231 \text{ N}$$

and then substitute this result into the first equation
to find F_p:

$$F_p = T \sin(30.0°) = 116 \text{ N}$$

EXAMPLE 8-5

A spider weighing 1.50×10^{-3} N is supported by two threads
attached to the ceiling, as shown in Fig. 8-5. One thread, A,
makes an angle of 45.0° with the vertical, the second thread, B,
makes an angle of 30.0° with the vertical. What are the
magnitudes of the tensions in the two threads?

SOLUTION

The forces acting on the suspended spider are the two tensions
$\vec{T}_A$ and $\vec{T}_B$ and the weight $\vec{w}$. The tensions have magnitudes
T_A and T_B and act in the
directions of the respective
threads, while $\vec{w}$ is
downward with magnitude
1.50×10^{-3} N.

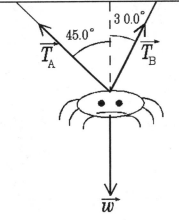

Figure 8-5

Example 8-5: Free-body
diagram of a spider
suspended from threads.

As in the previous problem,
taking x- and y-directions to
be horizontal to the right
and vertically upward, the
equations of equilibrium are

$$\sum F_x = 0, \text{ therefore}$$
$$T_B \sin(30.0°) - T_A \sin(45.0°) = 0$$

$$\sum F_y = 0, \text{ therefore}$$
$$T_B \cos(30.0°) + T_A \cos(45.0°) - 1.50 \times 10^{-3} \text{ N} = 0$$

This differs from the previous example in that the two
unknown forces T_A and T_B occur in <u>both</u> equations. The
method of approach is to use one equation to express one
unknown in terms of the other, and then substitute that
relation into the other equation. In this example, from the first
equation above, we have

$$T_{\mathrm{B}} = T_{\mathrm{A}} \, \frac{\sin(45.0°)}{\sin(30.0°)}$$

Substituting this relation into the second equation gives

$$T_{\mathrm{A}} \left(\frac{\sin(45.0°)}{\sin(30.0°)} \cos(30.0°) + \cos(45.0°) \right) = 1.50 \times 10^{-3} \, \mathrm{N}$$

which yields

$$T_{\mathrm{A}} = 7.77 \times 10^{-4} \, \mathrm{N}$$

Then, from the relation above between T_{B} and T_{A}, we find $T_{\mathrm{B}} = 1.10 \times 10^{-3}$ N.

8.4   Dynamics

We now generalize the treatment to include dynamics, i.e., the occurrence of an acceleration $\vec{a}$ due to a non-zero resultant force $\vec{F}_{\mathrm{R}}$. In all the examples in this chapter that we shall illustrate (as well as problems you will be asked to solve), the force and acceleration will be assumed constant and therefore the kinematic equations developed in Chapter 7 for constant acceleration will be applicable (e.g., $v = v_0 + at$, Eq. [7-7]). While these conditions do not in general hold for most applications in modern science and engineering, the methods we shall illustrate supply the foundation for treating more complex situations which are beyond the scope of this book.

Figure 8-6

Example 8-6: A toboggan pulled across a horizontal surface.

EXAMPLE 8-6

A person pulls a 10.0-kg toboggan across a horizontal surface with a force of 15.0 N directed at an angle of 37.0° above the horizontal, as shown in Fig. 8-6. What is the magnitude of the acceleration of the block? Neglect friction.

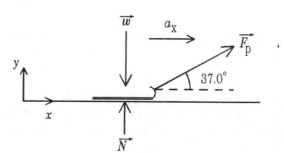

SOLUTION

The free-body diagram for the toboggan shown in Fig. 8-6 includes, besides the pulling force of magnitude 15.0 N denoted $\vec{F}_p$, the two other forces acting on the toboggan, namely its weight $\vec{w}$ and the normal force $\vec{N}$ of the ground. We choose positive x- and y-directions to be horizontal toward the right, in the direction of the toboggan's motion, and vertically upward. As we expect from common experience, the toboggan will not normally be lifted vertically, hence there is no vertical acceleration and the vertical component of the resultant force must be zero. The toboggan moves only horizontally, so the acceleration we seek is the component a_x in the x-direction. The components of Newton's second law, Eq. [8-4], become

$$\sum F_x = ma_x$$
$$\sum F_y = 0$$

Since only the pulling force $\vec{F}_p$ has both non-zero x- and y-components, denoted $F_{p,x}$ and $F_{p,y}$, respectively, these equations become

$$F_{p,x} = ma_x$$
$$F_{p,y} + N - w = 0$$

The horizontal component of $\vec{F}_p$ is $F_{p,x} = 15.0 \cos(37.0°)$ = 12.0 N, and then solving the first of the above equations for a_x yields

$$a_x = \frac{F_{p,x}}{m} = \frac{12.0 \text{ N}}{10.0 \text{ kg}} = 1.20 \text{ m/s}^2$$

Although the equation involving the y-components is not actually required in this solution (it yields the magnitude of the normal force N), we have included it to illustrate the resolution of forces.

EXAMPLE 8-7

A person of mass 70.0 kg skis down a snow-covered hill inclined at 20.0° from the horizontal. Assuming the skis are well waxed so that the friction between them and the snow is negligible, what is the magnitude of the acceleration of the skier?

SOLUTION

In this problem, the acceleration of the skier is along the direction of the incline, and therefore it is convenient to choose *x*- and *y*-directions parallel and perpendicular to the incline, rather than horizontal and vertical. With this choice of coordinate system, and because there is again no acceleration of the skier in the *y*-direction, the component form of Newton's second law is the same as in the previous example (i.e., $\sum F_x = ma_x$, $\sum F_y = 0$). The FBD for this question is shown in Fig. 8-7, where the inclination angle (20.0°) is denoted θ for generality.

Figure 8-7

Example 8-7: Free-body diagram for a skier on a frictionless incline.

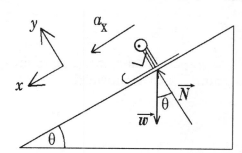

Only two forces are acting on the skier in the absence of friction, the weight $\vec{w}$ acting downward and the normal supporting force $\vec{N}$ of the incline. (Since these forces are not parallel, it should be obvious that the resultant force on the skier is non-zero, regardless of the choice of coordinate system.) Noting that only $\vec{w}$ has both non-zero *x*- and *y*-components in this case, the equations of Newton's second law are

$$w_x = ma_x$$
$$N + w_y = 0$$

Noting the angles in Fig. 8-7, one sees that w_x and w_y are related to the weight *w* and angle θ by

$$w_x = w \sin(\theta)$$
$$w_y = -w \cos(\theta)$$

[8-9]

where the negative sign in the second line indicates that $\vec{w}$ has a negative *y*-component. Using $w = mg$ and combining the two equations for w_x, we get

$$a_x = \frac{w_x}{m} = \frac{mg \sin(\theta)}{m}$$
$$= g \sin(\theta)$$
$$= (9.80 \text{ m/s}^2)\left[\sin(20.0°)\right]$$
$$= 3.35 \text{ m/s}^2$$

Notice that the mass m cancels from the final result, which means that any skier, regardless of his or her mass, slides down a frictionless incline with an acceleration of magnitude $g \sin(\theta)$. Also, as in the previous example, we did not have to use the y-component equations, which simply express consistency with Newton's first law. Note that the latter equations combine to give

$$N = mg \cos(\theta)$$

which reduces to $N = mg$ for a horizontal surface ($\theta = 0$). This result still holds when friction is included, as will be discussed later.

EXAMPLE 8-8

What is the apparent weight of an 800-N man standing in an elevator (Fig. 8-8) that is ascending with an upward acceleration of 3.20 m/s²?

SOLUTION

Here we are back to considering a one-dimensional problem, but one with several interesting implications. What is meant by *apparent weight*? We identify this with the force that the passenger's feet exert on the floor of the elevator. This follows by analogy with the equilibrium case in a stationary frame of reference, where a person's weight is normally measured. As we've seen, the true weight is the downward gravitational force $\vec{w}$ of Earth on the person. In equilibrium, a person is supported by an equal and opposite upward force $\vec{N}$ exerted by the floor. As mentioned earlier, $\vec{w}$ and $\vec{N}$ are not an action–reaction pair. The *reaction* to $\vec{N}$ is the force exerted by the person's feet on the floor, which at *equilibrium* has the same magnitude as $\vec{w}$. If the person is standing on a weigh scale, it is the latter downward force exerted by the person which is registered by the scale.

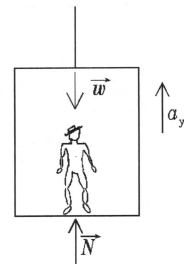

Figure 8-8

Example 8-8: A man in an ascending elevator.

Now consider the passenger in the accelerating elevator. We wish to find the force exerted on the elevator floor by the passenger's feet. By Newton's third law, this has the same magnitude N as the normal force $\vec{N}$ exerted <u>by</u> the floor <u>on</u> the passenger. Thus, we are really looking for N. The free-body diagram for the passenger is shown in Fig. 8-8, where the positive y-direction is chosen to be vertically upward. The only forces acting on the passenger are $\vec{N}$ and the downward gravitational force $\vec{w}$. We have added to the figure an arrow representing the upward acceleration $a_y = 3.20$ m/s² of the passenger due to the motion of the elevator. Hence Newton's second law gives

$$\sum F_y = N - w = ma_y$$

or, rearranging

$$N = w + ma_y$$

We are given the passenger's weight, $w = 800$ N, rather than his mass m, but the latter is readily determined from $w = mg$. Solving this for m and substituting into the previous equation, we finally get

$$N = w\left(1 + \frac{a_y}{g}\right) = 800 \text{ N} \left(1 + \frac{3.20 \text{ m/s}^2}{9.80 \text{ m/s}^2}\right)$$
$$= 1060 \text{ N}$$

Thus, the apparent weight, whose magnitude is given by N, is significantly greater than the true weight of 800 N.

During most of its motion, a typical elevator is not accelerating but moves at constant speed. An elevator does accelerate during the intervals between starting from rest and reaching its final speed, or between moving at constant speed and coming to a stop. You have probably noticed the sensation of extra "weight" when an elevator accelerates upward. Note that an upward acceleration, $a_y > 0$, such as in the previous example, also occurs when an elevator is initially moving <u>downward</u> and then comes to rest.

The expression for N derived in the previous example can also be applied when the acceleration a_y is <u>negative</u>. This occurs either when an elevator starts from rest and moves downward or is initially moving upward and then comes to rest. In these cases of negative, i.e., downward acceleration, the apparent weight given by N is <u>less</u> than the true weight of the person. In the extreme case that $a_y = -g$, i.e., when the elevator is in free fall (due to a snapped cable!), we see that $N = 0$ and the passenger seems to be *weightless*. This effect can also be produced in a high-altitude jet plane which undergoes a rapid dive, or in a spacecraft orbiting Earth (for reasons discussed later in Section 9.2). Although we won't go into this here, the physiological effects of prolonged apparent weightlessness is an active area of current medical research.

8.5   Friction

Friction has already been mentioned several times, and described as a force which opposes the relative sliding motion of two objects whose surfaces are in contact with each other. Friction plays a major role in everyday life. The ability to walk or run requires a sufficient amount of friction between the soles of shoes and the ground to prevent the foot from slipping backward or forward. Likewise, friction between the tires and the road is necessary to drive or turn an automobile or bicycle, and is of equal importance for bringing a vehicle to a stop. Friction also exists within an animal's body, particularly at the surfaces between joints and between other structures (e.g., tendons and bones) that slide over one another. In these cases, friction can be a severe impediment to proper movement of the structures and can also lead to wear.

To describe friction quantitatively, let us consider the example of an object (e.g., a book) resting on a horizontal surface, as in Fig. 8-9. Suppose a horizontal force $\vec{F}$ is applied in an attempt to move the object; this force could be measured by a spring scale, for example. Usually it is found that the object does not move until the magnitude, F, of the applied force exceeds a certain critical value F_{max}. For all $F < F_{max}$, the object remains at rest, which implies that $\vec{F}$ is exactly balanced by the opposing *static friction force* $\vec{F}_s$, i.e., $F_s = F$. Notice this tells us that the magnitude of the friction force F_s is <u>not</u> constant, but continually adjusts itself in response

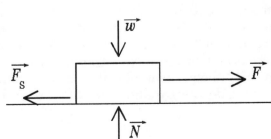

Figure 8-9

A force $\vec{F}$ attempting to move an object at rest on a surface must overcome the static friction force $\vec{F}_s$.

to the applied force. Once the critical value F_{max} is reached, the object breaks loose and starts to slide along the surface, indicating that the horizontal forces are no longer in equilibrium. Thus, F_{max} equals the maximum magnitude that the friction force F_s can attain, i.e., $F_s \leq F_{max}$.

What determines the value of F_{max}? Experience tells us that this depends on how tightly the object and the supporting surface are pressed together, as well as on the natures of the two contacting surfaces, e.g., whether they are rough or slippery. Experiments show that, for a large variety of surfaces in contact, F_{max} is proportional to N or,

$$F_{max} = \mu_s N \qquad \text{[8-10]}$$

where N, the magnitude of the normal force exerted on the object, represents the force pressing the two contacting surfaces together. Recall that the object exerts an equal but opposite force of magnitude N on the supporting surface. The proportionality factor μ_s is known as the *coefficient of static friction*; the term "static" is employed because so far we are dealing with surfaces which are not in relative motion. The value of μ_s depends on the nature and composition of <u>both</u> contacting surfaces. Rough or sticky surfaces have larger values of μ_s than do smooth or slippery surfaces. Typical values of μ_s range from 0.1 to 1. Note that because μ_s is a proportionality constant between two force magnitudes, it has no dimensions or units.

The relation Eq. [8-10] is only an approximate representation of what is actually a very complex phenomenon. Even a surface which appears completely smooth at a macroscopic level is usually rough at a molecular level, and might be pictured as an irregular mountain range with jutting peaks and deep valleys (Fig. 8-10). When two such surfaces are pressed together, the "mountains" of one surface can become locked in the "valleys" of the other, these being held together by attractive intermolecular forces (which are fundamentally electrical in nature); this is the origin of friction. Relative sliding motion of the surfaces can only come about when the applied force reaches a certain threshold value which is able to overcome the intermolecular forces binding the surfaces together.

Figure 8-10

Schematic illustration of two surfaces in contact, at the microscopic level, illustrating the origin of friction.

Once the object starts to slide over the supporting surface, it is usually found that the magnitude of the friction force <u>decreases</u>. This is again in agreement with common experience: e.g., it takes more force to start a sled moving than to keep it moving. Experiments show that the magnitude of the force of *kinetic friction* F_k ("kinetic" referring to the fact that the surfaces are now moving relative to each other) usually obeys a relation similar to Eq. [8-10], namely

$$F_k = \mu_k N \qquad \textbf{[8-11]}$$

where the proportionality constant μ_k is called the *coefficient of kinetic friction*. Consistent with the fact that F_k is usually less than F_{max}, one finds that $\mu_k < \mu_s$. As is true of Eq. [8-10], Eq. [8-11] is not always exactly valid, particularly for complicated surfaces like those of plastics or biomaterials, but will serve as a reasonable approximation that we shall use in the remainder of this chapter.[6] Kinetic friction forces may also depend on the relative speeds of the moving objects, but we shall ignore this complication.

These concepts of friction are best illustrated by several examples.

EXAMPLE 8-9

You are attempting to drag a 600-N crate across the floor by pulling horizontally on a rope attached to it. To start moving the crate, you find that you must pull with a horizontal force of magnitude 250 N. Once the crate starts to move, however, you can keep it moving at constant speed with only a 220-N force. What are the coefficients of static and kinetic friction between the crate and the floor?

SOLUTION

Figure 8-11 shows the FBD for the crate, with the pulling force denoted $\vec{F}_p$ and the friction force denoted $\vec{F}_f$. Both in the state of rest and while the crate is moving at constant speed, the crate is at equilibrium and the components of the net force in the x- and y-directions are zero. Hence,

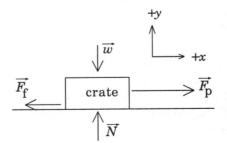

Figure 8-11

Example 8-9: Free-body diagram for a crate being pulled across the floor.

---

6. Equations [8-10] and [8-11] also break down for very large values of N.

$$\sum F_x = F_p - F_f = 0$$

therefore $F_f = F_p$

$$\sum F_y = N - w = 0$$

therefore $N = w = 600$ N

To just start the crate moving, we are given that $F_p = 250$ N. The friction force in this case has its maximum possible magnitude $F_f = F_{max} = \mu_s N$. Hence, the relation $F_f = F_p$ becomes

$$\mu_s N = 250 \text{ N}, \text{ therefore } \mu_s = \frac{250 \text{ N}}{600 \text{ N}} = 0.417$$

Once the crate is moving, the pulling force is reduced to a value $F_p = 220$ N, while the friction force becomes kinetic friction, $F_f = \mu_k N$. Hence we find

$$\mu_k N = 220 \text{ N}, \text{ therefore } \mu_k = \frac{220 \text{ N}}{600 \text{ N}} = 0.367$$

In the previous example, the normal force $\vec{N}$ of the floor on the crate is exactly equal (in magnitude) to the weight of the crate. This, of course, is because $\vec{N}$ and $\vec{w}$ are the only forces acting vertically on the crate and so they must be balanced for vertical equilibrium. This is also consistent with our description of $\vec{N}$ as representing the force which presses the two surfaces together. In that example, one intuitively associates the "pressing" force with the weight of the crate. This equivalence of N and w is not always true, however. Let's consider a variant of the previous example.

EXAMPLE 8-10

Suppose now you try to pull the crate in Example 8-9 with the rope making an angle of 30.0° above the horizontal. With what force must you pull to keep the crate moving at constant speed?

SOLUTION

The FBD for the crate is now that in Fig. 8-12. The motion at constant speed again implies that the forces are balanced horizontally, and that the friction force is to be identified with the kinetic friction $\mu_k N$. Now the equilibrium conditions in the x- and y- directions are:

$$\sum F_x = F_p \cos(30.0°) - F_f = F_p \cos(30.0°) - \mu_k N = 0$$

$$\sum F_y = N - w + F_p \sin(30.0°) = 0$$

These are two simultaneous equations for the two unknowns N and F_p. Let us solve the second equation for N,

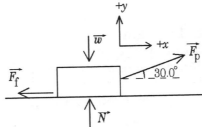

Figure 8-12

Example 8-10: Free-body diagram for a crate being pulled at an angle above the horizontal.

$$N = w - F_p \sin(30.0°)$$

This shows that the normal force (which measures the total force pressing the two surfaces together) is <u>less</u> than the weight of the crate, reduced by an amount equal to the vertical component of the <u>upward</u> pulling force F_p. Substituting this expression for N into the first equation gives

$$F_p \cos(30.0°) - \mu_k \left[w - F_p \sin(30.0°) \right] = 0$$

which can now be solved for F_p:

$$F_p = \frac{\mu_k w}{\left[\cos(30.0°) + \mu_k \sin(30.0°) \right]}$$

$$= \frac{0.367 \times 600 \text{ N}}{\left[\cos(30.0°) + 0.367 \sin(30.0°) \right]}$$

$$= 210 \text{ N}$$

where we used values for w and μ_k from the previous example.

Next, let us illustrate the effects of friction in several dynamics problems.

EXAMPLE 8-11

Let's go back to the skier we studied in Example 8-7, but now suppose the wax has worn off and there is a non-zero coefficient of kinetic friction between the skis and snow, $\mu_k = 0.150$. What now is the magnitude of the acceleration of the skier?

SOLUTION

Figure 8-13

Example 8-11: Free-body diagram for an object sliding down an incline in the presence of friction.

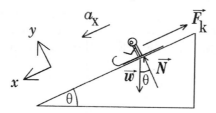

As earlier, we take x- and y-axes parallel and perpendicular to the slope. The FBD for this problem, shown in Fig. 8-13, differs from that in Fig. 8-7 only by inclusion of the kinetic friction force $\vec{F}_k$, which here acts up along the slope, opposite to the direction of the skier's motion. As in Example 8-7, the balance of forces perpendicular to the incline gives

$$N = -w_y = mg\cos(\theta)$$

Newton's second law for the components of the forces parallel to the slope now becomes

$$ma_x = w_x - F_k$$

or, substituting $w_x = mg\sin(\theta)$ (see Example 8-7) and $F_k = \mu_k N$,

$$ma_x = mg\left[\sin(\theta) - \mu_k\cos(\theta)\right]$$

from which we get

$$\begin{aligned} a_x &= g\left[\sin(\theta) - \mu_k\cos(\theta)\right] \\ &= 9.80 \text{ m/s}^2\left[\sin(20.0°) - 0.150\cos(20.0°)\right] \\ &= 1.97 \text{ m/s}^2 \end{aligned}$$

As in Example 8-7, this result is independent of the skier's mass. Notice that the acceleration in this case is much less than in the absence of friction (3.35 m/s²). If the angle of inclination were reduced somewhat, or the coefficient of friction were slightly increased, the forces parallel to the incline would become balanced and the skier would then descend with constant speed (see Exercise 8-16). A slight variant of this question is to determine the conditions required—e.g., the maximum value of the static friction coefficient for a given angle of inclination—for the skier to start moving without giving himself an initial push. In this case, one considers the equilibrium conditions when the static friction force just reaches its maximum value (see Exercise 8-17).

Similar considerations arise in analyzing the importance of friction between a person's shoe and the ground during walking, mentioned earlier.

EXAMPLE 8-12

You are walking along a wood floor. During the walking process, your heel strikes the floor with a force $\vec{F}$ at an angle θ to the vertical. If $\mu_s = 0.500$ for normal shoe leather on wood, what is the maximum value that θ can be and still ensure that you do not slip?

SOLUTION

Figure 8-14 illustrates the forces present when the shoe heel strikes the ground during walking. The force $\vec{F}$ generally exceeds the weight of the person, due to contributions from muscular forces which act in moving the leg. The foot also experiences a normal force $\vec{N}$ and static friction force $\vec{F}_s$ due to the ground. In order for the heel not to slip during this process, these forces must be balanced.

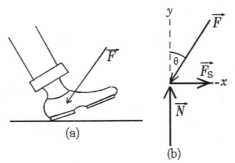

(a)

(b)

Figure 8-14

Example 8-12: (a) When the foot strikes the ground during walking, it experiences a force $\vec{F}$ exerted by the leg as shown; (b) Free-body diagram showing all forces on the foot.

Hence, resolving forces along the x- and y-axes indicated,

$$\sum F_x = -F \sin(\theta) + F_s = 0 \quad \text{or} \quad F_s = F \sin(\theta)$$
$$\sum F_y = N - F \cos(\theta) = 0 \quad \text{or} \quad N = F \cos(\theta)$$

The condition that the static friction force should satisfy $F_s \le \mu_s N$ leads to the relation

$$F \sin(\theta) \le \mu_s F \cos(\theta)$$

or

$$\tan \theta \le \mu_s$$

For given μ_s, this puts an upper limit on how large the step angle θ can be so that the foot does not slip. If $\mu_s = 0.500$, we find $\theta_{max} = \tan^{-1}(0.500) = 26.6°$.

Thermal Energy Produced by Friction

Whenever kinetic friction is acting, thermal energy (colloquially called "heat") is generated. A familiar example of this is the heat produced by rubbing the palms of one's hands together vigorously. The origin of the thermal energy will be discussed in Section 8.7 under conservative and non-conservative forces. In machines, heat may have harmful effects, such as causing burned-out electrical contacts and bearings. Likewise, in body joints, heat is damaging because proteins are easily denatured by it. Another possible damaging effect of friction is the overextension of muscles trying to force motion of joints. The application of lubricants such as oils to overcome friction is familiar in machinery. Such lubricants maintain a fluid film between the solid surfaces which prevents them from coming into molecular contact. (Frictional forces also exist between layers of a fluid, albeit much reduced compared to solid surfaces. This is discussed in connection with *viscosity* of fluids in Chapter 12.) The surfaces of body joints are bathed in a lubricating liquid known as synovial fluid, which is believed to play a role in reducing joint friction, although its importance in this role is still under considerable debate.

8.6 Conservation of Momentum

Collisions between moving objects or between moving and stationary objects seem to happen far too frequently. When two bodies collide, they exert very large and complicated forces on each other for a very short time. Such a situation is difficult to analyze using Newton's laws directly. To understand collision phenomena, a quantity known as *momentum* and the law of *conservation of momentum* (which we will see can be derived from Newton's laws) are introduced. Using these concepts, we can describe collisions of balls with baseball bats (or tennis rackets, golf clubs or pool cues), collisions between automobiles (and see why the occupants of the smaller vehicle are more likely to be injured), and the "collision" between a person's hipbone and floor when the person trips and falls. Several other phenomena, such as rocket propulsion, can also be explained by momentum conservation.

The momentum $\vec{p}$ of a body of mass m moving with velocity $\vec{v}$ is defined as

$$\vec{p} = m\vec{v} \qquad\qquad \text{[8-12a]}$$

This is a product of the scalar m with the vector $\vec{v}$, and therefore is also a vector. In terms of its components in two dimensions, Eq. [8-12a] is

$$p_x = mv_x \text{ and } p_y = mv_y \qquad\qquad \text{[8-12b]}$$

The significance of momentum can be seen by rewriting Newton's second law $\sum \vec{F} = m\vec{a}$ using the fact that $\vec{a} = \mathrm{d}\vec{v}/\mathrm{d}t$,

$$\sum \vec{F} = m\frac{\mathrm{d}\vec{v}}{\mathrm{d}t} \qquad \text{[8-13a]}$$

We will assume in this text that the mass of the body remains constant, although there are some situations where this need not be true. For constant m, we note that

$$m\frac{\mathrm{d}\vec{v}}{\mathrm{d}t} = \frac{\mathrm{d}}{\mathrm{d}t}(m\vec{v}) = \frac{\mathrm{d}\vec{p}}{\mathrm{d}t}$$

hence Eq. [8-13a] can be expressed as

$$\sum \vec{F} = \frac{\mathrm{d}\vec{p}}{\mathrm{d}t} \qquad \text{[8-13b]}$$

This is a useful rewriting of Newton's second law, and indeed is the form adopted in most advanced treatments of mechanics. For example, this equation indicates that a large net force is required to bring an object with large momentum $|\vec{p}|$ to rest in a given time interval $\mathrm{d}t$. The magnitude of the object's momentum could be large either because it is moving with high speed $|\vec{v}|$ (e.g., a bullet) or because it has a very large mass m (e.g., a large truck).

If the resultant force $\sum \vec{F}$ acting on a body is zero, then Eq. [8-13b] shows that $\mathrm{d}\vec{p}/\mathrm{d}t = 0$, i.e., the momentum of the body does not change with time. This is the law of conservation of momentum. Of course, so far this is nothing more than Newton's first law, since if the net force on a single body is zero then the object's acceleration is zero, and so its velocity $\vec{v}$ (and hence momentum $\vec{p} = m\vec{v}$) is constant. It is in generalizing this result to collisions involving two or more bodies that the momentum conservation law comes into its own. If two objects are colliding, then we have to consider the total momentum of the system, i.e., the vector sum of the individual momenta. If we label the two bodies as 1 and 2, the total momentum $\vec{p}$ is given by

$$\vec{p} = \vec{p}_1 + \vec{p}_2 \qquad \text{[8-14]}$$

Taking the time derivative of this expression, and using Newton's second law for each body 1 and 2, we get

$$\frac{\mathrm{d}\vec{p}}{\mathrm{d}t} = \vec{F}_1 + \vec{F}_2 \qquad \text{[8-15]}$$

where $\vec{F}_1$ and $\vec{F}_2$ are the net forces acting on bodies 1 and 2, respectively. Now let us suppose, for simplicity, that the net force on each body is due only to its interaction with the other body, as when they collide with each other. Then $\vec{F}_1$ and $\vec{F}_2$ form an action–reaction pair and thus, by Newton's third law, satisfy $\vec{F}_1 + \vec{F}_2 = 0$. Hence we again find that $\mathrm{d}\vec{p}/\mathrm{d}t = 0$, and the total momentum of the two bodies remains constant. This result is also true if other forces are acting on each body in addition to those which they exert on each other, as long as the vector sum of these *external forces* is zero.

Applied to a collision between two objects, the momentum conservation law tells us that the total momentum after the collision equals the total momentum before the collision. We can write this as the vector equation

$$\vec{p}_1 + \vec{p}_2 = \vec{p}'_1 + \vec{p}'_2 \qquad \text{[8-16a]}$$

or

$$m_1\vec{v}_1 + m_2\vec{v}_2 = m_1\vec{v}'_1 + m_2\vec{v}'_2 \qquad \text{[8-16b]}$$

where the quantities on the right-hand side of these equations, denoted by the prime symbol (′), are those present after the collision. Let's now illustrate these relations with several examples.

EXAMPLE 8-13

A car and a truck collide head on. Their masses and velocities just before impact are indicated in Fig. 8-15. As a result of the collision, the two vehicles become entangled with each other and move away as one combined mass. What is the velocity of the combined wreckage after the collision?

Figure 8-15

Example 8-13: A collision between a car and a truck moving in opposite directions.

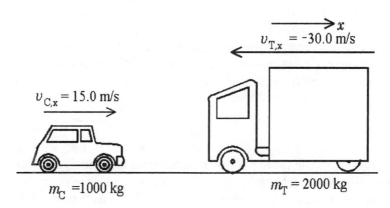

SOLUTION

Instead of numerals "1" and "2," we'll designate the car and truck by symbols "C" and "T," respectively. In this example, the car (m_C = 1000 kg) and truck (m_T = 2000 kg) are initially moving toward each other along a straight line, hence the total initial momentum lies along this line. We'll call this the x-axis, with the positive direction being that in which the car is moving. By conservation of momentum, the total momentum of the combined mass after the collision must also be along this axis, hence we need to consider velocity components only along x. Equation [8-16b] then becomes

$$m_C v_{C,x} + m_T v_{T,x} = (m_C + m_T) v'_x$$

where we have used the fact that the two vehicles move as one after the collision, with a total mass ($m_C + m_T$) and a common final velocity denoted v'_x. All the v's in this equation represent velocity components and not speeds, and thus can have either positive or negative values. With our choice for the positive x-direction, $v_{C,x}$ = 15.0 m/s and $v_{T,x}$ = –30.0 m/s. Solving this equation for v'_x, we find

$$v'_x = \frac{m_C v_{C,x} + m_T v_{T,x}}{(m_C + m_T)} = \frac{\left[(1000\,\text{kg} \times 15.0\,\text{m/s}) - (2000\,\text{kg} \times 30.0\,\text{m/s})\right]}{(1000\,\text{kg} + 2000\,\text{kg})}$$

$$= -15.0\,\text{m/s}$$

The minus sign in this result tells us that the final velocity of the wreckage is toward the left, the direction in which the truck was initially moving.

As stated earlier, the momentum conservation law is strictly valid only if the net external force on the colliding objects is zero. In the last example, this is clearly true regarding forces in the vertical direction, since neither vehicle has a vertical acceleration, and hence the vertical components of the forces (i.e., gravity and the normal forces due to the road) must be balanced. One might worry about friction forces between the vehicles' tires and the road, which are necessary for the forward motion of the vehicles. Usually the magnitudes of these forces are very small compared with the interaction forces between the vehicles when they collide, and can safely be neglected. This is no longer true in the time following the collision, when friction will eventually bring the two vehicles to rest. Therefore, in problems involving momentum conservation, the momenta and velocities of the colliding objects should be strictly those which occur <u>just before</u> and <u>just after</u> the collision. This proviso becomes less of a concern when the external frictional forces are indeed very small, such as when the colliding objects are sliding over an ice surface or, better yet, are floating freely in the zero-gravity environment of outer space.

EXAMPLE 8-14

A 60.0-kg astronaut is floating motionless in outer space. She is carrying a 5.00-kg space hammer. She throws the hammer away from her with a speed of 6.00 m/s. What is the speed of the astronaut after she throws away the hammer?

SOLUTION

This example could be said to describe the opposite of a collision, since initially the two objects (here the astronaut and hammer) are joined together, and then they move apart. Since the initial momentum is zero, the final velocities $v_{A,x}'$ and $v_{H,x}'$ of the astronaut and hammer, respectively, satisfy

$$m_A v_{A,x}' + m_H v_{H,x}' = 0$$

Hence,

$$v_{A,x}' = -\frac{m_H}{m_A} v_{H,x}' = -\left(\frac{5.00\,\text{kg}}{60.0\,\text{kg}}\right)(6.00\,\text{m/s})$$
$$= -0.500\,\text{m/s}$$

Therefore, the astronaut moves away with a speed of 0.500 m/s in the opposite direction (indicated by the negative sign) to that of the hammer.

Similar considerations to those in the last example explain the recoil velocity of a gun when it fires a bullet as well as the motion of a small boat when it is initially stationary in water and a person then tries to walk from one end to the other. A related effect occurs when a fish swims forward by imparting rearward momentum to water on flapping its tail. The same idea of momentum conservation explains the propulsion of a jet plane or rocket. When the jet fires its engines, hot gases are ejected at high speed from its base, so to conserve momentum, the jet must move off in the opposite direction. This process is a little more complicated than the ones we've analyzed so far because the total mass of the jet decreases in time as its fuel is consumed, but the dominant principle is still that of momentum conservation. When necessary, squids use the same principle of "jet propulsion" to move themselves rapidly, by squirting out a stream of water from a tube in their heads.

Now let us go back to the car–truck collision of Example 8-13, but generalize the motion to being two-dimensional.

EXAMPLE 8-15

This time, the car and truck collide at an angle of 30.0° as shown in Fig. 8-16. The initial speeds of the car and truck, denoted v_C and v_T, are the same as in Example 8-13. What now is the final velocity of the tangled wreckage?

SOLUTION

The basic equation to analyze is still Eq. [8-16b], but now we must take account of the vector character and resolve this equation into x- and y-components, with the directions of axes as indicated on Fig. 8-16.

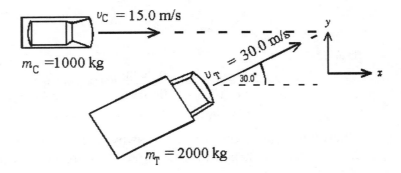

Figure 8-16

Example 8-15: A collision between a car and a truck at an angle of 30°.

The components of the final velocity after the collision are denoted v_x' and v_y'. Conservation of the x-component of momentum gives, as in Example 8-13,

$$m_C v_{C,x} + m_T v_{T,x} = (m_C + m_T)v_x'$$

where $v_{C,x} = v_C$ and $v_{T,x} = v_T \cos(30.0°)$. Hence,

$$v_x' = \frac{\left[1000\,\text{kg} \times 15.0\ \text{m/s} + 2000\,\text{kg} \times 30.0\ \text{m/s} \times \cos(30.0°)\right]}{3000\,\text{kg}}$$

$$= 22.3\,\text{m/s}$$

The treatment of momentum conservation in the y-direction is analogous:

$$m_C v_{C,y} + m_T v_{T,y} = (m_C + m_T) v'_y$$

but $v_{C,y} = 0$ while $v_{T,y} = v_T \sin(30.0°)$. Hence,

$$v'_y = \frac{\left[2000\,\text{kg} \times 30.0\,\text{m/s} \times \sin(30.0°)\right]}{3000\,\text{kg}} = 10.0\,\text{m/s}$$

We can now obtain the magnitude and direction of the final velocity $\vec{v'}$,

$$|\vec{v'}| = \sqrt{(v'_x)^2 + (v'_y)^2}$$
$$= \sqrt{(22.3\,\text{m/s})^2 + (10.0\,\text{m/s})^2}$$
$$= 24.4\,\text{m/s}$$

$$\theta' = \tan^{-1}\left(\frac{v'_y}{v'_x}\right) = \tan^{-1}\left(\frac{10.0\,\text{m/s}}{22.3\,\text{m/s}}\right)$$
$$= 24.2°$$

where θ' is measured relative to the positive x-direction.

Notice that, in the previous example as well as in any other examples involving two-dimensional motion, the x- and y-components of momentum are <u>individually</u> conserved. This reflects the basic <u>vector</u> character of momentum.

So far, we have assumed that the colliding objects stick together after the collision (or, in the inverse situation, stick together before they are separated). Such collisions are called *completely inelastic*. Much more commonly, colliding objects go their separate ways after the collision, as in the case of colliding billiard balls, for example. Such collisions are either *elastic* or *inelastic*. In an elastic collision, the total kinetic energy (to be discussed in Section 8.7) is also conserved during the collision. In a general inelastic collision, the total kinetic energy after the collision is less than that before the collision, the difference being converted to other forms of energy (e.g., heat and sound). For inelastic (other than completely inelastic) or elastic collisions, calculations are generally much more difficult than those described here. Conservation of momentum alone is not sufficient to solve the problems, unless one is provided with additional information as in the following example.

EXAMPLE 8-16

Two hockey pucks of equal mass m collide head-on. Initially, one puck has a speed v, while the other puck is stationary. After the collision, the first puck is stationary. What is the speed of the second puck?

SOLUTION

This is again a one-dimensional problem. The initial momentum is along the direction (chosen to be the x-direction) of the first puck, $p_x = m_1 v_{1,x} = mv$. The total momentum along x after the collision is just due to the second puck, $p_x' = m_2 v_{2,x}'$. Equating p_x and p_x' then gives $v_{2,x}' = v$. This is illustrated in Fig. 8-17.

Before collision

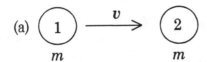

After collision

Figure 8-17

Example 8-16: Movement of two pucks (a) before and (b) after collision.

In the above example, we are told that the first puck is at rest after the collision. A slight generalization of this question is given in Problem 8-38 at the end of this chapter, for the case where the first puck does <u>not</u> collide head-on with the second puck but rather hits it slightly off to one side. In that case, both pucks move away after the collision at different angles with respect to the x-axis.

8.7   Work and Energy

Work

Although the concept of *energy* is familiar to all of us, defining this term is not so easy. We encounter several forms of energy in this text: gravitational, electrical, chemical, nuclear, etc. In this section, we will define precisely what is meant by *mechanical energy*, and discuss its relationship to several other forms of energy. We will see that the various forms of energy can be converted from one type to another, according to the principle of *conservation of energy*. For example, animals execute motion via muscular forces at the expense of chemical energy derived from the food they eat. By using up this energy, they are able to perform various feats, or, in other words, they do *work*. Our first step, then, is to define what is meant by work.

Figure 8-18

The definition of work W done by force $\vec{F}$: $W = F \, \Delta r \cos(\theta)$.

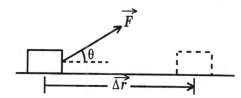

Everyone probably remembers from high-school physics that "work = force × distance," but this is valid only for one-dimensional situations. More precisely, work is defined as follows: suppose a constant force $\vec{F}$ is applied to an object while it is undergoing a displacement $\Delta \vec{r}$ in a straight line, where the vectors $\vec{F}$ and $\Delta \vec{r}$ have an angle θ between them (Fig. 8-18). The work W done by the object is defined as the product of the magnitude of the displacement $\Delta r = |\Delta \vec{r}|$ and the component $F_r = F \cos(\theta)$ of the force in the direction of the displacement,

$$W = F_r \, \Delta r = F \, \Delta r \cos(\theta) \qquad \textbf{[8-17]}$$

Since work has dimensions of force times length, the corresponding SI unit for work is the newton-metre (N·m), a combination which is given the name "joule" (J).

We will frequently be concerned with evaluating the work done <u>by</u> and <u>against</u> gravity, as in the following example.

EXAMPLE 8-17

In a "biceps curl" exercise, a weight lifter raises a dumbbell of mass 10.2 kg a vertical distance of 0.250 m.

(a) How much work is done by gravity on the dumbbell?

(b) Neglecting any acceleration, how much work is done by the weight lifter on the dumbbell?

SOLUTION

(a) We'll assume that the dumbbell is raised vertically in a straight line (although, as will be seen later, this is not necessary). The displacement of the dumbbell has a magnitude $\Delta y = 0.250$ m in the upward (positive) vertical direction. For later purposes, we write this as $\Delta y = y_2 - y_1$, as in Fig. 8-19. The force of gravity $\vec{w} = m\vec{g}$ on the dumbbell acts downward, and thus the angle θ between the displacement and the force is 180°. Hence, using Eq. [8-17], the work done by gravity is

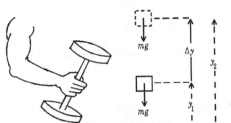

Figure 8-19

Example 8-17: Work done by gravity when an object (e.g., the dumbbell) undergoes a vertical displacement Δy.

$$W_{grav} = mg(y_2 - y_1)\cos(180°) \qquad \text{[8-18]}$$
$$= -mg(y_2 - y_1)$$

which has the value

$$W_{grav} = -10.2\,\text{kg} \times 9.80\,\text{m/s}^2 \times 0.250\,\text{m}$$
$$= -25.0\,\text{J}$$

Notice this is <u>negative</u>, since the displacement and force (weight) are in opposite directions.

(b) If there is no acceleration of the dumbbell, the upward force exerted on it by the weight lifter must exactly balance its weight. Hence, the work done by the weight lifter is <u>positive</u>, with the same magnitude as W_{grav}, i.e., 25.0 J.

As seen in the previous example, work (which is a *scalar* quantity; see Box 1-1) can be either positive or negative. More generally, work is negative when the component of the force is in the opposite direction to that of the displacement, i.e., $90° < \theta \leq 180°$ in Eq. [8-17]. The work W is zero if the force $\overrightarrow{F}$ is perpendicular to the displacement $\Delta \overrightarrow{r}$, or if there is no movement of the object, i.e., $\Delta r = 0$. For example, suppose the weight lifter in Example 8-17 is holding up the dumbbell at a <u>constant</u> height: although still exerting a force to overcome the dumbbell's weight, he is not doing any work since the displacement is zero. Similarly, if he walks across the floor while holding the dumbbell, he does no work on the dumbbell since the force he exerts is vertically upward while the displacement is horizontal. These examples show that "work" in a mechanical sense does not necessarily agree with its common everyday meaning.

Work and Kinetic Energy

So far, we have defined work only by Eq. [8-17]. The physical significance of work emerges from the following principle, known as the *work–energy theorem*:

The total work done on an object by all forces acting on it equals the change in the object's kinetic energy.

For an object of mass m moving with speed v, the *kinetic energy K* is defined as

$$K = \tfrac{1}{2}mv^2 \qquad\qquad \textbf{[8-19]}$$

Mathematically, the work–energy theorem is then

$$W_{TOT} = \Delta K \qquad\qquad \textbf{[8-20]}$$
$$= \tfrac{1}{2}mv_2^2 - \tfrac{1}{2}mv_1^2$$

In Eq. [8-20], v_1 and v_2 denote the initial and final speeds of the object, respectively, during the interval that the work W_{TOT} is being done on it.

We emphasize that W_{TOT} is the total work done by <u>all</u> the forces acting on the object. This can be evaluated either by calculating the work done by each separate force present, then adding up the quantities of work algebraically, or by calculating the <u>resultant</u> of all forces acting and then determining the work done by this single resultant force.

Now we shall give a proof of the work–energy theorem, restricted for simplicity to the case of motion in a straight line under a constant resultant force $\vec{F}_R$. For concreteness, let the direction of motion be along the x-axis. Denoting the component of the net force on this axis as $F_R \cos(\theta) = F_{R,x}$, where θ is the angle between $\vec{F}_R$ and the x-axis, the total work done during a displacement Δx is

$$W_{TOT} = F_{R,x}\, \Delta x$$

But from Newton's second law, Eq. [8-4], $F_{R,x} = ma_x$, where a_x is the object's acceleration along the x-axis. Hence

$$W_{TOT} = ma_x \Delta x$$

Finally, we can re-express the quantity $a_x \Delta x$ using the one-dimensional kinematic equation Eq. [7-9], $v^2 = v_0^2 + 2a\Delta x$ (replacing v_0 and v by v_1 and v_2, respectively), to get

$$W_{TOT} = \tfrac{1}{2}m(v_2^2 - v_1^2)$$

as in Eq. [8-20]. Although this proof is restricted to the conditions mentioned above, it can be extended to show that the work–energy theorem is valid for motion along any curved path in space and for a net force which varies with position.

Before discussing the theorem any further, we should first verify that the unit of kinetic energy is the same as that for work. From the definition Eq. [8-19] ($K = \tfrac{1}{2}mv^2$), the SI unit for kinetic energy is clearly $kg \cdot (m/s)^2$. This is indeed equivalent to N·m (or J), since $1\ N = 1\ kg \cdot m/s^2$.

As seen above, the derivation of the work–energy theorem hinges on Newton's second law, and really can be considered simply as a consequence of that law. One can rationalize the theorem as follows. If a non-zero resultant force $\vec{F}_R$ acts to produce a displacement of some object, then $\vec{F}_R$ does work on the object. But the action of the net force should also be accompanied by an acceleration of the object, hence a change in its velocity. Therefore one should expect that there is a connection between the total work done on an object and its change in velocity. This connection is expressed by the work–energy theorem. Note, however, that the kinetic energy K depends only on an object's speed, and the work–energy theorem contains no information about possible changes in the direction of an object's velocity.

EXAMPLE 8-18

A person pulls a sled of mass 5.19 kg across the ground, by exerting a force of 25.6 N inclined at 35.0° above the horizontal, as shown in Fig. 8-20. The coefficient of kinetic friction between the sled and ground is 0.502. If the sled is moved 2.75 m horizontally, (a) what is the work done by each of the four forces acting on the sled; (b) what is the total work done; and (c) what is the final speed of the sled, assuming that it starts from rest?

Figure 8-20

Example 8-18: A sled pulled at an angle above the horizontal.

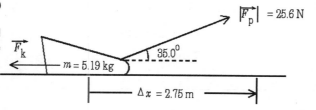

SOLUTION

(a) The free-body diagram for this example is the same as that in Fig. 8-12 for Example 8-10 (apart from the value of the inclination angle). Note that the normal force $\vec{N}$ and weight $\vec{w}$ are perpendicular to the displacement of the sled, hence these two forces do no work. Non-zero work is done by both the applied pulling force $\vec{F}_p$ and kinetic friction $\vec{F}_f = \vec{F}_k$. The work done by $\vec{F}_p$ is

$$W_{F_p} = F_p \cos(\theta)\,\Delta x$$
$$= (25.6 \text{ N})\big[\cos(35.0°)\big](2.75 \text{ m})$$
$$= 57.7 \text{ J}$$

Noting that the direction of $\vec{F}_k$ is exactly opposite to the displacement (i.e., the angle between them is 180°), the work done by $\vec{F}_k$ is

$$W_k = -F_k \Delta x$$
$$= -\mu_k N \Delta x$$
$$= -\mu_k \big[mg - F_p \sin(\theta)\big]\,\Delta x \quad \text{(see Example 8–10)}$$
$$= -0.502\big[5.19 \text{ kg} \times 9.80 \text{ m/s}^2 - 25.6 \text{ N} \sin(35.0°)\big](2.75 \text{ m})$$
$$= -49.9 \text{ J}$$

(b) Hence, the total work done is

$$W_{\text{TOT}} = W_{F_{\text{p}}} + W_{\text{k}}$$
$$= 57.7\,\text{J} - 49.9\,\text{J} = 7.8\,\text{J}$$

(c) Denoting the final speed of the sled as v, the work–energy theorem gives

$$W_{\text{TOT}} = \Delta K = \tfrac{1}{2}mv^2 - 0$$

Therefore
$$v = \sqrt{2W_{\text{TOT}}/m}$$
$$= \sqrt{2(7.8\,\text{J})/5.19\,\text{kg}}$$
$$= 1.7\,\text{m/s}$$

Note that Example 8-18 could have been solved by applying Newton's second law to find the acceleration of the sled and then using kinematic equations to find the final velocity. Of course, these are precisely the steps used in our derivation of the work–energy theorem. The latter theorem eliminates the intermediate step of finding the acceleration when all we wish to determine is the final speed. There can also be circumstances—e.g., when the net force and hence acceleration are <u>not</u> constant—for which Newton's laws cannot be readily solved, but the total work and hence the change in kinetic energy can still be calculated.

We saw in the previous example that the work done by kinetic friction is <u>negative</u>. This will be true of friction in general, since it always resists the relative movement of an object over a surface and hence acts in a direction opposite to its displacement.

Gravitational Potential Energy: Conservation of Energy

In Example 8-17, Eq. [8-18], we derived the work done by gravity on an object of mass m when it goes from an initial vertical height y_1 to another height y_2, $W_{\text{grav}} = -mg(y_2 - y_1)$. Although in the example the object of mass m increased in height, i.e., $y_2 > y_1$, one can easily check that Eq. [8-18] also applies if the object decreases in height, i.e., $y_1 > y_2$. In the latter case, the work done by gravity would be <u>positive</u>.

Furthermore, although derived assuming that the object moved in a vertical straight line, Eq. [8-18] is <u>independent</u> of the path followed by the object. For example, suppose the mass m is sliding down an inclined plane between vertical heights y_1 and y_2, as pictured in Fig. 8-21.

Figure 8-21

Calculating the work done by gravity on an object sliding down an incline: $W_{grav} = mg \sin(\theta)\, \Delta r$.

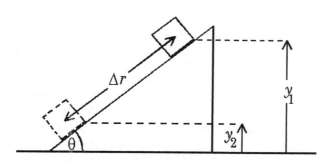

Recalling from Eq. [8-9] that the component of the object's weight parallel to the incline is $mg \sin(\theta)$, then the work done by gravity in this case is

$$W_{\text{grav}} = mg\, \sin(\theta)\Delta r \qquad \textbf{[8-21]}$$

where Δr is the magnitude of the object's displacement along the incline. But one can see from Fig. 8-21 that $\Delta r \sin(\theta) = y_1 - y_2$, and hence Eq. [8-21] can be written as $W_{\text{grav}} = mg(y_1 - y_2)$, which is identical to Eq. [8-18]. More generally, using calculus methods, it can be shown that Eq. [8-18] describes the work done by gravity when mass m follows an <u>arbitrary</u> curved path between vertical heights y_1 and y_2, such as, for example, during projectile motion.

Now let's apply the work–energy theorem Eq. [8-20] ($W_{\text{TOT}} = \Delta K = \frac{1}{2}mv_2^2 - \frac{1}{2}mv_1^2$) to a case where the only work being done is due to gravity, i.e., $W_{\text{TOT}} = W_{\text{grav}}$. This applies if an object is in free fall, or is undergoing projectile motion, or is sliding along an incline as in Fig. 8-21, provided that friction forces are negligible in the latter case. (Note that the normal force of such an incline does no work on the object since it is perpendicular to the object's displacement.) Then, combining Eqs. [8-18] and [8-20], we have

$$mg(y_1 - y_2) = \tfrac{1}{2}mv_2^2 - \tfrac{1}{2}mv_1^2$$

which can be re-arranged to

$$\tfrac{1}{2}mv_1^2 + mgy_1 = \tfrac{1}{2}mv_2^2 + mgy_2 \qquad \textbf{[8-22]}$$

If we define the *gravitational potential energy U* for mass m at vertical height y as

$$U = mgy \qquad \textbf{[8-23]}$$

then the previous equation is equivalent to

$$K_1 + U_1 = K_2 + U_2 \qquad \text{[8-24a]}$$

or

$$E \equiv K + U = constant \qquad \text{[8-24b]}$$

where the total *mechanical energy* E of mass m is defined as the sum of its kinetic and gravitational potential energies. Equation [8-24] shows that the total mechanical energy of mass m is constant, having the same value in its initial state 1 and final state 2 (as well as in all intermediate states). This is our first instance of the principle of conservation of energy.

If kinetic energy K is considered to be the energy associated with an object due to its motion, then gravitational potential energy U may be considered the energy of an object due to its vertical height. Note that, so far, we have implicitly been measuring the vertical coordinate y relative to an origin $y = 0$ on the ground. But the choice of origin in the definition Eq. [8-23] of U is arbitrary. This really follows from Eq. [8-18] for W_{grav}, since the latter involves only the height <u>difference</u> $y_1 - y_2$. A shift in the origin $y = 0$ would add a constant term to U in Eq. [8-23], but as long as the same shift in origin is made for all states of mass m, such a constant will cancel between both sides of Eq. [8-24a].

Let's apply Eq. [8-22] or [8-24a] to an object such as a ball of mass m which is initially held stationary ($v_1 = 0$) at some height y_1. The initial total energy E is only gravitational potential energy, $E = U_1 = mgy_1$. On being released, the ball accelerates toward the ground under the action of gravity, thereby gaining speed and kinetic energy while losing potential energy since its height y decreases. (Equivalently, gravity does positive work on the ball and its kinetic energy increases by the work–energy theorem.) When it reaches the ground ($y_2 = 0$), moving with speed v_2, all of its initial potential energy has been converted to an equal amount of kinetic energy $K_2 = \frac{1}{2}\,mv_2^2$. We can say that at its initial height y_1, the ball had the <u>potential</u> to develop kinetic energy. Alternatively, we can think of potential energy as energy which is <u>stored</u> in an object, ready to be converted to kinetic energy when it is released from rest.

As mentioned previously, we can also apply the energy conservation relations Eqs. [8-22] or [8-24] to projectile motion, as in the following example.

Figure 8-22

Example 8-19: A basketball player making a free-throw shot at the basket.

EXAMPLE 8-19

A basketball player makes a free-throw shot at the basket, as shown in Fig. 8-22. The ball's speed is 7.50 m/s when it is released from the player's hands, and its height is 2.17 m above the floor. What is the ball's speed as it goes through the hoop, which is 3.14 m above the floor?

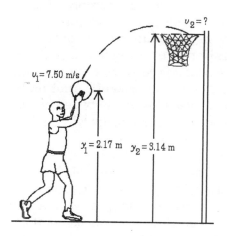

SOLUTION

We use the conservation of energy relation, Eq. [8-22]. The fact that the ball may initially rise and then drop down is unimportant in this equation, since only the vertical displacement $y_2 - y_1$ really matters. From Eq. [8-22], we have $\frac{1}{2}mv_1^2 + mgy_1 = \frac{1}{2}mv_2^2 + mgy_2$. Dividing each term by the mass m, and re-arranging:

$$v_2^2 = v_1^2 + 2g(y_1 - y_2)$$

Hence,

$$v_2 = \sqrt{v_1^2 + 2g(y_1 - y_2)}$$
$$= \sqrt{(7.50 \text{ m/s}^2) + 2(9.80 \text{ m/s}^2)(2.17 \text{ m} - 3.14 \text{ m})}$$
$$= 6.10 \text{ m/s}$$

Note that v_2, the final <u>speed</u> of the ball, is less than the initial speed v_1 since the final elevation y_2 is greater than the initial elevation y_1 and therefore some of the ball's initial kinetic energy has been converted to potential energy. Also notice that we have not been concerned with the directions of the initial and final velocity (presumably the final velocity in the y-direction has a downward component), since these directions do not enter the conservation of energy condition.

Conservative and Non-Conservative Forces

The conservation of energy principle derived in the previous subsection considered only the case of potential energy due to gravity. However, several other types of forces have an associated potential energy, and Eq. [8-24] also holds true if U includes one or more of these energy terms. An example is the *elastic potential energy* of a spring, already encountered in Chapters 1 and 2. By compressing or stretching a spring, potential energy can be stored in the spring, ready to be converted into kinetic energy on releasing one end of the spring. This effect is seen during the operation of a spring board or diving board: the potential energy stored in the board is converted into the kinetic energy of the jumper or diver during the latter's takeoff. Another force which produces potential energy is the *electric force* between charged particles. The electric force and potential energy will be discussed in Chapter 15.

One feature that all these forces (i.e., gravity, elastic, electric), which are known as *conservative forces*, have in common is that the work done by the force in displacing an object between an initial and final point depends only on the coordinates of those points and not on the path followed by the object in moving between the points. We have already seen this in the case of gravitational potential energy.

All other types of forces are said to be *non-conservative*. The clearest example of such a force is the kinetic friction force $\vec{F}_k$. The key difference from conservative forces is that the work done by a non-conservative force in displacing an object <u>does</u> depend on the path followed by the object. We have already noted that $\vec{F}_k$ always acts in the direction opposite to an object's motion and hence always does negative work. For instance, when an object (such as the skier in Example 8-11) slides down an incline, $\vec{F}_k$ acts upward along the incline while if the object is initially moving upward along the incline, $\vec{F}_k$ will be acting downward. The force of friction does negative work in both directions and hence depends on the path of motion (unlike gravity).[7]

The forces applied by a person to push or lift an object are usually non-conservative. While in some particular instances such a force may

---

7.   Another aspect illustrated by the skier in Example 8-11 is that the work done by friction would depend on the total length of the skier's downhill path and not simply, as in the case of gravitational work, on the difference between his initial and final positions. For example, by zig-zagging down the hill, the skier can significantly increase the total path length and hence magnitude of the friction work, although the skier's initial and final positions may be the same as if he skied straight down the hill.

behave like a conservative force, usually that force "adjusts" itself to the circumstances and its associated work is path-dependent.

Let us separate the total work W_{TOT} done on an object into two parts, that done by all conservative forces present, W_c, and that done by all non-conservative forces, denoted W_{nc}. That is, $W_{TOT} = W_c + W_{nc}$. Then it can be shown that, on re-arranging terms, the work–energy theorem Eq. [8-20] becomes

$$W_{nc} = \Delta K + \Delta U \equiv \Delta E \qquad \text{[8-25]}$$

where ΔU is the change in potential energy due to all conservative forces acting on the object. If there are no non-conservative forces present, $W_{nc} = 0$ and Eq. [8-25] becomes $\Delta E = 0$, i.e., the total mechanical energy is constant, as in Eq. [8-24]. More generally, Eq. [8-25] shows that any work done by non-conservative forces produces a change in the mechanical (kinetic plus potential) energy of the object.

We have learned that the work W_k done by kinetic friction is always negative. If friction is the only non-conservative force acting, then $W_{nc} = W_k$ and hence the change in mechanical energy according to Eq. [8-25] is negative. That is, the final mechanical energy is less than the initial energy. One may ask, where has the missing energy gone? The answer is that it has been converted into *thermal energy* of the contacting surfaces. As discussed in Section 8.5, friction is due to attractive binding forces between the molecules of both contacting surfaces. When the surfaces move relative to each other, the molecules are perturbed and end up vibrating more vigorously. This results in an increase in the mechanical energy of the vibrating molecules, which is signaled by an increase in the temperature of the surfaces. We can express this as

$$W_k = -\Delta E_{th} \qquad \text{[8-26]}$$

where ΔE_{th} is the change in thermal energy of the whole system. Since W_k is negative, Eq. [8-26] is consistent with an <u>increase</u> in thermal energy. Using Eq. [8-26], Eq. [8-25] (identifying $W_{nc} = W_k$) can be rewritten as

$$\Delta E = 0 \qquad \text{[8-27]}$$

where the energy E now includes both mechanical and thermal energy, i.e.,

$$E = K + U + E_{th} \qquad \text{[8-28]}$$

EXAMPLE 8-20

A crate of mass $m = 5.00$ kg slides down a ramp of length 2.00 m and inclined at an angle of 30.0°, as shown in Fig. 8-23. The crate starts from rest at the top of the ramp and experiences a constant friction force of magnitude $F_k = 6.50$ N. Use energy methods to determine the speed of the crate when it reaches the bottom of the ramp.

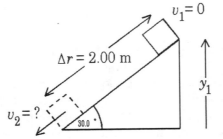

Figure 8-23

Example 8-20: A crate sliding down an incline.

SOLUTION

We use Eqs. [8-27] and [8-28], expressing the fact that the total (mechanical plus thermal) energy stays constant. The initial energy E_1 of the crate is just due to its gravitational potential energy, i.e.,

$$E_1 = mgy_1$$

where we measure the y-coordinate from the bottom of the ramp. The final energy E_2 consists of the crate's kinetic energy $\frac{1}{2}mv_2^2$ plus the thermal energy E_{th} generated (since only <u>changes</u> in energy really matter, we can take $\Delta E_{th} = E_{th}$). Hence, setting $E_1 = E_2$,

$$mgy_1 = \frac{1}{2}mv_2^2 + E_{th}$$

The thermal energy E_{th} is equal to the <u>magnitude</u> of the friction work $F_k \Delta r$, where $\Delta r = 2.00$ m is the length of the incline. Using $y_1 = \Delta r \sin(30.0°)$, we therefore have

$$mg\Delta r \sin(30.0°) = \frac{1}{2}mv_2^2 + F_k \Delta r$$

Hence, solving for v_2,

$$v_2 = \sqrt{2\left[mg\sin(30.0°) - F_k\right]\Delta r/m}$$

$$= \sqrt{2\left[5.00\text{ kg} \times 9.80\text{ m/s}^2 \times \sin(30.0°) - 6.50\text{ N}\right](2.00\text{ m})/5.00\text{ kg}}$$

$$= 3.79\text{ m/s}$$

We have seen that, when friction forces are acting, by including the generated thermal energy in our definition of the total energy E, the latter stays constant. More generally, by introducing additional kinds of energy, it is found that the principle of conservation of energy always holds for any isolated system:

Energy cannot be created or destroyed, but only transformed from one type to another.

Going back to Eq. [8-25], where E includes only mechanical energy, let's consider the situation where the work W_{nc} is provided by an applied force, such as due to a person lifting a mass from a lower to a greater height. In that case, W_{nc} is positive and the total mechanical energy of the mass <u>increases</u>. This could come about only if the person expends some internal energy E_I to drive the muscles which do the work of lifting. That is, $W_{nc} = -\Delta E_I$, where $\Delta E_I < 0$. Again, we could then rewrite Eq. [8-25] in the form of Eq. [8-27], where the total energy E now consists of the mechanical energy of the mass and the internal energy of the person. Therefore, we find again that the total energy of the system (in this case, the mass being lifted and the person doing the lifting) is conserved. This is a universal principle, and applies to all other possible types of energy, including sound energy (Chapter 2), nuclear energy (Chapter 6), electrical energy (Chapter 15), and chemical energy (not covered in this text).

Food Energy

We mentioned at the beginning of this section that the energy which enables animals to do muscular work (what we called E_I in the previous paragraph) ultimately is derived from *food energy*. This is basically chemical energy, i.e., energy stored in chemical bonds, which is released when the food is digested inside the body.

Traditionally, food energy has been measured in units of *calories* (cal) rather than joules. One calorie is the amount of thermal energy required to raise the temperature of one gram of water by 1°C. This is equivalent to 4.186 J. A more common traditional unit for food energy is the (food) Calorie (note the capital C), equal to 10^3 cal. Nowadays, this is more often referred to as a kilocalorie (kcal).

It is also useful to introduce one other term, namely *power P*. This is defined as the time rate of change of energy, i.e.,

$$P = \frac{\Delta E}{\Delta t}$$ [8-29]

where Δt is the time interval over which the energy change ΔE occurs (this definition applies to <u>any</u> type of energy change). The SI unit of power is the joule per second (J/s), which conventionally is called a *watt* (W).[8]

EXAMPLE 8-21

A 70.0-kg person runs up the stairs, of height 20.0 m, in 12.0 s. What power is required to do this?

SOLUTION

The total energy expended by the person (neglecting changes in his speed) equals his change in gravitational potential energy ΔU,

$$\Delta U = mg\Delta y = 70.0 \, \text{kg} \times 9.80 \, \text{m/s}^2 \times 20.0 \, \text{m}$$
$$= 1.37 \times 10^4 \, \text{J}$$

The power required is then

$$P = \frac{\Delta U}{\Delta t} = \frac{1.37 \times 10^4 \, \text{J}}{12.0 \, \text{s}} = 1.14 \times 10^3 \, \text{W}$$

When resting, the average person typically uses up food energy at the rate of 100 W. This is the basic metabolic rate, required for driving processes such as breathing, blood circulation, and body heating. Any activities of the body beyond this require additional energy. For example, walking at a rate of 5 km/h requires an additional 200 W of power. However, like any other engine, the human body is not 100% efficient at converting food energy into useful mechanical energy or work. For most people, the overall efficiency is about 25%, that is, only 25% or one-quarter of the food energy released can be converted to useful work, while the remaining 75% is lost as heat.

---

8.   James Watt (1736–1819), Scottish engineer.

EXAMPLE 8-22

A 70.0-kg person requires about 240 kcal of food energy to walk 4000 m on level ground. How many more kilocalories of food energy are required to walk 4000 m along a hill inclined at 30.0° to the horizontal? Assume that 4 J of food energy are required for an output of 1 J of useful mechanical energy.

SOLUTION

In terms of energy, the difference between the level walk and the walk up the hill is due to the increase in gravitational potential energy during the latter. The vertical distance moved by the person is $\Delta y = (4000 \text{ m}) [\sin(30.0°)] = 2000 \text{ m}$. Hence the additional mechanical energy required is

$$\Delta U = mg\Delta y = 70.0 \, \text{kg} \times 9.80 \, \text{m/s}^2 \times 2000 \, \text{m}$$

$$= 1.37 \times 10^6 \, \text{J}$$

The required food energy ΔE_F is four times this, i.e.,

$$\Delta E_F = 4 \times \Delta U = 5.48 \times 10^6 \, \text{J}$$

Converting to kcal (1 kcal = 4186 J),

$$\Delta E_F = 5.48 \times 10^6 \, \text{J} \times \frac{1 \, \text{kcal}}{4186 \, \text{J}} = 1.31 \times 10^3 \, \text{kcal}$$

Exercises

8-1 What is the magnitude of the horizontal force that will stop, in 4.0 s, a 5.0-kg mass sliding on a frictionless floor if the mass has an initial speed of 12 m/s?

8-2 A sprinter of mass 78.0 kg accelerates out of the starting blocks with a horizontal acceleration of magnitude 14.0 m/s^2. How much horizontal force must the sprinter exert on the starting blocks during the start to produce this acceleration? Which object exerts the force that propels the sprinter: the blocks or the sprinter himself?

8-3 A person applies a constant horizontal force of 85.0 N to a block of ice on a smooth horizontal floor. Neglect friction. The block starts from rest and moves 11.5 m in 4.50 s.

   (a)   What is the mass of the block of ice?

   (b)   If the person stops pushing at the end of 4.50 s, how far does the block move in the next 4.50 s?

8-4 When a karate expert breaks a brick, the velocity of his hand changes from (typically) 12 m/s downward to essentially zero in a time of only 3.0×10^{-3} s.

   (a)   What is the acceleration (assumed constant) of the hand?

   (b)   If the mass of the hand is 0.66 kg, what are the magnitude and direction of the resultant force on it? What object exerts almost all of this force on the hand?

8-5 A parachutist relies on the drag force of her parachute to reduce her acceleration toward the ground. If she has a mass of 61.0 kg and her parachute supplies an upward force of 342 N, what is her actual acceleration?

8-6 At the surface of Mars, the acceleration due to gravity is $g = 3.72$ m/s^2. A person weighs 590 N at the surface of Earth.

   (a)   What is the person's mass on Earth's surface?

   (b)   What are the mass and weight of the person on the surface of Mars?

8-7 What would be the weight of an 80.0-kg person on an asteroid if the asteroid's mass is 7.00×10^{10} kg and its radius is 175 km?

8-8 A crow of mass 1.10 kg sits on a telephone wire midway between two poles 50.0 m apart. The wire, assumed weightless, sags by 0.098 m. What is the magnitude of the tension in the wire?

8-9   An albatross can glide horizontally at constant velocity for short distances near the surface of the water. If it has a mass of 8.50 kg, what is its weight and what is the vertical upthrust exerted on it by the air?

8-10 It is found during a gale that the mooring cable of a captive weather balloon is at an angle of 20.0° to the vertical, and that the tension in the cable is 2.00×10^3 N. Find the magnitude of

(a)   the horizontal force exerted by the wind, and

(b)   the buoyant upthrust experienced by the balloon. (Ignore the weight of the balloon and of the cable.)

8-11 Find the magnitude of the tension in each cord A, B, and C as shown in diagram (a) and (b) if the weight of the suspended object is w.

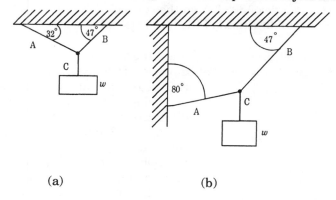

(a)   (b)

8-12 A 200-N wagon is to be pushed at constant speed up a ramp inclined at 35.9° above the horizontal. Neglect friction. If the force pushing the wagon is parallel to the incline, calculate the magnitude of this force.

8-13 A student of mass 66.0 kg stands on a bathroom scale in an elevator. As the elevator starts moving, the scale reads 550 N. What is the acceleration (magnitude and direction) of the elevator?

8-14 A block of ice is released from rest at the top of a 5.00-m long ramp and slides to the bottom in 1.68 s. What is the angle between the ramp and the horizontal? Neglect friction.

8-15 A horse pulls an old-fashioned cutter along a level icy surface at a constant speed of 15.0 m/s. The cutter and its load have a total mass of 250 kg. If the coefficient of kinetic friction between the cutter and the icy surface is 0.110, what is the magnitude of the horizontal pulling force provided by the horse?

8-16 A skier descends at constant speed a snow-covered hill inclined at 20.0° from the horizontal. What is the coefficient of kinetic friction between the skis and snow?

8-17 A skier wishes to descend a snow-covered hill with inclination angle of 15.0°. What is the maximum value that the coefficient of static friction between the skis and snow can be so that the skier is able to start moving without giving himself an initial push?

8-18 A girl on a toboggan slides down a hill inclined at 29.0° from the horizontal. Starting from rest, she travels 11.5 m in 2.50 s. What is the coefficient of kinetic friction between the toboggan and the snow?

8-19 Two freight cars are rolling toward one another on a level track. The first has mass 40.0 tonnes and is traveling at 2.00 m/s. The second has mass 125 tonnes and is traveling at 1.50 m/s. The two cars couple together after colliding. Find the final speed and the direction in which they move. Neglect friction. (1 tonne = 1000 kg)

8-20 A 1.00-kg duck is flying horizontally at 20.0 m/s when it is seized by a 0.80-kg hawk diving down at 30.0 m/s. The hawk is coming in from behind at an angle of 30.0° from the vertical. What is the velocity (magnitude and direction) of the birds just after contact?

8-21 In the Canada–Russia hockey series, the Canadian G. Howl and the Russian V. Blastov collided at centre ice. Howl was traveling across the rink at 10.0 m/s and Blastov met him at an angle (as shown in the accompanying diagram) at 15.0 m/s. Howl has a mass of 100 kg and Blastov 90.0 kg. Upon collision, they locked in an embrace of international friendship. With what speed and in what direction did this amiable pair travel after their meeting? Neglect friction with the ice.

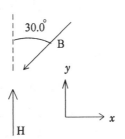

8-22 A jumper of mass 75.0 kg throws two 2.50-kg weights backwards while airborne with speeds of 5.00 m/s to increase his linear speed. If his initial speed while airborne was 10.0 m/s, calculate his final speed.

8-23 You and a friend are playing catch with a football on a field covered with a sheet of ice. Assume there is negligible friction between your feet and the ice. Your friend throws you the 0.400-kg football so that it is traveling horizontally at 13.0 m/s. Your mass is 70.0 kg, and you are initially at rest.

   (a)   If you catch the ball, with what speed do you and the ball move afterward?

   (b)   If, instead, the ball hits you and bounces off your chest, so that afterwards it is moving horizontally at 7.50 m/s in the opposite direction, what is your speed after the collision?

8-24 A girl pulls a toboggan of mass 4.90 kg up a hill inclined at 26.3° to the horizontal. The vertical height of the hill is 28.0 m. Neglecting friction between the toboggan and snow, determine how much work the girl must do on the toboggan to pull it at constant velocity up the hill.

8-25   A driver applies the brakes on a car traveling on a level road with speed v_0, so that the brakes lock and the tires slide rather than roll.

(a)   Use the work–energy theorem Eq. [8-20] to give an equation for the stopping distance of the car in terms of v_0, the acceleration of gravity g, and the coefficient of kinetic friction μ_k between the tires and the road.

(b)   The car stops in a distance of 89.5 m if $v_0 = 90.0$ km/h. What is the stopping distance if $v_0 = 60.0$ km/h? (Assume that μ_k remains the same.)

8-26   A girl constructs a simple pendulum by tying a stone to a string of length 1.85 m. She lifts the stone so that the string makes an angle of

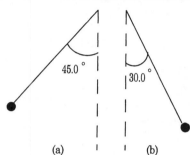

(a)   (b)

45.0° with the vertical [(a) in the diagram] and then releases it from rest. The stone swings down and then up. Use conservation of energy to find the speed of the stone when the string makes an angle of 30.0° with the vertical [(b) in the diagram].

8-27   Surplus energy from an electric power plant can be temporarily stored as gravitational potential energy by using this surplus energy to pump water from a river into a reservoir at some altitude above the level of the river. If the reservoir is 250 m above the level of the river, what mass of water must be pumped in order to store 2.00×10^{13} J of energy?

8-28   The brakes are applied to a car traveling at 87.0 km/h. The wheels lock and the car skids to a halt in 49.0 m. The magnitude of the kinetic friction force between the car's tires and the road is 7.45×10^3 N.

(a)   How much thermal energy is produced during the skid?

(b)   Use conservation of energy to determine the mass of the car.

8-29   A basketball player has a power output of 3.10×10^2 W. How long (in minutes) must she play in order to have a total energy output of 4.00×10^5 J?

8-30   How much power is produced when a 65.0-kg person performs a vertical jump of 36.0 cm height in 0.840 s?

8-31   When an insect such as a flea jumps, the energy is provided not by muscle alone, but also by an elastic protein that has been compressed like a spring. If a flea of mass 2.15×10^{-7} kg jumps vertically to a height of 67.0 mm, and 75.0% of the energy comes from elastic potential energy stored in the protein, determine the initial amount of stored elastic potential energy.

8-32 A 59-kg woman is doing push-ups. During each one, she raises the centre of mass of her body by 0.26 m. How many push-ups must she do in order to use at least 1.0 kcal of food energy (over and above the energy required for her to rest for an equal amount of time)? Neglect the energy used in lowering her body, and assume an efficiency of 25% in converting food energy to usable energy.

Problems

8-33 To perform a standing vertical jump, an athlete initially crouches down and then abruptly straightens his body, exerting a force on the ground (due to muscle contraction) in excess of his weight. By Newton's third law, the ground exerts an equal and opposite force on the athlete. The athlete's body then accelerates upward and rises above the ground. If an athlete weighing 870 N attains a jump height of 1.15 m, and the time of the jump before his feet leave the ground is 0.290 s, what is the average force he applies to the ground?

8-34 During a rescue operation, a man of mass 70.0 kg hangs by two inextensible cables A and B inclined at 60.0° to the horizontal (see diagram).

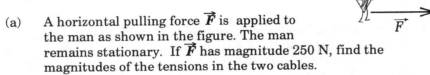

(a) A horizontal pulling force $\vec{F}$ is applied to the man as shown in the figure. The man remains stationary. If $\vec{F}$ has magnitude 250 N, find the magnitudes of the tensions in the two cables.

(b) Determine the magnitude of $\vec{F}$ for which the cable B would become slack, i.e., its tension becomes zero.

8-35 Blocks A and B in the diagram below weigh 95.0 N and 20.0 N, respectively. The coefficient of static friction between block A and the surface on which it rests is 0.32. The system is in equilibrium.

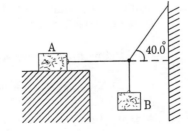

(a) Find the magnitude of the friction force exerted on block A.

(b) If everything else remains the same, what would be the maximum weight of block B for which the system could remain in equilibrium?

8-36 Three people are attempting to push a large table in the *x*-direction as shown in the diagram below. Two of them push with horizontal forces $\vec{F}_1$ and $\vec{F}_2$ of magnitudes and directions indicated in the figure. What is the magnitude and

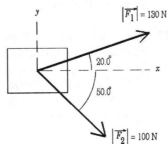

direction of the <u>smallest</u> force that the third person should exert?

8-37 A 2.00-kg brick is given an initial speed of 4.00 m/s up an inclined plane starting from a point 2.00 m from the bottom as measured along the plane. The plane makes an angle of 30.0° with the horizontal, and the coefficient of kinetic friction is 0.200.

(a)   How far up along the incline will the brick travel? Use energy methods.

(b)   After reaching its highest point, the brick will start to slide back down the incline. With what speed will it reach the bottom of the inclined plane?

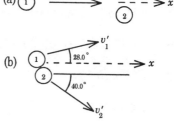

8-38 Hockey puck 1 is originally traveling at 25.0 m/s in the x-direction on a smooth icy surface [(a) in the diagram]. It strikes, slightly off-centre, another puck, 2, which was initially at rest. Puck 1 is deflected at 28.0° from its original direction of motion, while puck 2 moves off at 40.0° from this direction [(b) in the diagram]. Both pucks have the same mass.

(a)   Calculate the speed of each puck just after the collision. Neglect friction with the ice.

(b)   What fraction of the total kinetic energy is lost during the collision?

8-39 Reconsider the collision of the hockey players G. Howl and V. Blastov in Exercise 8-21, assuming that they approach each other with the same velocities as in that exercise. However, instead of locking together, now they move separate ways after the collision. Immediately after the collision, Howl is moving at 3.00 m/s opposite to his original direction of motion. Neglect friction with the ice.

(a)   What is Blastov's velocity (magnitude and direction) immediately after the collision?

(b)   Calculate the change in total kinetic energy of the players.

8-40 A donkey is used to raise buckets of water from a well. He is walking around a horizontal circle of radius 2.00 m at a speed of 3.00 km/h, turning the arm of a crank which raises the buckets of water. By this means, 100 kg of water is raised a distance of 2.00 m every second. Using energy methods, calculate the magnitude of the force exerted by the donkey on the arm of the crank.

Answers

8-1 15 N

8-2 1.09×10^3 N; the blocks

8-3 (a) 74.8 kg (b) 23.0 m

8-4 (a) 4.0×10^3 m/s^2 upward
(b) 2.6×10^3 N upward; brick

8-5 4.19 m/s^2 downward

8-6 (a) 60.2 kg (b) 60.2 kg, 224 N

8-7 1.22×10^{-8} N

8-8 1.4×10^3 N

8-9 83.3 N, 83.3 N

8-10 (a) 684 N (b) 1.88×10^3 N

8-11 (a) $T_A = 0.69\,w$, $T_B = 0.86\,w$,
$T_C = w$
(b) $T_A = 1.1\,w$, $T_B = 1.6\,w$,
$T_C = w$

8-12 117 N

8-13 1.47 m/s^2 downward

8-14 21.2°

8-15 270 N

8-16 0.364

8-17 0.268

8-18 0.125

8-19 0.652 m/s in direction of car 2

8-20 21 m/s, 33° below horizontal

8-21 3.66 m/s, 14.1° below –x-axis

8-22 11.0 m/s

8-23 (a) 0.0739 m/s (b) 0.117 m/s

8-24 1.34×10^3 J

8-25 (a) $\Delta r = \dfrac{v_0^2}{2g\mu_k}$ (b) 39.8 m

8-26 2.40 m/s

8-27 8.16×10^9 kg

8-28 (a) 3.65×10^5 J (b) 1.25×10^3 kg

8-29 21.5 min

8-30 273 W

8-31 1.06×10^{-7} J

8-32 7

8-33 2.32×10^3 N

8-34 (a) $T_A = 646$ N, $T_B = 146$ N
(b) 396 N

8-35 (a) 23.8 N   (b) 26 N

8-36 32.1 N in y-direction

8-37 (a) 1.21 m   (b) 4.53 m/s

8-38 (a) $v_1' = 17.3$ m/s,
$v_2' = 12.7$ m/s
(b) 0.263

8-39 (a) 7.64 m/s at 11.0° above
–x-axis
(b) -1.20×10^4 J

8-40 2.35×10^3 N

9 Mechanics of Biological Systems: Rotational Motion

9.1 Introduction

In the previous two chapters, we have analyzed translational (or linear) motion of objects. This is the only type of motion that occurs for an object which can be considered a point particle. In addition to translational motion, extended objects (or non-point particles) can also undergo *rotational motion*. In purely rotational motion, different points of the body move in concentric circles about a stationary axis through the body. The most general type of motion of an extended body is a combination of translational motion of the body's *centre of mass* (defined later in Box 9-1) and of rotational motion about an axis through the centre of mass.

There are many common examples of rotational motion. A bicycle or automobile wheel rotates about its axis, as do the blades of a circular saw and the rim of a ferris wheel. The human body performs (mainly) rotational motion in throwing a ball or swinging a hammer or lifting a dumbbell during muscle-building exercises. During walking or running, the swinging of the limbs is (mainly) rotational motion.

Due to its similarity to rotational motion of extended objects, we will first discuss the topic of *uniform circular motion*. This describes the motion of any object (point particle or otherwise) which moves in a circular path at constant speed.

9.2 Uniform Circular Motion

An example of an object undergoing uniform circular motion is a ball attached to the end of a string which you are twirling about your head in a horizontal circle at constant speed. Other examples of uniform circular motion are a car rounding a curve of constant radius at constant speed and a satellite moving in a circular orbit. Although the <u>speed</u> of such an object is constant, the <u>direction</u> of its velocity vector is continually changing and therefore the object has a non-zero acceleration, which is called *centripetal acceleration* and is denoted $\vec{a}_c$.

Figure 9-1 shows the path of a particle undergoing uniform circular motion, indicating the position vector $\vec{r}$ and instantaneous velocity vector $\vec{v}$ at several regularly spaced intervals along this path. Notice that at each point, the position $\vec{r}$ and velocity $\vec{v}$ are <u>perpendicular</u> to each other and the velocity is tangent to the circle. This last fact is consistent with the discussion in Chapter 7 that, for an object moving along a curved path, the instantaneous velocity $\vec{v}$ is tangent to the curve at every point along the path.

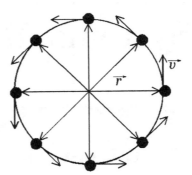

Figure 9-1

Position $\vec{r}$ and velocity $\vec{v}$ of an object (●) at several intervals during uniform circular motion.

We now present a trigonometric argument for deriving the magnitude and direction of the centripetal acceleration $\vec{a}_c$, starting from the definition of acceleration (Eq. [7-18])

$$\vec{a}_c = \lim_{\Delta t \to 0} \frac{\Delta \vec{v}}{\Delta t} \qquad \textbf{[9-1]}$$

Figure 9-2a shows a particle in uniform circular motion as it moves between points P_1 and P_2 on the circumference of a circle of radius r in a small time interval Δt. The position vector of the particle changes from $\vec{r}_1$ to $\vec{r}_2$, with $\vec{r}_2 - \vec{r}_1 = \Delta \vec{r}$. The corresponding velocity vectors are denoted $\vec{v}_1$ and $\vec{v}_2$ and their difference $\vec{v}_2 - \vec{v}_1 = \Delta \vec{v}$. To evaluate the latter, we redraw the vectors $\vec{v}_2$ and $-\vec{v}_1$ in Fig. 9-2b and place them "head to tail." This shows that $\Delta \vec{v}$ points approximately toward the centre of the circle, a condition which becomes exact as $\Delta t \to 0$. Since $\vec{a}_c$ is in the same direction as $\Delta \vec{v}$ in this limit, according to Eq. [9-1], we conclude that the direction of $\vec{a}_c$ is <u>toward the centre of the circle</u> at every point along the object's path.

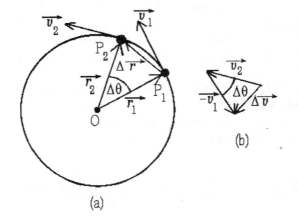

(a)

(b)

Figure 9-2

a) Uniform circular motion; b) Diagram showing that, as $\Delta \theta$ approaches zero, $\Delta \vec{v} = \vec{v}_2 - \vec{v}_1$ is directed from P_1 toward the centre O of the circle.

Now that the direction of $\vec{a}_c$ has been established, its magnitude can be determined. First we note that the angles labeled $\Delta \theta$ in Fig. 9-2a,b are equal. This follows from the fact that $\vec{v}_1 \perp \vec{r}_1$ and $\vec{v}_2 \perp \vec{r}_2$. Since

$|\vec{r_1}| = |\vec{r_2}| \equiv r$ and $|\vec{v_1}| = |\vec{v_2}| \equiv v$, the triangle OP_1P_2 in Fig. 9-2a and that drawn in Fig. 9-2b are similar. Since ratios of corresponding sides are equal,

$$\frac{|\vec{\Delta v}|}{v} = \frac{|\vec{\Delta r}|}{r}$$

and therefore

$$\frac{|\vec{\Delta v}|}{\Delta t} = \frac{v}{r} \frac{|\vec{\Delta r}|}{\Delta t}$$

Taking the limit $\Delta t \to 0$, we obtain

$$a_c = \lim_{\Delta t \to 0} \frac{|\vec{\Delta v}|}{\Delta t} = \frac{v}{r} \lim_{\Delta t \to 0} \frac{|\vec{\Delta r}|}{\Delta t} \qquad \text{[9-2]}$$

But the speed of the moving particle is given by

$$v = \lim_{\Delta t \to 0} \frac{|\vec{\Delta r}|}{\Delta t} \qquad \text{[9-3]}$$

and hence Eq. [9-2] becomes

$$a_c = \frac{v^2}{r} \qquad \text{[9-4]}$$

In summary, an object moving at constant speed v along a circular path of radius r has a centripetal acceleration $\vec{a_c}$, whose direction at each instant is toward the centre of the circle and whose magnitude is given by $a_c = v^2/r$. This result can be generalized to an object moving with non-constant speed along a circular path. The centripetal acceleration a_c as given above still describes the component of the object's acceleration in the direction toward the centre of the circle. However, if the speed changes, then the object also has a non-zero component of acceleration in the direction tangent to the circle, i.e., parallel to its motion. We shall not discuss this generalization any further here, but shall come back to related ideas when describing rotational motion in Section 9.4.

Equations [9-3] to [9-4] are often expressed in terms of other quantities—in particular, the frequency of the circular motion or the corresponding period T. The frequency f denotes the number of revolutions made per unit time, while the period is the time for one revolution. These are related as in Eq. [1-11],

$$T = 1/f \qquad \text{[9-5]}$$

In a time T, the object travels a distance equal to the circumference $2\pi r$ of the circle, hence the speed v is given by

$$v = 2\pi r/T = 2\pi r f = r\omega \qquad\qquad \textbf{[9-6]}$$

where we have defined the angular frequency $\omega = 2\pi f$.

Therefore, Eq. [9-4] can be rewritten in the equivalent forms

$$a_c = 4\pi^2 r f^2 = 4\pi^2 r/T^2 = \omega^2 r \qquad\qquad \textbf{[9-7]}$$

EXAMPLE 9-1

A penny is placed flat on a phonograph turntable at a distance of 10.0 cm from the centre. If the turntable is playing at 33⅓ rpm (revolutions per minute), what is the speed of the penny and what is the magnitude of its centripetal acceleration?

SOLUTION

We <u>assume</u> that the penny does not slide off the rotating turntable, but remains at the fixed distance $r = 10.0$ cm = 0.100 m from the centre. We are given that the frequency of revolution is $f = 33⅓$ rpm. Although not essential, it is conventional to express f in the number of cycles or revolutions per second. In these units,

$$f = 33\,⅓\,\frac{\text{rev}}{\text{min}} \times \frac{1\,\text{min}}{60\,\text{s}} = 0.556\,\frac{\text{rev}}{\text{s}}$$

(This is often written simply as 0.556 s^{-1}, since a "revolution" or "cycle" is a pure dimensionless number and has no physical unit.) Then from Eq. [9-6],

$$v = 2\pi r f = 2\pi \times (0.100\ \text{m}) \times (0.556\ \text{s}^{-1})$$
$$= 0.349\,\text{m/s}$$

while from Eq. [9-4],

$$a_c = v^2/r = (0.349\ \text{m/s})^2 / 0.100\ \text{m}$$
$$= 1.22\,\text{m/s}$$

Centripetal Force

Like all other motions, uniform circular motion is governed by Newton's second law (Chapter 8). Thus, the centripetal acceleration $\vec{a}_c$ of a particle of mass m must be the consequence of a net force $\Sigma \vec{F} = \vec{F}_c$ acting toward the centre of the circle, with constant magnitude $F_c = ma_c$. We call this the *centripetal force*, which generally will be the resultant of several basic forces such as we've already learned about. In the example of an object on a string which is whirled in a circle around your head, the centripetal force $\vec{F}_c$ is provided by the tension of the string, which constantly pulls the object toward the centre of the circle. If the string breaks, it no longer exerts a force and the object will fly off at a tangent to the circle (as required by Newton's first law). In the case of a satellite in circular orbit about a planet, the force "pulling" on the satellite is the gravitational attraction between the satellite and planet. An analogous explanation applies to planets which are themselves in orbit about a star.

What is the origin of the centripetal force acting on the penny on a turntable in Example 9-1? The answer is that the centripetal force is due to friction between the coin and the turntable surface. If no friction were acting to inhibit relative motion of the coin and surface, then the penny would slide off the surface in a tangential direction when the turntable rotates. Similarly, friction between a car's tires and the road is the main source of the centripetal force on the car when it is rounding a curve. In the absence of such friction, the car would tend to keep moving in a straight line, skidding out of the circular path, a not uncommon occurrence![1]

As the preceding discussion indicates, there are many possible physical sources of centripetal force. Whatever its origin, the magnitude of the centripetal force on an object is related to the object's speed v and the radius r of the curved path by

$$F_c = mv^2/r \qquad\qquad \textbf{[9-8]}$$

Other types of forces which can act as a centripetal force are illustrated in the following examples.

---

1. Highway curves are often banked at an angle to prevent skidding even on slippery roads: in this case, part of the centripetal force is due to the normal force of the road on the car.

EXAMPLE 9-2

A bird of mass 0.500 kg dives and then pulls out of the dive by flying in a portion of a vertical circle of radius 15.0 m as shown in Fig. 9-3. At the bottom of the circular arc, the bird's speed is 25.0 m/s.

(a) What is the magnitude of the net upward force acting on the bird?

(b) What is the magnitude of the upward lift exerted on the bird by the air?

Figure 9-3

Example 9-2: A bird pulling out of a dive.

15.0 m $\overrightarrow{F}_{\text{air}}$

$v = 25.0$ m/s

$m\overrightarrow{g}$

SOLUTION

(a) At the bottom of the circular arc, the centre of the circle is upward from the bird. The net upward force is then equal to the centripetal force, hence its magnitude is given by Eq. [9-8],

$$F_c = \frac{mv^2}{r} = \frac{0.500 \text{ kg} \times (25.0 \text{ m/s})^2}{15.0 \text{ m}} = 20.8 \text{ N}$$

(b) We have to relate the net force $\overrightarrow{F}_c$ to the actual physical forces acting on the bird. The latter are the downward force of gravity $m\overrightarrow{g}$ and the upward lift exerted by the air, $\overrightarrow{F}_{\text{air}}$. (There may also be a horizontal drag force exerted by the air, but this does not affect the vertical resolution of forces.) Hence, considering components of the forces in the vertical direction,

$$F_c = \sum F = F_{\text{air}} - mg$$

Thus, solving for F_{air} and using the result of part (a),

$$F_{\text{air}} = F_c + mg = 20.8 \text{ N} + 0.500 \text{ kg} \times 9.80 \text{ m/s}^2$$
$$= 25.7 \text{ N}$$

EXAMPLE 9-3

A certain amusement park ride consists, as shown in Fig. 9-4a, of compartments which rotate at fixed distance R around a central vertical shaft. At a sufficiently high rotation frequency f, a rider will be pinned against the wall of the compartment by friction and will not fall out even if the floor of the compartment is opened. Find the minimum frequency f_{min} for this to happen, if the coefficient of static friction between the compartment wall and the rider is μ_s.

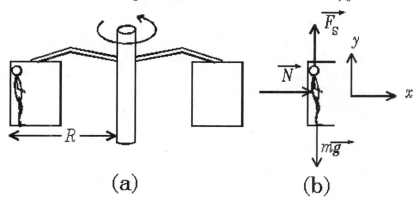

(a)   (b)

Figure 9-4

Example 9-3: Friction pinning a rider against the wall in an amusement park ride.

SOLUTION

The free-body diagram for a rider is shown in Fig. 9-4b where y denotes the vertical direction and x is the direction from the compartment to the axis of rotation along the central shaft. The real forces on the person are the weight $m\vec{g}$, the friction force $\vec{F}_s$ between the person and the wall, and the normal force $\vec{N}$ of the compartment wall. Note that the latter acts in the x-direction, i.e., toward the axis of rotation, and is the only force acting along x. Hence, $\vec{N}$ can be identified as the centripetal force, and the acceleration along x equals the centripetal acceleration $\vec{a}_c$. When the rider is pinned to the wall, there is no motion in the vertical direction, i.e., the component of the net force in this direction is zero. Therefore Newton's laws are

$$\sum F_y = 0, \text{ therefore } F_s - mg = 0$$
$$\sum F_x = ma_c, \text{ therefore } N = ma_c = m(4\pi^2 Rf^2)$$

Combining these relations with that for static friction,

$$F_s \le \mu_s N$$

then gives

$$mg \leq \mu_s m(4\pi^2 R f^2)$$

Cancelling the mass m of the rider, we find

$$g \leq \mu_s(4\pi^2 R f^2)$$

or

$$f \geq \frac{1}{2\pi}\sqrt{\frac{g}{\mu_s R}}$$

The <u>minimum</u> angular speed for which the rider is supported without falling out of the compartment is given by the lower bound of this relation, $f_{min} = (1/2\pi)\sqrt{g/(\mu_s R)}$. For instance, if $R = 5.0$ m and $\mu_s = 0.20$, this yields $f_{min} = 0.50$ s^{-1}, corresponding to one revolution every two seconds.

The work done by a force was discussed in Chapter 8. Although there we considered only the case of straight-line motion under a constant force, the relation between force and work can be generalized (using calculus techniques) to motion along curved paths under non-constant forces. One idea carries through straightforwardly to uniform circular motion: since the centripetal force is, at all times, directed toward the centre of the circle and hence is always perpendicular to the instantaneous displacement of an object, the work done by the centripetal force is zero.

Centrifugal "Force" and Centrifuges

Suppose you are the rider in Example 9-3. As required by Newton's third law, the normal force $\vec{N}$ exerted by the compartment wall against your body is accompanied by the reaction force of your body against the wall. Subjectively, you would feel as if you are being pressed against the wall, by some force which is trying to throw you outward away from the centre. A related sensation is encountered when you go around a corner rapidly in a car; you seem to be thrown away from the centre of curvature, as if acted on by a *centrifugal* ("fleeing from the centre") force. However, it is important to recognize that <u>there is no such centrifugal force</u>! In the absence of any forces, an object's centre of mass will always tend to move in a straight line at constant speed, due to Newton's first law. Therefore, an external centripetal force is required to move the object along a circular path. In the case of a car going around a corner, friction between you and the seat provides the centripetal force acting on the lower part of your body, but the upper part of your body tries to follow a straight, tangential path and thus seems to be "pushed" outward until it is restrained by seat belts or by the inside wall of the car.

Figure 9-5

Operation of a centrifuge, showing how a particle initially at A will continue to move in a straight line until it reaches the bottom of the test tube at B.

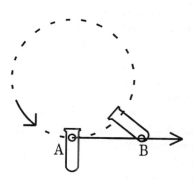

A practical application of these ideas is provided by a *centrifuge*, which is a device used to separate different substances in a solution or suspension according to their densities. The samples being analyzed are contained in test tubes which are rotated very rapidly in a horizontal circle, as shown in Fig. 9-5. A suspended particle in the solution which is initially at point A near the top of the test tube will tend to move in a straight line while the tube is being rotated, until it comes into contact with the bottom of the test tube at point B. Thereafter, it will be constrained by the test-tube wall (mainly through the normal force of the wall, similar to Example 9-3) to keep moving in the circular path followed by the whole tube, and thus it stays at the bottom of the tube.

If you were an observer who could remain stationary relative to the tube while it is rotating, you would observe the particle accelerating outward from the centre of the circle until it reaches the bottom, and thus you might imagine that it is being pushed outward by a fictitious *centrifugal force*. The apparent need to invent this fictitious force is due to the fact that such a rotating observer is accelerating relative to Earth, and therefore is in a *non-inertial frame of reference*. In an *inertial* frame of reference, i.e., to someone outside the centrifuge, there is no need to invoke the idea of a centrifugal force.

A more detailed discussion of the real forces acting on a particle in a centrifuge, in particular due to the effects of the surrounding liquid, will be given in Chapter 13.

9.3 **Rotational Equilibrium: Torque**

In dealing with rotational motion of extended objects, we shall make an idealization which in most practical cases is a good approximation. This is to assume the object is a *rigid body*, one which does not change in size or shape. If we consider different points in a rigid body, then the relative distances between all such points remain constant, independent of the object's motion.

Although the general subject of this section is rotational <u>motion</u>, in this section we shall analyze the conditions required for an object to be in *rotational equilibrium*, that is, to be in a state either of rest or of motion with constant angular velocity. We recall that, for an object to be in translational equilibrium, the vector sum of the forces acting on it must be zero (see Chapter 8). For an extended object, an additional requirement must be satisfied to ensure that it has no tendency to rotate. This is the condition that the sum of the *torques* exerted by the forces about any point in the body must be zero. Let's illustrate what this means, and the definition of "torque," with a simple example.

Figure 9-6a shows two children, of masses m_A and m_B, sitting on a teeter-totter. The teeter-totter itself is hinged to the *pivot* (or fulcrum), which provides the axis about which it is free to rotate. The distances of the children to the pivot are denoted d_A and d_B, respectively. Our "rigid body" system in this case consists of the two children together with the teeter-totter beam.

The forces acting on this body are indicated in Fig. 9-6b. For the moment, we will neglect the weight of the beam itself, hence the forces are the downward weights having magnitude $m_A g \equiv F_A$ and $m_B g \equiv F_B$ and the upward force of magnitude R exerted by the pivot on the beam. To be in *translational equilibrium*, the total weight $(m_A + m_B)\vec{g}$ must be balanced by the pivot force $\vec{R}$. To maintain *rotational equilibrium*, another condition is required,

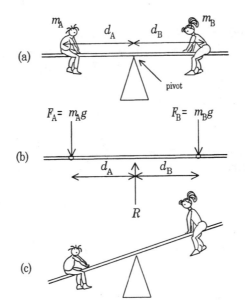

Figure 9-6

Two children on a teeter-totter, illustrating an application of torque.

otherwise one side of the teeter-totter will swing downward and the other side will swing upward, as shown in Fig. 9-6c. You may recall from childhood experience that to maintain rotational balance, the heavier child has to sit closer to the pivot axis than does the lighter child; i.e., if $m_A > m_B$, then $d_A < d_B$. The exact relationship between these quantities, as proven by experiment, is

$$F_A d_A = F_B d_B \qquad \text{[9-9]}$$

The relation Eq. [9-9] can be rewritten as $d_A/d_B = F_B/F_A = m_B/m_A$, hence $d_A < d_B$ if $m_A > m_B$, as expected.

Although we stated that Eq. [9-9] is "proven by experiment," it can also be derived from Newton's laws together with the constraint that all parts of the system are connected as a *rigid body* and thus always maintain the same relative distances. However, to show this requires a somewhat lengthy mathematical excursion which we shall forego.

Let us now rewrite Eq. [9-9] in a form which applies more generally to other rotational equilibrium problems. We define the torque τ produced by a force of magnitude F as the product of F and the *perpendicular distance* $d_\perp$ from the axis of rotation to the *line of action* of $\vec{F}$; that is

$$\tau = \pm F d_\perp \qquad\qquad [9\text{-}10]$$

The definition of torque also requires specifying the sign in Eq. [9-10]. We shall use the following convention:

τ is positive if the force $\vec{F}$ by itself would produce counterclockwise rotation about the rotation axis. Otherwise, τ is negative.

With this convention, the torque τ_A due to the force $\vec{F}_A \equiv m_A\,\vec{g}$ in Fig. 9-6b is

$$\tau_A = F_A\, d_A$$

while the torque due to $\vec{F}_B \equiv m_B\,\vec{g}$ is

$$\tau_B = - F_B\, d_B$$

Now one notes that Eq. [9-9] can be rewritten as

$$\tau_A + \tau_B = 0 \qquad\qquad [9\text{-}11]$$

The generalization of this relation for any body to be in rotational equilibrium is that the algebraic sum of all the torques acting on the body must be zero, i.e.,

$$\sum \tau = 0 \qquad\qquad [9\text{-}12]$$

From the definition, Eq. [9-10], we see that torque has the dimension of force times distance, hence its SI unit[2] is the newton-metre (N·m).

---

2.   Unfortunately, this unit is the same as that for "work," but a torque is <u>not</u> work and so N·m must <u>not</u> be replaced by J in this case.

It is important to keep in mind that the torque produced by a force depends not only on the magnitude of the force but also on the distance from the axis about which it acts. The perpendicular distance $d_\perp$ is called the *moment arm*, while the torque itself is sometimes called the *moment* of the force. From the example above, we can conclude that torque measures the effectiveness of a force in causing a body to rotate. The fact that the effectiveness increases (for a fixed magnitude of the force) when the moment arm $d_\perp$ increases is illustrated by trying to close a heavy door. It is a lot easier to move the door if you push it near the door knob than if you try to push it near the hinges. In the particular case that $d_\perp = 0$ (e.g., you try to push the door right at the hinges), then Eq. [9-10] gives $\tau = 0$. That is, a force cannot exert a torque, and hence has no effect on rotation, if its line of action passes through the axis of rotation. This is the reason we are able to neglect the pivot force $\vec{R}$ in solving the rotational equilibrium problem in Fig. 9-6.

The moment arm $d_\perp$ is the perpendicular distance from the axis of rotation to the line of action of the force. The moment arms in Figs. 9-6a and 9-6b are immediately seen by inspection to be the distances d_A and d_B. But what about the situation shown in Fig. 9-6c? This case is redrawn in Fig. 9-7. Now the perpendicular distances of the lines of action of $\vec{F}_A$ and $\vec{F}_B$ to the pivot axis are the distances denoted d_A' and d_B', which are no longer equal to the distances d_A, d_B from the point of application of each force to the pivot. Thus the torques due to $\vec{F}_A$ and $\vec{F}_B$ are now given by (remembering the sign convention)

$$\tau_A = F_A d_A'$$
$$\tau_B = -F_B d_B'$$

This illustrates the more general type of situation. Usually the appropriate perpendicular distances $d_\perp$ (e.g., d_A', d_B') can be determined by elementary geometric arguments. In terms of the angles ϕ_A and ϕ_B illustrated in Fig. 9-7 and the original distances d_A, d_B, we find

$$\tau_A = F_A d_A \sin(\phi_A)$$
$$\tau_B = -F_B d_B \sin(\phi_B)$$

[9-13]

For the example shown in Fig. 9-7, it happens that $\phi_B = \phi_A$ and thus $\sin(\phi_B) = \sin(\phi_A)$. Substituting the expressions in Eq. [9-13] for τ_A and τ_B into the torque balance relation Eq. [9-11], the angle-dependent terms cancel from both sides and the latter relation becomes identical to Eq. [9-9]. In other examples, such a cancellation of the angles won't necessarily occur.

Figure 9-7

Calculation of the moment arms for the teeter-totter case in Fig. 9-6(c).

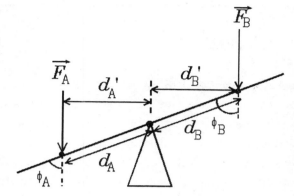

We have introduced the fundamental condition for rotational equilibrium, namely that the sum of the torques acting on a body equals zero (Eq. [9-12]) and have described two alternative but equivalent methods for calculating the torque due to a force $\vec{F}$, given by Eqs. [9-10] and [9-13]. The general conditions for equilibrium of a body also must include, of course, that for translational equilibrium, $\sum \vec{F} = 0$. In terms of components,

$$\sum F_x = 0 \text{ and } \sum F_y = 0$$

(As previously, we shall restrict attention to problems involving at most two dimensions.) For instance, in the example of Fig. 9-6, the latter conditions tell us that the upward reaction force $\vec{R}$ of the pivot on the teeter-totter has magnitude $R = (m_A + m_B)g$.

An important aspect of problem solving in the topic of translational and rotational equilibrium is the fact that <u>you can choose any point in space as the pivot point through which the axis of rotation is located</u>, in order to calculate the torques acting on a body. The physical basis for this property is simply that, if the object is not rotating, then it is not rotating about <u>any</u> axis. One other aspect in this connection has to be emphasized. Although the choice of pivot point is arbitrary, once a point is chosen then the same point must be used to calculate <u>all</u> the torques in any given problem. Usually, a judicious choice for the pivot point results in significant simplification of the problem, as will be illustrated in the following examples.

One final aspect to be dealt with in discussing the equilibrium properties of an extended body is how to account for the weight of the body (see Box 9-1).

Box 9-1   Centre of Mass

In the teeter-totter example discussed so far, we neglected the weight of the teeter-totter beam itself and took account only of the children's weights, assuming the latter were concentrated at the points where they sit. If the weight of the beam wasn't negligible, how could we include it? Because it is an *extended* object, its weight doesn't act at a single point but is distributed over the entire body. However, without going into details, one can show that the torque due to the weight of an extended body can be evaluated by assuming that the entire downward force of gravity is concentrated at a single point, its *centre of mass* (cm). The torque is then given by

$$\tau_{\text{grav}} = \pm mg \, d_{\perp}^{(\text{cm})}$$

where m is the total mass of the object and $d_{\perp}^{(\text{cm})}$ is the moment arm associated with the centre of mass. (The choice of sign in this equation follows the usual convention for calculating torque.)

(There are actually two distinct points: "centre of mass" and "centre of gravity." Strictly, the centre of gravity should be used for calculating the torque due to gravity, but in a constant and uniform gravitational field, the two points coincide.)

For a body of arbitrary shape and mass distribution, determining the location of the centre of mass can be a rather laborious process. For objects with a uniform mass distribution and a high degree of symmetry, having an obvious *geometric centre*, the centre of mass is located at the centre. This applies to a uniform sphere, cube, cylinder, rectangular plate, or circular disk. For the problems in this text, the location of the centre of mass will be stated unless it is obvious because of symmetry.

We shall now describe several examples illustrating the principles of this section and their application to biomechanics. As has been true in much of our previous discussions, the best way to begin solving a mechanical equilibrium problem is to draw a free-body diagram.

EXAMPLE 9-4

Figure 9-8 shows a human arm lifting a weight such as a dumbbell, where the forearm makes a 90° angle with the upper arm. The forces acting on the forearm are the downward weight $\vec{w}$, the tension $\vec{T}$ in the biceps muscle (assumed to be acting vertically upward), and the support reaction force of the upper arm acting at the elbow joint, $\vec{E}$. Suppose that $w = 150$ N. What are the magnitude T of the muscle force and the magnitude and direction of the elbow force $\vec{E}$ to maintain the forearm in equilibrium? The distances between the elbow joint and the point of application of $\vec{T}$, and between the elbow and the weight $\vec{w}$, are indicated in Fig. 9-8. Neglect the weight of the forearm itself.

Figure 9-8

Example 9-4: Maintaining a forearm in equilibrium.

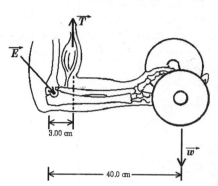

SOLUTION

The free-body diagram of this system is shown in Fig. 9-9, where the relevant distances are labelled d_T and d_w (= 3.00 cm and 40.0 cm, respectively). In this problem, as in most examples to follow, there are three unknowns, namely T and the magnitude and direction of $\vec{E}$. The latter could alternatively be represented by its horizontal (x) and vertical (y) components. In Fig. 9-8, we have made a guess as to the direction of $\vec{E}$. There is no need to worry whether this guess is qualitatively correct, as the solution of the equilibrium equations will tell us the actual direction.

Figure 9-9

Example 9-4: Free-body diagram.

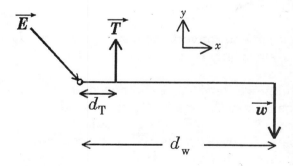

The equilibrium conditions are $\sum \tau = 0$ and $\sum F_x = 0$, $\sum F_y = 0$. In principle, we could examine these in any order, but it usually turns out to be simplest to first consider the torque condition $\sum \tau = 0$, with an appropriate choice of the pivot point. While we have said that the choice of pivot point is arbitrary, it is usually possible to choose that point so that the lines of action of one or more of the unknown forces pass through it and hence give no contribution to the total torque. The best choice of pivot point in the present example is at the elbow joint, since this choice eliminates the contribution of the elbow force $\vec{E}$, which really involves two unknowns. The torque condition Eq. [9-12] then becomes

$$\tau_T + \tau_w = 0$$

or

$$Td_T - wd_w = 0$$

where we have used the fact that the moment arms associated with $\vec{T}$ and $\vec{w}$ are just the indicated distances d_T and d_w, and have also inserted the appropriate signs of their torques. (Note that $\vec{T}$ acting alone would produce counterclockwise rotation of the forearm about the elbow joint, hence its torque is positive, while $\vec{w}$ alone would cause clockwise rotation and hence has a negative torque.) The last equation can now be solved for T,

$$T = \frac{d_w}{d_T}w = \left(\frac{40.0 \text{ cm}}{3.00 \text{ cm}}\right)(150 \text{ N}) = 2.00{\times}10^3 \text{ N}$$

(Note that we did not have to convert the distances d_w and d_T to SI units of metres, since only their ratio is involved.)

Having found the unknown force T, we can now determine the remaining unknowns E_x and E_y from the translational equilibrium conditions $\sum F_x = 0$, $\sum F_y = 0$. It should immediately be apparent, however, that $\vec{E}$ has no component in the horizontal (x) direction, since neither of the other two forces $\vec{T}$ and $\vec{w}$ have components in this direction. Hence $\vec{E}$ acts only in the vertical or y-direction, rather than at a non-zero angle to the vertical direction as assumed in Figs. 9-8 and 9-9. The value of the component E_y is obtained from the condition

$$\sum F_y = T - w + E_y = 0$$

Therefore,

$$E_y = w - T = 150\,\text{N} - 2.00 \times 10^3\,\text{N} = -1.85 \times 10^3\,\text{N}$$

The negative sign in this answer for E_y means that $\vec{E}$ points vertically downward.

In the previous example, it turned out that all of the forces acted along parallel lines, so that only one force balance condition, $\sum F_y = 0$, had to be considered. This won't always be the case. A generalization of this example is given in Exercise 9-12 at the end of this chapter, where the upper arm is thrust forward so that it no longer makes a right angle with the forearm. In this case, one must make use of both conditions $\sum F_x = 0$ and $\sum F_y = 0$ to find the force $\vec{E}$ exerted by the upper arm at the elbow. We shall describe analogous examples in the following.

EXAMPLE 9-5

Figure 9-10

Example 9-5: Beam supported by a cable.

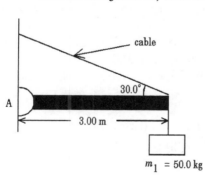

A uniform horizontal beam of length 3.00 m is attached to a wall at the joint A, and is also supported by a cable, as shown in Fig. 9-10. A mass of 50.0 kg is suspended from the end of the beam. If the mass of the beam is 10.0 kg, find the magnitude of the tension in the cable and the force (magnitude and direction) exerted on the beam by the joint A.

SOLUTION

This is a slightly more complicated problem than the preceding one. A properly labeled free-body diagram is of paramount importance, and this is shown in Fig. 9-11. The forces acting on the beam are the tension $\vec{T}$ in the cable, the weight $\vec{w}_1 = m_1\vec{g}$ of the suspended mass and the weight $\vec{w}_2 = m_2\vec{g}$ of the beam itself, and finally the supporting force $\vec{F}_A$ of the joint. As in the previous example, both the magnitude and direction of $\vec{F}_A$ are initially unknown, so we draw it at some assumed angle θ_A to the horizontal. However, as will be pointed out at the end, it is possible to deduce some information about the direction of $\vec{F}_A$ without detailed calculation.

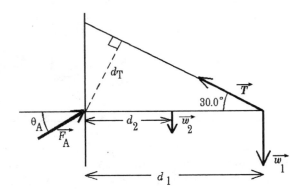

Figure 9-11

Example 9-5: Free-body diagram.

As in the previous example, let us first apply the torque condition $\sum \tau = 0$. Again, the pivot to use is arbitrary, but it is most convenient to take torques about the "natural" joint at A, since then the force $\vec{F}_A$ has no contribution:

$$\sum \tau = T d_T - d_1 w_1 - d_2 w_2 = 0$$

where d_T, d_1, d_2 are the moment arms associated with $\vec{T}$, $\vec{w}_1$, and $\vec{w}_2$, respectively, as shown in Fig. 9-11. The distance d_1 is clearly the length of the beam, $d_1 = 3.00$ m. Assuming the mass of the beam is uniformly distributed, its centre of mass is located at its midpoint, hence $d_2 = 1.50$ m. The moment arm d_T for the cable tension, the perpendicular distance from A to the line of action of $\vec{T}$, is indicated in the diagram and is easily found from geometry to be $d_T = d_1 \sin(30.0°)$. Hence we can solve the previous equation for the magnitude of the tension T,

$$T = (d_1 w_1 + d_2 w_2) / d_T$$

$$= \frac{(3.00\,\text{m} \times 50.0\,\text{kg} + 1.50\,\text{m} \times 10.0\,\text{kg})9.80\,\text{m/s}^2}{3.00\,\text{m} \times \sin(30.0°)}$$

$$= 1078\ \text{N}$$

$$= 1.08 \times 10^3\,\text{N} \text{ (to three significant digits)}$$

Now that T has been determined, we can use the remaining translational equilibrium conditions, $\sum F_x = \sum F_y = 0$, to find the x- and y-components of the supporting force $\vec{F}_A$, where the x- and y-directions are horizontal to the right and vertically upward, respectively.

$$\sum F_x = F_{A,x} + T_x = 0$$
$$\sum F_y = F_{A,y} + T_y - w_1 - w_2 = 0$$

The *x*- and *y*-components of $\vec{T}$ are easily deduced to be $T_x = -T\cos(30.0°)$ and $T_y = T\sin(30.0)$, hence one finds

$$F_{A,x} = -T_x = T\cos(30.0°) = 1078[\cos(30.0°)] = 934 \text{ N}$$

$$F_{A,y} = w_1 + w_2 - T_y = 60.0\,\text{kg} \times 9.80\,\text{m/s}^2 - 1078\,\text{N}[\sin(30.0°)] = 49.0 \text{ N}$$

From these, the magnitude and direction of $\vec{F}_A$ are found by familiar methods:

$$F_A = \sqrt{F_{A,x}^2 + F_{A,y}^2} = 935\,\text{N} \text{ and } \theta_A = 3.00°,$$

where θ_A is shown in Fig. 9-11.

The fact that both $F_{A,x}$ and $F_{A,y}$ turn out to be positive is consistent with the direction initially assumed for $\vec{F}_A$. However, as mentioned earlier, this guess has no effect on the solution of the problem, and is shown only for illustrative purposes. The fact that $\vec{F}_A$ is pointing to the right, i.e., that $F_{A,x} > 0$, could have been deduced immediately from the FBD since it must offset the one other force having a horizontal component, namely the tension $\vec{T}$, whose horizontal component is toward the left. Since there are three other forces with vertical components, namely w_1, w_2, and T_y, the sign of $F_{A,y}$ is not immediately obvious by inspection alone. This example is similar to some biomechanical situations that occur (see Box 9-2).

Box 9-2 Forces Supporting the Head of a Cow

A biomechanical problem closely related to Example 9-5 concerns the forces supporting the neck of a quadrupedal animal such as a cow (see accompanying figure). The spinal column is analogous to the beam and supports the head of the animal, which is analogous to the suspended weight $\vec{w}_1$. The so-called *nuchal ligament* (1) plays a role similar to that of the cable, in supporting the spinal column and head. Finally, the spinal column (3) is attached to the neck plate (2), analogous to the wall joint A in Example 9-5. The analog of the joint force $\vec{F}_A$ is a (nearly horizontal) *compressive* force exerted by the neck plate on the spinal column. Analyzing the forces $\vec{T}$ and $\vec{F}_A$ is relevant to determining the properties, e.g., strengths, of the ligament and spinal column that are required for the animal to function and thrive.

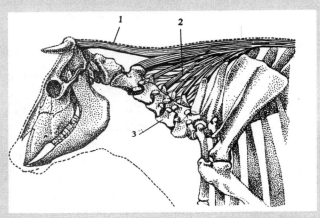

The nuchal ligament (1) and neck plate (2) support the spinal column (3) and head of a cow.

Related considerations are important in designing devices such as braces, splints and traction apparatus used to correct a variety of orthopedic problems in humans.

We'll give one more example illustrating how to handle static equilibrium problems.

EXAMPLE 9-6

A woman is bending over with her spine at an angle of 35.0° from the horizontal to lift a container weighing 260 N. Her upper body weighs 440 N and has a centre of mass C located 0.302 m from the vertebral joint denoted O, as shown in Fig. 9-12. The perpendicular moment arm from the container's weight to O is 0.380 m. Suppose that the muscles in her back which maintain the posture are a perpendicular distance of 1.50 cm from the joint, and act in a direction parallel to the line OC. Calculate the magnitude F_m of the force exerted by the

Figure 9-12

Example 9-6: Forces exerted on a person bending over.

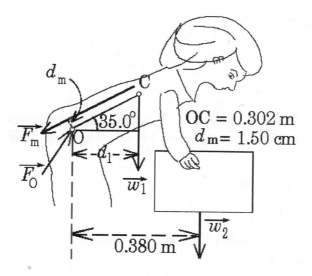

muscles and the force $\vec{F}_O$ (magnitude and direction) exerted by the joint on the vertebrae.

$OC = 0.302$ m
$d_m = 1.50$ cm

SOLUTION

We'll denote the magnitudes of the weights of the upper body and the container as w_1 (= 440 N) and w_2 (= 260 N), respectively. The lines of action of these forces and that of the muscles are indicated in Fig. 9-12. The force $\vec{F}_O$ is applied directly to the vertebral joint O, and we have drawn it in an assumed direction.

To find F_m, we can apply the torque condition Eq. [9-12] choosing O for our pivot. In this case, the unknown force $\vec{F}_O$ does not contribute to the torque condition, which becomes

$$\sum \tau = F_m d_m - w_1 d_1 - w_2 d_2 = 0$$

Note the signs associated with each term. The moment arms d_m (= 0.0150 m) and d_2 (= 0.380 m) are explicitly given, while d_1 is found from the diagram and given distance OC = 0.302 m to be d_1 = (0.302 m) cos(35.0°) = 0.247 m. Hence,

$$F_m = \frac{(w_1 d_1 + w_2 d_2)}{d_m} = \frac{(440 \, \text{N} \times 0.247 \, \text{m} + 260 \, \text{N} \times 0.380 \, \text{m})}{0.0150 \, \text{m}}$$

$$= 1.383 \times 10^4 \, \text{N}$$

To prevent round-off error in subsequent calculations, one extra digit has been kept in the value for F_m. To three significant figures, $F_m = 1.38 \times 10^4$ N.

The force $\vec{F}_O$ exerted by the joint can now be found by applying the translational equilibrium conditions. We'll take the x- and y-axes as horizontal to the right and vertically upward, respectively. For the x-direction,

$$\sum F_x = F_{O,x} + F_{m,x} = F_{O,x} - F_m \cos(35.0°) = 0$$

therefore $\quad F_{O,x} = F_m \cos(35.0°) = 1.133 \times 10^4 \, \text{N}$

Thus the horizontal component of $\vec{F}_O$ acts towards the right, opposite to that of $\vec{F}_m$, as could have been deduced from the figure. For the y-direction,

$$F_y = F_{O,y} + F_{m,y} - w_1 - w_2 = 0$$

therefore $\quad F_{O,y} = w_1 + w_2 + F_m \sin(35.0°)$

$$= 440 \, \text{N} + 260 \, \text{N} + 1.383 \times 10^4 \, \text{N} \sin(35.0°)$$

$$= 8.63 \times 10^3 \, \text{N}$$

The positive value of $F_{O,y}$ means that it is acting <u>upward</u>. Again, this could have been deduced immediately from the figure, since all the other forces have vertically downward components. Finally, the magnitude and direction of $\vec{F}_O$ are found from

$$F_O = \sqrt{F_{O,x}^2 + F_{O,y}^2} = \sqrt{(1.133 \times 10^4 \, \text{N})^2 + (8.630 \times 10^3 \, \text{N})^2} = 1.42 \times 10^4 \, \text{N}$$

$$\theta_O = \tan^{-1}\left(\frac{F_{O,y}}{F_{O,x}}\right) = \tan^{-1}\left(\frac{8.630 \times 10^3 \, \text{N}}{1.133 \times 10^4 \, \text{N}}\right) = 37.3°$$

where θ_O is measured with respect to the positive x-axis. Therefore $\vec{F}_O$ acts in a direction nearly parallel to the line OC.

The most noteworthy aspects of the preceding example are the large magnitudes of both the muscle force, F_m, and the joint force, F_O, relative to the combined weights $(w_1 + w_2) = 700 \, \text{N}$. Such strong forces produce tremendous stresses on the muscles and backbone, and are a common cause of lower back pain. The large value of F_m is due not only to the magnitudes of w_1 and w_2 but also to the <u>small</u> value of the moment arm d_m relative to d_1 and d_2. This is just a slightly more complicated version of the idea illustrated by our earlier discussion of the teeter-totter, namely that the ratio of the distances of the two children from the pivot is inversely proportional to their relative masses. Several other anatomical applications of these ideas are contained in the problems at the end of this chapter.

Gravitational Stability

We conclude this section with a further look at the significance of the centre of mass (see Box 9-1). The location of the centre of mass within a body has a crucial effect on the gravitational stability of the body. To illustrate this, consider an object resting on a horizontal surface, as in Fig. 9-13a. Suppose the object is then tilted slightly as in Fig. 9-13b. In this case, the centre of mass (cm) is still located over the base of the object, and the line of action of its weight is to the left of the corner touching the surface. The weight now produces an unbalanced torque on the object which will act to return the object to its original upright position. In Fig. 9-13c, however, the object is given a larger tilt such that the cm is no longer located over the object's base and is to the right of the pivot. In this case, the unbalanced torque due to its weight will cause the object to tip over onto its side, resulting in Fig. 9-13d.

Figure 9-13

Stability of an object against tilting (cm = centre of mass).

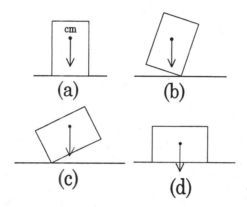

Imagine repeating these steps starting with the situation in Fig. 9-13d. Here the area of the object's base in contact with the surface is larger, while its centre of mass is closer to the base. It is not hard to see that it must be given a considerably larger initial tilt than before in order to "tip over" and return to its original upright position in Fig. 9-13a.

This illustrates that the location of a standing object's centre of mass with respect to its base determines its gravitational stability. To remain stable, its centre of mass must be located above the area bounded by its supporting base. In general, the lower the centre of mass and the larger the area of support, the more stable is the object against being overturned. The stability of animals that walk erect on two legs, such as humans and birds, depends on having relatively large feet. A standing human is least stable when his or her feet are close together: stability is gained by moving the feet apart so that the base is broadened. Of course, a person should not lean too far forward without other means of support (such as a cane or crutch), otherwise the person will suffer the same fate as the object in Fig. 9-13c. Four-legged animals have a larger area of support bounded by their legs, hence are very stable and require only small feet or hooves.

Finally, we'll mention that those ideas also apply to the stability of trees against being overturned by the action of a strong wind. If the wind bends the tree trunk so that its centre of mass is displaced laterally by a significant distance, then the tree is in danger of snapping or being uprooted. Of course, other factors also play a role, such as the width and stiffness of the trunk and the strength with which it is rooted in the soil.

9.4   Rotational Kinematics

To analyze rotational motion, we shall first deal with rotational kinematics, which is concerned with how to characterize such motion in terms of displacements, velocities, and accelerations. Let us consider a rigid body rotating about a fixed axis which passes through a point (denoted O) in the body, as shown in Fig. 9-14a. Suppose that the rotation axis is perpendicular to the page, and hence any given point P in the object will move in a plane parallel to the plane of the page. The distance OP is fixed (since the body is assumed to be rigid), and therefore, as the body rotates, the locus of P moves along a circular path centred at O. This path is indicated by the dashed curve in Fig. 9-14a. We can characterize the rotational position of point P by means of the angle θ between the line OP and some fixed (i.e., non-rotating) axis in the plane of rotation which passes through O, such as the x-axis indicated in Fig. 9-14a.

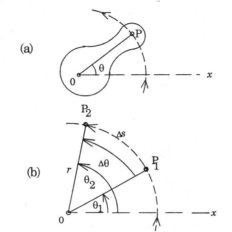

(a)

(b)

Figure 9-14

(a) Definition of the angular position of a point P in a rotating rigid body;
(b) Relation between angular displacement Δθ and linear displacement Δs.

Now suppose that the point P rotates from an initial position P_1 to another position P_2 corresponding to a change in angle from θ_1 to θ_2 (see Fig. 9-14b). The actual distance moved by the point P is the *arc length* Δs measured along the circular path traced out by this point. It is clear that Δs is proportional to the *angular displacement* $\theta_2 - \theta_1 = \Delta\theta$, while for a fixed value of Δθ, the magnitude of Δs increases in proportion to the distance OP ≡ r. Therefore

$$\Delta s = r\Delta\theta \qquad \textbf{[9-14a]}$$

which can be rewritten as

$$\Delta\theta = \frac{\Delta s}{r} \qquad \textbf{[9-14b]}$$

For the reasons described in Box 1-2, Δθ in Eq. [9-14] must be expressed in units of radians.

At any instant, different points in the rotating body may make different angles θ with respect to the fixed x-axis. Depending on how far each point is from the rotation axis, different points may move different distances Δs in a given time interval. However, since the body is rigid, <u>all</u> points undergo the <u>same</u> angular change $\Delta\theta$ in the same time interval.

Rotational kinematics is concerned with how the angular position θ of any point in a rotating body changes with time. The *angular velocity* ω characterizes the rate of change of θ and is defined as

$$\omega = \frac{d\theta}{dt} \qquad \textbf{[9-15]}$$

That is, ω is the time derivative of θ. If angles are measured in units of radians and time in seconds, then ω has units of radians per sec (rad/s). Another common unit for angular velocity is revolutions per minute (rev/min or rpm). Since 1 rev = 2π rad, the conversion between the two units is

$$1\,\text{rev/min} = 1\frac{\text{rev}}{\text{min}} \times \frac{1\,\text{min}}{60\,\text{s}} \times \frac{2\pi\,\text{rad}}{1\,\text{rev}} = \frac{2\pi}{60}\frac{\text{rad}}{\text{s}}$$

Since all points in a rotating rigid body undergo the same angular change in a given time interval, all points of the body have the <u>same</u> angular velocity ω.

If the angular velocity of a body changes during its motion, then it has an *angular acceleration* α. The latter is defined in terms of the rate of change of ω, analogous to the way translational acceleration is related to translational velocity:

$$\alpha = \frac{d\omega}{dt} \qquad \textbf{[9-16]}$$

The usual unit for α is rad/s². Just as all points in a rotating body have the same angular displacement and angular velocity, all points have the same angular acceleration.

Constant Angular Acceleration Kinematics

Now let us consider the case of objects rotating at constant angular acceleration. We can derive equations for angular position and velocity by the same methods used in Section 7.2, in the section on nonuniform motion, for straight-line motion with constant acceleration. Choosing the axis for measuring angles so that at the initial time $t = 0$, a point in the body has angular position $\theta_0 = 0$, while the initial angular velocity is ω_0, the angular position θ and velocity ω at any later time t are given by the following equations. Beside each equation for angular variables, we show the corresponding equations from Section 7-2 for straight-line motion with constant acceleration.

$$\omega = \omega_0 + \alpha t \qquad \left[v = v_0 + at \right] \qquad \textbf{[9-17a]}$$

$$\theta = \omega_0 t + \tfrac{1}{2}\alpha t^2 \qquad \left[x = v_0 t + \tfrac{1}{2}at^2 \right] \qquad \textbf{[9-17b]}$$

$$\omega^2 = \omega_0^2 + 2\alpha\theta \qquad \left[v^2 = v_0^2 + 2ax \right] \qquad \textbf{[9-17c]}$$

$$\theta = \left(\frac{\omega_0 + \omega}{2} \right) t \qquad \left[x = \left(\frac{v_0 + v}{2} \right) t \right] \qquad \textbf{[9-17d]}$$

The relations for the angular variables in Eq. [9-17a] to [9-17d] are seen to be completely analogous to those for straight-line motion, and can be obtained from the latter by simply replacing each straight-line quantity by its rotational analog.

Relating Angular and Linear Variables

It is useful to relate the angular kinematic variables of a rotating body to corresponding "linear" variables at a particular point in the body. Here we use the terms *(linear) velocity* and *(linear) acceleration* for the familiar quantities $\vec{v}$ and $\vec{a}$ as discussed in connection with purely translational motion. (Note that "linear" does not necessarily refer to motion in a straight line; it could just as well occur along a curved path, as in Section 9.2.) We have already discussed one connection between "angular" and "linear" variables, namely in Eq. [9-14] relating the distance moved Δs by a point in a body to the angular displacement $\Delta\theta$. The corresponding linear velocity of the point P in Fig. 9-14 is a vector $\vec{v}$ which at each instant is tangent to the circular path traced out by that point, such as was shown earlier in Figs. 9-1 and 9-2. The magnitude of $\vec{v}$ is the linear speed v, given by

$$v = \lim_{\Delta t \to 0} \frac{\Delta s}{\Delta t} = \frac{ds}{dt}$$

Substituting Eq. [9-14a] for Δs and noting Eq. [9-15], we find

$$v = \lim_{\Delta t \to 0} r\frac{\Delta\theta}{\Delta t} = r\frac{d\theta}{dt} = r\omega \qquad \textbf{[9-18]}$$

Note that Eq. [9-18] relating v and ω is the same as Eq. [9-6] for uniform circular motion. In deriving the last result, we have used the fact that the distance r of point P from the rotation axis in Fig. 9-14 is constant. Similar reasoning shows that the angular acceleration α and *tangential acceleration* a_t, defined as

$$a_t = \frac{dv}{dt}$$

are related by

$$a_t = r\alpha \qquad\qquad [9\text{-}19]$$

Just as the relation Eq. [9-14] is valid only when the angular displacement $\Delta\theta$ is expressed in radians, Eqs. [9-18] and [9-19] are valid only when ω and α are expressed in units of rad per unit time and rad/(unit time)^2, respectively.

We call a_t the tangential acceleration because it is in fact the component of the acceleration vector $\vec{a}$ at point P in the direction parallel to the instantaneous velocity $\vec{v}$, i.e., in the tangential direction. Note that a_t equals the rate of change of the speed of point P and (like α) could be either positive or negative. If the object is rotating with constant angular velocity ω, and hence with constant speed v, then $a_t = 0$. In this case, individual points in the object are undergoing uniform circular motion (see Section 9-2). Just as for any point object in uniform circular motion, the acceleration $\vec{a}$ at point P in a rotating rigid body also has a component of magnitude $a_c = v^2/r$ (Eq. [9-4]) in the direction from P toward the axis of rotation, which is due to changes in the <u>direction</u> of $\vec{v}$. The relation $a_c = v^2/r$ still holds even when ω (and hence v) is not constant. Then the total linear acceleration $\vec{a}$ of a point in a rotating rigid body is obtained by vector addition of $\vec{a}_t$ and $\vec{a}_c$. In the rest of this chapter, however, we shall not concern ourselves with the "centripetal" component $\vec{a}_c$, since the "centripetal forces" which keep the point P moving on a circular path are implicitly taken into account by the assumption that the body is rigid.

Finally, while the quantities $\Delta\theta$, ω, and α are the same for all points in a rotating body, we see that the linear variables Δs, v, and a_t depend on the distance r between the point P of interest and the axis of rotation, and, in particular, are proportional to r.

EXAMPLE 9-7

The wheel of an exercise bicycle increases its angular velocity from $\omega_0 = 0$ to $\omega = 12.5$ rad/s in 3.00 s.

(a) What is the wheel's angular acceleration, assuming this to be constant?

(b) How many revolutions did the wheel make during the 3.00 seconds?

(c) If the radius of the wheel is 0.760 m, what is the final linear speed of a point on the wheel rim?

SOLUTION

(a) Using Eq. [9-17a],

$$\alpha = \frac{\omega - \omega_0}{t} = \frac{(12.5 - 0)\text{rad/s}}{3.00 \text{ s}} = 4.17 \text{ rad/s}^2$$

(b) We could use either Eq. [9-17b] or [9-17d] to find the final angular position θ. Using the latter,

$$\theta = \tfrac{1}{2}(\omega_0 + \omega)t = \tfrac{1}{2}(0 + 12.5) \text{ rad/s } (3.00 \text{ s})$$
$$= 18.8 \text{ rad}$$

Converting to revolutions,

$$\theta = 18.8 \text{ rad} \times \frac{1 \text{ rev}}{2\pi \text{ rad}} = 2.99 \text{ rev}$$

(c) The linear speed is obtained from Eq. [9-18]

$$v = r\omega = 0.760 \text{ m} \times 12.5 \text{ rad/s}$$
$$= 9.5 \text{ m/s}$$

Notice that the radian unit is omitted from the last result, since it is dimensionless (Box 1-2).

9.5 Rotational Dynamics: Moment of Inertia

Now we shall study the dynamic principles that relate rotational motion to the forces acting on a rigid body. First we will consider only pure rotational motion of an object about a fixed axis. We have already seen in Section 9.3 on torque that rotational equilibrium of a rigid body requires the net torque $\tau_{net} = \sum\tau$ about any point to be zero. This is the rotational analog of Newton's first law. It should come as no surprise to learn that, if the net torque exerted on a body is <u>not</u> zero, then the body will develop an angular acceleration α. We shall state without going through the proof that the precise relationship between τ_{net} and α is

$$\tau_{net} = I\alpha \qquad\qquad \textbf{[9-20]}$$

which is the rotational analog of Newton's second law.

The new quantity I which appears in Eq. [9-20] is known as the *moment of inertia*. This plays the role for rotational motion analogous to that of mass for translational motion. To define this quantity, let us imagine that our rigid body is composed of several small point-like

Figure 9-15

A rigid object, such as the bicycle wheel of Example 9-7, represented as a collection of point masses.

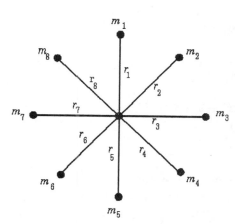

particles with masses $m_1, m_2, \ldots$ at fixed perpendicular distances $r_1, r_2, \ldots$ from the axis of rotation, shown schematically in Fig. 9-15. Here the axis of rotation is assumed to be perpendicular to the plane of the figure. For such a collection of masses, the moment of inertia is given by

$$I = \sum_i m_i r_i^2 \qquad [\textbf{9-21}]$$

where the summation is over all particles $i = 1, 2, \ldots$ in the body. Clearly, from this definition, the SI unit of I is kg·m².

Figure 9-16

Example 9-8: Four point particles held by rigid rods.

EXAMPLE 9-8

A body consists of four point particles held by rigid rods of negligible mass about an origin O, with masses and distances indicated in Fig. 9-16. Calculate: (a) the moment of inertia of the object about the x-axis, I_x; (b) the moment of inertia about the y-axis, I_y; and (c) the moment of inertia I_z about the z-axis perpendicular to the x–y plane and passing through the origin O.

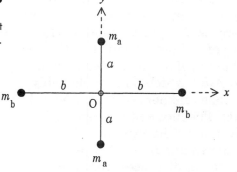

SOLUTION

(a) The two masses that lie on the x-axis do not contribute to I_x since the distances from the axis to these masses are zero. The perpendicular distances of the other two masses from the x-axis are both equal to a. Hence

$$I_x = m_a a^2 + m_a a^2 = 2m_a a^2$$

(b) By analogous arguments,

$$I_y = 2m_b b^2$$

(c) For the axis perpendicular to the *x–y* plane passing through O, it is clear that

$$I_z = 2m_a a^2 + 2m_b b^2$$
$$= I_x + I_y$$

The last example demonstrates several key features of the moment of inertia. The first is that the moment of inertia is not a single unique parameter of any body, but depends on the location of the rotation axis. A given body generally will have different values of *I* for rotation about different axes. Secondly, one sees that, for a given set of masses comprising the body, the moment of inertia increases when the distances of any masses from the axis of rotation increase. In terms of Eq. [9-20], the greater the moment of inertia, the greater is the net torque τ_{net} required to impart a given angular acceleration to an object.

In practice, most rigid bodies of interest are not a collection of point masses but rather have a continuous distribution of mass. Such a body can be regarded as the limit of an infinite number of infinitesimal mass elements. In this limit, the sum in Eq. [9-21] becomes an integral, which is straightforward to evaluate for highly symmetrical objects. Table 9-1 lists values for the resulting moments of inertia about the indicated axes of several such objects, assuming the mass is uniformly distributed through each body. In all cases, *I* is proportional to the total mass *M* of the object and to the square of a characteristic length. For example, we see that the moment of inertia of a thin-walled hoop or hollow cylinder of radius *R* is given by MR^2, which follows plausibly from Eq. [9-21] since all infinitesimal mass elements are located at the same distance *R* from the rotation axis. The moment of inertia of a solid disk or cylinder is ½ MR^2, hence smaller than that of a hollow disc or cylinder of the <u>same</u> total mass *M* and radius *R*. This is because most of the mass in the solid case is distributed at distances from the rotation axis smaller than *R*.

Now we return to the rotational form of Newton's second law, Eq. [9-20] ($\tau_{net} = I\alpha$), and illustrate the use of this equation for analyzing rotational dynamics.

Table 9-1   Moments of Inertia for Various Rigid Bodies

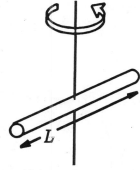

Long thin rod with axis
through centre

$I = \frac{1}{12} ML^2$

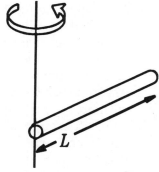

Long thin rod with axis
through one end

$I = \frac{1}{3} ML^2$

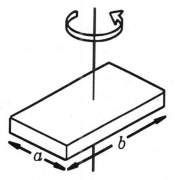

Rectangular plate

$I = \frac{1}{12} M(a^2 + b^2)$

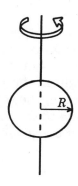

Solid sphere

$I = \frac{2}{5} MR^2$

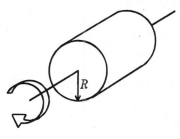

Solid cylinder or disk

$I = \frac{1}{2} MR^2$

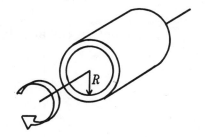

Thin-walled hoop or
hollow cylinder

$I = MR^2$

EXAMPLE 9-9

A wheel of radius $R = 1.50$ m and total mass $M = 10.0$ kg is mounted on a fixed axle as shown in Fig. 9-17. A cable is wrapped several times around the wheel and is pulled with a force of magnitude $F_p = 97.0$ N. Assume that the cable unwinds without slipping, turning the wheel. Through what angle does the wheel turn in 1.00 minute if it starts from rest? Neglect friction between the wheel and axle.

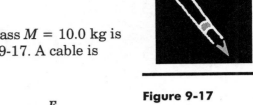

Figure 9-17

Example 9-9: A cable turning a wheel.

SOLUTION

This example combines Eq. [9-20] with the kinematic considerations of Section 9.4. We first solve τ_{net} for the angular acceleration α. In the absence of friction, and because the weight of the wheel acts through its centre of mass (i.e., through the rotation axis), the only torque contributing to τ_{net} is due to the force of magnitude F_p pulling on the cable. This torque is given by $F_p R$. Hence

$$\alpha = \frac{F_p R}{I} = \frac{F_p R}{\frac{1}{2}MR^2} = \frac{2F_p}{MR}$$

where we used Table 9-1 for the moment of inertia of a solid disc. The value of α is therefore found to be

$$\alpha = \frac{2(97.0\,\text{N})}{(10.0\,\text{kg})(1.50\,\text{m})} = 12.9\,\text{rad/s}^2$$

We obtain the angular displacement using Eq. [9-17b],

$$\theta = \omega_0 t + \frac{1}{2}\alpha t^2 = 0 + \left(\frac{12.9}{2}\,\text{rad/s}^2\right)(60\,\text{s})^2$$

$$= 2.32 \times 10^4\,\text{rad}$$

9.6   **Rotational Kinetic Energy**

In Section 8.7, we learned that any object in translational motion has kinetic energy $K = \frac{1}{2}mv^2$. A rigid body undergoing pure rotational motion also has kinetic energy, which we can evaluate by going back to the picture of the body as a collection of point particles and summing the kinetic energies of all these individual particles,

$$K = \frac{1}{2} \sum_i m_i v_i^2$$

where v_i is the linear speed of particle i. From Eq. [9-18], $v_i = r_i \omega$, where r_i is the perpendicular distance of particle i from the rotation axis and ω is the angular velocity of the body, which is the same for all points in the body. Hence the expression above for K becomes

$$K = \frac{1}{2} \sum_i m_i r_i^2 \omega^2 = \frac{1}{2} \left(\sum_i m_i r_i^2 \right) \omega^2$$

$$= \frac{1}{2} I \omega^2$$

[9-22]

where we used the definition of I in Eq. [9-21].

Equation [9-22] is analogous to the expression for translational kinetic energy, with ω in place of v and the moment of inertia I in place of mass m. The rotational kinetic energy in Eq. [9-22] is <u>not</u> a new form of energy, but is just ordinary kinetic energy for the particular case of rigid-body rotation.

We can use Eq. [9-22] in combination with the conservation of energy considerations of Section 8.7.

EXAMPLE 9-10

A uniform rigid rod of length L and total mass M is free to rotate about a pin through one end of the rod. The rod is initially held at rest in a horizontal position and then released (Fig. 9-18). Neglect friction between the rod and pin.

(a) What is the angular speed of the rod when it reaches its lowest position?

(b) What are the linear speeds of the centre of mass and of the free end of the rod at its lowest position?

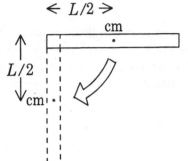

Figure 9-18

Example 9-10: A rigid rod rotating about a pin through one end.

SOLUTION

(a) Initially the rod is at rest and has no kinetic energy. After it is released and has rotated to the vertical position shown in the figure, it will have gained kinetic energy $\frac{1}{2}I\omega^2$. Relative to its original position, the centre of mass (cm) of the rod has dropped a vertical distance $L/2$. Therefore, in its initial state, the rod has potential energy $Mg(L/2)$ relative to the potential energy of its final state, which we set equal to zero. Hence, from conservation of energy,

$$Mg(L/2) = \frac{1}{2}I\omega^2$$
$$= \frac{1}{2}(\frac{1}{3}ML^2)\omega^2$$

where we used Table 9-1 for the moment of inertia of the rod. Therefore,

$$\omega^2 = 3g/L$$
$$\text{and} \quad \omega = \sqrt{3g/L}$$

(b) The centre of mass is located at distance $L/2$ from the rotation axis, hence in its final state,

$$v_{cm} = (L/2)\omega = \frac{1}{2}\sqrt{3gL}$$

whereas the end of the rod is at distance L from the axis, hence

$$v_{end} = L\omega = 2v_{cm}$$

The preceding example could not be solved easily using $\tau_{net} = I\alpha$, since in this case the net torque on the rod (due to its weight $M\vec{g}$ acting vertically downward through the centre of mass) changes as the rod rotates, and hence the angular acceleration α is <u>not</u> constant.

Rolling Objects

So far, we have been describing pure rotational motion, in which the axis of rotation remains fixed in space. The most general motion of a rigid body, however, consists of both translational and rotational contributions. Any such motion can be represented as a combination of translational motion of the centre of mass and rotation of the body about an axis through the centre of mass. The total kinetic energy of the body is then given by

$$K = \tfrac{1}{2}Mv_{cm}^{2} + \tfrac{1}{2}I_{cm}\,\omega^{2} \qquad \textbf{[9-23]}$$

Figure 9-19

A wheel which rolls without slipping across a surface <u>has</u> centre of mass velocity $\vec{v}_{cm}$ and angular velocity ω about the centre of mass.

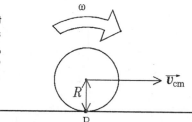

where v_{cm} is the translational speed of the centre of mass and I_{cm} is the moment of inertia about an axis through the centre of mass.

Let us consider a wheel which rolls without slipping across a flat surface, illustrated in Fig. 9-19. In this situation, it can be shown that v_{cm} and the angular velocity ω about the centre of the wheel are related by

$$v_{cm} = R\omega \qquad \textbf{[9-24]}$$

where R is the radius of the wheel. One notes that Eq. [9-24] is similar to Eq. [9-18] ($v = r\omega$) relating ω to the linear speed of a point in a rotating body with a <u>fixed</u> axis of rotation. In fact, Eq. [9-24] is most easily derived by going into the frame of reference of the rotating wheel. In this frame, the rotation axis is fixed and the surface contacting the wheel moves <u>backward</u> with speed v_{cm} relative to the centre of the wheel. The outside edge of the wheel has this same speed, which indeed is related to the angular speed ω by Eq. [9-18] with $r = R$ and $v = v_{cm}$.

Using Eq. [9-24] to express ω in terms of v_{cm}, the kinetic energy in Eq. [9-23] becomes

$$K = \tfrac{1}{2}Mv_{cm}^2 + \tfrac{1}{2}I_{cm}\left(\frac{v_{cm}}{R}\right)^2 \qquad \textbf{[9-25]}$$

$$= \tfrac{1}{2}Mv_{cm}^2\left(1 + \frac{I_{cm}}{MR^2}\right)$$

This equation applies to any round object which rolls without slipping along a flat surface. The proviso "without slipping" means that point P on the wheel that contacts the surface in Fig. 9-19 is instantaneously at rest relative to the surface; otherwise, Eq. [9-24] would not be valid. This condition in turn requires that there is a sufficiently large friction force between the wheel rim and surface. However, because there is no slippage, this friction force (really, a static friction force) does no work on the rolling object and plays no role in energy conservation.

EXAMPLE 9-11

A round object (such as a ball or disc) is released from rest at the top of an incline of height h and then rolls down the incline without slipping (Fig. 9-20). Note from Table 9-1 that the moment of inertia about the centre of mass of the object can be written as $I_{cm} = cMR^2$, where c is a numerical constant depending on the object's shape. Use conservation of energy to determine the centre-of-mass speed of the object when it reaches the bottom of the incline.

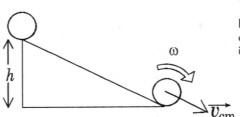

Figure 9-20

Example 9-11: A round object rolling down an incline without slipping.

SOLUTION

As in Example 9-10, we equate the initial potential energy of the object, Mgh (relative to the potential energy in the final position), with the final kinetic energy given by Eq. [9-25],

$$Mgh = \tfrac{1}{2}Mv_{cm}^2\left(1 + \frac{I_{cm}}{MR^2}\right)$$

Cancelling a factor of M on both sides of this equation, and substituting $I_{cm} = cMR^2$, we get

$$gh = \tfrac{1}{2}v_{cm}^2 \,(1+c)$$

or

$$v_{cm} = \sqrt{\frac{2gh}{(1+c)}}$$

The remarkable feature of the result for v_{cm} in Example 9-11 is that it doesn't depend on either the mass M or radius R of the object, but only on the shape-dependent factor c. For example, all solid cylinders have $c = \tfrac{1}{2}$, giving

$$v_{cm} = \sqrt{4gh/3}$$

and therefore all such cylinders attain the same final speed v_{cm} and roll down the incline in the same amount of time, regardless of their size or mass. Likewise, all thin-walled hollow cylinders have $c = 1$, resulting in

$$v_{cm} = \sqrt{gh}$$

again independent of size or mass. We see that any solid cylinder has a larger value of v_{cm} than any hollow cylinder, and hence travels down the incline in a shorter time. The physical explanation is that objects with smaller values of c produce relatively smaller amounts of rotational kinetic energy and thus have more kinetic energy available for translation.

9.7   Angular Momentum

Some of the most striking aspects of rotational motion are explained by the concept of *angular momentum*. Just as we have learned that every rotational quantity (i.e., ω, α, τ, I) is the analog of some quantity characterizing translational motion, angular momentum is the rotational analog of momentum. Strictly speaking, angular momentum is a vector quantity, as is true in a general sense of angular velocity, acceleration and torque. However, we can ignore this vector nature when dealing with rotation about a fixed axis of symmetry. In that case, the angular momentum L of a rotating body is defined by

$$L = I\omega \qquad\qquad \textbf{[9-26]}$$

where I and ω are the moment of inertia and angular velocity, respectively. This definition is analogous to that of (linear) momentum,

$\vec{p} = m\vec{v}$. The SI unit of angular momentum following from this definition is kg·m²/s.

Taking the time derivative of Eq. [9-26] and assuming that the moment of inertia I remains constant, we obtain

$$\frac{dL}{dt} = I\frac{d\omega}{dt}$$

Using Eqs. [9-16] ($\alpha = d\omega/dt$) and [9-20] ($\tau_{net} = I\alpha$), this becomes

$$\frac{dL}{dt} = \tau_{net} \qquad\qquad \text{[9-27]}$$

which is just another way of expressing $\tau_{net} = I\alpha$ (Eq. [9-20]). However, as we shall discuss, there are some situations in which the moment of inertia of a body does not stay constant. In these cases, Eq. [9-27] is still valid but Eq. [9-20] is not.

Conservation of Angular Momentum

If the net torque τ_{net} on the system is zero, Eq. [9-27] tells us that the angular momentum L is constant, independent of time. This is the principle of *conservation of angular momentum*, which will be the focus of this subsection. If the moment of inertia I is constant, constant angular momentum L implies constant angular velocity ω via Eq. [9-26], i.e., the system is in rotational equilibrium. But most animal bodies, including those of humans, are not perfectly "rigid" and rather are <u>multisegmented</u>, capable of changing body shape and thereby changing the moment of inertia value even when the rotation axis is unchanged. In such a case, if the moment of inertia changes from an initial value I_1 to a different value I_2, conservation of angular momentum[3] requires that the angular velocity change from ω_1 to ω_2, where ω_1 and ω_2 are related by

$$I_1\omega_1 = I_2\omega_2 \qquad\qquad \text{[9-28]}$$

Perhaps the most familiar example of conservation of angular momentum is that of a figure skater performing a spin (Fig. 9-21). The skater begins by going into a slow spin with her arms and one leg

---

3. The forces causing changes to an animal's body shape and moment of inertia are usually *internal forces* acting between different parts of the body. By Newton's third law, these internal forces as well as their corresponding torques add to zero. Hence, as long as the net torque due to *external* forces is zero, we can consider the angular momentum $L = I\omega$ to remain constant during a change in shape.

Figure 9-21

A figure skater increases her angular velocity from (a) ω_1 to (b) ω_2 by drawing in her arms and legs.

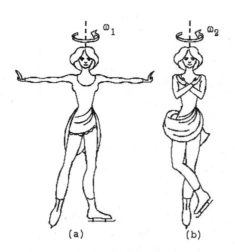

extended from her body, thus having a large initial moment of inertia I_1 and small angular speed ω_1. By drawing in her arms and leg rapidly, her moment of inertia decreases to a significantly smaller value I_2. Neglecting any external torques such as due to friction with the ice, the product $I\omega$ must stay constant and hence her angular speed increases to a new value ω_2, which may be quite large.

EXAMPLE 9-12

A figure skater rotating at 0.500 rev/s pulls in her arms and one leg as in Fig. 9-21, so that her moment of inertia about a vertical axis through her body is reduced by 62.0%.

(a) What is her new angular velocity?

(b) By what factor does her rotational kinetic energy increase?

SOLUTION

(a) Since the moment of inertia I decreases by 62.0%, the final value of I is (100.0% − 62.0%) = 38.0% of the original value, i.e.,

$$I_2 = 0.380\,I_1$$

By applying Eq. [9-28] ($I_1\omega_1 = I_2\omega_2$),

$$\omega_2 = \frac{I_1}{I_2}\,\omega_1 = \frac{\omega_1}{0.380} = 2.63\,\omega_1$$

therefore $\omega_2 = 2.63 \times 0.500\,\text{rev/s} = 1.32\,\text{rev/s}$

(b) The ratio of the final to initial rotational kinetic energies is

$$\frac{K_2}{K_1} = \frac{\frac{1}{2}I_2\omega_2^2}{\frac{1}{2}I_1\omega_1^2} = \frac{(I_2\omega_2)\omega_2}{(I_1\omega_1)\omega_1}$$

But $I_2\omega_2 = I_1\omega_1$, so the ratio becomes

$$\frac{K_2}{K_1} = \frac{\omega_2}{\omega_1}$$

which has the value of 2.63 according to part (a). Hence the rotational kinetic energy increases by a factor of 2.63 (the same factor by which the angular velocity increases).

One might well ask, given that any external torque on the figure skater in the above example was neglected, how could her kinetic energy increase? The answer is that the increase in kinetic energy is provided by expenditure of internal energy due to the effort of moving her arms and legs.

The principle of conservation of angular momentum is also used by other athletes, such as high divers performing somersaults. After jumping from a diving board with a small angular speed, a diver can reduce his moment of inertia and hence increase his rate of spin by curling into a tight tucked position. Just before entering the water, the diver straightens out his body so that his angular speed returns to a small value.

Exercises

9-1 A lab technician is centrifuging a blood sample at a rotation rate of 3.00×10^3 rpm (revolutions per minute). If the radius of the circular path followed by the sample is 0.150 m, determine the speed of the sample.

9-2 The blood sample in a centrifuge experiences a centripetal acceleration which is 6.00×10^3 times larger than the acceleration due to gravity. How many revolutions per minute is the sample making, if it is located at a distance of 5.00 cm from the axis of rotation?

9-3 To produce artificial gravity in a space laboratory, the laboratory should rotate. Suppose the laboratory has the shape of a ring of radius 1.00 km, and its frequency of rotation is adjusted so that the centripetal acceleration on the ring is equal in magnitude to the acceleration due to gravity on Earth. What is that frequency of rotation, the period of rotation, and the speed of a person standing in the ring?

9-4 A stone of mass 0.950 kg is attached to one end of a string and is whirled in a horizontal circle of radius 0.850 m. The string will break if its tension exceeds 575 N. What is the maximum speed that can be attained by the stone without breaking the string?

9-5 A car of mass 1.10×10^3 kg goes around an unbanked highway curve of radius 320 m at a speed of 27.0 m/s. What is the minimum value of the coefficient of static friction between the car's tires and the road that will enable the car to go around the curve without sliding? If the curve were "sharper", i.e., its radius were smaller, would your answer increase or decrease?

9-6 A downhill skier of weight 510 N reaches point A in the following diagram, where a jump is necessary. Point A is at the bottom of a circular

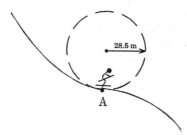

28.5 m

A

arc of radius 28.5 m. When she reaches this point, just before leaving the ground, the skier's velocity is horizontal with magnitude 15.0 m/s. What is the magnitude of the normal force acting on the skis at point A?

9-7 A 100-kg man stands 1.00 m from the end of a 4.00-m long scaffold which weighs 750 N. What are the magnitudes T_1 and T_2 of the tensions in the ropes supporting the scaffold? (See diagram.)

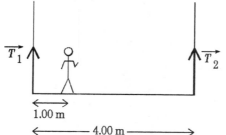

$\vec{T}_1$

$\vec{T}_2$

1.00 m

4.00 m

9-8   A teeter-totter is 20 kg in mass. Assume it is a uniform plank of length 2.00 m. The pivot point is placed 0.20 m from the midpoint. What weight must be placed on the short end (see diagram) to balance the teeter-totter?

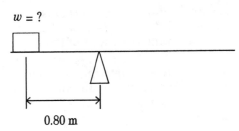

$w = ?$

0.80 m

9-9   A uniform plank of length 6.00 m and weight 230 N rests on two supports, one located beneath the left end and the other beneath a point 1.50 m from the right end of the plank (see diagram). To what distance x can a person weighing 480 N walk on the overhanging part of the plank before it just begins to tip?

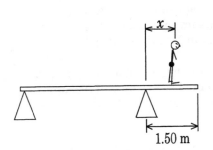

x

1.50 m

9-10 The following diagram shows a person's arm held elevated away from the body. The arm is supported in this position by the deltoid muscle, which is attached to the upper arm bone midway between the joint O and the centre of mass of the arm at point C. Suppose the weight of the arm is w. If the arm makes an angle θ with respect to the vertical, while the angle between the muscle and the upper arm bone is ϕ, find the magnitude T of the tension in the deltoid muscle.

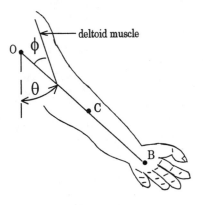

deltoid muscle

O
ϕ
θ
C
B

9-11   The following diagram shows a person doing a static knee exercise. The lower leg is kept at an angle of 22.0° from the vertical, and the mass m

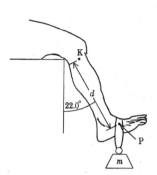

suspended from the foot is 12.0 kg. Assume that the weight $\vec{w}$ of mass m pulls down on the foot at point P, which is located at a distance $d = 0.600$ m from the knee joint K. Determine the torque exerted by $\vec{w}$ about the knee joint K. Assume that a counterclockwise torque is positive.

9-12   Suppose the arm in Example 9-4 (Fig. 9-8) is thrust forward so that the upper arm makes an angle of 30.0° with the vertical, while the forearm stays horizontal. Determine the new values of the magnitude T of the tension in the biceps muscle and of the magnitude and direction of the force $\vec{F}$ exerted by the upper arm bone on the forearm. Assume that the tension in the biceps muscle acts in a direction parallel to the upper arm.

9-13   Two people start at the same place and walk around a circular lake in opposite directions. One person has an angular speed of 1.67×10^{-3} rad/s, and the other has an angular speed of 3.33×10^{-3} rads. How long in minutes will it be before they meet?

9-14   A wheel rotates about its axis with a constant angular acceleration of 2π rad/s². If its initial angular velocity is 600 rpm,

   (a)   what is its angular velocity in rpm and rad/s at the end of one minute?

   (b)   how many revolutions does the wheel make during this minute?

9-15   A figure skater spins at 1.50 rev/s. Her finger tip is 0.900 m from the rotation axis. What is the linear speed of her finger tip?

9-16   A discus thrower is spinning at a rate of 10.0 rad/s when he releases the discus. The thrower with his outstretched arm is modeled as a rigid body, with the distance from the axis of rotation to the discus being 0.850 m.

   (a)   What is the linear speed of the discus when it is released?

   (b)   What is the angular acceleration of the thrower, assumed constant, if it took 1.75 s to reach the release speed when starting from rest?

9-17 A flywheel of radius 0.300 m starts from rest and accelerates with a constant angular acceleration of 0.500 rad/s². Determine the tangential acceleration and magnitude of the centripetal acceleration at a point on the wheel rim (a) at the start, and (b) after it has turned through 240°.

9-18 The bit of a dentist's high-speed drill accelerates from rest to an angular speed of 3.20×10^4 rad/s. During this process, the drill bit turns through 2.13×10^4 rad. Assuming a constant angular acceleration, how long would it take the bit to reach its maximum angular speed of 7.55×10^4 rad/s, starting from rest?

9-19 A rotating wheel of radius 0.300 m slows down with a constant tangential acceleration (at the wheel rim) of magnitude 15.0 m/s². At time $t = 4.00$ s, a point on the wheel rim has a tangential speed of 45.0 m/s.

(a) Calculate the wheel's constant angular acceleration.

(b) What are the angular velocities of the wheel at $t = 4.00$ s and $t = 0$?

(c) Through what angle did the wheel turn between $t = 0$ and $t = 4.00$ s?

9-20 Four small spheres of mass 0.250 kg each are arranged on a square of side-length 0.380 m, connected by rigid rods of negligible weight (see diagram).

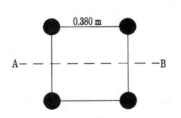

Calculate the moment of inertia of the system

(a) about an axis through the centre of the square, perpendicular to its plane

(b) about an axis which bisects two opposite sides of the square, i.e., along the line AB in the diagram for this exercise.

9-21 The following diagram shows three pairs of insect wings. The rotation axis for each wing is denoted AB. If each wing has the same total mass (distributed uniformly over the wing) and thickness, which <u>one</u> of the following statements describing the moment of inertia I about the axis AB is correct?

(a) I of the wings shown in (i) is larger than that of both (ii) and (iii).

(b) I of the wings in (ii) is larger than that of both (i) and (iii).

(c) I of the wings in (iii) is larger than that of both (i) and (ii).

(d)   *I* of all three wings (i), (ii), and (iii) is the same.

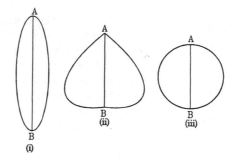

9-22 A person having moment of inertia 31.5 kg·m² is spinning about a vertical axis. If the person's angular speed reduces at a constant rate from 4.25 rad/s to 2.15 rad/s in 6.50 s, what is the magnitude of the frictional torque acting on the person's body?

9-23 The following diagram shows a model for the motion of the human forearm when throwing a dart. The forearm can rotate about an axis at the elbow joint, due to the force $\vec{T}$ exerted by the triceps muscle. The forearm has moment of inertia 0.0700 kg·m² (including a contribution from the dart) about the rotation axis, and dimensions shown in the figure. Assume that the force $\vec{T}$ acts perpendicular to the forearm. Find the magnitude of $\vec{T}$ required to give the dart a tangential speed of 5.50 m/s in 0.150 s, starting from rest. (Ignore effects of gravity or of any frictional forces.)

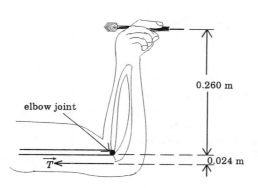

9-24 A grindstone of radius 28.0 cm is rotating at 20.0 rad/s. An axe is held against the grindstone with a normal force of magnitude 13.0 N. The coefficient of kinetic friction between the axe and grindstone is 0.750. If the moment of inertia of the grindstone is 0.420 kg·m²,

(a)   how long will it take the grindstone to stop rotating?

(b)   how much work is done by friction to bring the grindstone to rest? *Hint*: Use the work–energy theorem.

9-25 A uniform solid sphere of mass *M* and radius *R* rolls from rest down a hill of height 20.0 m. What is the sphere's speed when it reaches the bottom of the hill? (The moment of inertia of the sphere is $I = 2/5\,MR^2$ from Table 9-1.)

9-26   A cable is wrapped around a solid cylinder of mass M and radius R. The free end of the cable is tied to a mass m, which is then released from rest

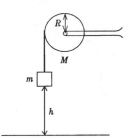

at a height h above the floor (see diagram). As the mass falls, the cable unwinds without slipping, turning the cylinder. Find expressions for the speed of the falling mass and the angular velocity of the cylinder just as the mass hits the floor. Neglect the weight of the cable and any friction between the cylinder and the axle about which it rotates. Use energy methods.

9-27   A figure skater is rotating at 0.640 rev/s.

(a)   What is her angular velocity in rad/s?

(b)   If she now pulls her arms in so that her moment of inertia is reduced by ⅓, what is her new angular velocity?

9-28   A diver jumps off a board with his arms straight up and legs straight down, so that his moment of inertia about his rotation axis is 19.0 kg·m². He then tucks into a tight ball, decreasing his moment of inertia to 4.60 kg·m². While tucked, he makes two complete revolutions in 1.30 s. If he hadn't tucked in, how many revolutions would he have made in a time of 1.60 s?

9-29   A ride in a playground consists of a large turntable of radius 2.00 m which rotates freely about a fixed vertical axis. A child of mass 19.0 kg is standing on the turntable right at its outer edge, and the system is initially rotating at an angular speed of 0.500 rev/s. The child then moves to a position 1.00 m from the centre of the turntable. What does the turntable's angular speed become? The moment of inertia of the turntable by itself is 400 kg·m². Assume that the child can be treated as a point particle.

Problems

9-30   A ball is tied to a cord and set in rotation in a vertical circle. Prove that the tension in the cord at the lowest point exceeds that at the highest point by six times the weight of the ball.

Hint: Consider Newton's second law for the ball at both the upper and lower points, as well as conservation of energy.

9-31   For cars traveling at a certain speed, a highway curve can be banked at just the right angle so that the cars are able to negotiate the curve without requiring friction. In the following diagram, the radius of the highway curve is 500 m.

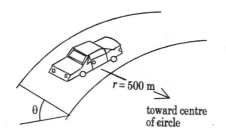

(a) Which force(s) constitute(s) the centripetal force on the car?

(b) For a car traveling with speed 100 km/h, what is the proper banking angle θ in the absence of friction?

9-32 A carnival ride consists of chairs that are swung at constant speed in a horizontal circle by 13.0-m long cables attached to a vertical rotating pole, as shown in the following diagram. Each cable makes an angle of 45.0° with the vertical. The total mass of a chair and its occupants is 240 kg.

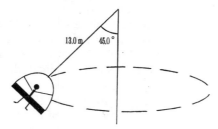

(a) What is the magnitude of the tension in the cable attached to the chair?

(b) What is the speed of the chair? Does this answer depend on the mass of the chair and its occupants?

Hint: Draw a free-body diagram for the chair, resolve all the forces on the chair along vertical and horizontal axes, and consider the components of Newton's second law along both axes.

9-33 The uniform gate in the following diagram has a mass of 51.0 kg. Assume that the force exerted on the gate by the upper hinge acts in a horizontal direction. Calculate the magnitude and direction of the supporting forces exerted by each hinge.

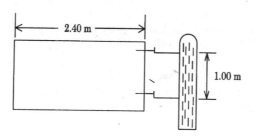

9-34 The system shown in the following diagram is in static equilibrium. The mass of the strut S is 100 kg.

(a) Find the largest mass m that can be supported if the maximum magnitude of the tension in the cable C is 2.00×10^4 N.

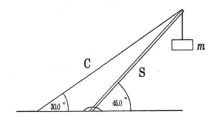

(b) For this case, calculate the magnitude and direction of the force exerted on the strut by the pivot.

9-35 When a person stands on tiptoe and is at equilibrium (see part (a) of diagram), the situation can be represented by the free-body diagram in part (b) of the diagram. This assumes that the floor exerts an upward force of magnitude F on the toe, the *Achilles tendon* exerts a tension of magnitude T in the direction indicated, and the leg bone exerts a force $\vec{F}_L$ on the foot at the ankle joint. In terms of F, find

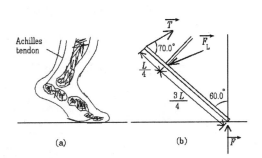

(a) the tension T in the Achilles tendon;

(b) the x- and y-components of the force $\vec{F}_L$.

9-36 A uniform ladder 10.0 m long rests against a frictionless vertical wall with its lower end 6.00 m from the wall. The ladder weighs 400 N. The coefficient of static friction between the ladder and the ground is 0.400. A man weighing 800 N climbs slowly up the ladder.

(a) Determine the magnitude of the frictional force between the ladder and the ground when the man has climbed up 3.00 m along the length of the ladder.

(b) How far along the length of the ladder can the man climb before the ladder starts to slip?

Hint: You may first want to determine the maximum magnitude of the frictional force that the ground can exert on the ladder.

9-37 A constant net torque of 20.0 N·m is exerted on a pivoted wheel for 10.0 s, during which time the angular velocity of the wheel increases from 0 to 100 rpm. The external part of the torque is then removed, and the wheel is brought to rest by friction in its bearings in 100 s. Calculate

(a)   the moment of inertia of the wheel

(b)   the magnitude of the frictional torque

(c)   the total number of revolutions made by the wheel.

9-38 A yo-yo is made from two uniform solid discs, each of mass M and radius R, connected by a light axle of radius b (see diagram). A string is wound several times around the axle and then held stationary while the yo-yo is released from rest, dropping as the string unwinds. Find expressions for the magnitudes of the downward acceleration of the yo-yo and of the tension in the string. Neglect the weight of the light axle.

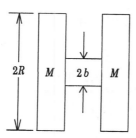

9-39 Consider the system sketched in the following diagram. The pulley has radius R and moment of inertia I. The rope does not slip over the pulley. The coefficient of kinetic friction between block A and the tabletop is μ_k. The system is released from rest and the block B descends. Use energy methods to find an expression for the speed of block B as a function of the distance h that it has descended. The masses of the blocks A and B are m_A and m_B, respectively. Neglect the mass of the rope and assume that it does not stretch while the blocks are moving.

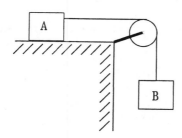

9-40 A turntable of radius 10.5 cm and mass 200 g is rotating at a constant angular velocity of 33$\frac{1}{3}$ rpm. There is no frictional torque and the turntable motor is turned off. A 100-g ball of putty is dropped from a negligible height onto the outer edge of the turntable. What is the change in kinetic energy of the system? Treat the turntable as a uniform solid disc and the ball of putty as a point particle.

Answers

9-1 47.1 m/s

9-2 1.04×10^4 rev/min

9-3 0.0158 s^{-1}, 63.5 s, 99.0 m/s

9-4 22.7 m/s

9-5 0.232; increase

9-6 921 N

9-7 $T_1 = 1.11 \times 10^3$ N, $T_2 = 620$ N

9-8 49 N

9-9 0.719 m

9-10 $T = 2w \dfrac{\sin(\theta)}{\sin(\phi)}$

9-11 -26.4 N·m

9-12 2.31×10^3 N , 2.18×10^3 N at 58.0° below horizontal

9-13 20.9 min

9-14 (a) 4.20×10^3 rpm, 440 rad/s
(b) 2.40×10^3 rev

9-15 8.48 m/s

9-16 (a) 8.50 m/s
(b) 5.71 rad/s^2

9-17 (a) $a_t = 0.150$ m/s^2, $a_c = 0$
(b) $a_t = 0.150$ m/s^2,
  $a_c = 1.26$ m/s^2

9-18 3.14 s

9-19 (a) -50.0 rad/s^2
(b) 150 rad/s, 350 rad/s
(c) 1.00×10^3 rad

9-20 (a) 0.0722 kg·m^2
(b) 0.0361 kg·m^2

9-21 (b)

9-22 10.2 N·m

9-23 411 N

9-24 (a) 3.08 s (b) -84.0 J

9-25 16.7 m/s

9-26 $v = \sqrt{\dfrac{2gh}{\left(1 + \dfrac{M}{2m}\right)}}$, $\omega = \dfrac{v}{R}$

9-27 (a) 4.02 rad/s (b) 6.03 rad/s

9-28 0.596 rev

9-29 0.568 rev/s

9-31 (a) the horizontal component of the normal force of the road on the car
(b) 8.95°

9-32 (a) 3.33×10^3 N
(b) 9.49 m/s; no

9-33 600 N to the right (upper hinge); 781 N upward and to the left at 39.8° above horizontal (lower hinge)

9-34 (a) 697 kg
(b) 2.48×10^4 N, 45.8° above horizontal

9-35 (a) 2.76 F
(b) $F_{L,x} = -2.12\,F$,
  $F_{L,y} = -2.78\,F$

9-36 (a) 330 N (b) 5.5 m

9-37 (a) 19.1 kg·m^2
(b) 2.00 N·m
(c) 91.7 rev

9-38 $a = \dfrac{g}{\left(1 + \dfrac{R^2}{2b^2}\right)}, T = M\,\dfrac{R^2}{b^2}\,a$

9-39 $\sqrt{\dfrac{2gh(m_B - \mu_k m_A)}{m_A + m_B + I/R^2}}$

9-40 -3.36×10^{-3} J

10 Elasticity and Scaling

10.1   Introduction

In Chapters 8 and 9 we studied the effects of forces on objects, in particular the relationship between forces and the motion of bodies. It was assumed that any extended body was either perfectly rigid or else composed of several jointed segments, each of which is perfectly rigid (Section 9.7). This is an idealized model which assumes that a body does not stretch, compress or bend when acted on by forces. In practice, all materials deform to some extent under the action of applied forces. The ability of a solid material to undergo deformation without breaking or becoming permanently deformed is termed its *elasticity*. The structural elements of animal bodies—bones, muscles and tendons—vary greatly in their ability to withstand applied forces without fracture or tear. The same is also true of plant bodies. In this chapter, we shall introduce a few simple concepts which enable us to predict what deformations occur when forces are applied to real bodies.

The skeleton of an animal must be able to support the weight of the animal, otherwise it may fracture under compression. This is one factor which is believed to impose upper limits on the size that any land animal can attain. The study of the effects on body structure and function due to variations in size among otherwise similar organisms uses the method of *scaling*, a type of dimensional analysis which forms the second topic of this chapter.

10.2   Elasticity: Tensile Stress and Strain

The simplest type of elastic behaviour to describe is the stretching of a rod, bar, wire, bone, or tendon when its ends are pulled. Figure 10-1 shows a bar with uniform cross-sectional area $\mathcal{A}$ subjected to a pair of equal and opposite forces $\vec{F}$ and $-\vec{F}$ pulling on its ends. The bar is said to be under *tension*: this is the same concept as described in connection with the tension of ropes and cables in Chapter 8. The *tensile stress* (often just *stress*) σ on the bar is defined as the ratio of the magnitude of $\vec{F}$ to the cross-sectional area $\mathcal{A}$, that is[1]

$$\sigma = \frac{F}{\mathcal{A}} \qquad\qquad \textbf{[10-1]}$$

Clearly, the SI unit of stress is the newton per square metre (N/m^2).

As a result of being under tensile stress, any solid object will be stretched from an initial length ℓ_0 to a new length $\ell_0 + \Delta\ell$, as shown in Fig. 10-1. The fractional change in length of the object is called the *tensile strain* (often just *strain*) and is denoted ε, that is

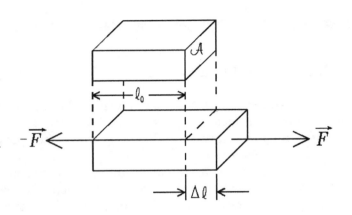

Figure 10-1

Under tensile stress, an object's length increases by an amount $\Delta\ell$.

$$\varepsilon = \frac{\Delta\ell}{\ell_0} \qquad\qquad \textbf{[10-2]}$$

Again, note that Eq. [10-2] is just a definition of tensile strain. Since ε is a ratio of two lengths, it is dimensionless.

The same definitions of stress and strain apply if the equal and opposite forces push inward rather than outward on the object. In such a case, the bar is said to be under *compression*. The corresponding stress, $\sigma = F/\mathcal{A}$, is now called the *compressive stress* while the corresponding strain, $\varepsilon = \Delta\ell/\ell_0$, is called the *compressive strain*. Of course, in this case the bar will decrease rather than increase in length, so that $\Delta\ell$ is interpreted as the <u>magnitude</u> of the length change.

---

1.   Although the bar in Fig. 10-1 is acted on by <u>two</u> equal and opposite forces, stress is defined with respect to just the single common magnitude *F* of those forces. Two equal and opposite forces must act on the bar because otherwise the net applied force would be non-zero, resulting in an acceleration of the bar.

It is usually observed that, for sufficiently small stress, the stress and strain are proportional. This relationship is known as *Hooke's law*[2] and is written as

$$\sigma = Y\varepsilon \qquad\qquad \textbf{[10-3a]}$$

or, equivalently,

$$\frac{F}{\mathcal{A}} = Y\,\frac{\Delta\ell}{\ell_0} \qquad\qquad \textbf{[10-3b]}$$

where the proportionality constant Y is known as *Young's modulus*. Since the strain is dimensionless, the dimensions and units of Y are the same as those of stress.

The relationship Eq. [10-3] basically states that the harder you squeeze or pull on an object, the more it compresses or stretches. Young's modulus Y can be considered a measure of the strength or stiffness of a material: a material with a large value of Y is relatively unstretchable or incompressible compared with a material having a smaller value of Y. (We will assume here, as is often but not always found, that Young's moduli for tensile and compressive deformations have the same values.) Table 10-1 lists values of Young's modulus for a number of materials, including several biomaterials. Note the wide range of these values, spanning materials which would be considered very "soft" or "pliant" to very "hard" or "rigid."

Table 10-1   Young's Modulus for Various Materials

Material	Y (N/m²)
Rubber	$(0.7\text{–}2)\times 10^7$
Wood	$(3\text{–}7)\times 10^7$
Silk	4×10^9
Glass	6×10^{10}
Steel	2×10^{11}
Blood vessel	$(1.2\text{–}4)\times 10^5$
Tendon	$(2\text{–}10)\times 10^7$
Cartilage	$(1\text{–}4)\times 10^7$
Teeth	1.5×10^{10}
Bone	$(1\text{–}2)\times 10^{10}$

2.   In Chapter 1 (Eq. [1-1]), Hooke's law was introduced for springs in a different form: $F_x = -kx$, which also shows the proportionality between applied force F_x and elongation x.

EXAMPLE 10-1

A strip of tissue 5.00 cm long with a cross-sectional area of 1.00×10^{-5} m² is cut from the wall of an aorta. This material has Young's modulus of 2.00×10^{5} N/m². A mass m is attached to one end of the strip and the other end is fixed to the ceiling so that the strip is suspended vertically (Fig. 10-2). What value of m will cause a 0.600-cm elongation of the strip? (Neglect the weight of the strip itself.)

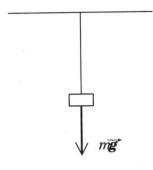

$m\vec{g}$

Figure 10-2

Example 10-1: A mass suspended from a strip of aortic tissue.

SOLUTION

As shown in Fig. 10-2, the mass m exerts a downward force $m\vec{g}$ on the strip. At equilibrium, the supporting ceiling necessarily exerts an equal and opposite upward force of magnitude mg on the strip. Hence the magnitude of the force F entering Eq. [10-3b] is $F = mg$. Given Y, the original length ℓ_0 and elongation $\Delta\ell$ of the strip, this equation becomes

$$\frac{mg}{A} = Y\,\frac{\Delta\ell}{\ell_0} = 2.00 \times 10^{5}\,\frac{N}{m^2}\left(\frac{0.600\,cm}{5.00\,cm}\right)$$

$$= 2.40 \times 10^{4}\,\frac{N}{m^2}$$

Notice that ℓ_0 and $\Delta\ell$ were left in their original units of centimetres, since only their ratio is required. Knowing the area A, we can solve the last equation for the suspended mass m:

$$m = (2.40 \times 10^{4}\,N/m^2)\,\frac{A}{g}$$

$$= (2.40 \times 10^{4}\,N/m^2)\,\frac{1.00 \times 10^{-5}\,m^2}{9.80\,m/s^2}$$

$$= 2.45 \times 10^{-2}\,kg$$

Elasticity and Plasticity

In introducing Hooke's law, Eq. [10-3a], we stated that it holds only for "sufficiently small stress." Whenever Hooke's law is obeyed, a graph of stress σ vs. strain ε is a straight line of slope equal to Young's modulus Y. However, beyond a certain limiting value of the stress, depending on the material, the stress and strain are no longer proportional. Typical graphs of stress vs. strain are shown in Fig. 10-3a,b. In each case, the straight-line portion ends at a point labeled "a" which is called the *proportional limit*. Beyond this point, the stress–strain curve is not a straight line and <u>Hooke's law is not obeyed</u>. Beyond the proportional limit, different materials can exhibit quite different stress–strain graphs (Fig. 10-3a,b).

Each graph in Fig. 10-3 is labeled by two other points. Point b is called the *elastic limit* or *yield point*, and the region of the graph between the origin and point b is called the *elastic range*. Within this

Figure 10-3

Two possible stress–strain relationships.

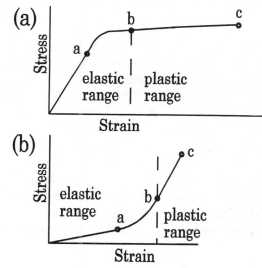

range, if the stress applied to a material is removed, the material will relax back to its original length. Often a deformation occurring in this range is completely *reversible*, that is, the stress–strain curve is exactly retraced when the stress is gradually reduced from the value at point b. Many materials, however, including most biomaterials, follow a slightly different stress–strain curve when the stress is decreased from point b to zero. This effect is known as *elastic hysteresis*. One uses the term *resilience* to describe how closely matched are the stress–strain curves for increasing and decreasing stress: in a material with 100% resilience, these curves are identical.

The region of the stress–strain graph beyond point b is called the *plastic range*. When the stress applied to a material is in this range, the material does not return to its original length when the stress is removed. On reducing the stress gradually, the material usually exhibits a quite different stress–strain curve which does not return to the origin but rather ends at a positive strain ε when the stress σ becomes zero (not shown in Fig. 10-3). This means that the material has undergone a permanent deformation.

On further increasing the stress, eventually the *fracture point* (or *breaking point*) c is reached, which has an obvious meaning. The value of the stress at this point is the maximum stress that can be applied without fracturing the material, and is known as the *breaking stress* or (for tensile forces) the *tensile strength* of the material. Some materials may have comparable Young's moduli but quite different values of the breaking stress. The fracture point c is also characterized by the corresponding value of the strain ε, which indicates the maximum deformation a material can withstand without breaking. For a stretching deformation, the limiting value of the strain at the fracture point is called the *extensibility*. In some materials, the fracture strain is much greater than the strain at the elastic limit (Fig. 10-3a), that is, the material can be stretched significantly before breaking. Such a material is said to be *ductile*. In contrast, a *brittle* material is one for which the fracture strain is relatively close to that at the elastic limit (Fig. 10-3b).

Biomaterials

We see that solid materials exhibit a wide diversity of responses to applied stresses. This is especially true of biomaterials. Depending on their role, biological tissues must be capable of resisting either tensile or compressive stresses. Tissues often undergoing tensile strain are the tendons, ligaments, muscles, blood vessels and nerves of animal bodies. Tissues often under compression are bones, vertebrae and joint cartilage. Ultimately, the elastic properties of a material are dependent on the chemical composition and the arrangement of, as well as intermolecular forces between, its constituent molecules. In rods or fibres of biomaterials, the molecular processes which occur during stretching or compression may be quite complicated, such as the uncoiling of molecular chains or the breaking of cross-linked chemical bonds.

Most structural components of biological systems are *composite materials* assembled from two or more chemical species. This is often important for enabling a tissue to respond to a variety of mechanical stresses. As an example, consider the nuchal ligament along the top of the neck of cows and other ungulates, described earlier in Box 9-2. It must be strong enough to support the head of the animal when grazing. It should also stretch readily when at first subjected to a sudden stress but then should "tighten up" before the strain becomes excessive, therefore acting as a biomechanical shock absorber.

These desired properties are exhibited by the stress–strain graph (solid curve) in Fig. 10-4. The two major structural components of the ligament are the fibrous proteins *collagen* and *elastin*. When each component is isolated (e.g., by the action of suitable enzymes), collagen and elastin produce quite different stress–strain curves shown by the dashed lines in Fig. 10-4. At first, collagen stretches easily with very little applied stress and then its stress–strain curve rises sharply. Under small stresses, it is mainly the elastin component, a "natural rubber" with a large extensibility, that provides the major resistance to stress. Analysis at the microscopic level indicates that the collagen proteins have a helical molecular arrangement and are essentially slack at low values of the stress. The collagen fibrils do not play a significant role until the ligament has stretched enough to "take up the slack." We shall see in Chapter 11 that the same basic mechanism, depending on the different elastic properties of collagen and elastin, controls the behaviour of arterial walls under swelling and plays a major role in determining the susceptibility of an artery to develop an aneurysm.

Figure 10-4

Stress–strain diagram for the nuchal ligament.

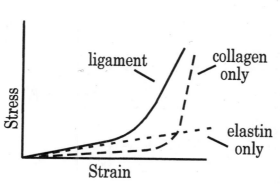

In combination with other structural proteins as well as with water, minerals, and carbohydrates, collagen is a major component of all connective tissue in animals, including bone, muscle, skin and blood vessels. On the other hand, the main structural component responsible for the tensile strength and rigidity of plants is *cellulose*, a polymeric sugar (or polysaccharide).

One further complication often arises in classifying the stress–strain behaviour of biomaterials. This is the fact that often the strain resulting from an applied stress does not develop instantaneously (as we have tacitly assumed up to now) but rather is time-dependent. For example, a fibre may stretch slowly in response to a stress, and return slowly to its original size and shape when the applied force is removed. Also, quite often the strain produced depends on the *rate* at which the stress is applied. Materials with this behaviour are called *viscoelastic*, exhibiting some of the characteristics of fluids, which will be discussed further in Chapter 12.

10.3   **Elasticity: Shear Stress and Strain**

Besides tension and compression, a solid material can be subjected to another type of stress, which is called *shear stress*.[3] The simplest example of this is illustrated in Fig. 10-5a, where a pair of equal and opposite forces of magnitude F_S are applied tangentially to opposite surfaces of a body. Analogous to Eq. [10-1], we define the *shear stress* σ_S on the body as

$$\sigma_S = \frac{F_S}{\mathcal{A}} \qquad\qquad \textbf{[10-4]}$$

where $\mathcal{A}$ is the area of each surface on which the forces act. This type of stress will result in a deformation of the body pictured in Fig. 10-5b, in which the opposite surfaces are displaced parallel to each other while their lengths remain nearly constant. In contrast to tensile or compressive strain, in which molecules of the material are pulled apart or pushed together, during shear, layers of molecules are forced to slide past each other. The *shear strain* ε_S is defined as

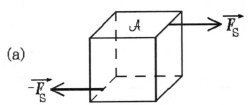
(a)

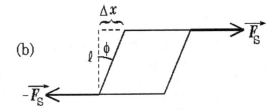
(b)

Figure 10-5

Deformation of a body under shear stress:
(a) before deformation;
(b) after deformation.

$$\varepsilon_S = \frac{\Delta x}{\ell} = \tan\phi \qquad\qquad \textbf{[10-5]}$$

where Δx is the relative parallel displacement of the two surfaces and ℓ is the vertical distance between them. The angle ϕ is defined in Fig. 10-5b. In practice, Δx is usually much smaller than ℓ and hence the deformation angle ϕ is very small, so that $\tan\phi \approx \phi$ and the shear strain is simply equal to ϕ, where the latter is expressed in units of radians.

For sufficiently small shear stress, σ_S and ε_S obey a form of Hooke's law analogous to Eq. [10-3], namely

$$\sigma_S = G\varepsilon_S \qquad\qquad \textbf{[10-6a]}$$

---

3.   One other type of deformation, bending, actually consists of simultaneous tensile and compressive deformations.

or

$$\frac{F_s}{\mathcal{A}} = G\,\frac{\Delta x}{\ell} \approx G\,\phi \qquad\qquad \textbf{[10-6b]}$$

where G is known as the *shear modulus*. Like Young's modulus Y, G has the same dimensions and units as stress. Usually, the value of G for a given material is one-third to one-half the value of Y. This means that most materials resist tensile or compressive stresses better than they resist shear stresses. In animal bodies, shear stresses often occur at joints, in the spinal vertebrae, and in the arm and leg bones.

EXAMPLE 10-2

Between each pair of vertebrae of the spine is a disc of cartilage of thickness 0.17 cm and radius 1.1 cm. Suppose the bottom of the disc is held fixed while a shearing force of magnitude 25 N is applied to the top surface, as shown in Fig. 10-6. If the top surface moves a distance of 4.5 µm relative to the bottom surface, what is the shear modulus of cartilage?

SOLUTION

Figure 10-6

Example 10-2: Disc of cartilage between spinal vertebrae.

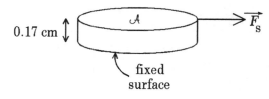

0.17 cm

fixed surface

Although only the shearing force applied to the top surface of the disc is shown in Fig. 10-6, this force is necessarily accompanied by an opposing force of the same magnitude on the bottom surface, to prevent overall motion of the disc. We rewrite Eq. [10-6] to solve for G:

$$G = \frac{(F_s/\mathcal{A})}{(\Delta x/\ell)} = \frac{F_s\,\ell}{\mathcal{A}\,\Delta x}$$

Substituting the values for F_s, ℓ, Δx and $\mathcal{A} = \pi r^2$ where r is the radius of the disc gives

$$G = \frac{(25\ \text{N})(1.7\text{x}10^{-3}\ \text{m})}{\pi(1.1\text{x}10^{-2}\ \text{m})^2(4.5\text{x}10^{-6}\ \text{m})}$$

$$= 2.5\text{x}10^7\ \text{N/m}^2$$

Most of the qualitative discussion of stress–strain behaviour given earlier also applies to shear stress and strain, so we won't go into further details. Instead, let's analyze a particular type of shear deformation which commonly occurs. This is known as *torsion*, which develops when a shear force of magnitude F_s is applied tangentially to the circumference at one end of a cylindrical rod (such as a bone) while the other end is held fixed, as pictured in Fig. 10-7. The corresponding shear stress is given by $\sigma_S = F_S/\mathcal{A}$, where $\mathcal{A}$ is the area of the circular end of the rod, and will cause the rod to twist around its long axis. Such a deformation occurs when a leg bone is twisted (for example, due to a skiing accident), often resulting in spiral or torsional fracture of the bone.

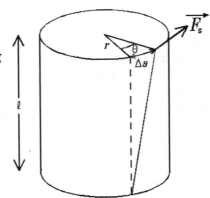

Figure 10-7

Torsional deformation of a solid cylinder.

Due to the stress, a point on the circumference of the top end of the rod is rotated through a distance Δs, subtending an angle θ as indicated in Fig. 10-7. Lower regions of the rod are twisted by smaller amounts, and at the bottom there is no rotation at all. The dashed and solid lines on the outer surface of the cylinder in Fig. 10-7 indicate how the amount of twisting varies from the top to the bottom of the rod. The shear strain at the outer surface of the cylinder is $\varepsilon_S = \Delta s/\ell$, where ℓ is the length of the rod. At the vertical axis through the centre of the rod, however, the shear strain is zero. Consequently, there is an <u>average</u> shear strain between the centre and outer surface of the rod given by

$$(\varepsilon_S)_{av} = \frac{(0 + \Delta s/\ell)}{2} = \frac{\Delta s}{2\ell}$$

This is not completely rigorous, but the final result is correct. Using $(\varepsilon_S)_{av}$ for the strain in Eq. [10-6a], we have

$$\sigma_S = G(\varepsilon_S)_{av}$$

or

$$\frac{F_S}{\mathcal{A}} = G\frac{\Delta s}{2\ell} \qquad \qquad \textbf{[10-7]}$$

The area $\mathcal{A}$ is that of the circular top end of the rod, $\mathcal{A} = \pi r^2$ where r is the radius of the rod. The previous equation can then be written as

$$F_S = \frac{\pi r^2 G \Delta s}{2\ell}$$

Multiplying both sides of this equation by r, and using $\Delta s = r\theta$, we finally get

$$\tau_S = \frac{\pi r^4 G \theta}{2\ell} \qquad \qquad \textbf{[10-8]}$$

where $\tau_S = F_S r$ is the *torque* exerted about the top surface of the rod. Equation [10-8] is our basic result, relating the torsion angle θ to the twisting torque τ_S.

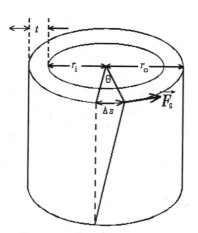

Figure 10-8

Torsional deformation of a hollow cylinder.

The derivation above applies to a filled or solid cylinder. We can also consider a hollow cylinder such as pictured in Fig. 10-8, which more closely approximates the leg bone of an animal. The thickness t and *average radius* $\bar{r}$ of the cylinder are given by

$$t = r_o - r_i \qquad \textbf{[10-9]}$$

and

$$\bar{r} = \frac{r_o + r_i}{2} \approx r_o \text{ for } t \ll r_o \qquad \textbf{[10-10]}$$

where r_i and r_o are the inner and outer radii of the cylinder, respectively. The area $\mathcal{A}$ must now be taken as that of the <u>solid</u> part of the top surface between r_i and r_o, namely

$$\begin{aligned}
\mathcal{A} &= \pi r_o^2 - \pi r_i^2 \\
&= \pi(r_o^2 - r_i^2) \\
&= \pi(r_o + r_i)(r_o - r_i) \\
&= 2\pi\bar{r}t
\end{aligned}$$

If the thickness t is much smaller than the mean radius $\bar{r}$, all parts of the top surface will be rotated through the same distance Δs, and hence the average strain in this case is simply $(\varepsilon_S)_{av} = \Delta s/\ell$. Equation [10-7] is then replaced by

$$\frac{F_S}{\mathcal{A}} = G\,\frac{\Delta s}{\ell}$$

Repeating the analysis following Eq. [10-7] using the current expression for area $\mathcal{A}$ and $r_o \approx \bar{r}$, we find

$$\tau_S = \frac{2\pi G\bar{r}^3 t\theta}{\ell} \qquad \textbf{[10-11]}$$

Equation [10-8] for a solid rod, or [10-11] for a hollow one, can be used to determine the shear modulus of the material making up a cylindrical rod, by applying a known torque τ_S and measuring the resulting twist angle θ.

EXAMPLE 10-3

During a skiing accident, one end of a person's lower leg bone is twisted by 8.0° relative to the other end. The leg bone can be approximated by a hollow cylinder of average radius 2.5 cm, thickness 0.60 cm, and length 0.35 m. What is the magnitude F_S of the force acting at a right angle to the tip of the ski (Fig. 10-9)

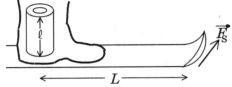

Figure 10-9

Example 10-3: Twisting force F_S on lower leg bone.

which caused this twist? The distance L from the tip of the ski to where the foot is attached is 1.25 m, and the shear modulus of bone is 1.0×10^{10} N/m².

SOLUTION

We use Eq. [10-11] to determine the twisting torque τ_S and hence the force F_S, where $\tau_S = F_S L$ (Fig. 10-9). Note that the twist angle, $\theta = 8.0°$, must be converted to radian units for use in Eq. [10-11]: $\theta = 8.0° \times \pi$ rad/180° = 0.14 rad.

$$F_S = \frac{2\pi G \bar{r}^3 t \theta}{L \ell}$$

$$= \frac{2\pi(1.0 \times 10^{10}\,\text{N/m}^2)(0.025\,\text{m})^3(0.60 \times 10^{-2}\,\text{m})(0.14\,\text{rad})}{(1.25\,\text{m})(0.35\,\text{m})}$$

$$= 1.9 \times 10^3\,\text{N}$$

EXAMPLE 10-4

A solid cylindrical rod and hollow cylindrical rod are both made of the same material and have the same length. Both rods have the same outer radius r_o, while the inner radius r_i of the hollow rod is 0.870 r_o. If one end of each rod is clamped while a weight w is suspended from the other end of each rod, as shown in Fig. 10-10, what is the ratio of the twist angles produced in the rods?

SOLUTION

Since both rods have the same outer radius, they experience the same twisting torque $\tau_S = wr_o$. Hence, equating the expressions for τ_S from Eqs. [10-8] and [10-11], we have

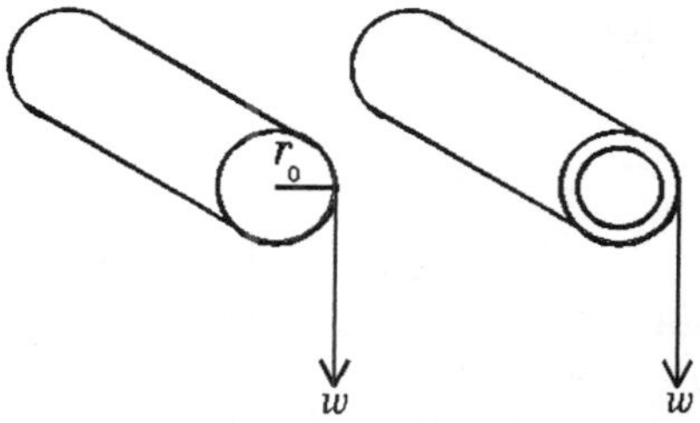

$$\frac{\pi r^4 G \theta_S}{2\ell} = \frac{2\pi G \bar{r}^3 t \theta_H}{\ell}$$

where the twist angles of the solid and hollow rods are denoted θ_S and θ_H, respectively. Cancelling the common factor $\pi G/\ell$ and re-arranging, we obtain

$$\frac{\theta_S}{\theta_H} = \frac{4\bar{r}^3 t}{r^4}$$

Figure 10-10

Example 10-4: Solid and hollow cylindrical rods.

The radius of the solid rod r is identical to r_o, while using Eq. [10-9] ($t = r_o - r_i$) and Eq. [10-10] [$\bar{r} = (r_o + r_i)/2$] and the fact that $r_i = 0.870\, r_o$, the mean radius and thickness of the hollow rod are found to be

$$\bar{r} = 0.935\, r_o \text{ and } t = 0.130\, r_o$$

Then the ratio of twist angles becomes

$$\frac{\theta_S}{\theta_H} = \frac{4(0.935\, r_o)^3 (0.130\, r_o)}{r_o^4} = 0.425$$

noting that the factors of r_o cancel.

Therefore, the twist angle of the solid rod is less than half that of the hollow rod, i.e., the solid rod has greater resistance to torsion. This is because the solid rod contains more material in its central region to be deformed.

Why Hollow Bones Are Good

A different outcome from that of the last example is obtained if both the solid and hollow rods have equal values of the *solid parts* of their cross-sectional areas rather than equal outer radii, that is, if $\pi r^2 = 2\pi \bar{r} t$. This is equivalent to saying that the rods have the same mass. For equal values of the applied torque, we still obtain the same relation between the twist angles as in Example 10-4, namely $\theta_S/\theta_H = 4\bar{r}^3/r^4$. For equal areas, this reduces to $\theta_S/\theta_H = \bar{r}/t$. Usually t is much smaller than $\bar{r}$ for a thin-walled hollow cylinder, and hence $\theta_S/\theta_H > 1$, i.e., the solid rod experiences a <u>greater</u> twist deformation. This is because both rods now contain the same mass of solid material, but the material near the central part of the solid rod is deformed little and therefore contributes little resistance to torsion. The greater torsion strength of a hollow rod compared with a solid rod of the same mass and length may be one reason why animals have evolved hollow rather than solid leg bones. This fact is certainly exploited in the bones of birds, which are thin hollow tubes with a high "strength-to-weight" ratio.

10.4 **Scaling**

Living organisms come in an enormous range of sizes, from the tiniest bacterium (length ≈ 0.3 μm) to blue whales (length ≈ 30 m) or the giant sequoia trees of California (height up to 100 m). One is tempted to ask if there are upper and lower limits to the size of organisms. To answer this question, we have to study how the size of an organism affects its interactions with its physical environment, which are strongly constrained by the laws of physics and chemistry.

The most important limiting factors on organism size are the strength of supporting structures (e.g., skeletons in animals, cell-wall tissue in plants), the demands on the muscular systems of animals for locomotion, and the need for energy intake (metabolism) via food and oxygen consumption. It is likely that lower limits to organism size are set primarily by metabolic needs and the accessibility of required nutrients as well as by the sizes of the macromolecules (e.g., proteins, DNA) present inside a biological cell. As yet there is no conclusive answer to whether an upper size limit to organisms exists. However, there is a very instructive body of arguments based on the ideas of *scaling*[4] which goes a long way toward suggesting answers to this question. Scaling is a type of dimensional analysis which describes how different parts of a system change size when the system as a whole changes size.

An important example of scaling believed to be related to the upper limits on the size of land animals is illustrated by the variation of leg-bone diameter with overall animal size. It was apparently first pointed out by Galileo that the leg bones of large mammals are disproportionately thicker, relative to overall body size, than those of smaller mammals. This is shown in Fig. 10-11, which compares the skeletons of a cat and an elephant, with both skeletons drawn to have approximately the same height. It is clear that, <u>relative</u> to the animal's height, the elephant's leg bones are significantly thicker than those of the cat. Why is this? Clearly, since the leg bones have to support the

---

4.   K. Schmidt-Nielsen, *Scaling: Why is Animal Size So Important?* (Cambridge University Press, Cambridge, 1984).

weight of the animal, the strength of these bones must increase as the weight increases. As we shall discuss in more detail later, the strength of a bone is proportional to its <u>cross-sectional</u> area. It will be shown that the disproportionate increase in leg-bone diameter with animal size follows from this fact with the use of simple scaling arguments, to which we now turn.

Figure 10-11

The skeletons of (a) a cat and (b) an elephant, drawn to have approximately the same height.

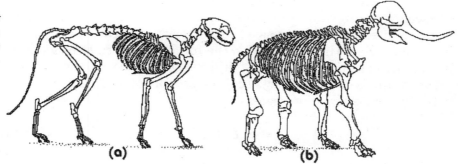

Geometric Scaling

Consider a set of geometrically similar (or *isometric*) objects, that is, objects of similar shape but different size. The simplest is a set of spheres. If R is the radius of a sphere, it has a surface area $\mathcal{A} = 4\pi R^2$ and volume $\mathcal{V} = 4/3\ \pi R^3$. Suppose we compare two spheres, one having double the radius of the other. The surface area of the larger sphere is $4\pi(2R)^2 = (2^2)\ 4\pi R^2$ while its volume is $4/3\ \pi(2R)^3 = (2^3)\ 4/3\ \pi R^3$, that is, the surface area of the larger sphere is $2^2 = 4$ times that of the smaller sphere and its volume is $2^3 = 8$ times that of the smaller sphere.

The same relative relationships between surface area and volume are obtained when comparing other isometric objects, such as the pair of cubes in Fig. 10-12. If each side of a cube has length L, the surface area of the cube is $\mathcal{A} = 6L^2$ (since it has six sides) while its volume is $\mathcal{V} = L^3$. Comparing two cubes whose sides have lengths L_1 and L_2, we see that the <u>ratios</u> of their surface areas and volumes obey $\mathcal{A}_2/\mathcal{A}_1 = (L_2/L_1)^2$ and $\mathcal{V}_2/\mathcal{V}_1 = (L_2/L_1)^3$. Again, if the ratio of lengths L_2/L_1 equals 2, the ratios of the surface areas and volumes are

Figure 10-12

Comparing two cubes of side lengths L_1 and L_2.

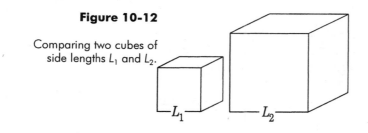

4 and 8, respectively. Of course, we could consider any other positive value for the length ratio L_2/L_1, not necessarily an integer, nor a value necessarily greater than 1.

These relationships generalize to any other set of isometric three-dimensional bodies, such as "cubical" objects whose side lengths are not all equal. They even apply to irregularly shaped objects such as animal bodies. The key point is that when the size of such an object is scaled up or down, all *linear dimensions* (e.g., length, width, and height) scale up or down in the same proportion. We could choose any one of these linear dimensions as a "characteristic length" L to represent the size of the object. Then the fundamental rules of scaling, when comparing isometric bodies of different size, are that the surface areas and volumes of such objects scale in proportion to L^2 and L^3, respectively. These are expressed as

$$\mathcal{A} = k_{\mathcal{A}} L^2 \propto L^2 \qquad \text{[10-12a]}$$

$$\mathcal{V} = k_{\mathcal{V}} L^3 \propto L^3 \qquad \text{[10-12b]}$$

where the proportionality constants $k_{\mathcal{A}}$ and $k_{\mathcal{V}}$ depend on the precise shape of the objects. However, usually these constants do not need to be specified (since they cancel when considering ratios, as in the comparison of cubes in the previous paragraph), and hence the relationships for $\mathcal{A}$ and $\mathcal{V}$ can be left as simple proportionalities.

The density, i.e., mass/volume, of tissue in most living organisms is close to that of water. It follows that body mass M is proportional to volume, and hence obeys the same scaling rule as in Eq. [10-12b], i.e.,

$$M \propto L^3 \qquad \text{[10-13]}$$

Two other important relationships follow from Eq. [10-12]. The ratio of surface area to volume is seen to obey

$$\mathcal{A}/\mathcal{V} \propto L^{-1} \qquad \text{[10-14]}$$

This illustrates the well-known fact that the surface-to-volume ratio decreases as the size of an object increases. Alternatively, Eq. [10-12b] can be expressed as $L \propto \mathcal{V}^{1/3}$, which on insertion in Eq. [10-12a] gives

$$\mathcal{A} \propto \mathcal{V}^{2/3} \qquad \text{[10-15a]}$$

Using the proportionality between mass and volume, Eq. [10-15a] could also be written as

$$\mathcal{A} \propto M^{2/3} \qquad \text{[10-15b]}$$

The relationships in Eqs. [10-15a] and [10-15b] are the most important ones to remember from this section.

Surface Area and Volume Effects

The reduction in surface area relative to volume with increasing size is probably the single most important factor in determining how size affects biological design. This should not be surprising, since contact between an organism and its surroundings occurs mainly via its surface, while its internal processes and structure depend mainly on its volume.[5] For example, the rate of heat loss from a body is proportional to its surface area. In a cold climate, an animal with a low surface-to-volume ratio can more easily conserve heat, which may explain why there are relatively few small mammals in arctic regions. On the other hand, to maintain a constant body temperature, heat loss through the surface must be balanced by heat production due to metabolic activity inside the body. If the rate of heat production were proportional to total body mass M, then large animals would tend to overheat. Hence the metabolic rates of animals do <u>not</u> scale in proportion to their masses but to a smaller power, approximately $M^{0.75}$. The fact that the exponent 0.75 is slightly larger than the value $2/3 \approx 0.67$ expected on the basis of the "surface rule," Eq. [10-15], has been the subject of considerable study and debate.

Microorganisms such as bacteria depend on the supply of oxygen and nutrients through transport across their membranes. In this case, a large surface-to-volume ratio is favourable. With increasing size, that ratio diminishes according to Eq. [10-14] and this may have a bearing on limiting the size of biological cells.

EXAMPLE 10-5

The rate of heat loss through the skin from a man of mass 80 kg is 85 W.

(a) What is the rate of heat loss from a child of mass 25 kg?

(b) Compare the <u>specific</u> rates of heat loss for the man and child, that is, the rates of heat loss per unit body mass.

5.   S. Vogel, *Life's Devices* (Princeton University Press, Princeton, 1988).

SOLUTION

(a) The rate of heat loss, i.e., *power P*, is proportional to the surface area, which in turn is proportional (Eq. [10-15]) to $M^{2/3}$. Hence, assuming the man (m) and child (c) are isometric, the ratio between their rates of heat loss is

$$\frac{P_c}{P_m} = \left(\frac{M_c}{M_m}\right)^{2/3}$$

Therefore

$$P_c = P_m \left(\frac{M_c}{M_m}\right)^{2/3}$$

$$= (85 \text{ W})\left(\frac{25}{80}\right)^{2/3}$$

$$= 39 \text{ W}$$

(b) The specific rate of heat loss is P/M. Hence for the man

$$\frac{P_m}{M_m} = \frac{85 \text{ W}}{80 \text{ kg}} = 1.1 \text{ W/kg}$$

while for the child

$$\frac{P_c}{M_c} = \frac{39 \text{ W}}{25 \text{ kg}} = 1.6 \text{ W/kg}$$

Therefore, the specific rate of heat loss is significantly larger for the child than for the adult.

Methods similar to those used in Example 10-5 can be applied to other quantities which scale in proportion to body surface area. For example, the amount of fur covering a mammal's body is proportional to the surface area times the thickness of the fur. If we compare two mammals of different size and assume that the fur thickness is the same for both animals, then the relative volumes or masses of fur will be in the same ratio as their surface areas (and hence in the ratio of their overall body masses to the power 2/3). However, it is often found that the fur thickness itself varies with body size, tending to increase with size among animals of the same species and thus better protecting larger animals against heat loss. This variation of fur thickness with size has to be taken into account in scaling arguments. Across different species of mammals, fur thickness is very much dependent on climate. For instance, large animals in warm climates tend to be hairless in order to facilitate heat loss.

Scaling of Bones and Skeletons

Now we are ready to analyze how the thickness of the leg bones of animals scale with overall body size, illustrated by the cat and elephant skeletons earlier in Fig. 10-11. As we stated, the strength of these bones must increase in proportion to the weight of the animal, otherwise they may fail under compression by crushing. How do we measure the "strength" of a bone? In Section 10.2, we described that the important parameter when a solid material is compressed is the *stress* on it, i.e., *the ratio of the compressive force to its cross-sectional area.*

It is plausible that the maximum compressive stress which can be supported by a bone equals the yield stress σ_b at the elastic limit (see Fig. 10-3), as any larger stress could produce a permanent deformation of the bone. If the corresponding maximum compressive force on the bone is denoted F_{max}, then $F_{max}/\mathcal{A}_B = \sigma_b$ or $F_{max} = \sigma_b \mathcal{A}_B$, where $\mathcal{A}_B$ is the cross-sectional area of the bone. This is the basis of the statement, made in the introduction to this section, that the strength of a bone, as measured by F_{max}, is proportional to its cross-sectional area. The bones of all animals are composed of essentially the same material and therefore have approximately the same yield stress σ_b. Hence, we can simply write $F_{max} \propto \mathcal{A}_B$, where the proportionality constant σ_b is independent of the animal species. Since the applied compressive force F_{max} is proportional to the animal's weight[6] and therefore to its mass M, we can conclude that

$$\mathcal{A}_B \propto M \qquad\qquad \textbf{[10-16]}$$

The cross-sectional area of a bone is $\mathcal{A}_B \approx \pi(d_B/2)^2$ where d_B is the bone diameter. Since, according to Eq. [10-13], the body mass M scales in proportion to L^3, where L is a characteristic linear dimension (e.g., height) of the body, Eq. [10-16] can be written as

$$d_B^2 \propto L^3$$

or

$$d_B \propto L^{3/2} \qquad\qquad \textbf{[10-17]}$$

---

6. Actually, the weight being supported by the legs is that of the whole body <u>minus</u> the weight of the legs, but it is reasonable to take this proportional to the total body weight w. Also, the compressive stress due to this weight is distributed over all leg bones, so that $F_{max} \propto w/n$ where n is the number of legs supporting the body (usually $n = 4$).

Hence, the bone diameter d_B does not scale in proportion to the characteristic length L but rather as $L^{3/2}$. This accounts for the disproportionate increase in bone diameter when a cat and elephant are compared. Were the bone diameter to scale with overall size according to geometric similarity, i.e., $d_B \propto L$, then its cross-sectional area and strength would increase only in proportion to L^2 while having to support a weight that increases proportional to L^3. This would inevitably lead to collapse of large animals.

In the analysis above, we have assumed that the leg bones of all animals are subjected to the maximum possible stress, the yield stress σ_b. This would allow for very limited margins of safety. In practice, the actual body weights supported by animal skeletons in static situations are much less (typically, by a factor of 10) than the limit set by the yield stress. On the other hand, when an animal is moving, the skeleton is subjected to much larger stresses during periods of acceleration and deceleration, stresses which can approach that at the yield point. This justifies gauging the strength of the bones by means of the yield stress σ_b. This is also consistent with the common occurrence of sprains and fractures in bones of athletes and racehorses during maximum performance.

All of the preceding analysis applies to terrestrial animals. Aquatic animals such as whales are very nearly neutrally buoyant in water, that is, their body weight is balanced by the buoyant force of water (Chapter 13). In these animals, the skeleton is not required to support the body weight and, consequently, does not scale out of proportion to body size. This is widely believed to be the main reason why whales can attain enormous sizes.

EXAMPLE 10-6

If all linear dimensions of an animal were increased by a factor of 5.0, by what factor should the leg-bone diameter be increased to achieve the same relative supporting strength?

SOLUTION

From Eq. [10-17], the bone diameter must scale proportional to $L^{3/2}$, where L is a typical linear dimension of the animal. Therefore, if L increases by a factor of 5.0, the bone diameter should increase by a factor of $5.0^{3/2} = 11.2$. (Note that the weight of the animal would increase by the enormous factor $5.0^3 = 125$, which also equals the scaling factor for the cross-sectional area of the legs.)

Let's consider some other consequences of the scaling relation Eq. [10-17] for bone thickness. Although derived specifically for the diameter of leg bones, Eq. [10-17] actually applies to all major skeletal bones. This is suggested by the cat and elephant skeletons shown in Fig. 10-11 and can be explained by the fact that the vertebrae and rib cage must also provide support for the body. The length of bones presumably should scale in proportion to the characteristic linear dimension L, hence the volume and mass of bones (proportional to the product of length and cross-sectional area) should scale as Ld_B^2. Using $d_B \propto L^{3/2}$, this becomes proportional to L^4. Expressing Eq. [10-13] ($M \propto L^3$) as $L \propto M^{1/3}$, we deduce that the total *skeletal mass* M_s scales with overall body mass as $M^{4/3}$, i.e., the skeletal mass increases out of proportion to body mass. Clearly, this would set a limit beyond which further increase in size becomes impossible, since the whole animal cannot be skeleton! (This is illustrated in Example 10-7.)

In reality, data for animal skeletons obey the scaling law $M_s \propto M^a$ where a ≈ 1.08 instead of 4/3 ≈ 1.33. Although this still means that skeletal mass increases with size more rapidly than does the total body mass, the rate of increase is significantly smaller than that predicted by our previous arguments, and remains within reasonable limits for even the largest animals.

The discussion above shows that the scaling laws for animal skeletons are not yet consistent with all facts. Alternative scaling theories have been proposed, most notably by McMahon,[7] although the latter still does not account for all observations. One feature which undoubtedly plays a role is the fact that animals of different sizes are <u>not</u> exactly isometric. For example, not all linear dimensions of the cat and elephant skeletons in Fig. 10-11 are in the same ratio, precisely because the leg bones of the elephant are relatively thicker than those of the cat. This non-isometric or *allometric* scaling of different body parts is, in fact, the general rule in biology. With variations in size, different parts of an organism do not change their "characteristic lengths" in the same proportion. In other words, changes in size affect the shape of organisms. Small degrees of non-isometry can lead to small but nonetheless significant deviations from the scaling laws developed in this chapter.

---

7.   T. McMahon, Size and Shape in Biology, *Science* **179**, 1201 (1973).

EXAMPLE 10-7

For a cat of total body mass $M = 5.0$ kg, it is found that 7.0% of this mass is due to skeleton, that is $M_s/M = 0.070$, where M_s is the skeletal mass. Assuming that skeletal mass varies with body mass as $M_s \propto M^{4/3}$, what percentage of body mass would be the skeletal mass of an elephant whose total body mass is 7000 kg?

SOLUTION

Dividing out a factor of mass M in the relation $M_s \propto M^{4/3}$, we find that the ratio of skeletal mass to total body mass varies as $M_s/M \propto M^{1/3}$. Now take the ratio of (M_s/M) for the elephant to that for the cat:

$$\frac{(M_s/M)_{ele}}{(M_s/M)_{cat}} = \left(\frac{M_{ele}}{M_{cat}}\right)^{1/3}$$

or

$$\left(\frac{M_s}{M}\right)_{ele} = \left(\frac{M_{ele}}{M_{cat}}\right)^{1/3} \left(\frac{M_s}{M}\right)_{cat}$$

Substituting values of the masses, this yields

$$\left(\frac{M_s}{M}\right)_{ele} = \left(\frac{7000}{5.0}\right)^{1/3} (0.070)$$
$$= 0.78$$

That is to say, 78% of the elephant's mass would be skeleton, an impossibly large percentage.

Dynamic Scaling

Our discussion above took account of both static and dynamic stresses on animal skeletons. However, other consequences of dynamics should be considered, since an animal must move if it is to survive. We have seen that larger animals must have heavier skeletons in order to support their bodies. But an increase in size puts greater demands on an animal's ability to move, and tends to reduce its agility. These effects appear to be more important in limiting the size of animals than do those considered previously.

To move itself, an animal must do work. The work required to move its body,[8] of mass M, a distance d is proportional to Md. In one step, the animal can move a distance on the order of its own size, and hence d is proportional to its characteristic linear dimension L. With M proportional to L^3 (Eq. [10-13]), the work required in one step is therefore $W \propto Md \propto L^4$. This work must be provided by the muscles in the animal's limbs. Just as the strength of a bone is proportional to its cross-sectional area, the maximum force F which can be exerted by a muscle is proportional to its cross-sectional area. (Compare a weight lifter's muscles with your own. The weight lifter's muscles are larger in cross-section, but not in length.) Assuming geometric similarity, then the area scales as L^2 while the distance moved in one step will again be proportional to L. Then the work that an animal could actually do is $W = Fd \propto L^3$. Thus we see that the work required to move an animal, $\propto L^4$, increases faster with increasing size than does the ability of the animal to provide the work, which clearly puts a limit on the size that can be attained.

Scaling arguments can also be used to analyze other aspects of animal motion such as running and jumping. It can be shown that animals of similar form but different masses ought to be able to jump to approximately the same height. As above, the maximum work that the muscles can provide is proportional to L^3. This work must lift a mass $M \propto L^3$ a distance h, hence the work required is proportional to L^3h. Consequently, h must be essentially the same value for all animals, regardless of their size. Of course, the picture is different if we examine the jump height h <u>relative</u> to the size L of the animal. Since h is constant, the relative height h/L varies inversely proportional to L. A flea can jump to a height more than 100 times its body height, while even the best human athlete cannot jump much higher than his or her own body height.

Using related arguments, it can be shown that the top speed on level ground of similarly constructed animals increases slightly with increasing mass. However, when running uphill, the extra energy cost favours smaller animals. Thus a horse can match a dog on level ground but quickly drops behind when a climb is encountered.

---

8. To move forward, an animal's feet must push backward on the ground with a horizontal force that matches the force of friction with the ground. The latter is proportional to the normal force, i.e., to the animal's weight. Hence, the force which the animal must exert on the ground to move itself forward is proportional to its weight or mass M.

We have discussed several applications of scaling to the biological design of organisms, but many other properties can be analyzed by similar tools. These include variations in size of individual organs (e.g., brains, heart and lungs); in heart rates and other aspects of blood circulation such as total blood volume and concentration of hemoglobin; in swimming and flying speeds; and last but not least the actual life span of organisms.

Exercises

10-1 A 79.0-kg mountaineer is suspended by a nylon rope of radius 4.50 mm. Under the weight of the mountaineer, the rope elongates by 9.00 cm from its original length of 26.0 m. For the nylon rope, calculate (a) the tensile strain, (b) the tensile stress, (c) the value of Young's modulus.

10-2 Given a tendon of cross-sectional area 3.00 mm², original length 0.650 m, and Young's modulus 2.00×10^7 N/m², what does its length become when a 1.10-kg mass is hung from it (a) on Earth, (b) on the Moon? (The gravitational acceleration on the Moon has magnitude 2.00 m/s².)

10-3 A wire whose cross-sectional area is 4.50 mm² is stretched by 0.120 mm by a certain weight. By how much will a wire of the same material and initial length stretch if its cross-sectional area is doubled and the same weight is attached?

10-4 How large a force is required to stretch a 2.50-cm diameter steel rod by 0.0100%? Young's modulus for steel is 2.00×10^{11} N/m².

10-5 A relaxed biceps muscle requires a force of 24.0 N to undergo an elongation of 2.80 cm. The same muscle under maximum tension requires a force of 480 N to experience the same elongation. Find Young's modulus for the muscle tissue under each of these conditions. Assume that the muscle is a uniform cylinder with an unstressed length of 0.190 m and cross-sectional area 12.0 cm².

10-6 When standing erect, a person's weight is supported chiefly by the larger of the two leg bones. Assuming this bone to be a hollow cylinder of 2.50-cm internal diameter and 3.50-cm external diameter, what is the compressive stress on the bone (in each leg) in the case of a person whose mass (excluding the legs) is 70.0 kg?

10-7 If the leg bones described in Exercise 10-6 have an unstressed length of 90.0 cm, (a) what is the amount of compression in each leg at the fracture point, and (b) what is the magnitude of the compressive force required to fracture the bone? Assume that Hooke's law remains valid up to the point of fracture, which occurs at a stress of 1.40×10^8 N/m². The value of Young's modulus for bone is 1.40×10^{10} N/m².

10-8 The humerus, which is the upper arm bone between the elbow and the shoulder joint, can be approximated as a 33.0-cm-long hollow cylinder with outer radius 1.05 cm and inner radius 0.430 cm. Suppose a gymnast whose arm bone has these dimensions does a one-arm handstand. Excluding the arm, the mass of the gymnast is 65.0 kg.

(a) What is the compressional strain of the humerus? Use the value of Young's modulus for bone given in Exercise 10-7.

(b) By how much is the humerus compressed?

10-9 A sample of bone in the form of a cylinder of cross-sectional area 1.40 cm^2 is loaded on its upper end by a mass of 10.0 kg. Careful measurement with a microscope reveals that the length decreases by 6.50×10^{-3}%. What is Young's modulus for this sample?

10-10 A 6.25-m long steel wire has a cross-sectional area of 0.0450 m^2. The stress at its proportional limit is 1.50×10^{-3} times its Young's modulus (see Table 10-1) while its breaking stress is 6.60×10^{-3} times its Young's modulus. The upper end of the wire is attached to a support and the wire hangs vertically, with various weights being attached to its lower end.

(a) How large a weight can be attached to the wire without exceeding the proportional limit?

(b) How much will the wire stretch under this load?

(c) What is the maximum weight that can be supported by the wire?

10-11 A block of gelatin has dimensions 40.0 mm by 40.0 mm by 20.0 mm when unstressed. A force of 0.245 N is applied tangentially to the upper surface, causing a 5.50-mm displacement relative to the lower surface (see diagram), which remains fixed.

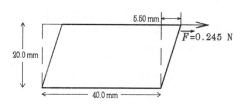

Find

(a) the shear stress

(b) the shear strain

(c) the shear modulus.

10-12   A bone sample in the shape of a cube, 0.300 m on a side, is subjected to two shearing forces, each of magnitude $F = 1.50 \times 10^6$ N (see diagram). Determine the angle ϕ. The shear modulus for bone is 1.00×10^{10} N/m².

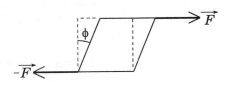

10-13   A torque of magnitude 170 N·m is applied to one end of a solid cylindrical steel shaft which is 3.50 m long and 2.60 cm in radius. The other end of the shaft is held fixed. Through how many degrees does the shaft twist? Take the shear modulus for steel to be 8.00×10^{10} N/m².

10-14   Which one of the following graphs represents the dependence of the twist-angle θ on the length ℓ of a solid rod of radius r, when the torque causing the twist is the same for all lengths?

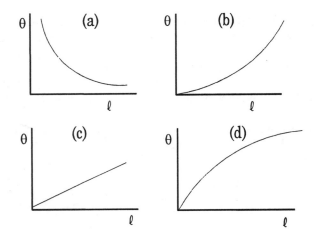

10-15   Equal twisting torques are applied to a solid cylindrical and hollow cylindrical rod of equal lengths and composed of the same material. The inner radius of the hollow rod is equal to the radius r of the solid rod, while the outer radius of the hollow rod is 1.200 r. What is the ratio of the twist angles developed in the rods?

10-16   If it takes 100 g of wool to knit a sleeveless sweater for a 15.0-kg child, how much wool is required for a similar sweater for a 120-kg man? Assume that all dimensions on the man are a simple multiple of those on the child, and that the thickness of the wool sweater is the same for both.

10-17 A 70.0-kg man and a 2.50-kg hairless dog go swimming. When they emerge, a 0.330-kg film of water adheres to the man's skin. How much water adheres to the dog's skin? What percentage of the mass of each is the water film? Assume that the bodies of the man and dog are isometric.

10-18 A small pizza of diameter 20.0 cm has a mass M_s. What diameter pizza would have twice this mass? Assume all pizzas have the same thickness.

10-19 A cell's oxygen requirement is proportional to its mass, but its oxygen intake is proportional to its surface area. If a cell is assumed to be a sphere of radius r, how does the ratio of oxygen intake to oxygen requirement vary as a function of r? Can a cell grow indefinitely and survive?

10-20 Two physically similar humans propose to compare the maximum weight they can lift. The taller of the two has a mass of 90.0 kg and can lift a maximum mass of 140 kg. The shorter has a mass of 60.0 kg. (Note that the forces exerted by muscles scale in proportion to their cross-sectional areas.)

(a) What is the maximum mass that can be lifted by the shorter person?

(b) What is the maximum mass lifted per kilogram of body mass for each person?

10-21 In a circus act, a pony and a horse are balanced on a teeter-totter. If the pony is 1.50 m tall and the horse is 2.50 m tall, what is the ratio of the distances

$$\frac{\text{pony to pivot point}}{\text{horse to pivot point}}?$$

10-22 If a giant species of ant were developed with all body measurements (excluding legs) 10.0 times as large as the common ant, how much larger would the leg diameters have to be in order to experience the same compressive stresses as in the common ant?

10-23 Go back to Example 10-7 comparing the relative skeletal masses of a cat and elephant, but use the scaling law $M_s \propto M^a$ (where $a = 1.08$) for the relation between skeletal mass M_s and total body mass M. With other numbers as in Example 10-7, what percentage of total body mass do you now predict for the skeletal mass of the elephant?

10-24 A large 4-legged animal has a body mass (excluding legs) of 350 kg. Model the legs of this animal as solid cylinders of radius 5.50 cm.

(a) Calculate the compressive stress on each leg due to the animal's main body weight.

(b) If a similarly shaped animal has a body mass (excluding legs) of 600 kg, what should be the radii of its legs in order to experience the same compressive stress as in part (a)?

Problems

10-25 The tibia is the lower leg bone (or shin bone) in a human.

(a) If the maximum strain it can experience before fracturing corresponds to a 1.00% change in length, what is the maximum force that can be applied to a tibia of cross-sectional area 2.90 cm^2? (Take Young's modulus to be 1.40×10^{10} N/m^2. This calculation is similar to that in Exercise 10-7(b).)

(b) Determine the maximum height from which a 75.0-kg person could jump and not fracture the tibia. Take the time between when the person's feet first touch the ground and when the person has stopped moving to be 0.0310 s. We do not recommend performing this experiment!

Hint: the maximum height will be that for which the normal force exerted by the ground on <u>each</u> lower leg, while the person is being brought to rest, equals the maximum force calculated in part (a).

10-26 A rod of length ℓ and mass m is supported at its ends by two wires of equal length (see diagram). The cross-sectional area of wire A is 1.00 mm^2 while that of wire B is 2.00 mm^2. Young's modulus for wire A is 2.40×10^{11} N/m^2 and that of wire B is 1.60×10^{11} N/m^2. At what point along the rod should a mass m (the same as the mass of the rod) be placed in order to produce (a) equal stresses in A and B, and (b) equal strains in A and B?

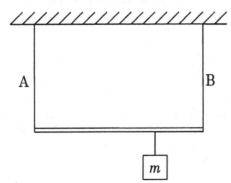

10-27 A square plate is 2.00 cm thick, has sides of length 8.00 cm, and has mass 0.450 kg. The plate material has shear modulus 1.60×10^7 N/m^2. One of the square faces of the plate rests on a flat horizontal surface, and the coefficient of static friction between the plate and the surface is 0.880. A force $\vec{F}_S$ is applied tangentially to the top surface of the plate, as in Fig. 10-5. Determine the maximum values of (a) the shear stress, (b) the shear strain, (c) the shear deformation Δx (see Fig. 10-5) that can be induced by the force $\vec{F}_S$ just before the plate begins to move.

10-28   A bone sample in the shape of a block is securely fastened to a horizontal tabletop. A force $\vec{F}$ of magnitude 2.50×10^3 N is applied to the top surface of the block as shown in the diagram. The values of Young's modulus Y and the shear modulus G for bone are: $Y = 1.40 \times 10^{10}$ N/m^2, $G = 9.00 \times 10^9$ N/m^2. Determine

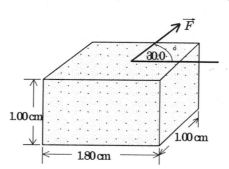

(a) the change in vertical dimension of the block

(b) the shear deformation of the block, i.e., the horizontal displacement of the top surface relative to the bottom surface.

10-29   A silo on a farm is constructed from a hemisphere of radius r on top of a cylinder of radius r and height h.

(a) Find expressions for both the volume and surface area of the silo as functions of h and r.

(b) If one wishes to double the volume but keep the ratio h/r constant, what are the new values for the height, radius, and surface area?

(c) If one wishes to double the volume while keeping r constant, what are the new values for the height and surface area?

10-30   We have seen that, when the body of an animal is scaled up in all dimensions by a certain factor α, the leg diameter must increase by the factor $\alpha^{3/2}$ in order to support the extra weight. Now suppose we wish to develop a super-pig in which all dimensions (<u>including</u> those of the legs) were scaled up by a factor of 2. How many extra legs would this super-pig require in order for each leg to experience the same compressive stress as the normal 4-legged pig before it was scaled up? Neglect the additional weight of the extra legs.

10-31   In earlier exercises, it was assumed that the <u>thickness</u> of the material covering a surface (i.e., the thickness of the wool in Exercise 10-16, of the water film in Exercise 10-17, of the pizza dough in Exercise 10-18) did not change when the other dimensions of the system were scaled. This will not always be the case.

A normal egg of mass 60.0 g has a shell of mass 2.00 g. Suppose a super-egg of mass 300 g is developed, which is found to have a shell of mass 7.60 g. What is the ratio

$$\frac{\text{shell thickness of super-egg}}{\text{shell thickness of normal egg}} \, ?$$

Assume that both eggs have a similar shape.

Answers

10-1 (a) 3.46×10^{-3}
(b) 1.22×10^7 N/m^2
(c) 3.52×10^9 N/m^2

10-2 (a) 0.767 m (b) 0.674 m

10-3 0.0600 mm

10-4 9.82×10^3 N

10-5 1.36×10^5 N/m^2,
2.71×10^6 N/m^2

10-6 7.28×10^5 N/m^2

10-7 (a) 9.00 mm
(b) 6.60×10^4 N

10-8 (a) 1.58×10^{-4}
(b) 5.21×10^{-5} m

10-9 1.08×10^{10} N/m^2

10-10 (a) 1.35×10^7 N
(b) 9.38 mm
(c) 5.94×10^7 N

10-11 (a) 153 N/m^2
(b) 0.275
(c) 557 N/m^2

10-12 1.67×10^{-3} rad

10-13 0.594°

10-14 (c)

10-15 θ(solid)/θ(hollow) = 1.06

10-16 400 g

10-17 0.0358 kg;
0.471% (man), 1.43% (dog)

10-18 28.3 cm

10-19 $\propto r^{-1}$; no

10-20 (a) 107 kg
(b) 1.56 (tall), 1.78 (short)

10-21 4.63

10-22 $10.0^{3/2} = 31.6$ times as large

10-23 12%

10-24 (a) 9.02×10^4 N/m^2
(b) 7.20 cm

10-25 (a) 4.06×10^4 N
(b) 56.4 m

10-26 (a) (5/6) ℓ from wire A
(b) (9/14) ℓ from wire A

10-27 (a) 606 N/m^2
(b) 3.79×10^{-5}
(c) 7.58×10^{-7} m

10-28 (a) 4.96×10^{-4} cm
(b) 1.34×10^{-3} cm

10-29

(a) $\mathcal{V} = \pi r^2 h + \dfrac{2\pi}{3} r^3$, $\mathcal{A} = 2\pi rh + 2\pi r^2$

(b) $r_2 = \sqrt[3]{2}\, r_1$, $h_2 = \sqrt[3]{2}\, h_1$, $\mathcal{A}_2 = 2^{\,2/3}\, \mathcal{A}_1$

(c) $h_2 = 2h_1 + \dfrac{2}{3} r_1$, $\mathcal{A}_2 = 2\mathcal{A}_1 - \dfrac{2\pi}{3} r_1^2$

10-30 4

10-31 1.30

11 Fluid Statics

11.1 Introduction

Materials are generally considered to fall into three classes termed "solids, liquids and gases." As has been shown in Chapter 10, a solid has elastic properties and reaches an equilibrium deformation (strain) when a stress is applied. Liquids and gases, however, never reach an equilibrium and continue to deform (flow) when a stress is applied. They are, for this reason, called *fluids*. Both solids and liquids are said to be forms of *condensed matter* that arise when the attractive forces between molecules of the material are relatively strong. In a gas, however, molecules interact relatively weakly and are further apart. The internal environment of living biological cells must be a liquid (an aqueous solution) for normal life processes such as the synthesis of proteins to occur. Sometimes, as in seeds, the fluidity of the environment is dramatically reduced so that the species can survive over winter and then regenerate in the spring. Liquids also carry nutrients to the various cells to fuel the metabolic processes and carry waste products away. Gases are also biologically valuable; air, for example, provides life-sustaining oxygen for metabolic processes and, for terrestrial organisms, the pressure to keep them in their proper shape.

A more careful description of matter shows that whether a material is in the solid, the liquid, or the gaseous phase depends on its temperature and pressure. At atmospheric pressure, water, for example, can exist in any of the three phases depending on the temperature. Since the mechanical properties of solids have been discussed in Chapter 10, this chapter will focus primarily on the basic properties of fluids, and describe why fluid properties such as pressure, surface tension, and viscosity are necessary to understand how living systems function. The chapter will also include a brief description of phase diagrams for simple systems such as water and carbon dioxide.

11.2 Pressure in Liquids

The molecules or atoms of a fluid are in random motion due to their temperature, and when these molecules or atoms collide with a surface they experience changes in momentum that generate a net force that is normal (perpendicular) to the surface. Therefore, a fluid exerts a *pressure* on any surface immersed within it. And, because of the random motions of the atoms or molecules, the pressure at any point in the fluid acts equally in all directions.

Pressure is defined as the magnitude of the normal force per unit surface area.

Dimensional analysis shows that the dimensions of pressure are

$$P = \left[\frac{\text{force}}{\text{area}} \right] = \frac{ML}{T^2} \times \frac{1}{L^2} = ML^{-1}T^{-2}$$

and the SI units of pressure are newtons per square metre (N/m²) or pascals (abbreviated Pa).[1]

Pressure that depends on the depth of fluid involved is called *hydrostatic pressure*. When we are swimming, for example, we feel a greater hydrostatic pressure at the bottom of the pool than we do near the surface. Hydrostatic pressure depends on the weight per unit volume of the fluid which is, in turn, the product of *density* of the fluid and the acceleration due to gravity(*g*). Pressure is a scalar quantity, and "weight" and "acceleration" here refer only to the magnitudes of these two vector quantities. Density (usually represented by the Greek lowercase letter rho [ρ]) is defined as mass per unit volume. Thus, the weight per unit volume is equal to ρg. For gases, the density can change considerably with depth (or altitude). For liquids, however, these changes are usually so slight that, within the limits required for life processes, they can be ignored. For example, if 1 m³ of water at 4°C were subjected to twice the normal atmospheric pressure, its volume would decrease to 0.99995 m³, and its density would increase to 1000.05 kg/m³. A gas, on the other hand, subjected to the same pressure change, would have its volume halved and its density doubled. The density of the gas would also change markedly with an increase or decrease in temperature, provided that the volume was allowed to change.

Liquids and solids have densities that are substantially greater than those of gases (Table 11-1). It is frequently convenient to express the

---

1.   Named for Blaise Pascal (1623–1662), French physicist.

density of a substance as the ratio of its density to that of water (1000 kg/m³). This ratio is called the *specific gravity* (s.g.) or relative density, and is dimensionless. Thus, a protein of specific gravity 1.33 would have a density of 1.33×10^3 kg/m³.

Table 11-1   Densities of Common Materials

Material	Density (kg/m³)
Dry air (20°C, 1 atm)	1.20
Water	1.00×10^3
Blood	1.05×10^3
Alcohol (ethyl)	0.792×10^3
Mercury	13.6×10^3
Ice (at 0°C)	0.92×10^3
Aluminum	2.70×10^3
Iron	7.87×10^3

$$\frac{dy}{dx}$$

To evaluate the variation of pressure with depth in a liquid at rest, consider Fig. 11-1. We can imagine that one compartment (or element of volume) of the liquid experiences a pressure from the remainder of the liquid. The diagram shows a volume element of the liquid having infinitesimal thickness dy and top (or bottom) surface area $\mathcal{A}$; the volume can be written as $\mathcal{A}dy$. Since pressure is defined as the magnitude of the normal force per unit surface area, there is a downward force $\vec{F}_{down}$ of magnitude $P_{upper}\,\mathcal{A}$ exerted on the upper surface of the volume element, and an upward force $\vec{F}_{up}$ of magnitude $P_{lower}\,\mathcal{A} = (P_{upper} + dP)\mathcal{A}$ on the lower surface, where dP represents the <u>increase</u> in pressure at the lower surface. The volume element also experiences a downward force, $\vec{F}_g = m\vec{g}$, which is its weight. Since mass is density times volume, then the magnitude of F_g is $(\rho\mathcal{A}dy)g$. If the volume element is at rest, then all the forces acting on it must sum to zero. If "up" is chosen to be positive and "down" negative,

$$-P_{upper}\,\mathcal{A} + (P_{upper} + dP)\mathcal{A} - \rho g\mathcal{A}\,dy = 0$$

and

$$dP = \rho g\,dy \qquad \textbf{[11-1]}$$

Since the variation of density with pressure in a liquid is negligible, then ρ is a constant and Eq. [11-1] can be

Figure 11-1

Forces on a fluid element.

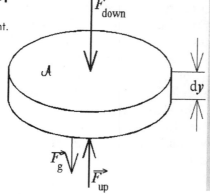

integrated between any two values of P and y to yield:

$$\int_{P_1}^{P_2} dP = \rho g \int_{y_1}^{y_2} dy$$

from which,

$$P_2 - P_1 = \rho g \, (y_2 - y_1)$$

Writing $\Delta P = P_2 - P_1$ for the pressure difference and $d = y_2 - y_1$ for the difference in depth,

$$\Delta P = \rho g d \qquad \qquad \textbf{[11-2]}$$

Note that the depth change, d, is the key variable that affects hydrostatic pressure in a liquid. If y_1 is taken at the surface of the liquid, then d becomes the depth from the surface. The shape of the container is irrelevant, and hydrostatic pressures are identical at all points having the same depth. Further, if the pressure at the surface, P_1, is the pressure due to the atmosphere, (i.e., $P_1 = P_{atm}$), then the pressure at y_1 is the *absolute pressure* (P_{abs}) at the depth d.

$$P_{abs} = P_{atm} + \rho g d \qquad \qquad \textbf{[11-3]}$$

Note that the pressure due to the atmosphere is transmitted through the liquid; that is, pressures are additive.

The pressure difference between two points (e.g., between the surface and depth d) is termed the *gauge pressure*. Very often, one of the points is at atmospheric pressure, and in this case the gauge pressure would represent the pressure above atmospheric pressure:

$$P_{gauge} = P_{abs} - P_{atm} \qquad \qquad \textbf{[11-4]}$$

Atmospheric pressure can change with both elevation and weather conditions.

Since pressure gauges register the difference in pressure (the gauge pressure) between two points, this difference (which can be positive or negative) will not be an absolute pressure unless (as rarely happens) one of the two points is at zero pressure. It is usually assumed that one of the points is at atmospheric pressure unless otherwise specified. Consider, for example, an automobile tire which is frequently filled to a pressure of approximately 2 atm (2×10^5 Pa). This means that the internal pressure is 2 atm more than the external pressure, which is typically 1 atm. Therefore, the gauge pressure inside the tire is 2 atm, but the total or absolute pressure inside is 3 atm.

Units of pascals (Pa) <u>must</u> be used in calculations involving equations such as Eq. [11-3] ($P_{abs} = P_{atm} + \rho g d$). However, pressure is frequently expressed in other units that are used for comparative purposes, but are not directly applicable for calculations. One standard atmosphere (1 atm) is the average atmospheric pressure at sea level, or 1.013×10^5 Pa. As another example, pressure may be quoted in millimetres of mercury or torr[2] since a common device used to measure pressure difference is a mercury *manometer* (Fig. 11-2). A manometer consists of a U-tube of glass containing a liquid; one end of the tube is connected to a chamber of gas on one end and to an external environment, often the atmosphere, on the other. Liquids that are commonly used in manometers are mercury, water, and oil. Basically, this device measures the gauge pressure equivalent to the hydrostatic pressure exerted by a column of the liquid of depth d. It may be converted to pascals using Eq. [11-2] ($\Delta P = \rho g d$). A list of pressure units and their conversion to pascals is shown in Table 11-2. Blood pressure is measured using a sphygmomanometer (see Box 11-1).

Figure 11-2

A mercury manometer measures pressure difference.

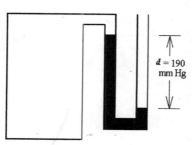

$d = 190$ mm Hg

Table 11-2   **Some Pressure Units and Their Equivalent in Pascals**

Pressure Unit	Equivalent in Pascals
1 mm mercury (Hg) at 20°C = 1 torr	132.8
1 standard atmosphere (atm)	1.013×10^5
1 bar	1.000×10^5

2.   Named for Evangelista Torricelli (1608–1647), Italian physicist, student of Galileo and inventor of the barometer.

Box 11-1   The Sphygmomanometer

The standard method for measuring blood pressure can be understood using Eq. [11-4]:

$$P_{gauge} = P_{abs} - P_{atm}$$

Through a person's arm runs the brachial artery with blood maintained at an absolute pressure P_{abs}. The surrounding tissue can be considered a fluid that transmits the atmospheric pressure P_{atm}. The artery remains inflated because $P_{abs} > P_{atm}$. If the atmospheric pressure could be increased to $P > P_{abs}$, then the artery would collapse and stop the blood flow.

Figure 11-3

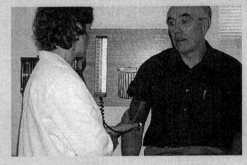

Measuring blood pressure. (Photo courtesy of Blair Fraser, M.D., and Anne Fraser, R. N.)

A bladder in the form of a "cuff" is wrapped around the arm (Fig. 11-3), and by means of a pump the pressure is increased until the blood flow is stopped. The pressure at which this occurs is measured by a pressure gauge of some sort and is a measure of the gauge pressure of the blood. The indication that flow has stopped is the cessation of the noise created by the turbulent flow of the blood when the artery is almost closed.

In practice, the cuff is inflated to a pressure above the cut-off. It is then slowly reduced and at some pressure the sound (as heard in a stethoscope) starts in spurts as the artery opens up periodically at the maximum (systolic) pressure exerted by the heart. When the cuff pressure has been reduced to the minimum (diastolic) pressure exerted by the heart, blood flows during the entire cardiac cycle, and the sound disappears. The systolic and diastolic pressures are recorded in millimetres of mercury, and for a healthy 20-year-old are of the order of 110 mm Hg and 70 mm Hg, respectively, usually quoted as 110/70.

EXAMPLE 11-1

The mercury-filled manometer in Fig. 11-2 indicates a difference in column heights of 190 mm. One side of the manometer is open to the atmosphere. What is the absolute pressure inside the chamber? (The density of mercury is given in Table 11-1.)

SOLUTION

Note that in this case, the level of the liquid is higher on the chamber side of the manometer than on the side open to the atmosphere. This indicates that the absolute pressure, P_{abs}, inside the chamber is less than atmospheric pressure and that the gauge pressure is –190 mm Hg.

The gauge pressure is determined in pascals by using $P_{gauge} = -\rho g d$, where the minus sign has been inserted because the gauge pressure is known to be negative. Thus,

$$P_{gauge} = -(13.6\times10^3 \text{ kg/m}^3)\times(9.80 \text{ m/s}^2)\times(190\times10^{-3}) \text{ m}$$
$$= -2.53\times10^4 \text{ Pa}$$

From Eq. [11-4], the absolute pressure inside the chamber is

$$P_{abs} = P_{atm} + P_{gauge}$$
$$= [1.013\times10^5 + (-2.53\times10^4)] \text{ Pa}$$
$$= 7.6\times10^4 \text{ Pa (or 0.76 atm)}$$

Remember that *in all cases,* if the gauge pressure is quoted as a column height (e.g., mm Hg), then before using it in any calculations it must first be converted to a physical pressure unit, using the expression

$$\boldsymbol{P_{gauge}} = \pm\rho g d.$$

EXAMPLE 11-2

What are the gauge and absolute pressures at a depth of 1.0 m in a pool of water?

SOLUTION

The gauge pressure at a depth of 1.0 m is the pressure difference a manometer would measure between that point and the surface. From Eq. [11-2] ($\Delta P = \rho g d$), this pressure difference is

$$\rho g d = (1000 \text{ kg/m}^3) \times (9.80 \text{ m/s}^2) \times (1.0 \text{ m})$$
$$= 9.8 \times 10^3 \text{ Pa}$$

From Eq. [11-4],

$$P_{abs} = P_{atm} + P_{gauge}$$
$$= (1.013 \times 10^5 + 9.8 \times 10^3) \text{ Pa}$$
$$= 1.11 \times 10^5 \text{ Pa}$$

Thus, the absolute pressure at this depth would be 1.11×10^5 Pa, or approximately 1.1 atm.

A swimmer becomes very much aware of this kind of increase in pressure described in Example 11-2 when submerging as little as 2 m, since the pressure difference across the eardrum increases from zero at the surface to a significant amount rather quickly. For example, if the pressure on the inside of the eardrum were 1.00×10^5 Pa and on the outside were 1.20×10^5 Pa then, since the area of the eardrum is approximately 0.66×10^{-4} m², there would be a net force on the eardrum of magnitude

$$F = P\mathcal{A} = 0.20 \times 10^5 \text{ N/m}^2 \times 0.66 \times 10^{-4} \text{ m}^2 = 1.3 \text{ N}$$

This is a lot of force; in fact, ten times that required to rupture the eardrum. The eardrum does not rupture, since the pressure inside the eardrum is also increased due to the water pressure acting over the body and to involuntary contraction of muscles in the respiratory system.

EXAMPLE 11-3   *The Hydrostatic Factor in the Human Circulatory System*

The change of pressure with depth of a liquid is called the *hydrostatic factor,* and is important in the vascular system of the human body. At the heart, the gauge systolic pressure is about 100 mm Hg. When the body is horizontal, the average arterial pressures in the brain and feet are approximately the same as at the heart, namely 100 mm Hg, because they are all at the same level. If the body is erect, the hydrostatic factor reduces the systolic pressure in the brain and increases that in the feet. Determine the systolic pressures in the (a) head and, (b) feet, if they are located 0.50 m above, and 1.3 m below, the heart, respectively. (Use Table 11-1 for the densities of mercury and blood.)

SOLUTION

First convert the systolic pressure at the heart to pascals:

P_{heart}

$= \rho_{mercury} \, gd$

$= (13.6 \times 10^3 \text{ kg/m}^3) \times (9.80 \text{ m/s}^2) \times (100 \text{ mm} \times 10^{-3} \text{ m/mm})$

$= 1.33 \times 10^4 \text{ Pa}$

(a) The arterial pressure difference between the head and the heart is

$P_{difference}$

$= -\rho_{blood} \, gd$

$= -(1.05 \times 10^3 \text{ kg/m}^3) \times (9.80 \text{ m/s}^2) \times (0.50 \text{ m})$

$= -5.1 \times 10^3 \text{ Pa}$

Therefore the systolic pressure in the head is

$$(1.33 \times 10^4 - 5.1 \times 10^3) \text{ Pa} = 8.2 \times 10^3 \text{ Pa}$$

This is about 60% of the pressure at the heart.

(b) Since the feet are located 1.3 m below the heart, then the pressure difference between the feet and the heart is

$P_{difference}$

$= (1.05 \times 10^3 \text{ kg/m}^3) \times (9.80 \text{ m/s}^2) \times (1.3 \text{ m})$

$= 1.3 \times 10^4 \text{ Pa}$

and the arterial pressure down at the feet is

$$(1.33 \times 10^4 + 1.3 \times 10^4) \text{ Pa} = 2.6 \times 10^4 \text{ Pa}$$

or almost double that at the heart. The potential for fainting and for swollen feet is closely related to these differences in arterial pressure.

11.3   Archimedes' Principle and Buoyancy

Any object, when placed in a fluid, appears to weigh less because of an upward force called the *buoyant force*.

> **Archimedes' principle states that an object, when placed in a fluid, experiences an upward (buoyant) force equal in magnitude to the weight of the fluid that the object has displaced.**

Therefore, any object when placed in a fluid experiences two forces:

a) the downward gravitational force due to the mass of the object
b) the upward buoyant force equal to the weight of the displaced fluid.

The downward gravitational force is $-mg = -\rho \mathcal{V} g$. The upward buoyant force (whose magnitude equals the weight of the displaced fluid) is then $m_f g$ where m_f is the mass of a volume $\mathcal{V}$ of the fluid (since this is the amount of fluid displaced by the object). Since $m_f = \rho_f \mathcal{V}$ where ρ_f is the density of the fluids, then the upward buoyant force is $\rho_f \mathcal{V} g$. Clearly, if the downward gravitation force is greater than the buoyant force, then the object will sink (as sediment) to the bottom of the container. If the reverse is true, the object will rise to the liquid surface and float there. If the sum of the two forces is zero, then the object is said to be neutrally buoyant and remains suspended in the liquid. The small particles in suspensions such as milk and latex paint are extremely close to being neutrally buoyant. The sedimentation of particles will be discussed more thoroughly in Chapter 13. A free-body diagram showing these forces acting on the object is shown in Fig. 11-4.

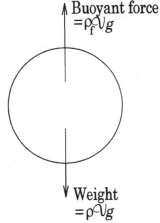

Buoyant force $= \rho_f \mathcal{V} g$

Weight $= \rho \mathcal{V} g$

Figure 11-4

Archimedes' principle.

Therefore, the "net" weight of the particle in the fluid is $(\rho - \rho_f)\mathcal{V}g$, and its gravitational potential energy is $(\rho - \rho_f)\mathcal{V}gd$.

EXAMPLE 11-4

A swimmer of mass 60 kg and density 0.97×10^3 kg/m³ is completely submerged and temporarily motionless at depth of 2 m. What is the net force on the swimmer and will the swimmer tend to sink further or to float upwards? Would this net force be any different if the swimmer were located at a depth of 20 m?

SOLUTION

$$F_{net} = (\rho - \rho_\ell)\mathcal{V}g = \rho\mathcal{V}g - \rho_\ell\mathcal{V}g = mg - \rho_\ell(m/\rho)g = mg(1 - \rho_\ell/\rho)$$
$$= 60 \text{ kg} \times 9.8 \text{ m/s}^2 \times (1 - 0.97) = 18 \text{ N}$$

This force is positive, so the swimmer will tend to drift upwards. The density of water (1×10^3 kg/m³) is essentially the same at all depths since water is incompressible. Therefore, the net force would be the same at all depths.

It is interesting to note that air also generates a buoyant force on objects, but because of the low density of air (see Table 11-1) it is almost 1000 times smaller than the buoyant force in water and is normally neglected. However, it is the source of lift for hot air balloons and helium-filled blimps. Further, unlike water, the density of air and the buoyant force decrease with increasing altitude. The next section discusses the density of gases such as air in more detail.

11.4 **Pressure in Gases**

Unlike the density of liquids, the density of gases is extremely dependent on temperature and pressure. Experimental studies by Robert Boyle[3] demonstrated that at constant temperature, the product of the absolute pressure of a fixed mass of gas and its volume is a constant, that is, $P\mathcal{V} = constant$. Sometimes this relationship, called *Boyle's law*, is written alternatively as

---

3. Robert Boyle (1627–1691), Anglo-Irish physicist and chemist.

$$P_1\mathcal{V}_1 = P_2\mathcal{V}_2 \qquad\qquad \textbf{[11-5]}$$

where the subscripts 1 and 2 refer to two different states of P and $\mathcal{V}$. Equation [11-5] shows that if the pressure is doubled (for example, by compressing the gas in a cylinder), the volume is halved. Since the same mass of gas is present in both states, then the density has doubled.

The product, $P\mathcal{V}$, varies linearly with temperature. This is shown in Fig. 11-5, in which $P\mathcal{V}$ is plotted for a fixed mass of gas as a function of temperature in degrees Celsius (°C). Amazingly, if one extrapolates the linear graph, there is an intercept at the temperature -273.15°C, at which point $P\mathcal{V} = 0$. At this temperature, called *absolute zero,* the gas molecules would lose all translational motion, and pressure would vanish. The equation that represents this behaviour is

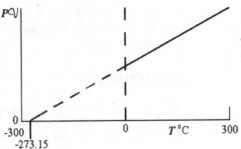

Figure 11-5

Variation of $P\mathcal{V}$ with temperature.

$$P\mathcal{V} = [constant](T_{\text{Celsius}} + 273.15) \qquad\qquad \textbf{[11-6]}$$

It is easy, however, to redefine the temperature scale such that a new "zero" is at absolute zero, but the degree steps remain the same as in the Celsius scale. This is the kelvin[4] temperature scale, T, with degree units of kelvin (K). Since $T = T_{\text{Celsius}} + 273.15$, then Eq. [11-6] becomes $P\mathcal{V} = [const]T$.

It is also known that the volume of a gas is directly proportional to the number of molecules of gas present. That is, if the number of moles of a gas is doubled, but the pressure and temperature are held constant, the volume will double. If n represents the number of moles of gas present, then Eq. [11-6] can be written as

$$P\mathcal{V} = n\text{R}T \qquad\qquad \textbf{[11-7]}$$

where the remaining constant, the *molar gas constant*, R, has a value of 8.315 J·mol^{-1}·K^{-1}.[5]

---

4. Baron William Thomson Kelvin (1824–1907), Scottish mathematician and physicist.
5. This implies that the unit of $P\mathcal{V}$ is the joule. This can be seen from the product of the units of P and $\mathcal{V}$: $(N/m^2)(m^3) = N\cdot m = J$.

Equation [11-7] is called the *ideal gas law* or the *equation of state for an ideal gas.* At normal temperature and pressure, such as that found near Earth's surface, most gases and gas mixtures obey the ideal gas law reasonably well. More complex equations of state that include effects due to intermolecular interactions and other factors have been proposed for more extreme temperature and pressure situations. Often, one is concerned with gas properties at *standard conditions*, which are defined to be a pressure of 1 atm and a temperature of 273.15 K (0°C). At standard conditions, the ideal gas law predicts that the volume of one mole of any gas would be 22.41 L or 22.41×10^{-3} m³.

EXAMPLE 11-5

A small marine organism carries with it a small air bubble that it uses to control its buoyancy. At the surface of a lake, the bubble has a radius of 0.50 mm. Determine the radius of the bubble at a depth of 10 m.

SOLUTION

The volume of the bubble at the surface is

$$\mathcal{V}_s = \frac{4}{3}\pi r^3 = \frac{4}{3}\pi\left(0.50 \times 10^{-3} \text{ m}\right)^3 = 5.2 \times 10^{-10} \text{ m}^3$$

The absolute pressure at the surface is atmospheric pressure, or 1.013×10^5 Pa. At a depth of 10 m, the absolute pressure is atmospheric pressure plus the hydrostatic pressure, $\rho g d$, at that depth:

$$P_{10m} = (1.013 \times 10^5 \text{ Pa}) + (1000 \text{ kg/m}^3 \times 9.80 \text{ m/s}^2 \times 10 \text{ m})$$
$$= 2.0 \times 10^5 \text{ Pa}$$

Using Boyle's law (Eq. [11-5]),

$$\mathcal{V}_2 = \frac{P_1 \mathcal{V}_1}{P_2} = \frac{\left(1.013 \times 10^5 \text{ Pa}\right)\left(5.2 \times 10^{-10} \text{ m}^3\right)}{2.0 \times 10^5 \text{ Pa}} = 2.6 \times 10^{-10} \text{ m}^3$$

and the new radius is

$$r_2 = \left(\frac{3}{4\pi}\mathcal{V}_2\right)^{\frac{1}{3}} = 4.0 \times 10^{-4} \text{ m} = 0.40 \text{ mm}$$

11.5   **Pressure in Earth's Atmosphere**

The ideal gas law also may be used to evaluate the density of a gas under conditions such as occur in Earth's atmosphere. Analysis of the molar fractions of dry air reveals it to consist of 78% nitrogen (N_2), 20% oxygen (O_2), 0.9% argon (Ar), 0.03% carbon dioxide (CO_2) and small amounts of xenon (Xe) and krypton (Kr). The atmospheric envelope, being compressible and influenced by the gravitational field of Earth, is relatively dense near the surface and gradually less so with increasing altitude. Normal sea level pressure is taken to be 1.013×10^5 Pa, but fluctuations of about 5% are not uncommon due to weather. At an altitude of 100 km, the pressure of the atmosphere decreases to almost a perfect vacuum.

Almost all living organisms on Earth above sea level exist at pressures between normal sea level pressure and approximately 5×10^4 Pa at the tree line in high mountains. At any altitude where the pressure is less than about 5×10^4 Pa (3000 m above sea level), exertion by humans is likely to result in breathing distress. Breathing pure oxygen will help but not indefinitely. At 14 000 m, even with 100% oxygen, barely enough oxygen will cross the alveolar walls in the lungs to maintain consciousness; beyond this, elevation pressurized systems are required. At 21 000 m the pressure is the same as that of the water vapour in the body. Any fluid will boil when its vapour pressure is the same as the prevailing atmospheric pressure, so actual boiling of body fluids would occur at this elevation in an unpressurized system, and death would occur very rapidly.

Determining a quantitative expression for pressure as a function of altitude is not difficult. In doing so, however, it is more meaningful to use the molecular equivalent of the ideal gas law. Since one mole of a gas contains Avogadro's number ($N_A = 6.022 \times 10^{23}$ mol^{-1}) molecules of gas, the universal gas constant per molecule is

$$k_B = \frac{R}{N_A} = 1.381 \times 10^{-23} \text{ J/K}$$

where k_B is known as *Boltzmann's constant*.[6] Using this definition, the equivalent molecular form of the ideal gas law, $P\mathcal{V} = nRT$ (Eq. [11-7]), can be written as

$$P\mathcal{V} = Nk_B T \qquad \qquad \textbf{[11-8]}$$

where $N = n\,N_A$ is the number of gas molecules present. However, like the density of any other substance, the density of gas is given by mass per

---

6.   Ludwig Edward Boltzmann (1844–1906), Austrian physicist.

unit volume. Since the mass of gas in a container (volume, $\mathcal{V}$) is nM, where n is the number of moles of gas and M is the molar mass,[7] then the density is described by

$$\rho_{\text{gas}} = \frac{nM}{\mathcal{V}} \qquad \textbf{[11-9]}$$

Solving Eq. [11-8] for $\mathcal{V}$ and substituting into Eq. [11-9] yields

$$\rho_{\text{gas}} = \frac{PM}{k_B N_A T} = \frac{Pm}{k_B T} \qquad \textbf{[11-10]}$$

where $m = M/N_A$ is the mass per molecule in kilograms.

While the density of air does change with altitude, if we consider only the incremental pressure change dP over a very small change in height dy, then over this tiny range, density can be assumed constant and Eq. [11-1], $dP = \rho g\, dy$, applies. Substitution of the expression for density from Eq. [11-10] into Eq. [11-1] gives

$$dP = -\frac{Pmg}{k_B T}\, dy$$

or

$$\frac{dP}{P} = -\frac{mg}{k_B T}\, dy \qquad \textbf{[11-11]}$$

where the (–) sign arises because atmospheric pressure decreases as the elevation y increases. This equation can be integrated from sea level, h_0 (where the pressure is P_0) to any elevation h_1 (where the pressure is P). To simplify the calculation we will assume that g and T are constant at all elevations. Indeed, within the elevation range of physiological interest, the temperature of the atmosphere is reasonably constant and the decrease of g is also quite small. The integration then becomes

$$\int_{P_0}^{P} \frac{dP}{P} = -\frac{mg}{k_B T} \int_{h_0}^{h_1} dy$$

with solution

$$\ln\!\left(\frac{P}{P_0}\right) = -\frac{mg}{k_B T}(h_1 - h_0)$$

or in exponential form

---

7.   The "molar mass" is the mass of one mole of a substance, i.e., the mass of Avogadro's number (6.022×10^{23} mol^{-1}) of molecules.

$$\frac{P}{P_0} = e^{-\frac{mg}{k_B T}(h_1 - h_0)} \qquad \textbf{[11-12]}$$

Equation [11-12] is known as the *barometric formula* and indicates an exponential decrease in pressure with elevation. The term $mg(h_1 - h_0)$ in the exponent is the additional gravitational potential energy of an air molecule at height h_1 above its potential energy at h_0. The product $k_B T$ (sometimes called the *thermal energy*) is also an energy term, so that the complete exponent is a dimensionless quantity as is required. A graph of P/P_0 vs. $h_1 - h_0$ is shown using linear and semilogarithmic vertical scales in Fig. 11-6.

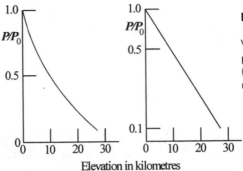

Figure 11-6

Variation of atmospheric pressure with altitude (logarithmic plot on the right).

Elevation in kilometres

From Eq. [11-10] it is clear that density and pressure are proportional to each other. Hence, using the barometric formula, but replacing the ratio of pressures with the ratio of densities, one can also determine the air density at any elevation:

$$\frac{\rho}{\rho_0} = e^{-\frac{mg}{k_B T}(h_1 - h_0)} \qquad \textbf{[11-13]}$$

EXAMPLE 11-6

Determine the absolute pressure and the air density in Earth's atmosphere at a height of 1000 m, as a percentage of the value at sea level. Assume the average mass of an air molecule is 4.8×10^{-26} kg and that the temperature is 25°C.

SOLUTION

$$\frac{P}{P_0} = e^{-\frac{mg}{k_B T}(h_1 - h_0)}$$

$$= e^{-\frac{4.8 \times 10^{-26}\,\text{kg} \times 9.8\,\text{m/s}^2}{1.381 \times 10^{-23}\,\text{J/K} \times (273.15 + 25)\,\text{K}}(1000 - 0)}$$

$$= 0.89$$

Therefore, the pressure (and the density) of air at 1000 m is 89% of that at sea level.

11.6 Phase Diagrams

The ideal gas law (Eq. [11-7]) can be rewritten in ratio form as

$$\frac{P\mathcal{V}}{nRT} = 1$$

and the ratio should equal unity at all temperatures and pressures. For any real gas there are relatively small variations from unity that can occur, especially at high pressures and low temperatures. These variations are caused by molecular interactions and by *excluded volume* effects, both of which are enhanced under these conditions because the molecules of gas are brought closer together. Excluded volume means that gas molecules occupy a volume into which other gas molecules cannot go, since that space is already occupied. If the temperature is lowered even further, there is eventually a point at which the ratio

Figure 11-7

Variation of the ratio (*P*$\mathcal{V}$/*nRT*) for a real substance.

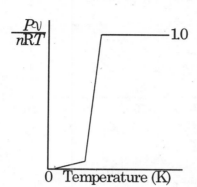

drops catastrophically to near zero (Fig. 11-7). This collapse is an indication of a *phase change* in which the gas has converted to a *condensed phase,* either a liquid or a solid. If the change is to a liquid then the system has passed through the *boiling point*; if the gas has solidified, the system has passed through the *sublimation point.*

The actual temperatures at which these phase changes occur depend on the pressure. For example, at sea level, the boiling point of water is 100°C, but on a mountain top, it could be much less. Similarly, when a phase transition takes place between the liquid phase and the solid phase, the temperature at which this occurs is called the *melting point.* An overview of this behaviour is apparent from a *phase diagram* that graphically displays the boiling point, sublimation point and melting point of a material as a function of pressure and temperature. Phase diagrams for water (H_2O) and carbon dioxide (CO_2) are shown in Fig. 11-8.

Note that the solid lines in Fig. 11-8 indicate where the phase transitions occur and whether they are solid↔liquid, solid↔gas, or liquid↔gas transitions. Along these lines the two adjacent phases can coexist (e.g., an ice–water mixture). Note also that there is a *triple point* at which all three phases can coexist. If one moves horizontally across the graph (which corresponds to constant pressure or an "isobar"), then at pressures above the triple point the system passes through all three phases. At pressures below the triple point, there are only two phases, solid and gas. This provides an interesting comparison between H_2O and CO_2 if one moves isobarically across the graph at one atmosphere of pressure. Clearly as the temperature is changed, H_2O goes from ice to water to gas (steam), whereas CO_2 moves directly from solid to gas. This is why *dry ice*, which is solid CO_2, is so named—it sublimates directly to gaseous CO_2 without passing through the liquid phase.

The phase diagrams also demonstrate the strong dependence of boiling point on pressure. From the H_2O phase diagram, the normal boiling point for water drops as the pressure is reduced and rises as the pressure is increased. The gas–liquid phase transition terminates at the *critical point*. Above this point the material cannot be characterized as either a liquid or a gas; it is simply a fluid. Surface tension, a phenomenon described in the following section, acts to separate one phase from another, but it too vanishes at the critical point.

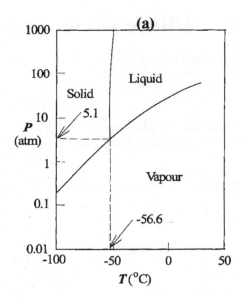

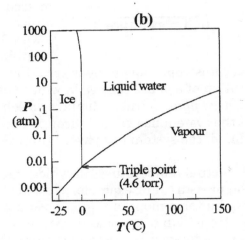

Figure 11-8

Phase diagrams for (a) carbon dioxide and (b) water.

11.7 **Surface Tension**

Small drops of water on wax paper form themselves almost into spheres, and the shapes would indeed be spherical if it were not for the force of gravity flattening them a bit. If you have watched films of astronauts in space, where the influence of gravity is negligible, you might have seen perfect spheres of water floating around the inside the cabin. The sphere has the particular distinction of having the least surface area for a given volume. Thus, the formation of spherical drops seems to indicate that the surface of the liquid has a higher energy than the bulk, and the liquid is trying to achieve a state of lowest energy by minimizing its surface area. This surface energy is associated with a property of liquids called the *surface tension*, and given the symbol γ.[8] Surface tension effects are very important in a number of biological phenomena including the transpiration of water in plants, the expansion of the lung alveoli in breathing (see Box 11-2), and the walking on water by many insects.

Since liquid surfaces, such as a water–air interface, tend to behave almost like a skin, some authors have drawn an analogy between surface tension and a rubber sheet. This is, unfortunately, a poor analogy since when the rubber sheet is stretched, no new material is brought to the surface. In a rubber sheet the old surface material is merely expanded, and each successive increment of surface area requires a greater force than the one before it. When a liquid surface is extended, it is done by creating new surface from molecules drawn from the bulk of the liquid. Thus, each successive increment in surface area requires the same force as the one before. The process can continue until there are no more interior molecules available to create new surface area.

---

8. γ is the lowercase Greek letter *gamma*.

Box 11-2   Surface Tension Effects in the Lungs

The most extensive surface of the body in contact with the environment is the moist interior surface of the lungs. To carry on the exchange of CO_2 and O_2 between circulating blood and the atmosphere in sufficient volume to sustain life requires one square metre of lung surface for each kilogram of body mass. For a normal adult this amounts to the area of a tennis court. The feat is accomplished by compartmentalizing the lungs into tiny air sacs called alveoli.

The surface tension in the outermost single layer of molecules of the film of tissue fluid that moistens the surface of the lungs accounts for a large percentage of the total elasticity of the alveoli. This elasticity must be overcome when we inhale and expand the surface area during breathing. A surface-active substance (surfactant) that has a much lower surface tension than water reduces the surface tension, brings about a more even distribution of pressure between the large and small alveoli, and reduces the overall pressure requirement, thereby decreasing the muscular effort required for respiration. Also, a lower surface tension permits a closer fit of the alveolar surface to the capillaries for maximum efficiency of gas transfer.

Specifically, during inhalation the surface area of the lungs is enlarged and the effective surface concentration of lung surfactant decreases (i.e., the same number of molecules is spread more thinly), leading to an increase in the surface tension. During exhalation the surface area of the lungs decreases, leading to an increase in concentration of surfactant and thus a decrease in surface tension. This decrease in surface tension stabilizes the airways in the lungs, preventing their collapse.

The surfactant in the lungs is a lipoprotein, that is, a compound molecule composed of protein and lipid constituents. In the lungs, the lipid has a polar head group which is hydrophilic (soluble in water) and long fatty acid chains which are hydrophobic (insoluble in water) and are believed to stick up out of the surface.

In *hyaline membrane disease* prevalent in premature babies, the lipid is absent. The molecules of the liquid lining have a high surface tension and attract each other. The alveoli collapse after exhalation and must be re-expanded. Thus, it becomes increasingly difficult for the infant to inflate its lungs. The baby often dies as a result of oxygen starvation and exhaustion. Researchers have been investigating the use of a lipid aerosol which has proved fairly successful in getting these children through the first 10 days or so until their body produces enough lipid surfactant.

While molecules in the bulk liquid are attracted equally in all directions by their neighbours and experience no net force, molecules at the surface are attracted by molecules beside them and by molecules within the bulk liquid, but negligibly by molecules from outside (since the number of molecules above the surface is small). Hence, work has to be done to pull molecules from the bulk, where they experience no force, to the surface, where they experience an attractive force back into the bulk. If work W is required to increase the surface area by $\Delta\mathcal{A}$, then the surface tension γ is

$$\gamma = \frac{W}{\Delta\mathcal{A}} \qquad\qquad \textbf{[11-14]}$$

Surface tension can alternately be expressed as force per unit length:

$$\gamma = \frac{F}{\ell} \qquad\qquad \textbf{[11-15]}$$

The reader should check that energy per unit area (J/m²) and force per unit length (N/m) indeed have equivalent dimensions of M/T^2.

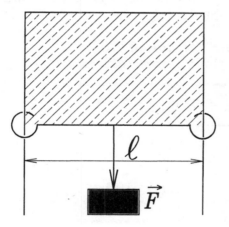

Figure 11-9

Measurement of the surface tension of a liquid film.

The force per unit length concept of surface tension can be visualized by the simple experimental apparatus shown in Fig. 11-9. A heavy wire frame with a moveable crossbar of light wire is drawn out of a soap solution, forming a film of the solution between the frame and the crossbar. Once completely out of the solution, the crossbar remains at rest, indicating that the film is exerting an upward surface tension force that counteracts the downward force, $\vec{F}$, due to the weight of the crossbar and the small mass attached. Unlike a sheet of rubber, the upward force is independent of the area of the film.

The upward force acts along <u>each</u> surface for the full length of the bar. Since the soap film has two surfaces (front and back, with bulk solution in between), then this force is $2\gamma\ell$. Since, at equilibrium, the magnitudes of the vertical forces sum to zero,

$$2\gamma\ell - F = 0$$

and thus, if F and ℓ are measured for this particular apparatus, the surface tension can be calculated as the ratio of F to the total length (2ℓ) of the film pulling up:

$$\gamma = \frac{F}{2\ell} \qquad\qquad \textbf{[11-16]}$$

A wide range of values of surface tension (Table 11-3) exists in nature.

Table 11-3   Surface Tension of Various Liquids at 20°C

Liquid	Surface Tension (N/m)
Water	72.8×10^{-3}
Blood plasma	50×10^{-3}
Lung surfactant	1×10^{-3}
Benzene	28.9×10^{-3}
Mercury	464×10^{-3}

Not surprisingly, surface tension is quite temperature-dependent; for water this effect is indicated in Table 11-4.

Table 11-4   Surface Tension of Water at Various Temperatures

Temperature (°C)	Surface Tension (N/m)
0	75.6×10^{-3}
20	72.8×10^{-3}
60	66.2×10^{-3}
100	58.9×10^{-3}

These values are obtained only with very pure water in very clean apparatus. Even small amounts of material in solution can cause considerable change. A 0.3% solution of acetic acid in water will decrease the surface tension by approximately 4%. Some materials decrease the surface tension (e.g., ammonium hydroxide), while others increase it (e.g., potassium hydroxide).

Surface Tension in Curved Surfaces

Interesting results of biological significance can occur when the effect of surface tension on curved surfaces is examined. Consider a soap bubble. We know we have to blow to create one, so the pressure must be greater inside than out. The bubble will reach equilibrium size when the tension in the liquid surface is sufficient to counteract the

Figure 11-10

Forces on one-half of a soap bubble (F_r = repulsive force; F_a = attractive force).

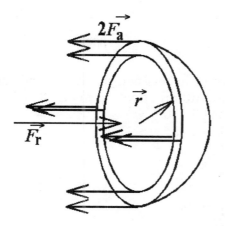

force exerted by the greater interior pressure. Figure 11-10 shows one-half of such a soap bubble.

If P is the internal absolute pressure, the excess internal pressure ($P - P_{atm}$) acts on the projected cross-sectional area of the bubble (πr^2) to create a net repulsive force of magnitude F_r that acts to push the bubble half to the right and tries to split the bubble apart. Thus,

$$P - P_{atm} = \frac{F_r}{\mathcal{A}} = \frac{F_r}{\pi r^2}$$

where $\mathcal{A}$ is the cross-sectional area (πr^2) of the bubble. To counteract this force, the surface tension exerts an attractive force to the left that tries to hold the bubble together. This force acts around the entire circumference ($2\pi r$) of the bubble where the two halves join. From Eq. [11-15] ($\gamma = F/\ell$), the magnitude of this attractive force is $F_a = \gamma\ell = \gamma 2\pi r$ for each of the inner and outer surfaces of the bubble. Considering both surfaces, equilibrium is obtained when $F_r - 2F_a = 0$. That is,

$$\left(P - P_{atm}\right)\pi r^2 - 4\pi\gamma r = 0$$

from which,

$$P - P_{atm} = \Delta P = \frac{4\gamma}{r} \qquad\qquad \textbf{[11-17]}$$

Figure 11-11

Connected soap bubbles: the bubble with a smaller radius will deflate to increase the size of the one with the large radius (dotted lines represent initial condition; solid lines represent final condition).

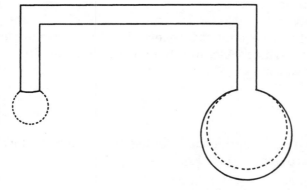

This indicates the surprising property of all bubbles: that the smaller the radius, the larger the internal pressure. If you blow two bubbles of different sizes and connect their interiors by a tube (Fig. 11-11), the smaller bubble will "deflate" and increase the size of the large one.

EXAMPLE 11-7

What is the gauge pressure difference between the inside and the outside of a soap bubble of radius 1.0 mm floating in air? The surface tension of the solution is 69×10^{-3} N/m.

SOLUTION

Using Eq. [11-17]:

$$P - P_{atm} = \frac{4\gamma}{r} = \frac{4 \times 69 \times 10^{-3} \text{ N/m}}{1.0 \times 10^{-3} \text{m}} = 2.8 \times 10^2 \text{ Pa}$$

If a bubble of air submerged in water were subjected to the same type of analysis as the soap bubble, there would be only one surface to generate the attractive surface tension force. As a result, the excess pressure of $4\gamma/r$ in Eq. [11-17] (for a two-surfaced bubble) becomes $2\gamma/r$ instead. In addition, the absolute pressure, P_{ext} external to the bubble can be substantially greater than atmospheric pressure, especially if it occurs at some depth in the liquid. Therefore, in this case,

$$P - P_{ext} = \Delta P = \frac{2\gamma}{r} \qquad \text{[11-18]}$$

EXAMPLE 11-8

(a) What pressure difference is required to blow a bubble 1.0 mm in radius just below the open surface of the same soap solution as used in Example 11-7?

(b) What additional pressure must be exerted if the bubble is blown 10 cm below the surface? Assume that the liquid's density is the same as that of water.

SOLUTION

(a) When blown in a bulk liquid, the bubble has only one surface. If it is blown at zero depth, then Eq. [11-18] gives

$$P - P_{ext} = \Delta P = \frac{2\gamma}{r} = \frac{2 \times 69 \times 10^{-3} \text{N/m}}{1.0 \times 10^{-3} \text{m}} = 1.4 \times 10^2 \text{ Pa}$$

(b) Use Eq. [11-2] to determine the additional pressure due to depth of liquid:

$$\Delta P = \rho g d$$
$$= (1000 \text{ kg/m}^3) \times (9.80 \text{ m/s}^2) \times (0.10 \text{ m}) = 9.8 \times 10^2 \text{ Pa}$$

Hence, the total pressure that must be exerted is

$$(9.8 \times 10^2 + 1.4 \times 10^2) \text{ Pa} = 1.1 \times 10^3 \text{ Pa}$$

As an aside, this same Eq. [11-18] would apply to a balloon at equilibrium, but the surface tension, γ, would have to be replaced by an elastic tension, T, that is not constant but increases as the balloon is more and more inflated.

The equation which we have derived for a sphere is a particular case of the more general equation for an ellipsoidal object (football shape) with two radii of curvature:

$$P - P_{\text{ext}} = \Delta P = \gamma\left(\frac{1}{r_1} + \frac{1}{r_2}\right) \qquad \textbf{[11-19]}$$

For the sphere, $r_1 = r_2$ and Eq. [11-19] reduces to Eq. [11-18]. In general, there is a term of the form tension/radius for each surface and each radius of curvature.

11.8 Capillarity

If small droplets of different liquids are placed on different surfaces and examined carefully, some droplets will appear extremely flattened whereas other are almost round beads. For water on (dirty) glass, the drop has the shape indicated in Fig. 11-12a, while for mercury on glass the shape is as in Fig. 11-12b.

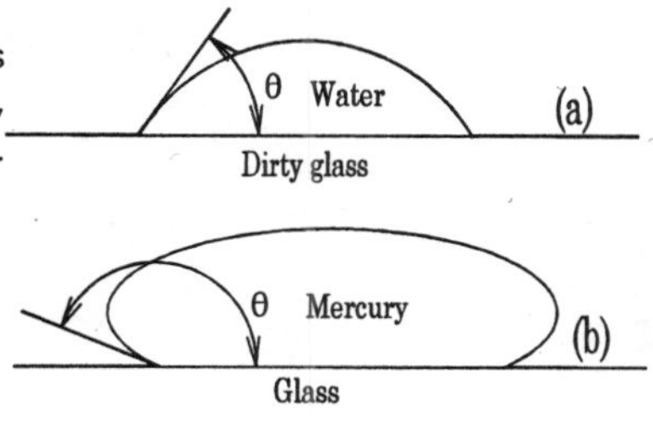

Figure 11-12

Angle of contact with glass of a droplet of (a) water, (b) mercury.

The different shapes arise because the liquid molecules have different affinities for the glass molecules. This behaviour is characterized by the *contact angle*, θ, which is shown in Fig. 11-12 for the two different droplets. The contact angle is a measure of the curvature of the liquid–vapour interface at a solid surface. For water on clean glass, $\theta \approx 0°$, whereas for water on dirty glass, $\theta \approx 30°$, and for mercury on glass, $\theta \approx 140°$. Liquids having contact angles between 0° and 90° are said to "wet" the solid surface. Liquids having contact angles between 90° and 180° are said "not to wet" the solid surface.

This wetting effect leads to the phenomenon of capillary rise (see Box 11-3). If a small tube of internal radius r is placed vertically in a liquid, the liquid will be observed to rise or fall in the tube depending on whether the contact angle is less than or greater than 90°. Consider the column of water at equilibrium in the glass tube in Fig. 11-13. The water has risen up the tube a distance y above the surface of the water in the container. This rise is due to the upward component of the surface tension force $\vec{F}$, which pulls all the way around the edge of the meniscus in the direction shown in Fig. 11-13. The magnitude of the upward component of this force is

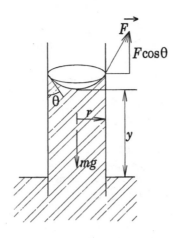

Figure 11-13

Capillary rise.

$$F_{up} = 2\pi r\gamma \cos(\theta)$$

The weight of the column of liquid in the tube provides a downward force of magnitude

$$F_{down} = -mg = -(\text{volume of the column}) \, \rho g = -\pi r^2 y \rho g$$

When the liquid in the column is at rest, these two forces sum to zero. Thus,

$$2\gamma \cos(\theta) = ry\rho g$$

and rearrangement of this equation yields an expression for the distance risen:

$$y = \frac{2\gamma \cos(\theta)}{\rho g r} \qquad\qquad \textbf{[11-20]}$$

Note that if θ is greater than 90°, $\cos(\theta)$ will be negative. The negative y-value that results represents the depression of the liquid in the tube below the surface of the liquid in the container. Such a capillary depression occurs for mercury in glass.

EXAMPLE 11-9

A glass tube of inner diameter 1.0 mm is inserted into (a) water and (b) mercury. What is the capillary rise (or depression)? Assume a contact angle of 0° for water on glass, and 140° for mercury on glass. Surface tensions and densities are given in Tables 11-3 and 11-1, respectively.

SOLUTION

Using Eq. [11-20]

(a) For water:

$$y = \frac{2\gamma\cos(\theta)}{\rho g r}$$

$$= \frac{2\,(72.8\times10^{-3}\text{ N/m})\cos(0°)}{(1000\text{ kg/m}^3)\times(9.80\text{ m/s}^2)\times(0.50\times10^{-3}\text{m})}$$

$$= 0.030\text{ m} = 3.0\text{ cm rise}$$

(b) For mercury:

$$y = \frac{2\gamma\cos(\theta)}{\rho g r}$$

$$= \frac{2\,(464\times10^{-3}\text{ N/m})\cos(140°)}{(1.36\times10^{4}\text{kg/m}^3)\times(9.80\text{ m/s}^2)\times(0.50\times10^{-3}\text{ m})}$$

$$= -0.011\text{ m} = 1.1\text{ cm depression}$$

Box 11-3   Capillary Rise in Plants

Capillarity plays an important role in the movement of water in plants. The vascular systems (xylem) of plants and trees can be considered as capillary in nature. Taking the radius of the xylem to be approximately 20 μm and assuming a contact angle of 0°, Eq. [11-20] predicts that water could rise to a height of about 0.75 m. This is sufficient to get water to the top of small plants but not trees. However, the walls of the xylem contain a network of passageways of quite small radius. These could easily support water columns in even the tallest trees.

11.9 Development of an Aneurysm

In discussions of blood flow that occur in Chapter 12, the blood vessel is treated as a solid pipe. However, the artery walls are actually elastic and possess quantities of the structural proteins elastin and collagen as well as contractile smooth muscle cells. The relative amounts of these components vary with distance from the heart. In the largest vessels, elastin and collagen are the most important contributors to the elastic properties of the wall. A graph of tension (T) in the wall versus radius (r) of such a vessel is shown in Fig. 11-14. The lower portion of the graph is principally due to the elastin component, while the more steeply rising portion is due mainly to the collagen component. The collagen fibrils have a very large Young's modulus (see Chapter 10) compared to elastin, but are essentially slack at smaller radii and do not become important until the vessel has stretched enough to "take up the slack."

In the smaller arteries, there is considerable smooth muscle, the active contraction of which can introduce a tension component in addition to the passive elastic components shown in Fig. 11-14. A number of interesting computations can be performed on the equilibrium of these various vessels. This section concerns one particular problem that can develop in the larger arteries where we do not have to worry about the complicating contribution from smooth muscle.

Earlier, the pressure difference across a flexible surface was discussed, and a general relationship between the curvature of the surface and the pressure gradient was provided (Eq. [11-19]):

$$P - P_{\text{ext}} = \Delta P = \gamma \left(\frac{1}{r_1} + \frac{1}{r_2} \right)$$

In the case of a blood vessel, the surface tension γ must be replaced by the elastic tension, T, in the wall of the vessel. Note that in an expanding bubble, γ would remain constant, but in an expanding artery the elastic tension would increase as the vessel radius increases. In Eq. [11-19], the radii describe the large and small curvatures associated with an ellipse. For a cylinder, $r_1 = r$ (the cylinder's radius), and $r_2 \to \infty$ since it corresponds to a straight line along the surface. Thus, one of the radii goes to infinity and its reciprocal to zero. Consequently, the equation can be rewritten for the blood vessel wall as:

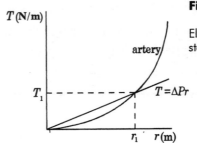

Figure 11-14

Elastic properties and stability of a normal artery.

$$\Delta P = T/r$$

$$\text{and } T = \Delta P\, r \qquad\qquad \textbf{[11-21]}$$

The ΔP represents the excess pressure inside the vessel relative to outside and is the quantity we call the blood pressure (see Box 11-1). If ΔP is constant, it can be represented as the slope of a straight-line graph relating the wall tension, T, to the vessel radius, r. An infinite number of combinations of radius and tension on this line satisfy Eq. [11-21]. The particular combination that actually occurs in a given vessel depends on the material properties of the vessel wall, specifically on its composition. Figure 11-14 simultaneously shows a plot of Eq. [11-21] and the tension-versus-radius curve for a real artery. Note that the artery expands easily at small radii, but stiffens at large radii as the "slack" in the collagen is taken up and the collagen starts to stretch. Note also that the point where the line and the curve cross indicates the stable radius and tension values for the artery in question.

This point of intersection (r_1, T_1) is the only point that satisfies both the condition that $T = \Delta Pr$ and also the tension-versus-radius data for the artery wall. Thus, r_1 is the stable radius, and T_1 is the stable elastic tension for a cylindrical vessel containing blood of pressure ΔP. If the blood pressure were to increase, the point of intersection would move up the curve since the slope of the straight line increases.

By way of comparison, Fig. 11-15 shows the stable radius and tension for a normal artery, and for one that has been weakened by disease or by aging. As would be expected, the stable radius for the weakened vessel is somewhat larger for the same blood pressure. In some—fortunately rare—instances the vessel wall can become so weakened that at peak blood pressure the point of intersection moves to extremely large values (Fig. 11-16) with the consequence that the blood vessel balloons or forms an *aneurysm* and sometimes bursts. This type of blowout can and does occur, and can be fatal.

Figure 11-15

Stability of a healthy and a weakened artery.

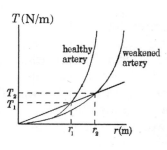

Figure 11-16

An aneurysm is produced when the point of intersection of *r* and *T* moves to extremely large values.

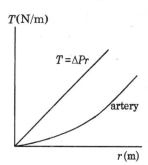

The potential for the development of an aneurysm can be determined quantitatively from the material properties of artery walls. The equation defining Young's modulus (Eq. [10-3]) can be rewritten in differential form for such a blood vessel and becomes

$$dT = Yt\,\frac{dr}{r}$$

where t is the thickness of the vessel wall and Y is its Young's modulus. The maximum rate of change of tension with respect to change of vessel radius is

$$\left(\frac{dT}{dr}\right)_{max} = \frac{Y_{max}t}{r}$$

However, from Eq. [11-21], $T/r = \Delta P$. Hence, $dT/dr = \Delta P$, and

$$\left(\frac{dT}{dr}\right)_{max} = \Delta P_{max}$$

Therefore,

$$\frac{Y_{max}t}{r} = \Delta P_{max} \qquad\qquad \textbf{[11-22]}$$

The measured value for Y in a healthy human iliac artery from a young adult is 7.0×10^5 N/m². Typical values for the artery radius, r, and thickness, t, are 0.35 cm and 0.07 cm respectively. Substitution of these values into Eq. [11-22] shows that the ΔP_{max} required for an aneurysm is 1.4×10^5 Pa (or 1000 mm of mercury), considerably more than 1 atm, whereas peak systolic pressure is normally only about 120 mm of mercury. The heart could never develop this large pressure, and an aneurysm cannot develop in a healthy vessel. While this is generally the case, unexpected aneurysms do sometimes develop in apparently normal individuals, especially at points where the arteries divide.

Exercises

11-1   A biologist uses a chamber (C) to study the effect of air pressure on small organisms placed inside. The pressure in the chamber is

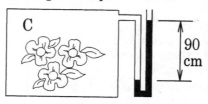

measured with a mercury-filled U-tube manometer open at one end to the atmosphere as shown in the diagram. For the case illustrated, what is the absolute pressure in the chamber? What is the gauge pressure in the chamber?

11-2   A typical human's systolic blood pressure is 120 mm of mercury. Express this in pascals. Is this an absolute or a gauge pressure?

11-3   A person 1.7 m tall stands on his head. Estimate by how much the hydrostatic pressure in his brain increases compared to its value when he is in the normal upright position.

11-4   A cube of wood of density 658 kg/m³ floats in a lake. What percentage of the wooden cube is submerged?

11-5   At a certain height above sea level, the gravitational potential energy and the thermal energy (k_BT) of the air molecules are the same. What is the pressure at that height?

11-6   Calculate the pressure in atmospheres at the top of Mount Everest whose peak is 10 km above sea level. Assume that the ratio between the weight of a single air molecule and its thermal energy is 1.0×10^{-4} m⁻¹.

11-7   The total surface area of the alveoli of the adult human lung is about 80 m². If this expands 5.0% on inhalation, how much work must be done to increase the area of the film lining the lungs if the surface tension of the liquid in the film is 1.0×10^{-3} N/m?

11-8   Sodium sulphate fertilizer solution has a surface tension of 7.3×10^{-2} N/m and a density of 1.5×10^3 kg/m³. Calculate the maximum diameter of xylem that would raise the solution to the top of a plant 50 cm tall. Assume that the contact angle is zero.

11-9   What is the tension in an arteriole wall if the mean blood pressure is 60 mm Hg and the radius is 0.010 cm?

Problems

11-10   If ice has a density of 0.92×10^3 kg/m³, what is the average volume occupied by one H_2O molecule in ice?

11-11   When a person is sucking hard, the gauge pressure in the lungs can be reduced to –80 mm Hg. Pure rum has a density of 0.92×10^3 kg/m³. What is the greatest height that rum can be sucked up a straw?

11-12   What are the gauge pressure and the absolute pressure in chamber A? Assume both manometer liquids to be water. The right-hand manometer is open at one end to the atmosphere, as shown.

11-13   Calculate the height in the atmosphere at which the atmospheric pressure is 0.10 times that at sea level. Assume that the average molar mass of the air molecules is 29 g/mol, and that the air is at a uniform temperature of 300 K.

11-14   The chamber below is being used for the study of algae. The chamber is a square box, 1.00 m on each side, and contains water at a depth of 30.0 cm. A mercury manometer is used to determine the pressure in 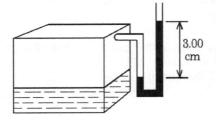 the chamber and registers a column height difference of 3.00 cm. One end of the manometer is open to the atmosphere. A small bubble of radius 0.0500 mm is located 20.0 cm below the surface of the water. Determine the gauge pressure and the absolute pressure inside the bubble.

11-15   Calculate the gauge pressure (relative to atmospheric pressure) inside a spherical bubble of radius 8.0×10^{-5} m located in a water tank 15 m below the surface of the water. On top of the water there is a 5.0 m layer of oil of density 0.50×10^3 kg/m³. The surface of the oil is open to the atmosphere.

11-16   A biologist has a sample of lung surfactant and wishes to measure its surface tension. She first determines that its density is 80% of the density of water, and that its contact angle with the lung tissue surface is zero. She places a glass capillary tube in a sample of water (the contact angle for the water–glass interface is also zero) and notes that the capillary rise is 0.10 m. She then places the same tube in surfactant and finds the rise to be only 0.020 m. What is the surface tension of the surfactant?

11-17 A capillary tube is dipped in water with its lower end 0.050 m below the water surface. Water rises in the tube to a height of 0.020 m above that of the surrounding liquid and the contact angle is zero. Then pressure is applied to blow bubbles out the end of the tube. What pressure difference is required to blow a hemispherical bubble at the lower end of the tube?

11-18 A liquid rises in a capillary as shown in the diagram below. Its density is 8.0×10^3 kg/m^3. What is (a) the gauge pressure at A? (b) the gauge pressure at B? Assume that the tube and liquid are open to the atmosphere.

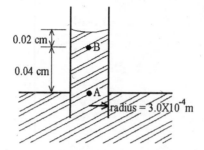

11-19 A capillary tube is placed in water and the water is observed to rise up in the tube to a height of 10 cm. The tube is then lowered further into the water so that only 8 cm of tube is above the water surface. Will the water:

(a) jet up like a fountain to a height of 2 cm above the top of the tube?

(b) flow over the top of the tube and run down the sides?

(c) rise to the top and stop?

Hint: As the water starts to flow upward at the top of the tube, what happens to the contact angle between the water and the tube?

Answers

11-1	2.2×10^5 Pa, 1.2×10^5 Pa	**11-11**	1.2 m
11-2	1.60×10^4 Pa, gauge	**11-12**	2.0×10^3 Pa, 1.03×10^5 Pa
11-3	1.7×10^4 Pa	**11-13**	2.0×10^4 m
11-4	65.8%	**11-14**	8.87×10^3 Pa, 1.10×10^5 Pa
11-5	3.7×10^4 Pa	**11-15**	1.7×10^5 Pa
11-6	0.37 atm.	**11-16**	1.2×10^{-2} N/m
11-7	4.0×10^{-3} J	**11-17**	6.9×10^2 Pa
11-8	4.0×10^{-5} m = 0.040 mm	**11-18**	(a) 0 (b) –31 Pa
11-9	0.80 N/m	**11-19**	(c)
11-10	3.3×10^{-29} m^3		

12 Fluid Dynamics

12.1 Introduction

The field of fluid dynamics is concerned with fluids in motion and with objects moving through fluids. For example, the process of breathing, the circulation of the blood, cytoplasmic streaming, and the movement of living creatures through air or water all involve fluid flow. The size of an artery or pipe and the speed of fluid flow through it are factors determining the nature of the flow. Similarly, an object's size and swimming speed are extremely important parameters that can lead to vastly different swimming mechanisms.

12.2 Streamline Flow

Most of the fluid-flow situations encountered in biological systems are of a fairly simple type where the fluid moves along smooth lines, known as "streamlines." This type of flow is known as *streamline flow* (sometimes called *laminar flow*). It can be continuous and constant, as in the flow of sap through the xylem of a tree, or *pulsatile*, as in heart-driven blood flow. The discussion in this section will be restricted to steady streamline flow. Pulsatile flow will be discussed in Section 12.7.

Figure 12-1

Streamline flow in (a) a straight tube, (b) a tube with variable radius.

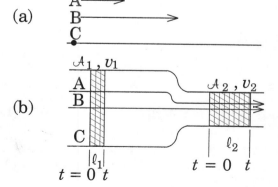

Figure 12-1a indicates a fluid, in this case a liquid, flowing through a tube at a low velocity. It is assumed that the liquid, density ρ, is incompressible: that there is no significant gain or loss of heat energy during flow and that the thermal motion of the molecules in the moving fluid is negligible relative to their flowing motion. If a small quantity of ink were released from a hypodermic needle inserted at A, the ink would be observed to sweep downstream along a streamline parallel to the sides of the tube. If the process were repeated at the centre of the tube at B, a similar streamline would form, but careful observation would reveal that this central streamline was traveling faster. Finally, if the ink was released right at the wall (at point C), it would remain essentially immobile and no streamline would be formed, indicating that the velocity decreases to zero there.

Obviously, the velocity is not constant at all points; instead a velocity distribution exists across the tube, having a value of zero at the walls and a maximum at the tube centre. An expression for this distribution will be developed shortly. However, it is often convenient to speak of the *average speed of flow*, v, of a liquid within a tube. This corresponds to the average speed of all the streamlines.

If the process is repeated with a tube that constricts or narrows as in Fig. 12-1b, the streamlines are forced to converge at the constriction. A careful observation would reveal that all the streamlines move more rapidly where the tube is narrowed, but that a similar velocity distribution is still present (i.e., the central streamlines still travel faster than those near the walls). The average flow speed is greater to the right of the constriction.

If the tube or pipe in Fig. 12-1b has rigid walls, then regardless of its shape, the rate at which the (incompressible) liquid enters one end must equal the rate at which it exits from the other end. This observation leads to a very useful conclusion. The same total volume of liquid must flow past any marked point in the wide or the narrow portions of the tube in the same time interval, t. That is, the volumes $\mathcal{A}_1\ell_1$ and $\mathcal{A}_2\ell_2$ are equal, where the $\mathcal{A}$'s refer to the cross-sectional areas of the wide and narrow parts of the tube and the ℓ's refer to the average distance traveled in time t. Since $\ell = vt$, then

$$\frac{\mathcal{A}_1\ell_1}{t} = \frac{\mathcal{A}_2\ell_2}{t}$$

and

$$\mathcal{A}_1 v_1 = \mathcal{A}_2 v_2 = \text{constant} \qquad \textbf{[12-1]}$$

This equation is known as *the equation of continuity*. Note that the product, $\mathcal{A}v$, has dimensions of volume per unit time (units: m³/s) and is defined as the as the *flow rate*, Q. Therefore,

$$Q = \mathcal{A}v \qquad \textbf{[12-2]}$$

Flow rate Q and flow speed v are, therefore, quantities with distinctly different meanings.

Sometimes in flow systems, situations are encountered in which a single pipe can divide (bifurcate) into two or more pipes such as when an artery branches. The reverse can also occur as when venules join to form veins. In all of these situations the total flow rate entering a bifurcation must equal the total flow rate leaving it. Thus, if n_1 tubes (of equal cross-sectional area, $\mathcal{A}_1$) each having an average flow speed, v_1, join and subsequently divide into n_2 tubes (of equal cross-sectional area, $\mathcal{A}_2$) each having an average flow speed v_2, the equation of continuity would be modified to:

$$Q = n_1 \mathcal{A}_1 v_1 = n_2 \mathcal{A}_2 v_2 = constant \qquad \textbf{[12-3]}$$

12.3 **Flow of an Ideal Fluid: Bernoulli's Theorem**

In an *ideal fluid*, no frictional forces are present to retard the flow. Therefore, an ideal fluid has no *viscosity*. While all real fluids are viscous, for most gases and many liquids, such as water, the viscosity is very small. It is, therefore, useful to consider flow behaviour in the

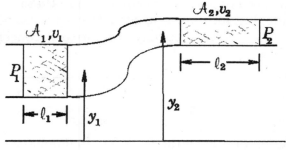

Figure 12-2

Bernoulli's law.

ideal situation. Consider the case in Fig. 12-2 where the tube changes not only in area but also in vertical elevation. The flow from region 1 to region 2 involves a change in gravitational potential energy (ΔU) of the liquid since region 2 is higher than region 1, and a change in the kinetic energy (ΔK) since, due to the equation of continuity, the average flow speed is greater in region 2 than in region 1. Consequently, there will be a total change of mechanical energy of

$$\Delta E = \Delta U + \Delta K \qquad \textbf{[12-4a]}$$

or

$$\Delta E = (mgy_2 - mgy_1) + (\tfrac{1}{2}mv_2^2 - \tfrac{1}{2}mv_1^2) \qquad \textbf{[12-4b]}$$

Some work must have been done to bring about this change in mechanical energy. Over the time interval t, the liquid in region 1 is pushed forward a distance ℓ_1 by a force of magnitude $F_1 = P_1 \mathcal{A}_1$, where P_1 is the pressure exerted by the fluid behind. The work, W_1, done by this force is $P_1 \mathcal{A}_1 \ell_1 = P_1 \mathcal{V}_1$, where $\mathcal{V}_1$ is the volume flowing past a point in region 1 in time t. Similarly, negative work is done by the backward force exerted on the liquid in region 2 by the fluid ahead; this work is $W_2 = -P_2 \mathcal{V}_2$. Since the flow rate, Q, is constant all along the tube, then $\mathcal{V}_1 = \mathcal{V}_2 = \mathcal{V}$ and the net work done is

$$W_1 + W_2 = (P_1 - P_2)\mathcal{V} \qquad \textbf{[12-5]}$$

This net work done is equal to the change in mechanical energy ΔE given by Eq. [12-4b]. Equating the right-hand sides of these equations, and replacing $\mathcal{V}$ with m/ρ, where m is the mass of liquid in the volume and ρ is the liquid's density, yields the expression

$$P_1 + \tfrac{1}{2}\rho v_1^2 + \rho g y_1 = P_2 + \tfrac{1}{2}\rho v_2^2 + \rho g y_2 \qquad \textbf{[12-6a]}$$

Equivalently,

$$P_1 + \tfrac{1}{2}\rho v_1^2 + \rho g y_1 = constant \qquad \textbf{[12-6b]}$$

Either form of the expression (Eq. [12-6a] or Eq. [12-6b]) is known as *Bernoulli's equation*.[1] Strictly speaking, the pressures used in these expressions should be absolute, but for most real cases they can also be gauge pressures since most situations involve only pressure <u>differences</u>.

EXAMPLE 12-1

Oil of density 850 kg/m³ flows in a tube 0.030 m in diameter at an absolute pressure of 1.60×10^5 Pa. At a smooth constriction the tube diameter reduces to 0.020 m and the pressure to 1.00×10^5 Pa. Calculate the rate of flow of oil in the tube. Assume the tube is horizontal.

SOLUTION

The Bernoulli equation describing the flow is Eq. [12-6a]:

---

1.   Daniel Bernoulli (1700–1782), Swiss mathematician.

$$P_1 + \tfrac{1}{2}\rho v_1^2 + \rho g y_1 = P_2 + \tfrac{1}{2}\rho v_2^2 + \rho g y_2$$

Since the tube is horizontal, y_1 equals y_2 and the expression simplifies with the elimination of the $\rho g y$ terms. It can be rearranged as

$$\tfrac{1}{2}(v_2^2 - v_1^2) = \frac{P_1 - P_2}{\rho} = \frac{(1.60 - 1.00)\times 10^5\,\text{Pa}}{850\,\text{kg/m}^3} = 70.6\ \text{m}^2/\text{s}^2 \qquad \textbf{[1]}$$

From the equation of continuity (Eq. [12-1]), $\mathcal{A}_1 v_1 = \mathcal{A}_2 v_2$, and since the cross-sectional area, $\mathcal{A}$, of a tube is πr^2, then

$$\pi\left(\frac{0.030\ \text{m}}{2}\right)^2 v_1 = \pi\left(\frac{0.020\ \text{m}}{2}\right)^2 v_2$$

and hence, $v_2 = 2.25\,v_1.$ Substituting this in Eq. [1] yields $v_1 = 5.9$ m/s.

Since $Q = \mathcal{A}_1 v_1$ then

$$Q = \pi\left(\frac{0.030\ \text{m}}{2}\right)^2 \times 5.9\ \text{m/s} = 4.2\times 10^{-3}\ \text{m}^3/\text{s}$$

EXAMPLE 12-2

By what percentage would the pressure drop in an artery as the blood (density 1.05×10^3 kg/m³) enters a region which has been narrowed down by an atherosclerotic plaque to a cross-sectional area only one-fifth of normal? Assume that the narrowing occurs within a very short distance, and hence any change in elevation will be negligible. Assume that the blood pressure in the normal artery is 100 mm Hg, and that the flow speed in this normal region is 0.12 m/s.

SOLUTION

The vessel is assumed to be fairly large so that the error introduced by ignoring the viscosity of the blood is not great; this allows the use of Bernoulli's equation. Designate the normal artery as region 1, and the narrow portion as region 2. Bernoulli's equation (Eq. [12-6a]) is

$$P_1 + \tfrac{1}{2}\rho v_1^2 + \rho g y_1 = P_2 + \tfrac{1}{2}\rho v_2^2 + \rho g y_2$$

Since $y_1 = y_2$, then the $\rho g y$ terms of Bernoulli's equation vanish, leaving

$$P_1 - P_2 = \frac{1}{2}\rho(v_2^2 - v_1^2) \qquad \qquad \textbf{[2]}$$

The equation of continuity (Eq. [12-1]) is $\mathcal{A}_1 v_1 = \mathcal{A}_2 v_2$.

Since $\mathcal{A}_2 = \mathcal{A}_1/5$, then $v_2 = 5\,v_1$, and substituting in Eq. [2],

$$P_1 - P_2 = 12\,\rho v_1^2 = 12 \times 1.05{\times}10^3\,\text{kg/m}^3 \times (0.12\ \text{m/s})^2 = 180\ \text{Pa}$$

To convert the normal arterial pressure ($P_1 = 100$ mm Hg) to pascals, use $P_1 = \rho g d$ (from Section 11.2):

$$P_1 = \rho g d = 1.36{\times}10^4\,\text{kg/m}^3 \times 9.80\ \text{m/s}^2 \times 0.100\,\text{m} = 1.33 \times 10^4\,\text{Pa}$$

Therefore,

$$\frac{\Delta P}{P} \times 100\% = \frac{180}{1.33{\times}10^4} \times 100\% = 1.4\%$$

Thus, the pressure decreases by 1.4% in the restricted area.

One of the most important consequences of Bernoulli's law is the counterintuitive fact that whenever a fluid flows faster, there is a <u>decrease</u> in pressure. As has just been shown in Example 12-2, the decrease in pressure can be calculated quantitatively for an incompressible liquid. Qualitatively, the same type of pressure drop occurs when a compressible fluid such as air is involved, and accounts for many phenomena ranging from the lift generated by a wing to the fact that baseballs that are hit down the first or third base lines often curve foul. This effect is often called the *Bernoulli effect*.

12.4 **Viscous Fluids**

Viscosity is the internal friction which occurs in a real fluid when there is flow. By flow we refer to either fluid motion such as in a tube or to the movement of the fluid around an object moving through it.

A quantitative understanding of viscosity is best achieved by applying the ideas of shear stress and strain, which were developed in Chapter 10. There, in the case of solids, it was seen that when a shear stress was applied, a corresponding shear strain resulted. In other words, a particular stress created a particular strain. However, if a similar fixed shear stress is applied to a fluid, the fluid flows and continues to flow as long as the stress is applied. That is, the strain continues to increase as long as the stress is present. This can be seen in Fig. 12-3.

Figure 12-3

Stress and strain in the shearing of a fluid.

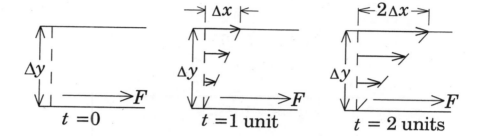

In this diagram a thin layer of liquid is placed between two plates, the top one moveable and the bottom one fixed. A force of magnitude F is applied parallel to the top plate (of area $\mathcal{A}$), making it move to the right. The shear stress σ_S is given by Eq. [10-4]: $\sigma_S = F/\mathcal{A}$. The shear strain (Eq. [10-5]) becomes $\varepsilon_S = (\Delta x/\Delta y)$, and Fig. 12-3 shows that this quantity is continually increasing as time progresses. It has been shown experimentally that the <u>rate</u> at which the strain increases depends linearly upon the stress. A greater stress will make the fluid flow faster. Therefore,

$$\frac{F}{\mathcal{A}} = \eta \frac{\Delta(\Delta x/\Delta y)}{\Delta t} = \eta \frac{\Delta}{\Delta t}\left(\frac{\Delta x}{\Delta y}\right) = \eta \frac{\Delta}{\Delta y}\left(\frac{\Delta x}{\Delta t}\right)$$

and since $(\Delta x/\Delta t)$ is the flow speed v, then,

$$\frac{F}{\mathcal{A}} = \eta \frac{\Delta v}{\Delta y} \qquad \textbf{[12-7]}$$

The quantity $(\Delta v/\Delta y)$ is called the *velocity gradient* and corresponds to the change in speed with position across the fluid. The proportionality constant η (lowercase Greek letter *eta*) that appears in Eq. [12-7] is called the *coefficient of viscosity* or more commonly the *viscosity* of the fluid. Its value depends on the composition of the fluid and on other parameters such as temperature and pressure (Table 12-1). Dimensional analysis of Eq. [12-7] shows that the dimensions of viscosity are $M \cdot L^{-1} \cdot T^{-1}$ (readers should prove it to themselves), and the standard units are N·s/m². An older, but still often used, unit is the *poise* named to honour Poiseuille, a 19th-century French chemist who investigated liquids (one poise = 10^{-1} N·s/m²).

Table 12-1   **The Viscosity of Various Fluids**

Fluid	Viscosity, η (N·s/m²)
Water at 20°C	0.00100 or 1 cP*
Water at 100°C	0.00028
Blood at 20°C	0.0045
Castor oil at 20°C	1.0
Air at 20°C	1.8×10^{-5}
Air at 100°C	2.1×10^{-5}

*1 cP = 1 centipoise = 10^{-2} poise

Table 12-1 shows trends that are typical of most liquids and gases. Note that the viscosity of a liquid (e.g., water) decreases with increasing temperature, whereas that of a gas increases. Viscosity in gases, where the molecules are relatively far apart most of the time, results from collisions arising from the thermal motion of molecules. Since the average molecular speed of random thermal motion increases with temperature, the viscosity of gases also increases with temperature (approximately as the square root of the absolute temperature). For liquids, in which the molecules are close together at all times, viscosity arises from the retarding effects of intermolecular forces that act to restrain the translational motion of the molecules. Rising temperature increases the kinetic energy of random thermal motion of the molecules and thus increases the relative fraction of molecules having sufficient kinetic energy to overcome the intermolecular forces, thereby effectively reducing the viscosity.

EXAMPLE 12-3

A glass microscope slide of dimensions 2.0 cm × 6.0 cm rests on a film of water of thickness 100 μm and temperature 20°C. It is found that a force of magnitude 1.20×10^{-4} N is required to move the slide horizontally with a speed of 1.0 cm/s. What is the coefficient of viscosity of the water at 20°C?

SOLUTION

Rearrange Eq. [12-7] to solve for η:

$$\eta = \frac{F \Delta y}{\mathcal{A} \Delta v} = \frac{(1.2 \times 10^{-4} \text{ N})(100 \times 10^{-6} \text{ m})}{[(2.0 \times 6.0) \times 10^{-4} \text{ m}^2](1.0 \times 10^{-2} \text{ m/s})} = 1.0 \times 10^{-3} \text{ N·s/m}^2$$

Note that $(\Delta y/\Delta v)$ was evaluated as the "change in position" $(100 \times 10^{-6} \text{ m})$ over the "change in velocity" $(1.0 \times 10^{-2} \text{ m/s})$.

Newtonian and Non-Newtonian Fluids

Fluids that obey Eq. [12-7] are called *Newtonian fluids*. These include gases and most of the common liquids such as water, alcohol, and liquid metals. Many fluid and quasi-fluid systems do not obey these equations and are said to be *non-Newtonian fluids*. *Thixotropic* substances are fluid when they are in motion and solid when they are not (e.g., wet sand and non-splatter paint). *Dilatant* substances are fluids for which the viscosity increases with increasing stress (e.g., suspensions of starch grains and Silly Putty).

From a biological point of view, one of the most important groups of these non-Newtonian fluids is the viscoelastic liquids. For such substances the expression for the shear stress contains both elastic (solidlike) and viscous (fluidlike) terms:

$$\frac{F}{\mathcal{A}} = G\varepsilon_s + \eta \frac{\Delta v}{\Delta y}$$ [12-8]

where G is the elastic shear modulus of the liquid (see Chapter 10). The first term represents the elastic (energy-storing) properties of the liquid whereas the second term represents its viscous (energy-losing) behaviour.

One of the most important of the viscoelastic liquids is *synovial fluid* that fills the cavities in the synovial joints, such as the knee, of mammals. It is similar to blood plasma but has less protein and contains a long polysaccharide, *hyaluronic acid*. It is this acid which seems to control the viscosity of the liquid. The viscosity of synovial fluid decreases as the stress and velocity gradient increase, which, of course, is necessary for smooth joint operation. Marked changes occur in the value of the viscosity of synovial fluid with the onset of certain diseases, in particular arthritis. Needless to say, there is at present a large research effort directed toward the investigation and characterization of such viscoelastic liquids, first to improve our understanding of their properties, and second to create suitable substitutes for the fluids in individuals suffering from joint disease.

12.5   **Flow of Viscous Fluids**

There are situations in which the effect of viscosity on flow behaviour must be considered. For flow in piping of small internal diameter, especially for a liquid with a large coefficient of viscosity, Bernoulli's law is no longer applicable. This is the case for the flow of blood in *capillaries*, which are the smaller vessels of the circulatory system. In the 19th century, Poiseuille, in an effort to measure the viscosity of blood, studied the flow of Newtonian fluids by measuring their flow rates in cylindrical tubes. When a liquid has a constant viscosity regardless of the tube size and regardless of the shear forces in the fluid, its Newtonian behaviour is confirmed. Blood is very close to being Newtonian down to very small tube radii, but some other liquids such as paints and molasses are very non-Newtonian, and their viscosity can change dramatically with different flow rates (shear rates).

The first step in the derivation of the equation for flow of a viscous fluid in a pipe of radius, R, and length, L, is to determine an expression that describes the velocity profile across the pipe. In Section 12.2 it was already indicated that the flow of a liquid is not constant across the pipe, but is maximum at the centre and approaches zero near the walls—this is a consequence of viscosity. Anyone who has waded across a small stream realizes that the current is much stronger at the centre of the stream. The idea here is to imagine that, in the pipe, the flow consists of thin concentric tubes or cylinders of fluid sliding over each other with the cylinders of fluid in the pipe centre moving faster than those near the walls (Fig. 12-4). This is the streamline flow discussed in Section 12-2.

The total flow rate, Q, will be the sum (the integral) of the flows in all these hypothetical cylinders. To begin, consider a very thin column of fluid of radius r at the very centre of the tube as shown. A force of magnitude F_1 exerted by fluid coming from behind pushes the column of fluid to the right, and a backward force of magnitude F_2 is exerted on the column by the fluid ahead. Thus, there is a net driving force of magnitude

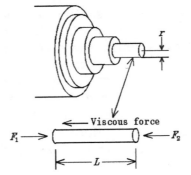

Figure 12-4

Viscous flow in a circular pipe.

$$F_\text{d} = F_1 - F_2 = (P_1 - P_2)\pi r^2$$

forcing fluid to the right, where P_1 and P_2 are the pressures related to F_1 and F_2. However, acting to oppose this motion is the frictional force between the column of fluid and the cylinder of fluid that surrounds it.

This is a viscous force (F_v) acting along the entire length, L, of the surface area of the column.

$$\frac{dy}{dx}$$

In order to determine the magnitude of this viscous force, Eq. [12-7] needs to be written in differential form as follows:

$$\frac{F_v}{\mathcal{A}} = \eta \frac{dv}{dy}$$

The area, $\mathcal{A}$, is the total contact area between the central column and the surrounding cylinder (both of length L), and is given by $\mathcal{A} = 2\pi rL$, which is the surface area of the column (excluding its ends). The quantity dv in this equation is the relative speed of the two surfaces, $v_{column} - v_{cylinder}$, but for simplicity it is left as dv. Since the important direction in this situation is radially outward, dy will be written as dr. Substituting for $\mathcal{A}$ and dy, and solving for the force (writing it as F_v since it depends on relative speed),

$$F_v = -2\pi\eta rL\frac{dv}{dr} \qquad \text{[12-9]}$$

The negative sign in front of the right-hand term in Eq. [12-9] arises because v decreases as r increases. Under conditions of steady flow, $F_d - F_v = 0$, and

$$(P_1 - P_2)\pi r^2 + 2\pi\eta rL\frac{dv}{dr} = 0$$

and rearranging leads to the expression for the *velocity gradient in a pipe*,

$$-\frac{dv}{dr} = \frac{P_1 - P_2}{2\eta L}r \qquad \text{[12-10]}$$

In this expression r has a minimum value of 0 at the centre of the pipe, and a maximum value of R, the inside radius of the pipe. Note that the velocity gradient varies linearly with r, and that it is 0 at the centre of the pipe and maximum near the walls. This is somewhat counterintuitive, since, as is shown quantitatively below, the velocity is greatest at the pipe centre and 0 at the walls. To obtain the expression for the velocity profile, it is necessary to rearrange Eq. [12-10] and integrate over the proper limits. The integral equation is,

$$-\int_v^0 dv = \frac{P_1 - P_2}{2\eta L}\int_r^R r\,dr$$

with the solution,

$$v = \frac{P_1 - P_2}{4\eta L}(R^2 - r^2) \qquad \text{[12-11]}$$

Equation [12-11] predicts that the velocity profile decreases parabolically from a maximum at the centre (where $r = 0$) to a minimum of 0 at the pipe wall (where $r = R$). By examining this function (Fig. 12-5), it is possible to determine why the velocity gradient, dv/dr, which is simply the slope of the velocity profile, is 0 at the pipe centre and a maximum at the pipe wall (see Eq. [12-10]). That is, near the walls, moving a small distance along r leads to a large change in v, whereas near the centre of the pipe, moving a small distance along r leads to only a small change in v. Clearly the shearing forces that would be experienced by objects carried along with the fluid are much greater near the walls. This is why a raft tends to lose direction and be rotated about when it is near the edge of a fast-moving stream, whereas it has relatively stable motion when it is in the centre of the stream.

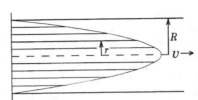

Figure 12-5

Velocity profile across a circular pipe carrying a viscous fluid.

To calculate the total flow rate, Q, carried along by the pipe it is necessary to sum all the individual flows, dQ, of each of the many sliding cylinders of fluid pictured in Fig. 12-4. Consider one of these which is viewed end-on in Fig. 12-6. The sliding cylinder is so thin that its wall thickness can be represented by dr. Therefore its cross-sectional area is $dA = 2\pi r\, dr$. Using an argument similar to that used in Chapter 2 (Fig. 2-8), the volume of fluid, dV, that the cylinder carries past a given point in the time interval dt is $dV = 2\pi rv\, dr\, dt$. Since $dQ = dV/dt$, then

$$dQ = 2\pi rv\, dr$$

and the total flow becomes

$$Q = \int_0^R 2\pi rv\, dr$$

Substituting the expression for v given by Eq. [12-11] and integrating yields *Poiseuille's law* (see Box 12-1):

$$Q = \frac{\pi R^4}{8\eta L}(P_1 - P_2) = \frac{\pi R^4 \Delta P}{8\eta L} \qquad \text{[12-12]}$$

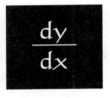

Figure 12-6

Cross-section of a circular pipe carrying a viscous fluid.

In this expression, the ratio $(P_1 - P_2)/L$ or $\Delta P/L$ is called the *pressure gradient*.

It is unusual for one variable, such as Q, to be dependent on the fourth power of another variable (R in this case). This means that, other factors remaining unchanged, a reduction in the radius to one-half the original value will result in a flow of one-sixteenth the original value.

This can be of major physiological significance in the circulation of the blood where reduction of arteriole radius due to smooth muscle contraction can change flow rates quite markedly. Such changes occur in response to direct nervous stimulation or to circulating sympathomimetic hormones such as adrenaline (epinephrine).

Even though Newtonian (viscous) fluids exhibit a parabolic velocity profile during streamline flow, they are still incompressible. Therefore, the equation of continuity still applies. However, it is important to emphasize that the "v" being referred to in the equation of continuity is the *average flow speed*, v_{av}, and, to avoid confusion, Eq. [12-2] could be written as

$$Q = Av_{av}$$ [12-13]

By substituting Eq. [12-13] into Eq. [12-12] and combining with Eq. [12-11] it can be shown that the flow speed in the centre of the pipe, $v_c = 2v_{av}$.

Box 12-1   When to Apply Bernoulli's Equation or Poiseuille's Law

It can be noticed that Poiseuille's law (Eq. [12-12]), and Bernouilli's equation (Eq. [12-6]) are in disagreement. Poiseuille's law predicts a pressure drop along a horizontal tube of constant diameter as a result of viscosity or fluid friction. Bernoulli's equation predicts no pressure drop under the same conditions since it deals with an ideal (and therefore frictionless) fluid. Sometimes there can be uncertainty as to which equation should be used. If the fluid is *viscous* and flowing through a tube which does not change in radius, then Poiseuille's law must be used. If there is a constriction or dilation in the vessel, then only Bernoulli's equation can be used even if some error is introduced by neglecting any viscous effects. These errors will be fairly small for low-viscosity liquids such as blood or water flowing in tubes of moderately large radius, but can be quite large with a very viscous liquid, and/or very small tubes.

EXAMPLE 12-4

An arteriole of diameter 2.5×10^{-5} m carries blood flowing at an average flow speed of 2.8×10^{-3} m/s. What is the pressure difference ΔP from one end to the other if the length of the arteriole is 5.0×10^{-3} m? What fraction of the total pressure drop in circulation does this ΔP represent?

SOLUTION

Combining the equation of continuity (Eq. [12-13]) and Poiseuille's law (Eq. [12-12]):

$$Q = \mathcal{A}v_{av} = \pi R^2 v_{av} = \frac{\pi R^4 \Delta P}{8\eta L}$$

and hence

$$\Delta P = \frac{8\eta L\, v_{av}}{R^2}$$

Using the viscosity of blood from Table 12-1, and other values as above,

$$\Delta P = \frac{8(4.5\times10^{-3}\,\text{N·s/m}^2)(5.0\times10^{-3}\,\text{m})(2.8\times10^{-3}\,\text{m/s})}{(1.25\times10^{-5}\,\text{m})^2} = 3.2\times10^3\,\text{Pa}$$

Normal average blood pressure in the large vessels (therefore approximately in the heart) is about 100 mm Hg. After the return to the heart, the pressure is approximately 0. Therefore the entire pressure drop of 100 mm Hg in the circulation is (from Section 11.2; also see Example 12-2):

$$\Delta P(\text{total}) = \rho g d$$
$$= 1.36\times10^4\,\text{kg/m}^3 \times 9.80\ \text{m/s}^2 \times 0.10\ \text{m}$$
$$= 1.33\times10^4\ \text{Pa}$$

Thus the drop in the arteriole is $\Delta P/P \times 100\%$ = $(3.2\times10^3\ \text{Pa}/1.33\times10^4\ \text{Pa}) \times 100\% \approx 24\%$ of the total drop in the circulation.

12.6   **Bolus Flow**

Many capillaries have a diameter of only 5 or 6 µm, whereas the erythrocytes (red blood cells) in the blood have a diameter of about 8 µm. Consequently the erythrocytes must deform to pass through the capillary. The cells occupy the entire cross-section of the interior of the vessel and pass through as a series of moving plugs with short sections of plasma trapped between them. This is known as *bolus flow* and is represented schematically in Fig. 12-7.

Figure 12-7

Bolus flow.

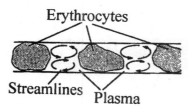

The flow in the trapped plasma sections becomes a specialized form of streamline flow. Rapidly moving streamlines down the centre catch up with an erythrocyte and are deflected to the outside where their speed, relative to the wall, becomes essentially zero. The following erythrocyte picks up these layers and forces them once more down the centre. The streamlines advance in a process that looks much like the tread on a bulldozer. Bolus flow has a decidedly beneficial result as the plasma sections are kept well stirred. This facilitates rapid movement of nutrients and waste products across the capillary walls.

12.7   **Pulsatile Flow**

In the larger arteries and the thoracic veins, the beating action of the heart creates a pulsatile flow in which the flow speeds are no longer steady. While the blood still advances in streamlines, the flow does not persist long enough for a parabolic velocity profile to develop. The velocity profile tends to be more uniform across the vessel, but its magnitude changes from one part of the cardiac cycle to another with the flow direction actually reversing at one point in the cycle. If the velocity profile is examined over an entire cycle, one gets the impression of a mass of fluid which advances four steps and then backs up one. With increasing distance from the heart, the pulses are gradually damped due to several factors, including the viscosity of the blood and the extensibility of the blood vessel walls. In the smaller, more distant arteries, if average values are used for the pressure gradient, Poiseuille's law is quite satisfactory for predicting the average flow rate. In the larger vessels, however, the observed flow is much less than that predicted by Poiseuille's law, and may be as little as one-fifteenth of the predicted value.

A graph of the instantaneous flow rate during pulsatile flow is shown in Fig. 12-8 for the femoral (thigh) artery of a dog. The heart rate is 2.75 s^{-1} (pulses per second) and the corresponding period (the inverse of the heart rate) of 0.36 s is represented on the horizontal axis as 360°. This representation is reminiscent of uniform circular motion, in which the process repeats itself every complete cycle. Notice that the flow reverses for part of the cycle even this far from the heart. The total accumulated flow during the cycle could be obtained by integrating the area under the curve Q, but the mathematical techniques are rather advanced. It is possible, however, to estimate the flow by approximating the positive and negative sections under the curve as triangles, for which area determinations are relatively easy (see Problem 12-16).

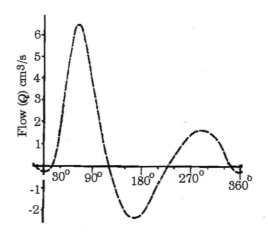

Figure 12-8

Pulsatile flow vs. phase (time) in the femoral artery of a dog.

12.8   Turbulent Flow

If an increasing pressure gradient is applied to a pipe or tube through which a Newtonian fluid is flowing, the flow rate increases linearly in accordance with Poiseuille's law, but only up to a certain limit. Beyond this point it takes progressively greater increments of pressure to produce increased flow (Fig. 12-9). In this region the flow becomes noisy and can be easily heard. In addition, the tube may even vibrate. If ink is injected so as to observe the flow pattern, the lines would be observed to swirl around and quickly disappear rather than advance in streamlines. The flow is said to be *turbulent*. In general, turbulent flow is inefficient for transporting fluids, and the associated noise can be disturbing. This is easily heard in operating a kitchen tap. At low flow rate the flow is silent, but at higher flow rate the resulting turbulent flow can be quite loud.

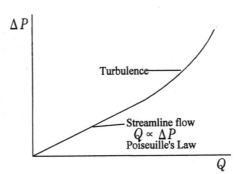

Figure 12-9

Onset of turbulent flow in a viscous liquid.

The energy required to move the fluid can be wasted since turbulence generates heat and, in addition, there can also be excessive wear and vibration on the pipes and pipe joints with increased chance of leaks developing.

The velocity profile in turbulent flow is extremely complex and difficult to analyze. However, it is possible to consider a time-averaged profile (Fig. 12-10). As compared to streamline flow, a time-averaged velocity profile indicates extremely large velocity gradients near the walls, but a more "flattened" profile near the centre of the tube. However, this is a time-averaged view; instantaneously the rapid swirling and mixing motions of turbulent flow can lead to large velocity gradients and high shear rates.

Figure 12-10

Velocity profiles in streamline and turbulent flow.

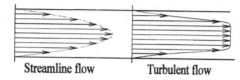

Streamline flow   Turbulent flow

Reynolds,[2] one of the first investigators of turbulent flow, discovered that a dimensionless parameter, now known as the *Reynolds number,* R_e , could be very useful in characterizing flows. The Reynolds number is defined by the equation,

$$R_e = \frac{\rho v_{av} D}{\eta}$$
[12-14]

where ρ is the fluid's density, v_{av} is the average flow speed, D is the diameter of the pipe, and η is the fluid's viscosity.

The flow of simple liquids such as water or blood usually changes from streamline to turbulent when the Reynolds number exceeds about 2000. However, this is not necessarily always the case. If the walls of the pipe are highly polished and all imperfections are removed, it is quite possible to have streamline flow still present even when the Reynolds number is as large as 16 000. Even in working models of the curved aorta with side branches, streamline flow has been obtained at $R_e \sim 16$ 000. However, the onset of turbulence is much like the onset of the process of crystallization, or of precipitate formation. If there is so much as one tiny imperfection, then what begins as a microscopic effect suddenly mushrooms throughout the system. Therefore, the proper view is that whenever there is a disturbance in the flowlines at $R_e > 2000$, it will usually amplify into full-scale turbulence, whereas if the disturbance occurs at $R_e < 2000$, it will usually die away and streamline flow will be restored.

2.   Osborne Reynolds (1842–1912), English engineer and physicist.

Some books define the Reynolds number using radius instead of diameter. In these cases, the limiting value for turbulence is changed from 2000 to 1000. If an R_e value is quoted in a book or set of tables, always check to find out whether diameter or radius was used in the calculation.

EXAMPLE 12-5

What is the peak Reynolds number in the abdominal aorta of the rabbit where $D = 3.0$ mm and the peak speed $v = 0.60$ m/s? If a disturbance is initiated, will it amplify into turbulence or will it die away? Use viscosity of blood as per Table 12-1, and density of blood as per Table 11-1.

SOLUTION

Using Eq. [12-14]:

$$R_e = \frac{\rho v_{av} D}{\eta} = \frac{1.05 \times 10^3 \, \text{kg/m}^3 \times 0.60 \, \text{m/s} \times 3.0 \times 10^{-3} \, \text{m}}{0.0045 \, \text{N·s/m}^2} = 420$$

Since R_e is less than 2000, the disturbance will die away.

EXAMPLE 12-6

Would one expect turbulence in a typical human aorta having a diameter of 0.020 m? Assume that blood has the same density as water and that its peak speed is 0.40 m/s for a person at rest.

SOLUTION

$$R_e = \frac{\rho v_{av} D}{\eta} = \frac{1000 \, \text{kg/m}^3 \times 0.40 \, \text{m/s} \times 0.020 \, \text{m}}{0.0045 \, \text{N·s/m}^2} = 1.8 \times 10^3$$

This is so close to the limiting value for turbulence that one cannot answer the question in a definite manner. Direct observation suggests that little if any turbulence occurs in the aorta even when the flow increases as in heavy exercise. There may be a little turbulence right at the peak of systole (ejection of blood from the heart), especially at the valves, but if so, it quickly dies out. In no other vessel is the Reynolds number this large. Consequently there is essentially no turbulence in the blood vessels of a healthy person with the exception of local disturbances at bifurcations (branching points), which quickly die out within a very short distance. However, when taking blood pressure using a sphygmomanometer (Box 11-1), the physician or nurse places a stethoscope to the inside of the elbow to hear the onset of turbulent blood flow as the constriction about the upper arm is gradually released. Once the arteries are fully opened, the sound disappears.

12.9 **Biological Ramifications of Fluid Dynamics**

Fluid Flow in Trees

The bulk of water movement in the trunks of trees is through the xylem tissue where long cells, typically 1 mm long and having radii from 20 μm to 200 μm (depending on the species), are joined more or less end-to-end to form conducting pathways that approximate cylinders. It might be expected that Poiseuille's law would apply in these vessels. Actual measurements are difficult to make, but do seem to be consistent with Poiseuille's law. In trees with vessels of 20 μm radius, the observed upward flow speed is 0.1 cm/s. This would require, if the viscosity is similar to that of water, a pressure gradient of 2×10^4 Pa/m. The observed pressure gradient is 3×10^4 Pa/m, which looks like rather poor agreement. However, if one accounts for the additional pressure gradient required to overcome gravity, the agreement is very good. Indeed, if the trunk is placed in a horizontal position, the expected 2×10^4 Pa/m is measured.

Blood Flow

If water, which is a Newtonian fluid, is made to flow through tubes of different radii, and Poiseuille's law is used to calculate the coefficient of viscosity, the same value is determined regardless of tube size. However, if the same experiment is repeated with blood, the blood apparently becomes less viscous as the tube radius decreases. This non-Newtonian behaviour (where fluid viscosity depends on the size of the tube) is almost undetectable for larger tubes, but as the radius decreases below 0.5 mm the phenomenon, known as the *Fahreus-Lindqvist effect*, can be quite significant (Fig. 12-11).

There is considerable disagreement as to the complete explanation of this non-Newtonian behaviour, and a number of phenomena may be involved. One of the more important factors is probably the fact that velocity gradients tend to cause the erythrocytes to tumble in toward the centre of the stream. Consequently, next to the walls a boundary layer develops in which the number of cells is significantly reduced so that the flow here is essentially that of the plasma which has a lower viscosity. The size of this boundary layer relative to the total cross-sectional area becomes relatively more important as the radius of the tube gets smaller.

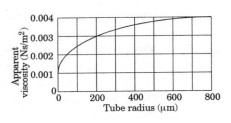

Figure 12-11

The Fahreus-Lindqvist effect: Variation of apparent viscosity with tube radius.

Often a large tube divides into two or more smaller tubes, such as frequently happens in the circulatory system. In this case, Poiseuille's law requires that to maintain the same total flow with similar pressure gradients, the total cross-sectional area of all n tubes after division is equal to $n^{1/2}$ times the area of the original tube. Thus a division into four smaller tubes would require twice the original area, and each of the new tubes must be one-half the original area. This is part of the explanation of why the total cross-sectional area of all the capillaries is much greater than the cross-section of the principal aorta.

Swimming

The physical description of how organisms swim includes the same concepts that were used in previous sections to describe fluids in motion. Indeed, it does not matter whether it is the fluid or the solid which experiences the net motion—the hydrodynamics are the same. Only the relative motion matters. However, it is interesting to see what the concepts of hydrodynamics require of the swimmer so that it can propel itself satisfactorily. There will be some surprises, especially for small motile cells such as certain bacteria, algae and spermatozoa.

One of the parameters that best characterizes the physics involved in the swimming process is the Reynolds number. When it was used earlier in Section 12.8 to describe the flow conditions in tubes, it was not mentioned where the equation that defines the Reynolds number came from or what it measured in any detail. Now, however, a better understanding of the Reynolds number and its significance is extremely important.

To begin, recall what it was like the last time you were swimming. In order to propel yourself forward you had to try to push water backward using your arms and legs. This backward motion of water, which opposes the force you apply to it, is often termed the "inertial" part of the opposing force. However, another part of the force you applied went into overcoming the frictional effects due to the viscosity of water. This is the viscous part of the opposing force. We define the quantity χ (lowercase Greek letter *xi*) as being the dimensionless ratio of the magnitudes of these two parts of the opposing force. Thus,

$$\chi = \frac{inertial\ force}{viscous\ force} \qquad \textbf{[12-15]}$$

The inertial force can be replaced by its equivalent (ma) where m refers to the mass of water moved and a refers to its acceleration. If the mass of water was initially at rest and achieves a final speed, v, then $a = v^2/2x$ (see Eq. [7-9]), in which x is the corresponding displacement (assuming $x_0 = 0$). Thus,

$$inertial\ force = \frac{mv^2}{2x}$$

The force required to move an object through a viscous liquid is linearly proportional to the speed of the object. Therefore, the viscous force can be replaced by $\mathcal{F}v$ where $\mathcal{F}$ is called the friction factor of the object. Albert Einstein showed that for spherical particles of radius r,

$$\mathcal{F} = 6\pi\eta r = 3\pi\eta D \qquad \textbf{[12-16]}$$

where η is the fluid's viscosity, and $D = 2r$ is now the diameter of the object instead of the diameter of a pipe as discussed earlier. The mathematical form of the friction factors for rod-shaped particles, ellipsoidal particles and dumbbell-shaped particles are much more complex, but in all cases, $\mathcal{F}$ is proportional to the product of the coefficient of viscosity and a parameter related to the size of the object. Therefore, the viscous force in Eq. [12-15] has the form

$$viscous\ force = C_0\,\eta D v \qquad \textbf{[12-17]}$$

where C_0 is a dimensionless constant (e.g., $C_0 = 3\pi$ for a sphere).

Substituting the expressions for inertial force and viscous force into Eq. [12-15],

$$\chi = \frac{mv}{2C_0 x\eta D}$$

This expression can be further simplified by noting that $m = \rho \mathcal{V}$, where ρ is the density and $\mathcal{V}$ is the volume of the moving fluid. The quantities $\mathcal{V}$ and x can be approximated as functions of D. That is,

$$\dot{x} = C_1 D \text{ and } \mathcal{V} = C_2 D^3$$

where C_1 and C_2 are constant parameters. In this approach, which is somewhat like the scaling methods discussed in Chapter 10, all terms having some power of length as their dimension can be expressed in terms of the object diameter D. Now χ becomes

$$\chi = \frac{C_2 \rho v D}{2 C_0 C_1 \eta}$$

All the dimensionless constants are then moved to the left-hand side and become the single dimensionless constant called the Reynolds number. Thus, R_e can be quantitatively expressed by the relationship

$$R_e = \frac{2\chi C_0 C_1}{C_2} = \frac{\rho v D}{\eta} \qquad \textbf{[12-18]}$$

However, qualitatively it is important to remember the origin of R_e, that is,

$$R_e = (constant)\frac{inertial\ force}{viscous\ force}$$

Through all this manipulation of constants and parameters, one sees that R_e is just a constant times the ratio of the inertial force to the viscous force. For any swimming organism, the hydrodynamic environment depends on the organism's size as described by R_e. The magnitude of R_e determines whether the inertial force or the viscous force has the greater influence on swimming behaviour.

Now let's see how Eq. [12-18] describes the way you swim. You can test this description the next time you are in a pool. When you want to propel yourself forward you extend your arms and legs sideways as much as you can on the power stroke in order to make D as large as possible. As well, to move faster, you move your arms and legs fairly rapidly (making v large). The effect of increasing both D and v is to dramatically increase R_e and hence increase the fraction of the total exerted force that goes into inertial effects. The result gives you sizeable forward motion, and, if you then keep your arms in close to the body (to minimize D), you can glide through the pool for a distance several times your body length. The more streamlined you make yourself, the further you glide. On the recovery stroke, as well, you bring your arms in close to minimize D and you move them more slowly. The lower value of R_e during recovery means that the inertial forces (that would tend to move you backward) are reduced.

These arguments also apply to fish. It is interesting to note the compromises which nature produces. Take, for example, a fish such as the sea raven (Fig. 12-12) that is known as a sprinter. Its oversized fins and tail give it rapid acceleration over short distances, but reduce its ability to glide. Hence it has the ability to dart quickly away from a predator, but is a poor long-distance swimmer. Other fish such as tuna that migrate long distances are more streamlined and have relatively small fins. Although they accelerate more slowly, they can glide smoothly and rapidly over long distances.

Figure 12-12

Profiles of gliding (skipjack tuna) and sprinting (sea raven) fish.

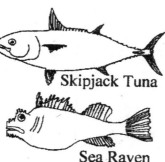

Skipjack Tuna

Sea Raven

Evaluation of R_e for large organisms such as dolphins yields values of 10^4 or higher. This is considerably above the turbulence limit discussed earlier. The dolphin has, however, a very smooth delicate skin to minimize turbulence so that under most swimming conditions the flow remains close to streamline. R_e for very small fish such as minnows is about 100. From the equations above, it is clear that for small fish, viscous effects become relatively much more important and limit their ability to glide.

Microscopic Swimmers

The case of microscopic swimmers such as motile bacteria, algae and spermatozoa is so interesting that we should deal with it separately. For such tiny particles, R_e is a very small quantity, typically 10^{-3} or less. Clearly, in this regime viscous forces become huge and inertial forces become negligible. The ability of the organism to glide is lost completely, since essentially all the driving force is used to overcome friction. Most microorganisms have developed motions that allow them to move along the direction of least friction. In other words, they are designed so that the friction factor associated with forward motion is much lower than the friction factor for sideways motion. Hence, wriggling or whipping motions tend to be translated into forward motions.

The cellular motions must be just right, however, for effective propulsion to occur. Symmetrical motions of a stiff fibre (like a stiff oar) would be useless, for example. Since the cell cannot glide, it would stop the instant the power stroke was completed. During the recovery stroke, the cell would move back to its initial starting point and make no net progress at all; instead, it would simply oscillate back and forth in one place. Microscopic motile cells are usually driven by slender flexible fibres called *flagella*. These flagella often beat in a helical wave or some related form of asymmetrical motion. Helical flagellar motion is characteristic of the spermatozoa of many creatures including mammals. Other motions, such as a type of breaststroke exhibited by cells such as the motile algae *Chlamydomonas reinhardii,* are less efficient. Although these cells bend the flagella in the recovery stroke (to try and reduce D), they are not totally successful and move backward during this part of the swimming cycle. These cells move by taking "two steps forward and one backward" as shown in Fig. 12-13.

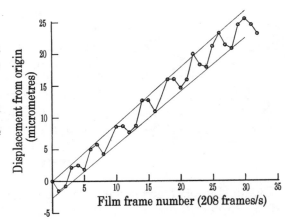

Figure 12-13

Displacement vs. time for a motile alga.

Exercises

12-1   Which of the following statements is not implied by Bernoulli's equation?

(a) In a horizontal tube, if the flow speed increases the pressure decreases.

(b) In a tube of variable cross-section, the product of flow speed times area is constant.

(c) The pressure increases with depth in the ocean.

(d) The total energy per unit volume of a fluid is constant.

(e) The kinetic energy may increase at the expense of the gravitational energy.

12-2   A large artery (radius r_1) branches into three smaller arteries of equal radii r_2. The branching occurs in such a way that the average flow speed of the blood is the same in each of the smaller arteries as in the larger one. For this to occur, what must be the value of r_2 relative to r_1?

12-3   Diagrams I and II represent schematically a human leg in the vertical and horizontal positions respectively. If the flow rate is to be the same in both cases, which of the following is true:

(a) The arterial pressure at the entry to the leg must be the same in both cases.

(b) The arterial pressure at the entry to the leg must be greater in I.

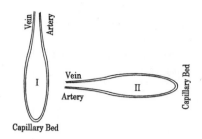

(c) The arterial pressure at the entry to the leg must be greater in II.

12-4   Referring to Poiseuille's law (Eq. [12-12]), if ΔP, η and L are held constant, sketch graphs of Q vs. R and Q vs. R^4. If R, ΔP, and η are held constant, sketch graphs of Q vs. L and Q vs. $1/L$.

12-5   Determine the Reynolds number associated with the swimming motion of a microscopic alga in water at 20°C if it has an effective diameter of 10 μm. The swimming speed of the alga is 100 μm/s.

12-6   Water at room temperature flows through a tube of internal diameter 2.0 cm at an average speed of 20 cm/s. Which one of the following statements about the flow is correct?

(a) The flow will probably be turbulent.

(b) The flow will undoubtedly be streamline.

(c) The flow will undoubtedly be turbulent.

(d) Any disturbances in the flow will quickly be dampened out.

(e) The flow will be turbulent if the viscosity of the liquid appears to decrease as the flow speed is increased from 0 to the final value of 20 cm/s.

Problems

12-7   An ideal fluid flows in the tube which constricts and drops as shown in the diagram below. What must h be in order that the pressure in the fluid at the bottom (P_2) equals the pressure at the top (P_1)? (Express h in terms of v_1 and g.)

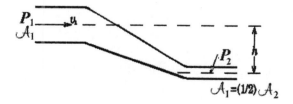

12-8   If the gauge pressure in the larger section of the horizontal cylindrical tube shown below is 2.00×10^5 Pa, then what is the expected gauge pressure in the smaller section? (Assume that the liquid is water and that its viscosity is negligible.)

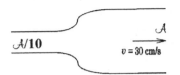

$\mathcal{A}/10$

$\xrightarrow{\mathcal{A}}$
$v = 30$ cm/s

12-9   The cross-sectional area of a pipe carrying water decreases linearly from $\mathcal{A}$ to $\frac{1}{2}\mathcal{A}$ over a distance of 0.15 m, as shown. If the speed of the water is 0.10 m/s and the gauge pressure is 50 Pa at the large cross-section, how high will the fluid be in the tube located 0.10 m downstream? (Assume for this problem that water has zero viscosity).

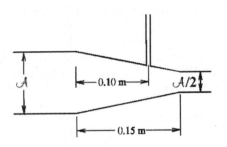

$\mathcal{A}$ |←0.10 m→| $\mathcal{A}/2$

|←— 0.15 m —→|

12-10   The blood in an artery of radius 5.0×10^{-3} m flows with a speed of 0.15 m/s. This artery subdivides into a large number of capillaries, each having radius 5.0×10^{-6} m. The flow speed in the capillaries is 5.0×10^{-4} m/s. Into how many capillaries does this artery divide? (Treat blood as an ideal fluid.)

12-11   Four small veins all of the same radius r_1 join together to form a larger vein of radius $r_2 = 5r_1$. If the average flow speed of the blood in each of the small veins is v_1, what is the average flow speed in the larger vein (in terms of v_1)?

12-12   Many clinical studies reveal that deposits in arteries effectively narrow the arterial passages. Compare a healthy artery and a diseased one whose cross-sectional area is only 0.600 of the healthy one. By what factor will the pressure gradient ($\Delta P/L$) have to change if the same volume of blood is to be carried by the diseased artery per unit time?

12-13 A viscous liquid flows out of a full tank through tubes of equal length, as shown in the diagram. If the upper tube has twice the radius of the lower tube, then determine the flow rate from the upper tube relative to that from the lower tube.

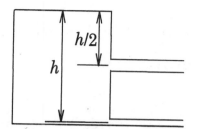

12-14 A patient is to be given an intravenous transfusion of blood. The blood is to flow from a bottle through a needle inserted into a vein in the patient's arm. The inside diameter of the 3.00-cm-long needle is 0.440 mm, and the required flow rate is 4.00 cm³ of blood per minute. How high should the bottle of blood be placed above the needle, assuming that the patient's blood pressure in the vein is 10.0 mm Hg and that the density of blood is 1.05×10^3 kg/m³.

12-15 Under the conditions of normal activity, an adult inhales about 1 L of air during each inhalation. With a watch, determine the time for one of your own inhalations (average several), and calculate the volume flux in cubic metres per second and the average flow speed of the air in metres per second through your trachea. The radius of the trachea in adult humans is approximately 10^{-2} m.

12-16 Figure 12-8 shows a graph of instantaneous flow rate vs. time for the pulsatile flow of blood in the femoral artery of a dog. As stated in Section 12.7, 360° on the horizontal axis represents the time for one pulse, which is 0.36 s.

(a) The first forward flow occurs from about 15° to 120°. During what time interval (in seconds) does this flow occur?

(b) The graph of the first forward flow is approximately triangular in shape. Determine the area of this triangle, using the height in cubic centimetres per second, and the base in seconds. Include units for this "area."

(c) The answer to part (b) represents the volume of blood that flows during the first forward flow. The flow from 120° to 230° is backward, and from 230° onward is forward again. The negative area from 120° to 230° is approximately the same size as the positive area from 230° onward, and hence these two flows contribute zero net flow. Hence, the answer to part (b) represents the volume of blood that flows during the entire pulse. The heart rate is 2.75 s⁻¹; what volume of blood flows in 1.0 min?

Answers

12-1 (b)

12-2 $r_2 = (3)^{-1/2} r_1 = 0.58 \, r_1$

12-3 (a)

12-4

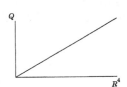

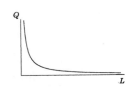

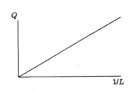

12-5 0.0010

12-6 (a)

12-7 $3v_1^2/(2g)$

12-8 1.55×10^4 Pa

12-9 0.0045 m

12-10 3.0×10^8

12-11 $4v_1/25$

12-12 $(\Delta P/L)_{\text{diseased}} = 2.78 \, (\Delta P/L)_{\text{healthy}}$

12-13 $Q_{\text{upper}} = 8 \, Q_{\text{lower}}$

12-14 1.08 m

12-15 approx. 2 to 3 m/s

12-16 (a) approx. 0.11 s

(b) approx. 0.36 cm^3

(c) approx. 59 cm^3

13 Thermal Motion of Molecules

13.1 Introduction

About 1827, Robert Brown[1] first observed the random chaotic motion of pollen grains in solution. The particles would sometimes change speed and direction without apparently colliding with anything. Brown was unable to explain the observations, and it was only a generation later that Maxwell[2] realized that collisions between the pollen grains and water molecules, which have their own similar but invisible chaotic motion, were influencing the motion of the pollen grains.

Brown's observations were among the first to show that molecules can have motion simply as a consequence of their temperature. While he observed translational motion in the pollen grains, it is now recognized that temperature can induce other motions of molecules as well, such as vibration and rotation. All these motions arise as a consequence of the continual exchange of *thermal energy* between a molecule and its surroundings. The quantitative analysis of the translational motion of molecules, called the *kinetic theory* of molecules, is the primary objective of this chapter.

13.2 The Boltzmann Equation and the Distribution of Thermal Energy

If a container contains identical molecules of a gas or a liquid, then even if the temperature is constant at all locations within the container, the molecules do not all have the same thermal energy. Instead, their energy is distributed, some molecules having energies

---

1. Robert Brown (1773–1858), Scottish botanist.
2. James Clerk Maxwell (1831–1879), Scottish mathematician and physicist.

greater than the average and some less. A similar distribution of energy can be observed by placing a number of marbles in a dinner plate and jiggling it gently back and forth. Some marbles exhibit rapid motion, while others are nearly still—but because of collisions the fast ones are sometimes slowed or stopped while others are speeded up. If the plate is jiggled more vigorously, then the overall motion of the marbles increases (that is, their average energy is increasing), but there are still some that are slow and others that are fast. The increased jiggling of the plate corresponds to an increased "temperature" of the marbles.

The "barometric formula" (Eq. [11-12]), derived in Chapter 11, is an excellent example of the distribution of the potential energies of air molecules. This equation

$$\frac{P}{P_0} = e^{\frac{-mg(h_2 - h_1)}{k_B T}}$$

when combined with Eq. [11-8], $P\mathcal{V} = N k_B T$, yields the expression

$$\frac{\dfrac{N k_B T}{\mathcal{V}}}{\dfrac{N_0 k_B T}{\mathcal{V}}} = e^{\frac{-mg(h_2 - h_1)}{k_B T}}$$

or

$$\frac{N/\mathcal{V}}{N_0/\mathcal{V}} = e^{\frac{-mg(h_2 - h_1)}{k_B T}} \qquad \text{[13-1]}$$

The quantity $N/\mathcal{V}$ is the number of air molecules per unit volume (e.g., per cubic metre) at elevation h_2, each having a potential energy mgh_2, and $N_0/\mathcal{V}$ is the same quantity at elevation h_1, where each molecule has potential energy mgh_1. The term $k_B T$ (remember "k_B" is Boltzmann's constant) in the denominator of the exponent is a measure of the thermal energy in the system. At high temperature (e.g., $T > 1000$ K), $k_B T$ becomes large relative to the numerator, and the right-hand side of Eq. [13-1] approaches unity. This would mean that N approaches N_0, indicating that the large thermal motion produces a distribution of molecules that is essentially uniform, i.e., independent of height. On the other hand, if $k_B T$ is small (e.g., at temperatures < 100 K) many fewer molecules would be found at elevation h_2.

A general way of describing the distribution of energy is through a probability function

$$\mathcal{P}_E = \frac{e^{-\frac{E}{k_B T}}}{\sum e^{-\frac{E_i}{k_B T}}} \qquad \text{[13-2]}$$

where $\mathcal{P}_E$ represents the probability that a molecule in a system has energy E or, equivalently, the fraction of molecules in a system that have energy E. This function, known as a *canonical distribution function*, is a fundamental equation in the discipline of *statistical thermodynamics*. The numerator of this expression corresponds to the particular energy state, E, whereas the denominator is a summation over all possible energy states that the molecules can have. If the summation is set equal to unity, then

$$\mathcal{P}_E = e^{-\frac{E}{k_B T}}$$

[13-3]

Figure 13-1

Exponential decay of energy-state probability with energy.

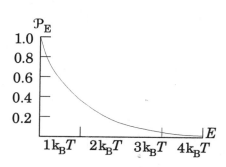

Figure 13-1 shows a plot of this function. The plot shows that it is more probable that the molecules of the system will have small energies than larger ones. The probability of having molecules with energies $E < k_B T$ is substantially greater than the probability of having energies $E > k_B T$. However, there are always a few molecules that will have energy many times $k_B T$.

If N molecules are present in the system, then the fraction of molecules having energy E_1 is

$$\mathcal{P}_{E_1} = \frac{N_1}{N} = e^{-\frac{E_1}{k_B T}}$$

[13-4]

and similarly, the fraction of molecules having energy E_2 is

$$\mathcal{P}_{E_2} = \frac{N_2}{N} = e^{-\frac{E_2}{k_B T}}$$

[13-5]

By taking a ratio of Eqs. [13-4] and [13-5],

$$\frac{N_2}{N_1} = e^{-\frac{E_2 - E_1}{k_B T}} = e^{-\frac{\Delta E}{k_B T}}$$

[13-6]

where $\Delta E = E_2 - E_1$.

Equation [13-6] is a general expression, sometimes known as the *Boltzmann equation*. It applies to all kinds of energy states (potential, kinetic, vibrational or rotational) that the molecules can have. It is identical in form to the modified barometric formula given in Eq. [13-1], which involves the gravitational potential energy states of the molecules.

EXAMPLE 13-1

A particular molecule at temperature T can occupy any one of three states, having energies of 0.27 k_BT, 0.88 k_BT, and 1.44 k_BT. If 10 000 of these molecules were placed in a container and held at temperature T, how many molecules would be in the highest energy state?

SOLUTION

From Eq. [13-2],

$$\mathcal{P}_{E_3} = \frac{N_3}{N} = \frac{e^{-\frac{E_3}{k_BT}}}{e^{-\frac{E_1}{k_BT}} + e^{-\frac{E_2}{k_BT}} + e^{-\frac{E_3}{k_BT}}}$$

$$\frac{N_3}{10\,000} = \frac{e^{-\frac{1.44k_BT}{k_BT}}}{e^{-\frac{0.27k_BT}{k_BT}} + e^{-\frac{0.88k_BT}{k_BT}} + e^{-\frac{1.44k_BT}{k_BT}}}$$

$$N_3 = 10\,000 \left(\frac{e^{-1.44}}{e^{-0.27} + e^{-0.88} + e^{-1.44}} \right)$$

$$N_3 = 10\,000 \left(\frac{0.237}{0.763 + 0.415 + 0.237} \right)$$

$$= 1.7 \times 10^3 \text{ molecules}$$

EXAMPLE 13-2

A gas in a container is at a constant temperature T. The gas molecules have two vibrational energy states whose energies are $E_1 = 0.60k_BT$ and $E_2 = 1.25k_BT$. If 1500 molecules are in state 1, how many molecules are expected to be in state 2?

SOLUTION

From Eq. [13-6],

$$\frac{N_2}{N_1} = e^{-\frac{E_2 - E_1}{k_B T}}$$

$$\frac{N_2}{1500} = e^{-\frac{(1.25 - 0.60)k_B T}{k_B T}} = e^{-0.65} = 0.52$$

$$N_2 = 1500 \times 0.52 = 7.8 \times 10^2 \text{ molecules}$$

13.3   Particles in Suspension: The Perrin Experiment

The distribution of potential energy among particles suspended in a liquid (water) was first demonstrated in 1890 by Perrin.[3] He reasoned that Robert Brown's dancing pollen grains should be governed by the same laws as applied to gas molecules. He argued that just as the number of molecules in the atmosphere decreases exponentially with elevation as described by the barometric formula, the number of particles in suspension should decrease exponentially from the bottom of the suspension to the top.

He set out to prove experimentally that this decrease existed. To do this he had to find particles that had just the right density. If they had exactly the same density as water, then in water they would be unaffected by gravity and all would have an effective gravitational potential energy of zero. On the other hand if they were less dense than water, they would all float to the top, and if they were significantly more dense they would all fall (*sediment*) to the bottom of the container. The ideal particles were those that had densities just slightly greater than water. He was able to achieve this requirement by making suspensions of gamboge (tree resin) particles of nearly uniform size (radius ~ 0.22 μm).

The experiment was carried out at 15°C (288 K) with these gamboge particles of volume $\mathcal{V} = 4.5 \times 10^{-20} \text{ m}^3$ and density $\rho = 1.20 \times 10^3 \text{ kg/m}^3$, suspended in water of density $\rho_\ell = 1000 \text{ kg/m}^3$ (subscript "ℓ" for "liquid"). After being thoroughly mixed, the suspension was placed in a vertical glass tube and allowed to remain undisturbed for many hours to

---

3.   Jean Baptiste Perrin (1870–1942), French physicist.

ensure that the system was at equilibrium. Then the number of particles at different heights in the tube was observed and counted with the aid of a microscope. Figure 13-2 illustrates the number of particles counted at successive 30-μm high "boxes" from the bottom of the tube.

		Relative number
d	↑30μm	12
c	↑30μm	22.6
b	↑30μm	47
a	↓	100

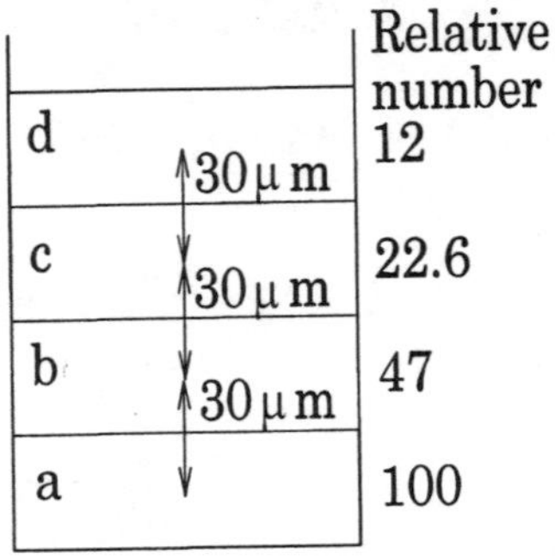

Figure 13-2

Perrin's experiment.

While Perrin predicted that the number of particles in the boxes should decrease with distance, h, from the bottom of the tube, he knew that the barometric formula as expressed by Eq. [13-1] had to be corrected to include the effects of buoyancy on the suspended particles. While the following derivation will be conducted explicitly for the comparison of the number of particles in box "d" relative to box "a" (Fig. 13-2), similar expressions can be obtained for all other comparisons. The Boltzmann expression relating the number of particles in box "d" (N_d) to box "a" (N_a) is

$$\frac{N_d}{N_a} = e^{-\frac{E_d - E_a}{k_B T}} \qquad \textbf{[13-7]}$$

where E_d and E_a are the respective gravitational potential energies. When used for comparison with experimental data, it is useful to convert this equation to the equivalent logarithmic form. Taking the natural logarithm of both sides of the expression yields

$$\ln\left(\frac{N_d}{N_a}\right) = -\frac{E_d - E_a}{k_B T} \qquad \textbf{[13-8]}$$

After a correction for buoyancy (Section 11.3 on *Archimedes' principle*) this function becomes

$$\ln\left(\frac{N_d}{N_a}\right) = -\frac{\mathcal{V}(\rho - \rho_\ell)g(h_d - h_a)}{k_B T} \qquad \textbf{[13-9]}$$

where h_d and h_a are the heights of the centres of boxes "d" and "a." If h_a is set to be zero for convenience, and if N_h represents the number of particles in a box centred at a distance h above this, then Eq. [13-9] can be written in a generic form

$$\ln\left(\frac{N_h}{N_0}\right) = -\frac{\mathcal{V}(\rho - \rho_\ell)gh}{k_B T} \qquad \textbf{[13-10]}$$

Figure 13-3

Semilogarithmic plot of
data in Fig. 13-2.

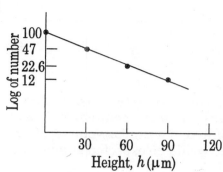

Equation [13-10] predicts that if the natural logarithm of the relative number of particles in each box (i.e., N_b/N_a, N_c/N_a, N_d/N_a) is plotted as a function of the box height h, then the graph should be a straight line with a negative slope of $-\mathcal{V}(\rho - \rho_\ell)g/(k_B T)$. Figure 13-3 demonstrates that this is precisely what is observed and that Perrin was correct in suggesting that particles in solution also obey kinetic theory.

EXAMPLE 13-3

Since all the other parameters in Eq. [13-10] have been given already in this section, use the data in Figure 13-2 to experimentally determine Boltzmann's constant ($T = 288$ K).

SOLUTION

Solving Eq. [13-10] for k_B yields

$$k_B = -\frac{\mathcal{V}(\rho - \rho_\ell)gh_d}{T \ln(N_d/N_a)}$$

$$= -\frac{4.5 \times 10^{-20} \text{ m}^3 \times (1.20 - 1.00) \times 10^3 \text{ kg/m}^3 \times 9.80 \text{ m/s}^2 \times 90 \times 10^{-6} \text{ m}}{(288 \text{ K}) \ln(12/100)}$$

$$= 1.3 \times 10^{-23} \text{ J/K}$$

This value for Boltzmann's constant is quite close to the accepted value of 1.381×10^{-23} J/K. If more data points had been taken, and a standard fitting procedure had been used to find the slope of the best-fit line to data such as that in Fig. 13-3, the result would have been even better.

EXAMPLE 13-4

In an experiment similar to Perrin's, 88 particles per unit volume are counted close to the bottom of a column, and 12 μm above this, 47 particles are counted in a similar volume. How many particles per unit volume would be expected at a height of 8.0 μm above the bottom?

SOLUTION

The phrase "in an experiment similar to Perrin's" in the question does not mean that the particles have the same volume and density as the ones in Perrin's experiment, nor that the temperature is the same. It means only that a number of particles (all having some particular volume and density) are mixed thoroughly in a liquid and allowed to come to equilibrium. The easiest way to deal with this is to combine all the parameters except h on the right-hand side of Eq. [13-10] into one constant, C:

$$\frac{\mathcal{V}(\rho - \rho_\ell)g}{k_B T} = C$$

Therefore,

$$\ln\left(\frac{N}{N_0}\right) = -Ch$$

Now, using the given information, determine C:

$$\ln\left(\frac{47}{88}\right) = -C(12\ \mu\text{m})$$

$$C = 0.0523\ \mu\text{m}^{-1}$$

Now find N at $h = 8.0$ μm:

$$\ln\left(\frac{N}{88}\right) = -(0.0523\ \mu\text{m}^{-1})(8.0\ \mu\text{m})$$

$$\frac{N}{88} = e^{-(0.0523\ \mu\text{m}^{-1})(8.0\ \mu\text{m})}$$

$$N = 58$$

13.4 Sedimentation

Perrin's experiment demonstrated that when conditions are just right, particles will distribute themselves exponentially in a liquid column. The "just right" means that the particle's mass and density must have magnitudes such that the right-hand side of Eqs. [13-9] or [13-10] is approximately unity. This means that the gravitational potential energy (*"m"gh*, where *"m"g* is the effective weight of the particle) is the same order of magnitude as $k_B T$. On the other hand, if the mass of the particle and/or its density is very large, then *"m"gh* can be many times greater than $k_B T$. In this case, thermal processes can never impart sufficient energy to raise the particles, and they will simply fall to the bottom of the liquid. Finally, there is the possibility that the particles are very small and/or their density is very close to that of the suspending medium. In this case the thermal energy is capable of equally distributing the particles everywhere in the solution by a process called *diffusion*. These differing situations are summarized in Table 13-1. This section deals with the first case, namely sedimentation. Diffusion will be discussed in the next section.

Table 13-1 **Effects of the Ratio *"m"gh/($k_B T$)* in the Boltzmann Equation**

Value	N/N_0	Result
"m"gh/($k_B T$) $\gg 1$	$N/N_0 \approx 0$	Particles sediment; thermal energy not sufficient to raise particles upward.
"m"gh/($k_B T$) ≈ 1	$0 < N/N_0 < 1$	Particles are exponentially distributed as in Perrin's experiment.
"m"gh/($k_B T$) $\ll 1$	$N/N_0 \approx 1$	Diffusion distributes particles everywhere in the solution.

Sedimentation refers to the settling of particles through the suspension due to gravitation. For example, if a sample of blood is allowed to stand quietly (after addition of a chemical that prevents coagulation), a clear, slightly yellowish zone will appear at the top of the container (Fig. 13-4). The demarcation line between the clear zone and the red material will gradually move downward. Microscopic examination reveals that the red material consists of erythrocytes

Figure 13-4

Red blood cells moving down through plasma in the process of sedimentation.

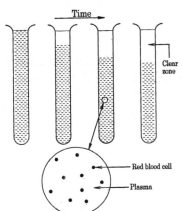

that are slowly moving down through the yellowish suspending medium, called *plasma*. The settling rate for an average human male is about 3 mm/h and for a female, 4 mm/h. In some disease conditions, this can increase drastically to around 100 mm/h, and different animal species often show marked differences in settling rates. A blood sample from a horse sediments much more rapidly than does a blood sample from a cow.

The erythrocyte normally has the shape of a biconcave disk as illustrated in Fig. 13-5, but for the calculations related to sedimentation it can be well represented as a sphere of radius $r = 3$ μm. As it falls through the plasma it experiences a retarding force F_v (upward) that is proportional to its downward speed v. Since the Reynolds number (Section 12.8) for particles of this size and speed is extremely small, an erythrocyte reaches its *terminal* (maximum) speed v_t almost instantly. Therefore, the magnitude of the retarding force is

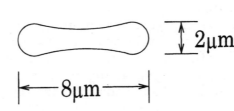

Figure 13-5

Dimensions of a red blood cell.

$$F_v = \mathcal{F} v_t \qquad \text{[13-11]}$$

The constant of proportionality in Eq. [13-11] is the friction factor $\mathcal{F}$, which is proportional to the viscosity of the medium. The friction factor for spherical particles was given in Chapter 12 (Eq. [12-16]):

$$\mathcal{F} = 6\pi\eta r$$

and hence Eq. [13-11] becomes

$$F_v = 6\pi\eta r v_t \qquad \text{[13-12]}$$

Gravitational and buoyant forces will also act upon sedimenting particles. From Section 11.3 the net gravitational (downward) force will be

$$F_d = (\rho - \rho_\ell)\mathcal{V}g$$

where ρ and ρ_ℓ are the densities of the sedimenting particle and the liquid medium, respectively. For spherical particles, $\mathcal{V} = 4\pi r^3/3$ and hence

$$F_d = \frac{4}{3}\pi r^3 (\rho - \rho_\ell)g \qquad \text{[13-13]}$$

At terminal speed, v_t, the upward and downward forces must sum to zero. Thus,

$$F_v - F_d = 0 \qquad \text{[13-14]}$$

and

$$6\pi r\eta v_t - \frac{4}{3}\pi r^3(\rho - \rho_\ell)g = 0 \qquad \text{[13-15]}$$

from which

$$v_t = \frac{2r^2 g}{9\eta}(\rho - \rho_\ell) \qquad \text{[13-16]}$$

EXAMPLE 13-5

What happens to the blood cells in a disease where the sedimentation rate suddenly increases from 4.0 to 100 mm/h? (The densities do not change significantly and direct experimentation shows negligible change of viscosity.)

SOLUTION

Since the densities and viscosity are constant, then from Eq. [13-16], the size r of the cells must be changing. Taking a ratio of the terminal speeds of cells from sick and healthy individuals yields (from Eq. [13-16]),

$$\frac{v_t(\text{sick})}{v_t(\text{healthy})} = \frac{r^2(\text{sick})}{r^2(\text{healthy})} = \frac{100 \text{ mm/h}}{4.0 \text{ mm/h}} = 25$$

$$\frac{r(\text{sick})}{r(\text{healthy})} = 5$$

The erythrocyte is a relatively rigid structure, so the only way the apparent radius can increase by a factor of five is for the cells to aggregate in groups. Microscopic examination of the blood from the diseased patients reveals that the erythrocytes are stacked in arrangements which resemble stacks of coins. The aggregates are called *rouleaux*.

13.5 **Centrifugation**

For the sedimentation of particles such as erythrocytes to be observed, the particles must be sufficiently large for them to sediment with an observable terminal speed. For these systems, it is also true (see Table 13-1) that the particle's gravitational potential energy greatly exceeds the thermal energy. If one wants to observe sedimentation of smaller particles, such as protein molecules whose typical radii are less than 0.1 μm, one would need to apply a much larger "gravitation-like" force. This can be achieved by spinning a tube containing the sample about a vertical axis in the rotor of a device called a *centrifuge* (see Section 9.2 under "Centrifugal 'Force' and Centrifuges").

In a centrifuge, the sample is spun rapidly at a radius R from the rotation axis of the rotor (Fig. 13-6). As the sample is rotated at a constant rate, the medium in which the sample is suspended experiences a centripetal acceleration that is directed inward toward the axis of rotation. The particles, because of their inertia, tend to move at a constant velocity in a straight line tangential to the circular path followed by the tube. As they do this, they move further from the rotation axis. To an outside observer the particles appear to be acted upon by a force

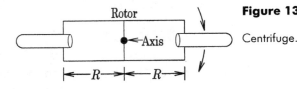

Figure 13-6

Centrifuge.

that moves them outward from the axis of rotation. This "fictitious" force, called the *centrifugal force*, plays the same role as does the gravitational force in Eq. [13-16], i.e., the centrifugal force causes the particles to undergo sedimentation. However, because the rotor is spinning rapidly (> 1000 Hz in "ultracentrifuges"), the centrifugal force F_c can be much greater than the gravitational force so that even very small particles tend to settle outward. At terminal outward speed v_t, the outward and inward forces must sum to zero, just as in gravitationally induced sedimentation. Therefore,

$$F_v - F_c = 0 \qquad\qquad [13\text{-}17]$$

If the rotor is turning at a frequency f, the centrifugal force on a spherical particle is (see Example 9-3),

$$F_c = \frac{4}{3}\pi r^3 (\rho - \rho_\ell)(4\pi^2 f^2 R)$$

After substituting this expression for F_c, and Eq. [13-12] for F_v, Eq. [13-17] becomes

$$6\pi\eta r v_t - \frac{4}{3}\pi r^3(\rho - \rho_\ell)(4\pi^2 f^2 R) = 0 \qquad \textbf{[13-18]}$$

and solving for v_t yields

$$v_t = \frac{8\pi^2 f^2 R r^2(\rho - \rho_\ell)}{9\eta} \qquad \textbf{[13-19]}$$

This expression holds only for spherical particles. It is possible to derive a more general expression for v_t for a particle of any shape (and friction factor $\mathcal{F}$). The result is:

$$v_t = \frac{4\pi^2 f^2 R}{\mathcal{F}}(m - m_\ell) \qquad \textbf{[13-20]}$$

where m is the mass of the particle and m_ℓ is the mass of liquid displaced by the particle.

From Eqs. [13-19] and [13-20] it is clear that v_t is proportional to the radius R of the centrifuge rotor and to the square of the frequency of rotation. As a consequence, identical particles would exhibit different terminal speeds from one instrument to another. A more commonly used parameter that is a property of the particle alone is the sedimentation coefficient s which is defined as

$$s = \frac{v_t}{4\pi^2 f^2 R} \qquad \textbf{[13-21]}$$

Because s involves dividing v_t by R and f^2, it is then a characteristic of the particle only, and is independent of the rotor frequency and radius of the centrifuge used. Combining Eqs. [13-20] and [13-21] relates s to the particle properties as follows:

$$s = \frac{m - m_\ell}{\mathcal{F}} \qquad \textbf{[13-22]}$$

or, for spherical particles, where $\mathcal{F}$ is known ($\mathcal{F} = 6\pi\eta r$) *and* $m = \rho \mathcal{V} = \rho(4\pi r^3/3)$,

$$s = \frac{2r^2}{9\eta}(\rho - \rho_\ell) \qquad \textbf{[13-23]}$$

The unit of the sedimentation coefficient is seconds (s). Sedimentation coefficients for a few common biological particles are given in Table 13-2.

Table 13-2   Some Sedimentation Coefficients

Substance	Sedimentation Coefficient, s (seconds)	Molar Mass (kg/mol)
Lysozyme	1.91×10^{-13}	14.4
Bovine serum albumin	5.01×10^{-13}	66.5
Bushy stunt virus	132×10^{-13}	10 700
Tobacco mosaic virus	193×10^{-13}	40 000

Molar mass M can be obtained from sedimentation information. By first multiplying both the numerator and the denominator of the right-hand side of Eq. [13-22] by Avogadro's number N_A and noting that $N_A \times m = M$,

$$s = \frac{M - N_A m_\ell}{N_A \mathcal{F}}$$
[13-24]

Since $m_\ell/m = \rho_\ell/\rho$, then $N_A m_\ell = N_A m \rho_\ell/\rho = M \rho_\ell/\rho$, and Eq. [13-24] becomes

$$s = \frac{M(1 - \rho_\ell/\rho)}{N_A \mathcal{F}}$$
[13-25]

Density-Gradient Centrifugation

Many biological macromolecules such as proteins, nucleic acids and polysaccharides have slightly different densities. Indeed, densities often vary between one protein and another. As a consequence, density can be used as a means of separating one macromolecular species from another. One technique that is ideal for this task is density-gradient centrifugation. In one common variation of this method, a solution of cesium chloride or sucrose is prepared in such a way that the

Figure 13-7

Density-gradient centrifugation: density increases toward the bottom of the centrifuge tube.

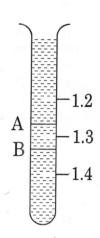

concentration increases down the centrifuge tube. As a consequence the density ρ_ℓ of this solution increases from the top to the bottom of the tube (Fig. 13-7).

Consider a mixture of polysaccharide A and protein molecules B of specific gravity $\rho_{ps} = 1.28$ and $\rho_{pro} = 1.32$, respectively. If this mixture is added to the top of the centrifuge tube and then rotated rapidly as before, the molecules will migrate toward the bottom with an initial terminal speed given by Eq. [13-19]. The two molecular specific gravities ρ_{ps} and ρ_{pro} are constant and are greater than the liquid's specific gravity ρ_ℓ at the top of the tube. As the molecules migrate down, however, ρ_ℓ gradually increases and at some point ρ_ℓ will equal 1.28 which is the specific gravity of the polysaccharide. At this point the polysaccharide stops because v_t for the polysaccharide goes to zero as a consequence of the factor $(\rho_{ps} - \rho_\ell)$ in Eq. [13-19] becoming zero. The protein continues further down the tube to the point where $\rho_\ell = 1.32$. Therefore, the two materials separate out as two bands, which can be removed separately.

The method can be made so sensitive that biomolecules such as DNA containing the normal isotope of nitrogen (^{14}N) can be separated from DNA that has been labeled with a heavier isotope of nitrogen (^{15}N). An interesting point is that the cesium chloride density gradient in the tube is relatively stable since the radii of the cesium and chloride ions are so small that their terminal speed is negligible relative to the protein or the polysaccharide. However, thermal energy can dissipate the concentration gradient of the cesium chloride over time via a process called diffusion, a topic that will be discussed in Section 13.8. Because of diffusion, density-gradient centrifugation experiments must be performed soon after the concentration gradient of cesium chloride has been prepared in the centrifuge tube.

EXAMPLE 13-6

Sedimentation coefficients are often determined by comparing the sedimentation rate of an unknown system with that of a standard (a particle whose sedimentation coefficient is well known). In a particular study, samples of an unknown protein and bovine serum albumin (BSA) are each dissolved in water in two centrifuge tubes. The two tubes are then spun together in a high-speed centrifuge. The terminal speed of the unknown protein is determined to be 1.43 times larger than that of the BSA. Both the unknown and BSA molecules are spherical and, from density-gradient centrifugation, both are found to have the same density. Determine the sedimentation coefficient, the radius (relative to the radius of BSA), and the molar mass of the unknown particle.

SOLUTION

Since both particles are spun at the same time in the same centrifuge, then the rotor frequency and radius are identical for both samples. From Eq. [13-21],

$$\frac{s(\text{unknown})}{s(\text{BSA})} = \frac{\dfrac{v_t(\text{unknown})}{4\pi^2 f^2 R}}{\dfrac{v_t(\text{BSA})}{4\pi^2 f^2 R}} = \frac{v_t(\text{unknown})}{v_t(\text{BSA})} = 1.43$$

From Table 13-2, the sedimentation coefficient of BSA is 5.01×10^{-13} s. Therefore,

$$s(\text{unknown}) = 1.43 \times 5.01\times10^{-13}\,\text{s} = 7.16\times10^{-13}\,\text{s}$$

From Eq. [13-23], a ratio can be established as follows:

$$\frac{s(\text{unknown})}{s(\text{BSA})} = 1.43 = \frac{\dfrac{2(r_{\text{unknown}})^2}{9\eta}(\rho_{\text{unknown}} - \rho_\ell)}{\dfrac{2(r_{\text{BSA}})^2}{9\eta}(\rho_{\text{BSA}} - \rho_\ell)} = \frac{(r_{\text{unknown}})^2}{(r_{\text{BSA}})^2}$$

Hence,

$$r_{\text{unknown}} = (1.43)^{1/2}\, r_{\text{BSA}}$$

The two proteins have identical densities, and mass $m = \rho\mathcal{V} = \rho(4\pi r^3/3)$ for spherical particles. Then, since molar mass $M = N_A m$,

$$\frac{M_{\text{unknown}}}{M_{\text{BSA}}} = \frac{(r_{\text{unknown}})^3}{(r_{\text{BSA}})^3} = (1.43)^{3/2} = 1.71$$

The molar mass of BSA is given in Table 13-2 as 66.5 kg/mol. Therefore,

$$M_{unknown} = 1.71 \times 66.5 \text{ kg/mol} = 114 \text{ kg/mol}$$

13.6   **Distribution of Kinetic Energy**

This chapter began with a discussion of how thermal energy becomes distributed over the possible energy states available to molecules. It was shown that the Boltzmann equation (Eq. [13-6]) could be used quantitatively to describe this distribution. The barometric equation and Perrin's experiment were particular cases that described how gravitational potential energy is distributed among particles.

The Boltzmann equation also determines the distribution of kinetic energies that particles can have. Kinetic energy involves motion, and its distribution is a consequence of collisions and other energy-exchange mechanisms among particles. If the molecules or particles are in the gaseous phase then the exchange occurs among the particles themselves, but if the particles are in a liquid medium the exchange can also occur with the liquid molecules. This following section is concerned with the information that the Boltzmann equation, when applied to the kinetic energy of particles, can tell us about the distribution of particle velocities, about the random walk process, and about diffusion (Brownian motion).

To begin, the kinetic energies associated with the x-components of the velocities of the particles will be considered. The kinetic energy of full three-dimensional motion will follow. If, from their motions in the x-direction, N_1 particles have kinetic energy $K_1 = \frac{1}{2}mv_{x1}^2$ then the number of particles N_2 expected to have kinetic energy $K_2 = \frac{1}{2}mv_{x2}^2$ can be determined by the Boltzmann equation (Eq. [13-6]). Thus,

$$\frac{N_2}{N_1} = e^{-\frac{K_2 - K_1}{k_B T}} = e^{-\frac{mv_{x2}^2 - mv_{x1}^2}{2k_B T}} \qquad \text{[13-26]}$$

If $v_{x1} = 0$, then the lower kinetic energy state is zero and

$$\frac{N}{N_0} = e^{-\frac{mv_x^2}{2k_B T}} \qquad \text{[13-27]}$$

where N is the number of particles having kinetic energy $\frac{1}{2}mv_x{}^2$ and N_0 is the number of particles having zero kinetic energy. This function is called a *Gaussian* function after Gauss.[4] A distribution such as this, where a quantity (N) depends on "e" raised to the negative square of another quantity (v_x) is called a *Gaussian distribution* or sometimes a *normal distribution*. When N is plotted as a function of velocity v_x, a "bell" shape results (Fig. 13-8).

In Fig. 13-8, the areas between the vertical lines indicate the fraction of particles having a velocity between v_x and $v_x + \Delta v_x$, where Δv_x is the separation of the lines. The graph indicates that the fraction of particles having extremely high-velocity components to either the right or the left approaches zero. This result is for velocity components in the

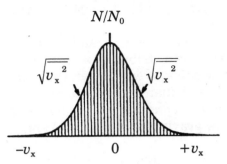

N/N_0

$\sqrt{\overline{v_x{}^2}}$ $\sqrt{\overline{v_x{}^2}}$

$-v_x$ 0 $+v_x$

Figure 13-8

Gaussian (normal) distribution of velocities.

x-direction only. Similar graphs would be obtained if the velocity components along the *y*- or *z*-axis were presented. The graph shows that more particles have zero velocity than any other single value. Stated another way, the greatest probability is for particles to be at rest.

In addition, for motions in the *x*-direction, since there are as many particles having velocity components to the right as to the left, then the average velocity is also zero. Therefore, in a Gaussian distribution of velocities, the *most probable velocity* is zero and the *average velocity* is zero! This will always be the case, regardless of the temperature. All that varying the temperature will accomplish is to broaden or narrow the "width" of the "bell." Clearly, therefore, when dealing with a Gaussian velocity distribution, most probable velocities and average velocities are of little interest.

However, for motion along the *x*-axis, the quantity in the numerator of the exponent of Eq. [13-27] is $v_x{}^2$, and it will have a mean value that is not zero since all values of $v_x{}^2$ must be positive even though v_x can be either negative or positive. Therefore, the *mean square velocity* is the key parameter of interest. As the temperature increases, the mean square velocity will also increase. Similarly, the *root mean square (rms) velocity*,

$$v_{x,\text{rms}} = \sqrt{\overline{v_x{}^2}} \qquad\qquad \textbf{[13-28]}$$

---

4.   Johann Karl Friedrich Gauss (1777–1855), German mathematician.

must also must be positive. Note that the line ("bar") above $v_x{}^2$ means "the average of" (see Box 13-1).

Box 13-1   Average (Mean) Quantities

There are several ways in which the average of a quantity can be indicated. If the quantity is x, then its average can be denoted as $\bar{x}$ (pronounced "x-bar"), x_{av} or $<x>$.

The positions of $+v_{x,rms}$ and $-v_{x,rms}$ are shown on the graph in Fig. 13-8. If the area under the Gaussian velocity distribution function that falls between these values is carefully measured (or evaluated by integration), it is found to be two-thirds of the total area under the curve. In other words, if the particles are at thermal equilibrium, two-thirds of them have velocity components between $+v_{x,rms}$ and $-v_{x,rms}$ and one-third would be moving at velocities greater than the former or less than the latter. Alternatively, one could state this in terms of probabilities by saying that a single particle has a two-thirds chance of moving at a velocity between $+v_{x,rms}$ and $-v_{x,rms}$ and a one-third chance of having a velocity outside this range.

The average kinetic energy associated with the motion along the x-axis is the thermal energy $k_B T$. Expressed mathematically, this is

$$\overline{\frac{1}{2}mv_x{}^2} = \frac{1}{2}k_B T \qquad \textbf{[13-29]}$$

or

$$\overline{mv_x{}^2} = m\overline{v_x{}^2} = k_B T$$

Equation [13-27] can, therefore, be re-expressed for x-axis motion in the form

$$\frac{N}{N_0} = e^{-\frac{v_x{}^2}{2\overline{v_x{}^2}}} \qquad \textbf{[13-30]}$$

where N is the number of particles having velocity component $\pm v_x$, N_0 is the number of particles having zero velocity along x, and the denominator of the exponent is the mean square velocity along x.[5] More discussion on mean square variables will be given in the following section.

The average kinetic energy for three-dimensional motion is just the sum of the average kinetic energies associated with the x-, y- and z-axes. Therefore,

$$\overline{\frac{1}{2}mv^2} = \overline{\frac{1}{2}mv_x^2} + \overline{\frac{1}{2}mv_y^2} + \overline{\frac{1}{2}mv_z^2} = \frac{3}{2}k_BT \qquad [13\text{-}31]$$

This expression is extremely important for relating the average kinetic energy of molecules of a gas, for example, to the thermal energy.

13.7   Random Walk Processes

Individual particles exhibit displacements as a result of their thermally driven velocities. This process is called *diffusion* or sometimes *Brownian motion*. Consider the displacements along the x-axis, as described in the previous section. Because an equal number of particles have velocity components to the right as to the left, then there must be as many displacements to the right as to the left. The average displacement along x must, therefore, be zero. And since the most probable velocity is zero, the most probable displacement must also be zero. As before, mean square values must be used. For motion along the x-axis, the *root mean square displacement* during time t is related to the root mean square velocity by $x_{\text{rms}} = v_{x,\text{rms}}t$. Squaring this equation gives

$$\overline{x^2} = \overline{v^2}t^2 \qquad [13\text{-}32]$$

An equation similar to Eq. [13-30] can be written for N/N_0, where N and N_0 now represent the number of particles experiencing a displacement $\pm x$ and the number of particles experiencing no displacement, respectively, over the time interval t:

$$\frac{N}{N_0} = e^{-\frac{x^2}{2\overline{x^2}}} \qquad [13\text{-}33]$$

---

5.   Note that the mean square velocity is <u>never</u> zero, unlike the square of the mean velocity which is <u>always</u> zero.

Figure 13-9

Gaussian (normal)) distribution of displacements.

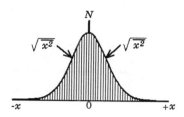

This is a Gaussian distribution of displacements (Fig. 13-9). As time increases, the magnitude of the root mean square displacement x_{rms} of the particles grows, and the distribution becomes broader and broader. However, the total area under the curve, which is a measure of the total number of particles, remains constant. If all the particles had started at the origin at $t = 0$, they would have a root mean square displacement $x_{rms(t_1)}$ after a time interval t_1, a larger root mean square displacement $x_{rms(t_2)}$ after a time interval t_2, and so on. The particles are *diffusing* away from the origin.

The diffusion of molecules is called a *random walk* process. To illustrate the phenomenon in simple terms consider the following

Figure 13-10

Displacements in a one-dimensional random walk.

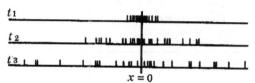

analogy. A gardener deposits a bushel of dry leaves at a position $x = 0$ at time $t = 0$ on a long street. The day is relatively calm with random breezes but no prevailing wind. These breezes and the random air currents induced by passing automobiles distribute the leaves up and down the street with approximately equal numbers in each direction. The positions of a sample of the leaves at three increasing times is shown in Fig. 13-10. This is exactly the phenomenon being described in Fig. 13-9. Note that the average displacement of the leaves is zero and the most probable displacement is also zero (as expected). While it is completely impossible to predict what one particular leaf (or particle) will do, it is possible to describe what will happen to the group.

Figure 13-11

A 300-s plot of the displacement of a motile *E. coli* bacterium swimming at about 2.0×10^3 cm/s. This is an excellent example of a random walk. If a copy is made of this page and 90-degree folds are made inward along the dashed lines, one can see the projections of the random walk on three faces of a cube.

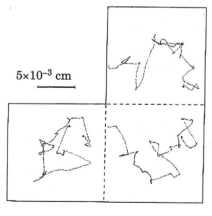

In the real diffusion situation, particles move due to thermal forces in their environment, not wind or other macroscopic currents. Nevertheless, some living systems also demonstrate random walks. Fig. 13-11 shows the random walk of a motile (swimming) bacterium. In gases and in solutions the amount of energy involved in the dynamic processes of translation and rotation is determined by temperature, whereas in the swimming cell the energy is provided by metabolic processes.

EXAMPLE 13-7

A scientist is measuring the random motion of 100 small particles in a long, very thin tube. With the aid of time-lapse photography she locates all particles at a given time and again 20 s later. She measures the displacements (all in the ±x-direction) and counts the number of particles that travel different distances from their starting points. Motion in one direction is arbitrarily called negative and in the opposite direction, positive. The following table is obtained.

Number of Particles (or Frequency of Occurrence, f_x)	Approximate Displacement, x (μm)
1	-30
6	-20
23	-10
40	0
23	+10
6	+20
1	+30

Determine:

(a) the mean displacement

(b) the mean square displacement

(c) the root mean square displacement

SOLUTION

The following table will aid in the solution.

Parameter	f_x	x	$f_x x$	x^2	$f_x x^2$
	1	-30	-30	900	900
	6	-20	-120	400	2400
	23	-10	-230	100	2300
	40	0	0	0	0
	23	10	230	100	2300
	6	20	120	400	2400
	1	30	30	900	900
Totals	100	0	0	2800	11 200

From the totals row of the table,

$N_{total} = \Sigma f_x = 100$, which is the total number of particles,
$\Sigma f_x x = 0$ µm is the sum of all the displacements and
$\Sigma f_x x^2 = 11\ 200$ µm^2 is the sum of all the displacements squared.

(a) The mean displacement is

$$\bar{x} = \frac{\Sigma f_x x}{N_{total}} = \frac{0\ \mu m}{100} = 0\ \mu m$$

(b) The mean square displacement is

$$\overline{x^2} = \frac{\Sigma f_x x^2}{N_{total}} = \frac{11200\ \mu m^2}{100} = 112\ \mu m^2$$

(c) The root mean square displacement is

$$x_{rms} = (112\ \mu m^2)^{1/2} = 10.6\ \mu m$$

Further analysis of Gaussian distributions of displacements along x (Fig. 13-9) would indicate that two-thirds of the particles experience displacements inside $\pm x_{rms}$ and one-third of the particles would experience displacements outside $\pm x_{rms}$. In addition, the number of dimensions in which the particles are able to move is also important. If the displacements are restricted to a two-dimensional (2-D) planar surface, as on a biological membrane, for example, then they would exhibit mean square displacements of

$$\overline{R^2_{2\text{-D}}} = \overline{x^2} + \overline{y^2} \qquad \text{[13-34]}$$

and if full three-dimensional motion is allowed, then

$$\overline{R^2} = \overline{x^2} + \overline{y^2} + \overline{z^2} \qquad \text{[13-35]}$$

13.8   The Diffusion Coefficient

More advanced statistical analyses of three-dimensional diffusive motion show that in solution, intermolecular collisions influence the mean square displacement and the mean square velocity with the result that

$$\overline{R^2} = 6\left(\frac{\overline{v^2}\tau}{3}\right)t = 6Dt \qquad \text{[13-36]}$$

where t is time and τ is the mean time between molecular collisions. The quantity

$$D = \frac{\overline{v^2}\tau}{3} \qquad \text{[13-37]}$$

is called the *diffusion coefficient*. The units of the diffusion coefficient are metres squared per second (m²/s). Some diffusion coefficients are given in Table 13-3.

Table 13-3   Some Diffusion Coefficients

Substance	Diffusion Coefficient, D (m²/s) in water at 20°C	M (kg/mol)
Water (H$_2$O)	2.00×10^{-9}	0.018
Sucrose	4.59×10^{-10}	0.342
Lysozyme	1.19×10^{-10}	14.1
Bovine serum albumin (BSA)	6.1×10^{-11}	66.5
Fibrinogen (human)	1.98×10^{-11}	330
Bush stunt virus	1.15×10^{-11}	1.07×10^4
Tobacco mosaic virus	4.4×10^{-12}	4.0×10^4

Any molecule in solution will exhibit diffusive Brownian motion. If we could only watch and measure the changes in velocity of one molecule, then it would be possible to learn a great deal about it. For instance, if the mean square velocity were always small then energy considerations would tell us that we were dealing with a large molecule. This is easy to recognize from Eq. [13-31]:

$$\overline{\frac{1}{2} m v^2} = \frac{3}{2} k_B T$$

For a constant thermal energy $k_B T$, if the particle mass m is large then its mean square velocity will be small. Thus, the degree of motion or diffusion of particles in a solvent decreases with increasing particle size. This property of diffusion is vital to the orderly functioning of biological cells. In fact, more than any other process, diffusion has determined basic properties such as the ultimate size of the cells.

Equation [13-36] $(\overline{R^2} = 6Dt)$ is very useful, especially when one is trying to determine whether or not many of the processes of the biological cell, such as transcription and translation, make sense in physical terms. For example, one might ask whether or not the rates of diffusion are sufficient to allow 50 amino acids per second to be assembled into protein at the ribosome. The following examples show that the times required by small particles to diffuse across a cell are very small. Indeed, there are only a few cases where the cell needs special machinery to transport molecules, and these cases are usually restricted to the transport of large macromolecules.

EXAMPLE 13-8

How long will it take a water molecule to diffuse 0.010 m through water?

This question is typical of the rather imprecise language that is frequently encountered in describing diffusion. The question should really be: "In a large sample of diffusing water molecules, how long will it take for the molecules to travel a root mean square distance of 0.010 m?" or equivalently, "How long will it take for one-third of a large number of water molecules to diffuse a distance of at least 0.010 m from their starting points?"

SOLUTION

The diffusion coefficient of water molecules in water near room temperature (from Table 13-3) is 2.0×10^{-9} m²/s. From Eq. [13-36],

$$t = \frac{\overline{R^2}}{6D} = \frac{(0.010 \text{ m})^2}{6 \times 2.0 \times 10^{-9} \text{ m}^2/\text{s}} = 8.3 \times 10^3 \text{ s} = 2.3 \text{ h}$$

This makes diffusion seem a very slow process, indeed. However, it should be remembered that within the confines of a living cell, the distances to be traveled are normally quite small. Consider now the same question, but changing the distance from 0.010 m to the length of a bacterial cell (e.g., 3.0 μm). Then

$$t = \frac{\overline{R^2}}{6D} = \frac{(3.0 \times 10^{-6} \text{ m})^2}{6 \times 2.0 \times 10^{-9} \text{ m}^2/\text{s}} = 7.5 \times 10^{-4} \text{ s}$$

The process in a real bacterial cell would not be quite this rapid since the cytoplasm will be about five times as viscous as water. As will be discussed shortly, this will decrease the diffusion constant to one-fifth of the value in water with the result that the time will be increased by a factor of five to 3.8×10^{-3} s or 3.8 ms. This is still very fast. Therefore, diffusion, while a slow process for long-distance transport, is a fast process within the confines of a biological cell.

EXAMPLE 13-9

Bovine serum albumin (BSA) has a diffusion coefficient (Table 13-3) of 6.1×10^{-11} m²/s in water. In the more viscous cytoplasm of a mammalian cell, D would be one-fifth of that value. Determine the time for BSA to "travel" a root mean square distance of 10 μm, the approximate length of a mammalian cell.

SOLUTION

Using Eq. [13-36],

$$t = \frac{\overline{R^2}}{6D} = \frac{5(10 \times 10^{-6} \text{ m})^2}{6 \times 6.1 \times 10^{-11} \text{ m}^2/\text{s}} = 1.4 \text{ s}$$

Stokes–Einstein Equation

The diffusion coefficient can be described in terms of the thermal energy k_BT and the friction factor $\mathscr{F}$ discussed in Sections 12.9 and 13.4. The relationship between these quantities is given by the *Stokes[6]–Einstein[7] equation:*

$$D = \frac{k_BT}{\mathscr{F}}$$

[13-38]

If written in the form $D\mathscr{F} = k_BT$, both sides of the expression have the units of energy. In this form the right-hand side is the thermal energy that drives diffusion, whereas the left-hand side is the energy dissipated to the environment during diffusion. The equation represents this energy exchange. For spherical particles $\mathscr{F} = 6\pi\eta r$ (Eq. [12-16]), and the Stokes–Einstein equation can be written as

$$D = \frac{k_BT}{6\pi\eta r}$$

[13-39]

Notice that D is proportional to the absolute temperature T, and inversely proportional to both the fluid viscosity η and the particle radius r.

Fick's Law

Another biologically important phenomenon involving the diffusion coefficient is the movement of molecules through semi-permeable membranes. Because of diffusion, particles tend to move from regions of high concentration to regions of low concentration, in an attempt to equilibrate the concentration everywhere. *Fick's law* states that J, the flux of particles (the net number of particles passing through an imaginary surface of unit area per unit time), is proportional to the concentration gradient $\Delta C/\Delta x$ between the regions. ΔC is the concentration difference across a distance Δx, and the greater the difference ΔC or the smaller the distance Δx, the greater will be the flux. In the limit of very small concentration differences over very small distances, $\Delta C/\Delta x$ is written as dC/dx. Thus, Fick's law is usually expressed by the relation

$$J = -D\frac{dC}{dx}$$

[13-40]

---

6.   Sir George Gabriel Stokes (1819–1903), British mathematician and physicist.
7.   Albert Einstein (1879–1955), German physicist.

where D is the diffusion coefficient of the diffusing molecules of interest. The negative sign in Eq. [13-40] indicates that the direction of the flux is down the concentration gradient, i.e., from high C to low C.

The diffusion coefficient is essentially independent of concentration although at higher concentrations it may be influenced by intermolecular interactions. Therefore, D is the same whether particles are uniformly spread through the solution or whether they are more localized, for example, on one side of a membrane. In the first case the concentration gradients dC/dx are microscopic and rapidly fluctuating. In the latter case they are macroscopic. Modern techniques such as laser-light scattering (a technique called "dynamic light scattering") measure diffusion coefficients via the microscopic rapidly fluctuating gradients. Older techniques rely on Fick's law, but these are still very instructive as is shown next.

Suppose two liquid compartments of volumes $\mathcal{V}_1$ and $\mathcal{V}_2$ are separated by a porous membrane of thickness Δx as in Fig. 13-12. The volume $\mathcal{V}_1$ contains a solution of molecules whose diffusion constant is to be measured. The initial concentration of the molecules on this side will be designated $C_{1,0}$ (the "0" subscript refers to time $t = 0$). The other volume $\mathcal{V}_2$ initially contains pure water so that its initial concentration of the molecules is $C_{2,0} = 0$. As time advances, the

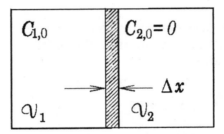

Figure 13-12

Diffusion across a semi-permeable membrane.

concentrations will gradually change as diffusion acts to equilibrate them as a result of the net movement of molecules from regions of high concentration to low concentration.

If the area of the membrane is $\mathcal{A}$, then the incremental flux J of particles through the membrane is defined as

$$J = \frac{1}{\mathcal{A}} \frac{dn}{dt} \qquad \text{[13-41]}$$

where n is the number of particles and t is time. Combining this definition with Fick's law (Eq. [13-40]) leads to

$$\frac{dn}{dt} = -D\mathcal{A} \frac{dC}{dx} \qquad \text{[13-42]}$$

If the concentrations on the two sides of the membrane at time t are represented by C_1 and C_2, then the concentration gradient dC/dx across the membrane can be approximated by $(C_1 - C_2)/\Delta x$, where Δx is the thickness of the membrane (Fig. 13-12), so that

$$\frac{dn}{dt} = -D\mathcal{A}\frac{C_2 - C_1}{\Delta x} \qquad \textbf{[13-43]}$$

This equation can be solved to obtain an equation which describes the concentration C_1 in volume $\mathcal{V}_1$ as a function of time. If the two volumes $\mathcal{V}_1$ and $\mathcal{V}_2$ are equal, the mathematical solution of Eq. [13-43] is (after several steps)

$$C_1 = \frac{C_{1,0}}{2} e^{-\frac{2D\mathcal{A}}{\Delta x \mathcal{V}_1}t} + \frac{C_{1,0}}{2} \qquad \textbf{[13-44]}$$

The general form of this solution can almost be guessed. If the concentration is $C_{1,0}$ on one side and zero on the other with two equal volumes, the one will decay in time from $C_{1,0}$ to $C_{1,0}/2$ and the other will grow from 0 to the same value. The reader should check that Eq. [13-44] gives $C_1 = C_{1,0}$ at time $t = 0$, and $C_1 = C_{1,0}/2$ as $t \to \infty$.

Since all the other variables in this expression can be known in advance, the equation can be solved for D. The value determined for D, however, corresponds to a value for the diffusion coefficient across the membrane. Usually what is required is not D across the membrane but D in pure solution. The latter can be determined by correcting the experimentally measured D for the percentage of the membrane area that actually consists of pores. Thus, if the area were 50 percent pores then the true diffusion coefficient would be twice that determined experimentally, and if the area were only 5 percent pores then the measured D must be increased by a factor of 20. This type of correction is valid (a) if the pores are straight and have a diameter significantly larger than that of the diffusing molecules, and (b) if there is no interaction between the membrane and the diffusing molecules.

EXAMPLE 13-10

The diffusion coefficient for potassium ions crossing a biological membrane 10 nm thick is $1.0 \times 10^{-16}\,\text{m}^2/\text{s}$. What number of potassium ions would move per second across an area 100 nm by 100 nm, if the concentration difference across the membrane is 0.50 mol/dm³ ? (1 dm³ = 10^{-3} m³ = 1 L)

SOLUTION

From Eq. [13-43]

$$\frac{dn}{dt} = -D\mathcal{A}\frac{C_2 - C_1}{\Delta x}$$

$$= -\frac{1.0 \times 10^{-16} \text{ m}^2/\text{s} \times (100 \times 10^{-9} \text{ m})^2 \times (-0.50 \times 10^3 \text{ mol/m}^3)}{10 \times 10^{-9} \text{ m}}$$

$$= 5.0 \times 10^{-20} \text{ mol/s}$$

$$= 5.0 \times 10^{-20} \text{ mol/s} \times 6.022 \times 10^{23} \text{ ions/mol}$$

$$= 3.0 \times 10^4 \text{ ions/s}$$

Determining Molar Mass from Diffusion Coefficient

If a diffusion coefficient has been measured for a spherical particle or molecule, then this information can be used to estimate the molar mass M of the particle. From Eq. [13-39], the radius r of the particle can be written as

$$r = \frac{k_B T}{6\pi\eta D}$$

and from the relation between the radius and volume of a sphere, $\mathcal{V} = (4/3)\pi r^3$, then

$$\mathcal{V} = \frac{4}{3}\pi\left(\frac{k_B T}{6\pi\eta D}\right)^3 \qquad \text{[13-45]}$$

In addition, $m = \rho\mathcal{V}$ and $M = N_A m$. Therefore

$$M = \frac{4}{3}\pi\rho N_A\left(\frac{k_B T}{6\pi\eta D}\right)^3 \qquad \text{[13-46]}$$

Molar masses calculated by this method will frequently be too large because of the presence of hydrophilic (water-loving) groups on their surfaces. Most soluble biological molecules will pick up a layer of bound water molecules equivalent to about 0.35 g of water per gram of biological molecule.

Even after corrections are made for bound water, errors often remain that can indicate how non-spherical the molecule is. For example, hemoglobin has an accepted molar mass of 68 kg/mol, but Eq. [13-46] yields 130 kg/mol. This is reduced to 89 kg/mol by correcting for bound water. The remaining difference suggests that the molecule is slightly non-spherical.[8] At the other extreme, tobacco mosaic virus (TMV) is a long thin rod of length 280 nm and diameter 15 nm. Its accepted molar mass is 4.0×10^4 kg/mol but, even after corrections for bound water, diffusion coefficient measurements yield a value of 1.60×10^5 kg/mol or four times larger than expected. This large discrepancy tells us that the molecule is very non-spherical, a fact that has been well demonstrated by other techniques. Thus, diffusion measurements can be a powerful source of information on structural properties of macromolecules.

As discussed above, Eq. [13-46] can be used to obtain molar masses, but to get accurate results requires a knowledge of the friction factor, which in turn requires knowledge of the shape of the system under investigation. The requirement for knowledge of the friction factor vanishes if both sedimentation coefficients and diffusion coefficients are measured. If Eq. [13-25] for sedimentation coefficient $(s = M(1 - \rho_\ell/\rho)/(N_A \mathcal{F}))$ is divided by Eq. [13-38] for diffusion coefficient $(D = k_B T/\mathcal{F})$, the friction factor disappears and one obtains

$$\frac{s}{D} = \frac{M(1 - \rho_\ell/\rho)}{N_A k_B T} \qquad \text{[13-47]}$$

from which

$$M = \frac{s N_A k_B T}{D(1 - \rho_\ell/\rho)} \qquad \text{[13-48]}$$

This approach, involving measurement of both s and D, provides one of the best methods for determining molar mass of biological particles of any shape. Molar masses obtained in this way are more precise than those obtained using only one of Eq. [13-25] or Eq. [13-46] for two reasons. First, one does not need to assume a spherical shape for the particle. Second, the molar mass in this case is of the particle, and no correction is needed for bound water.

---

8.   That is, it has a larger surface than does a sphere of the same volume.

13.9   **Osmotic Pressure**

Osmotic pressure is encountered when solvent molecules diffuse across semi-permeable membranes in response to concentration gradients of other molecules that are too large or have other properties (e.g., electric charge) preventing their passage through the membrane. Figure 13-13a represents the beginning of an ideal experiment in which water molecules can move freely in all directions including back and forth across the semi-permeable membrane. The volume on the left side (A) of the membrane contains only water, but the volume on the right side (B) of the membrane also contains molecules that are too large to flow through the membrane pores. Because some of the volume on side B is occupied by these molecules, there must be fewer water molecules per unit volume than on side A that contains only water. The water is free to flow everywhere and diffusion will try to equalize the water concentration everywhere, and hence there will be a net diffusion of water from side A to side B. The arrows in the figure show the direction of this water flow.

As flow occurs, however, volume is being added to side B and a hydrostatic pressure difference develops between the two sides because the liquid level is higher on side B than on side A. As the pressure builds, there is a greater and greater tendency to force water back to side A; at some point equilibrium is achieved and no further net flow of water occurs (Fig. 13-13b). The hydrostatic pressure difference between the two sides at equilibrium, $P_B - P_A = \rho g d$, is called the *osmotic pressure*, represented by Π (uppercase Greek letter *pi*).

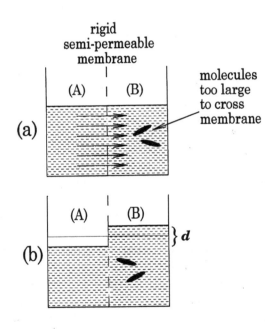

Figure 13-13

Osmosis: (a) at the beginning of the experiment; (b) at equilibrium.

Interestingly, water flow across the membrane does not necessarily have to occur. A piston or some other pressurizing device could be applied to side B to prevent any volume change. However, the pressure exerted by the piston would be the same as the hydrostatic pressure resulting from the water level difference described above.

The point has just been made that the reason that water moved from side A to side B was that initially the water was more concentrated in side A, and molecules tend to diffuse from regions where they are highly concentrated to regions where they are less concentrated. This implies that the energy E_A associated with side A is greater than the energy E_B associated with side B, since energy must come from somewhere to facilitate the movement of molecules and to create the osmotic pressure. This is indeed the case. Molecules have a greater energy when they are concentrated than when they are dispersed. The degree of organization of a system is related to a property known as *entropy*. When the entropy of a system is very low, the system is highly ordered, concentrated or confined, but when the entropy is high the system is disorganized, dilute and dispersed. The energy of a system increases when entropy decreases.

The energies E_A and E_B are the energies of the solvent molecules that are free to move across the membrane. Since these energies are of a single chemical component, they are more often termed *chemical potentials*, μ_A and μ_B, which really are chemical potential energies per mole. (Note: these are not gravitational potential energies, but potential energies due to entropy.) If the solvent molecules are water, the term *water potential* is often used instead of chemical potential. In many biology books, water potential is given the symbol ψ (uppercase Greek letter *psi*).

The osmotic pressure, which is the pressure difference between side A and side B, is related to μ_A and μ_B by the expression,

$$\Pi \mathcal{V}_m = \mu_A - \mu_B \qquad \textbf{[13-49]}$$

where $\mathcal{V}_m$ is called the partial molar volume of the solvent. This is the volume occupied by one mole of the solvent. One mole of water, for example, would occupy 0.018 dm^3 at room temperature.

The chemical potential difference between side A and side B, $\mu_A - \mu_B$, can be evaluated in terms of the physical properties of the solution. For relatively dilute solutions, the mole fraction of the solvent X_v (subscript "v" for sol<u>v</u>ent) and the solute X_u (subscript "u" for sol<u>u</u>te) are given by

$$X_v = \frac{n_v}{n_v + n_u} \quad \text{and} \quad X_u = \frac{n_u}{n_v + n_u} \qquad \textbf{[13-50]}$$

where n_v and n_u are, respectively, the number of moles of solvent and solute molecules that are present in the solution. Note that $X_u = 1 - X_v$. For side A, which is pure water, $X_v = 1$ and $X_u = 0$. If side B is relatively dilute, then the fraction of water molecules having chemical potential μ_A relative to the number having energy μ_B is X_v. Therefore, the Boltzmann equation becomes

$$X_v = e^{-\frac{\mu_A - \mu_B}{RT}} \qquad \text{[13-51]}$$

where R is the molar gas constant (8.315 J·mol^{-1}·K^{-1}). The denominator in the exponent of Eq. [13-51] is RT instead of the usual k_BT because μ_A and μ_B in the numerator are energies per mole instead of energies. Therefore, Boltzmann's constant k_B in the denominator has been multiplied by Avogadro's number N_A to give Boltzmann's constant per mole, which is the molar gas constant ($R = k_B N_A$).

The exponent in Eq. [13-51] is usually quite small, and hence the following approximation can be used: $e^{-x} \approx 1 - x$, where $x = (\mu_A - \mu_B)/(RT)$ in this case. Thus, Eq. [13-51] can be written approximately as

$$X_v = 1 - \frac{\mu_A - \mu_B}{RT} \qquad \text{[13-52]}$$

This equation can then be rearranged as follows

$$1 - X_v = X_u = \frac{\mu_A - \mu_B}{RT}$$

By replacing $\mu_A - \mu_B$ with $\Pi \mathcal{V}_m$ according to Eq. [13-49] and rearranging,

$$\Pi = \frac{RT}{\mathcal{V}_m} X_u \qquad \text{[13-53]}$$

Therefore,

$$\Pi = \frac{RT}{\mathcal{V}_m} \left(\frac{n_u}{n_v + n_u} \right) \qquad \text{[13-54]}$$

In dilute solution the number of moles of solute n_u is small relative to the number of moles of solvent n_v, and therefore $n_v + n_u \approx n_v$. Equation [13-54] becomes

$$\Pi = \frac{RT n_u}{\mathcal{V}_m \, n_v}$$

But in dilute solution the solvent accounts for essentially all the volume $\mathcal{V}$. Therefore, the denominator ($\mathcal{V}_m n_v$) can be written as $\mathcal{V}$, and then $n_u/\mathcal{V}$ is the concentration of the solute C, which is normally expressed in units of moles per cubic metre (mol/m³). Hence,

$$\Pi = RTC \qquad \text{[13-55]}$$

EXAMPLE 13-11

Consider a situation in which a semi-permeable membrane separates two compartments, each having volume of 1.00 m³. Compartment A contains 10 000 moles of water and compartment B contains 9 990 moles of water and 10.0 moles of a solute molecule that is too large to cross the membrane.

(a) Determine the difference in the mole fractions of water on the two sides, and show that it is identical to the mole fraction of the solute in compartment B.

(b) Determine the osmotic pressure difference between compartments A and B. The temperature is 20.0°C.

SOLUTION

(a) In compartment A, the mole fraction of water is $X_{vA} = (10\,000/10\,000) = 1$. In compartment B, the mole fraction of water is $X_{vB} = (9990/10\,000)$. The difference in the mole fractions of water is

$$X_{vA} - X_{vB} = \frac{10\,000}{10\,000} - \frac{9990}{10\,000} = \frac{10}{10\,000}$$

However, the mole fraction of solute in compartment B is

$$X_{uB} = \frac{10}{9990 + 10} = \frac{10}{10\,000}$$

Thus, $X_{vA} - X_{vB} = X_{uB}$, as required.

(b) The osmotic pressure difference between compartments A and B can be determined from Eq. [13-55]:

$$\Pi = RTC = \left(8.315\,\frac{J}{mol \cdot K}\right) \times (273.15 + 20.0)K \times \left(10.0\,\frac{mol}{m^3}\right) = 2.44{\times}10^4\,Pa$$

Instead of expressing solute concentration as the number of moles per unit volume, it is sometimes more convenient to express it as the mass of solute per unit volume: $\mathscr{C} = m/\mathcal{V}$. Since the concentration in moles per unit volume is $C = n/\mathcal{V}$ (where n is the number of moles of solute), and $n = m/M$ (where M is the molar mass of the solute), then $C = m/(M\mathcal{V}) = \mathscr{C}/M$. Hence, Eq. [13-55] ($\Pi = RTC$) can be written as

$$\Pi = \frac{RT\mathscr{C}}{M} \qquad\qquad \textbf{[13-56]}$$

EXAMPLE 13-12

What is the osmotic pressure of a solution of ovalbumin ($M = 44.6$ kg/mol) at 20.0°C if $\mathscr{C} = 1.00$ g/dm³? Calculate the height of a column of water that this osmotic pressure could raise.

SOLUTION

First, the concentration in grams per cubic decimetre must be converted to kilograms per cubic metre:

$$1.00\,\frac{\text{g}}{\text{dm}^3} \times \frac{1\,\text{kg}}{10^3\,\text{g}} \times \frac{10^3\,\text{dm}^3}{1\,\text{m}^3} = 1.00\,\frac{\text{kg}}{\text{m}^3}$$

Equation [13-56] is now used:

$$\Pi = \frac{RT\mathscr{C}}{M} = \frac{\left(8.315\,\dfrac{\text{J}}{\text{mol}\cdot\text{K}}\right) \times (273.15 + 20.0)\text{K} \times \left(1.0\,\dfrac{\text{kg}}{\text{m}^3}\right)}{44.6\,\dfrac{\text{kg}}{\text{mol}}} = 54.7\,\text{Pa}$$

This is a very dilute solution of a very high molar mass material, and the resulting osmotic pressure is very small. The height of the column of water that this osmotic pressure could raise can be calculated from

$$\rho g d = 54.7\,\text{Pa}$$

$$d = \frac{54.7\,\text{Pa}}{9.80\,\text{m/s}^2 \times 1000\,\text{kg/m}^3}$$

$$= 5.58 \times 10^{-3}\,\text{m}$$

$$= 5.58\,\text{mm}$$

Osmotic Pressure and Biological Cells

Solvent molecules such as water can not only pass through the semi-permeable membranes in these examples, but also through real biological membranes. When a molecular species can cross the membrane it is said to be *osmotically inactive.* In the case of biological membranes, osmotically inactive substances include water; dissolved gases such as oxygen, nitrogen and carbon dioxide; and small neutral molecules such as short-chain alcohols and glycerol.

Large molecules, charged organic molecules, and inorganic ions are prevented from crossing membranes and are, therefore, *osmotically active.* Their presence is vital for creating the concentration imbalances that cause osmotic pressure. Biological cells must have a positive internal pressure called *turgor pressure* (i.e., the inside of the cell is at higher pressure than the outside), and have special osmoregulatory processes to ensure this pressure is maintained.

Osmolarity

Osmotic pressure is a *colligative* quantity; that means that it depends on the number of solute species present and not on their size or particular composition. If the 10 moles of solute in compartment B of Example 13-11 dimerized (i.e., two molecules aggregated to form one larger molecule) to form 5 moles of solute, then C would be halved and the osmotic pressure would be halved, according to Eq. [13-55] $(\Pi = RTC)$. Similarly, if the solute were sodium chloride (NaCl), which ionizes in solution to form two ions Na^+ and Cl^-, then the concentration of solute particles would be twice the molar concentration and the osmotic pressure would be doubled.

To avoid confusion, the term *osmolarity* is used to describe the total number of osmotically active solute molecules per unit volume in a solution. Thus, for NaCl, the osmolarity would be twice the molarity. If a substance completely ionized into three parts (e.g., $CaCl_2 \rightarrow Ca^{++} + 2\,Cl^-$) then the osmolarity would be three times the molarity, etc.

Osmolality

At high concentrations, the osmolarity of the solution becomes a poor parameter for predicting osmotic pressures. The extensive solute–solute and solute–solvent interactions that take place at higher concentrations can invalidate the assumption of ideality that was required to evaluate expressions for osmotic pressure such as Eqs. [13-55] and [13-56]. One consequence of these interactions is that at high concentrations, the volume of a mixture may substantially deviate from the sum of the volumes of the components. Some very high molar mass polymers, especially those with an extended conformation in water (e.g., polyethylene glycol), will have an osmotic pressure considerably greater than expected from their concentration. In these molecules, each chain seems to act like several separate molecules. As well, the volume of a solution changes measurably with temperature due to thermal expansion. Osmolarity, which is the number of moles of solute per unit volume, is not generally, therefore, the best measure of concentration.

At high concentrations, an improved estimate of concentration is given by the solution *osmolality, C_{os},* which is the number of moles of solute per unit mass of solvent (expressed in units such as moles per kilogram). However, there is no simple relationship between osmolarity and osmolality for the reasons cited above. Scientists have now taken the approach that the osmolality of a solution must be inferred from experimental measurements of Π. Although this information can be obtained by experiments similar to that demonstrated in Fig. 13-13, newer, more accurate, devices such as *vapour pressure osmometers* are generally used. The empirical relationship they use is similar to Eq. [13-55] ($\Pi = RTC$):

$$\Pi = ZTC_{os} \quad \text{hence,} \quad C_{os} = \frac{\Pi}{ZT} \qquad \text{[13-57]}$$

where the constant Z plays the role of R except that the units are different ([kg atm]/[K mole]). An osmotic pressure of 22.4 atmospheres ($22.4 \times 1.013 \times 10^5$ Pa) corresponds to a 1.00 osmolal solution at 0°C. At 35°C the pressure would be 25.5 atm, but the osmolality remains the same (1.00 osmolal).

EXAMPLE 13-13

If 0.0300 moles of NaCl are dissolved in 1.00 dm³ (1.00 L) of water at 4°C, determine the osmolarity and the osmolality of the solution at (a) 20°C and (b) 70°C. The densities of water at 4°C, 20°C, and 70°C are 1.000×10^3, 0.998×10^3, and 0.972×10^3 kg/m³, respectively.

SOLUTION

The molarity of a solution is usually expressed in units of moles per litre, and since 0.0300 moles of NaCl are dissolved in 1.00 L of water, the molarity in this case is 0.0300. Since NaCl ionizes into Na^+ and Cl^-, the solution osmolarity will be twice the molarity. Therefore, at 4°C the solution is $2 \times 0.0300 = 0.0600$ osmolar. The density of water at 4°C is 1000 kg/m³ or 1.00 kg/L, so that the solution will also be 0.0600 osmolal, i.e., 0.0600 moles of solute ions per kilogram of solvent. The osmolality remains at 0.0600 osmolal at all the other temperatures, since the mass of solvent remains constant. However, the osmolarity will gradually fall as the volume of the solution increases by thermal expansion, that is, as the density decreases. The osmolarity values are as follows:

(a) At 20°C, 0.0600 osmolar × 0.998 = 0.0599 osmolar

(b) At 70°C, 0.0600 osmolar × 0.972 = 0.0583 osmolar

13.10 Osmotic Effects on Biological Cells

The semi-permeable properties of cell membranes serve to separate the *lumen* (the cell contents) from the extracellular environment. This suggests that osmotic pressure differences between the cell contents and the exterior might lead to water movements in or out of the cell, which could be harmful. Indeed, biological cells have elaborate systems to allow them to osmoregulate by pumping small compatible solute molecules (such as proline or betaine) across membranes in response to changing osmotic environments. In the absence of these regulating systems, cells can experience serious damage.

The osmolality of the lumen of a biological cell can vary over quite a large range. For erythrocytes, however, a typical value is 0.30 osmolal. If the external medium is also 0.30 osmolal, then the two solutions (i.e., the internal and the external solutions) are said to be *isosmotic*, that is, they have the same osmolality. As a result, there is no "net" flow of water in or out of the cell and its volume remains constant. Note that the actual solute composition of the two solutions need not be the same.

If a cell is placed in a solution and the cell does not shrink or swell, the solution is said to be *isotonic* for that particular type of cell. For example, a 0.30 osmolal solution of NaCl is isotonic for erythrocytes. If, instead, a cell is placed in a solution and the cell swells (or possibly bursts), the solution is *hypotonic*. Such a solution has an osmolality less than that of the lumen, and this leads to a net flow of water into the cell. In contrast, a *hypertonic* solution is one that has an osmolality greater than that of the lumen. In this case the net flow of water is out of the cell, causing its volume to shrink.

However, these simple relationships between the *tonicity* of a solution and its osmolality hold only if the solutes are osmotically active, that is, if the solute molecules cannot cross the membrane. Glycerol, for example, is not osmotically active because as a small neutral molecule it, like water, is relatively free to cross the membrane. One can prepare a 0.30 osmolal glycerol solution, but, as will be shown in Example 13-16, this is *not* an isotonic solution for a cell whose lumen is 0.30 osmolal. Sometimes, therefore, solutions can be isosmotic without being isotonic.

Consider the effects of placing red blood cells into a hypotonic medium of pure water (0.00 osmolal). Since the lumen is 0.30 osmolal, there will be a net flow of water into the cell. As the water enters, the increasing volume causes the shape of the cell to change from its normal shape of a biconcave disc (see Fig. 13-5) to evolve into a sphere. Volume can increase considerably without an increase in surface area as this shape change occurs. Since a sphere has the greatest volume for a fixed surface area, then continued influx of water and further volume increases must be accompanied by the stretching of the membrane. Eventually, the strain will become so great that the cell ruptures and the contents are released into the external medium. The membrane may then re-seal itself and can resume its usual biconcave shape and exist as a "ghost." The rupture has caused complete equilibration of the lumen with the medium. Up to the point at which the membrane has started to stretch, the cell behaves as an *ideal osmometer*. This means that if the cells were placed into a solution of different osmolality the volume would increase or decrease according to the equation

$$C_{os1} \mathcal{V}_1 = C_{os2} \mathcal{V}_2 \qquad \textbf{[13-58]}$$

where C_{os1} is the original osmolality of the lumen and medium (when the cell's volume is in equilibrium), $\mathcal{V}_1$ is the original volume, and C_{os2} is the osmolality of the lumen and medium after the cell has equilibrated at its new volume, $\mathcal{V}_2$.

Since there is no membrane stretching involved, Eq. [13-58] applies generally to all volume reductions arising from hypertonic effects. If red blood cells are placed in a concentrated solution of salt, they lose volume and eventually, if the osmotic change is sufficiently large, take on a crenated (wrinkled) appearance.

EXAMPLE 13-14

Consider a mammalian erythrocyte originally present in an isotonic 0.30 osmolal NaCl solution at 20°C. Since the membrane is impermeable to the ions, this establishes that the cell interior must have an osmolality of 0.30 as well. A small aliquot of this solution (also at 20°C) is injected into a much larger solution of 0.20 molar NaCl. Determine what happens to the cell.

SOLUTION

As shown in Example 13-13, since the solution is reasonably dilute and at room temperature, the osmolality of the solution will be extremely close to the osmolarity (which is twice the molarity in this case). Thus, the 0.20 molar NaCl is 0.40 osmolal. Initially the situation will be as in Fig. 13-14.

In this case water will leave the cells until the concentration of solute molecules inside increases to 0.40 osmolal. Let the initial volume of each cell be $\mathcal{V}_1$. Since the volume of the solution is huge relative to the cellular volume, the exiting water will negligibly influence the osmolality of the medium. The water will leave the cell until the osmolality of the lumen matches that of the medium.

Figure 13-14

Example 13-14: Conditions inside and outside a red blood cell in a 0.4 osmolal NaCl solution.

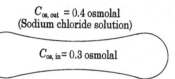

$C_{os,\,out} = 0.4$ osmolal
(Sodium chloride solution)

$C_{os,\,in} = 0.3$ osmolal

Let x represent the number of moles of solute entities in the cell. Initially,

$$\frac{x}{\mathcal{V}_1} = 0.30 \text{ osmolal}$$

Let $\mathcal{V}_2$ represent the volume of each cell after the water has left the cell. Then,

$$\frac{x}{\mathcal{V}_2} = 0.40 \text{ osmolal}$$

Divide the first equation by the second to eliminate the x's. Thus,

$$\frac{x / \mathcal{V}_1}{x / \mathcal{V}_2} = \frac{\mathcal{V}_2}{\mathcal{V}_1} = \frac{0.30}{0.40} = 0.75$$

Therefore, $\mathcal{V}_2 = 0.75\,\mathcal{V}_1$, i.e., the final volume is three-quarters of the original, and the final concentration of the lumen is 0.40 osmolal.

Alternatively, this result can be obtained using the ideal osmometer equation (Eq. [13-58]),

$$C_{os1}\mathcal{V}_1 = C_{os2}\mathcal{V}_2$$

$$0.30\ \mathcal{V}_1 = 0.40\ \mathcal{V}_2$$

$$\mathcal{V}_2 = 0.75\ \mathcal{V}_1$$

Therefore the final volume is three-quarters of the original, and the final concentration of the lumen is 0.40 osmolal.

EXAMPLE 13-15

If the erythrocytes in Example 13-14 were placed in a 0.10 molar NaCl solution, what would happen?

SOLUTION

The osmolality of this new medium is 0.20 osmolal, which is less than the original lumen osmolality of 0.30. Water would enter the cell until the initial interior osmolality was reduced to 0.20 osmolal. In this case applying Eq. [13-58] yields

$$0.30\ \mathcal{V}_1 = 0.20\ \mathcal{V}_2$$

$$\mathcal{V}_2 = 1.5\ \mathcal{V}_1$$

The volume is increased by 50 percent and the final internal concentration will be 0.20 osmolal. In fact, under these circumstances the cell is spherical and is just beginning to stretch as a consequence of this volume increase. If the new medium had been 0.15 osmolal, the volume would have had to double. However, the cell cannot tolerate this increase in volume and so would *lyse* (burst).

EXAMPLE 13-16

Suppose that the original cells in Example 13-14 are injected into a 0.30 osmolal solution of glycerol. What happens?

SOLUTION

The glycerol molecules are osmotically inactive which means that they can freely cross the cellular membrane. Figure 13-15 shows the initial situation. The behaviour of the cell is explained by the fact that not only water but also glycerol may cross the membrane. Consider what happens if the movement of water is initially neglected. The glycerol will tend to diffuse into the cell until the concentration of glycerol is the same inside as out (0.30 osmolal). But since the cell lumen already had an osmolality of 0.30 it is now raised to a total of 0.60 osmolal. But this means that the interior is more concentrated than the exterior (which remains at 0.30 osmolal). The situation is sketched in Fig. 13-16.

Figure 13-15

Example 13-16: Initial conditions inside and outside a red blood cell in a 0.3 osmolal glycerol solution.

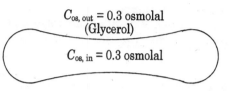

Figure 13-16

Example 13-16: After diffusion of glycerol.

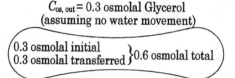

If the water is now permitted to move, it also crosses into the cell to reduce its concentration imbalance; the cell swells and bursts. Of course the water movement cannot be manipulated in this way. In reality, as soon as the glycerol starts to move in, the water follows as well. This inward movement of water will tend to dilute the concentration of the lumen. Because glycerol keeps moving in, the water can never reduce the total internal concentration to 0.30 osmolal, and so the cell keeps on swelling as more glycerol and more water enter. Finally it bursts. Note that this was a situation where the lumen and the medium were isosmotic, but not isotonic.

Exercises

13-1   In an experiment similar to Perrin's, 100 particles per unit volume were counted close to the bottom of a column, and 1.0 μm above this, 50 particles were counted in a similar volume. How many particles per unit volume would be expected at a height of 5.0 μm above the bottom?

13-2   In an experiment similar to Perrin's, the number density of particles at height 10.0 μm compared to the number density at the bottom is $10^{-4}:1$. What is the ratio of the number density at height 5.0 μm to the number density at the bottom?

13-3   Two spheres of the same size but different material have terminal speeds in water in the ratio of 9.0:1. If the slower sphere has a density 2.0 times that of water, what is the density of the faster sphere?

13-4   A centrifuge is a device which is used to speed up the sedimentation of particles by increasing the acceleration. How much faster will a particle sediment if the acceleration is increased by a factor of 10?

13-5   Suppose that the position of each molecule in a system of molecules could be observed as a function of time. At a certain time t, the displacement of each molecule relative to its displacement at $t = 0$ is determined. The following table gives data for the number of molecules found as a function of the x-component of the displacement.

No. of Molecules (thousands)	x-component of Displacement (m)
2	-0.045 to -0.035
4	-0.035 to -0.025
7	-0.025 to -0.015
9	-0.015 to -0.005
10	-0.005 to +0.005
9	0.005 to 0.015
7	0.015 to 0.025
4	0.025 to 0.035
2	0.035 to 0.045

For the x-component of the displacement, determine

(a) the average value
(b) the mean square value
(c) the rms value
(d) the most probable value

Sketch a histogram showing the distribution of molecules as a function of the x-component of displacement.

13-6   The diffusion coefficient of water in water is 2.0×10^{-9} m^2/s. How long will it take for one-third of the water molecules in a large population to diffuse at least 1.0 mm from their initial position?

13-7   A biological macromolecule has a diffusion coefficient of 1.5×10^{-11} m^2/s. If the position of each molecule in a solution were known at some instant, then 1.0 s later how far will approximately one-third of the molecules have traveled?

13-8   A spherical molecule has a diffusion coefficient of 5.0×10^{-11} m^2/s in water. How long will it take one-third of such molecules to diffuse a distance of at least 1.0 μm in an *E. coli* cell where the viscosity is 5.0 times as great as the viscosity of water?

13-9   Consider two different protein molecules which are both spherical, but one has a molar mass which is eight times the other. What is the diffusion coefficient in water of the larger molecule relative to the smaller?

13-10 Molecules can diffuse across living cells in very small fractions of a second. If the width of cells were increased by a factor of 10, then by what factor would the average time required for molecules to diffuse across them increase?

13-11 If the diffusion coefficient for CO-hemoglobin in water at 20°C is 6.2×10^{-11} m^2/s, and for CO-hemoglobin in an unknown liquid is 15.5×10^{-11} m^2/s, what is the viscosity of the unknown liquid?

13-12 A certain kind of spherical molecule is observed to diffuse through water so that one-third of the molecules cover at least 0.50 cm in a time of 10 h at 20°C. Compute the radius of the molecule.

13-13 The concentration of glucose (a sugar) outside a bacterial membrane of thickness 10 nm is found to be 0.50 mol/dm^3. Inside the cell the glucose concentration is 0.050 mol/dm^3. Measurement of the amount of glucose crossing the bacterial membrane yields a flux of 1.20 mol·m^{-2}·s^{-1}. Assuming that this glucose flux is due solely to diffusion, determine the diffusion coefficient of glucose. Assume that 5.0 percent of the bacterial membrane area is pores.

13-14 A human erythrocyte of surface area 140 μm^2 is placed in a solution of 0.30 molar glycerol. If the diffusion coefficient of glycerol across the membrane is 1.0×10^{-12} m²/s, what is the initial number of molecules crossing the cell wall per second? Assume the membrane is 10 nm thick.

13-15   Once inside the cell, the glycerol molecules in Exercise 13-14 have a diffusion coefficient of about 2.0×10^{-10} m²/s. Remember that the cell is a biconcave disc about 8 μm in diameter and about 1 μm thick in the centre. Approximately how long would it take molecules to diffuse from the membrane to the centre of the cell:

(a) if they entered at the periphery of the biconcave disc?

(b) if they entered at the centre of the biconcave disc?

13-16   The diffusion coefficient of a large protein molecule is found to be 2.0×10^{-13} m²/s in water at 20°C. Assuming it is spherical with a density of 1.3×10^3 kg/m³, compute its molar mass.

13-17   What is the effect of: (a) water of hydration, and (b) non-spherical shape on the diffusion coefficient of a molecule, and on the molar mass calculated from the diffusion coefficient?

13-18   Suppose that the erythrocyte in Exercise 13-14 is placed in an NaCl solution that is isotonic for the erythrocyte. What will be the net number of molecules crossing the cell wall per second?

13-19   Solution A contains 0.50 mole/dm³ sucrose (which does not ionize in water) and solution B contains 0.75 mol/dm³ of sodium chloride (which ionizes into Cl⁻ and Na⁺ ions). Equal volumes of solution A and solution B are mixed together at 4°C. Estimate the osmotic pressure of the mixture.

13-20   Two different solutions are prepared at a temperature of 4°C to have identical molarities of 0.500 mol/dm³. The first solution is composed of glucose (which does not ionize in water), the second is sodium chloride (which ionizes into Cl⁻ and Na⁺ ions). Compare (a) the osmolarity and (b) the osmolality of the two solutions at 60°C. (The density of water is 1000 kg/m³ at 4°C and 983 kg/m³ at 60°C.)

13-21   Erythrocytes having a lumen concentration of 0.30 osmolal are placed in an 0.35 osmolal NaCl solution.

(a) Do they swell or shrink?

(b) By what percentage does the volume of each cell change?

13-22   A 0.22 osmolal NaCl solution is isotonic for a type of cell having a membrane that is impermeable to Na⁺ and Cl⁻. A cell of this type is placed in another NaCl solution, and the cell's volume increases by 34%. What is the osmolality of the solution?

Problems

13-23   In an experiment similar to Perrin's, 100 particles per unit volume were counted close to the bottom of a column, and 1.0 cm above this 50 particles were counted in a similar volume. If the experiment were repeated on the moon where $g = 1.6$ m/s^2, how many would you expect to find 1.0 cm up (assuming there were 100 particles per unit volume near the bottom as on Earth)?

13-24   A bacterium has a mass of 2.0×10^{-15} kg and a density of 1.050×10^3 kg/m^3. When submerged in water (density 1000 kg/m^3), what would be its apparent weight?

13-25   Sand is defined as mineral particles of diameter 0.05×10^{-3} m to 2.0×10^{-3} m; clay is defined as mineral particles of diameter less than 2.0×10^{-6} m.

(a) Assuming that the particles are spherical and have identical densities, what is the ratio of the terminal speeds of the largest sand and largest clay particles, when sedimenting in water?

(b) How long will it take the largest clay particles, assumed spherical, to settle 0.010 m in water? The density of clay minerals is about 2.6×10^3 kg/m^3.

13-26   A block of ice is held 2.0 m below the surface of the water and then released. What is the magnitude of its initial acceleration? If its mass is 2.0 kg and its frictional factor is 3.0 kg/s, what is the terminal upward speed? (Use 920 kg/m^3 for the density of ice.)

13-27   A spherical protein of molar mass 4.5×10^6 g/mol and density 1.33×10^3 kg/m^3 is placed in an ultracentrifuge tube filled with water at a distance of 14 cm from the rotation axis. The centrifuge then spins at 1.2×10^3 Hz.

(a) What is the terminal speed of the protein?

(b) How long will it take for the protein to sediment a distance of 1.5 mm?

13-28   Two species of spherical protein molecules are being compared. The surface area of one is 2.0 times that of the other. What is the ratio of their diffusion coefficients?

13-29   A student working 3.0 m away from you in a chemistry lab performs an experiment which creates a bad odour. It takes a while for the smell to get to you (assume that air currents are negligible). You are able to move 2.0 m further away from the source but the smell catches up with you in another 60 s. What is the diffusion coefficient in air for the molecules you smell?

Hint: the root mean square distance and the time must be measured from the source of the diffusing molecules.

13-30   Data from measurements on Infectious Pancreatic Necrosis Virus (IPNV), a trout virus, are listed below. In all cases the measurements were performed using water at 20°C as the solvent. Electron microscopic pictures indicate that the virus particle is spherical. Estimate the ratio of hydrated molar mass/dry molar mass for this virus using the following data:

Diffusion coefficient $D = 6.67 \times 10^{-12}$ m²/s
Sedimentation coefficient $s = 435 \times 10^{-13}$ s
Density $\rho = 1.42 \times 10^3$ kg/m³

Answers

13-1	3	**13-15**	(a) 1×10^{-2} s
13-2	10^{-2}:1		(b) 2×10^{-4} s
13-3	1.0×10^4 kg/m³	**13-16**	4.1×10^{12} g/mol
13-4	10 times	**13-17**	Both decrease the diffusion coefficient and increase the calculated molar mass.
13-5	(a) 0 m (b) 3.89×10^{-4} m²		
	(c) 1.97×10^{-2} m (d) 0 m		
13-6	83 s	**13-18**	zero
13-7	at least 9.5×10^{-6} m	**13-19**	2.3×10^6 Pa
13-8	1.7×10^{-2} s	**13-20**	(a) 0.492 osmolar for glucose, 0.983 osmolar for sodium chloride (b) 0.500 osmolal for glucose, 1.000 osmolal for sodium chloride
13-9	$D_{large} = 0.5\, D_{small}$		
13-10	100		
13-11	4.0×10^{-4} N·s/m²	**13-21**	(a) shrink (b) decrease by 14%
13-12	1.9×10^{-9} m	**13-22**	0.16 osmolal
13-13	5.3×10^{-10} m²/s		
13-14	2.5×10^{12} molecules/s	**13-23**	89
		13-24	9.3×10^{-16} N

13-25   (a) 1.0×10^6 (b) 2.9×10^3 s   **13-28**   1.4

13-26 0.85 m/s^2; 0.57 m/s   **13-29** 0.044 m^2/s

13-27   (a) 7.1×10^{-5} m/s (b) 21 s   **13-30**   2.2 ($M_{hydr} = 1.2 \times 10^8$ g/mol; $M_{dry} = 5.4 \times 10^7$ g/mol)

14 Heat and Heat Flow in Biological Systems

14.1   Introduction

Fish are poikilotherms (cold-blooded creatures) and normally their body temperatures would be expected to be essentially that of their environment. This is true for most fish, which consequently are rather sluggish at low temperatures and more active at higher temperatures. Some ocean fish, however, such as the bluefin tuna, are powerful swimmers and, while still poikilotherms, have developed a special mechanism for keeping their main swimming muscles at a temperature several degrees higher than ambient temperature. Even warm-blooded animals exhibit special characteristics that depend on their size and location. Large animals that are found near the tropics, such as elephants, have huge ears to help dissipate heat, whereas mammals whose habitat is nearer the poles have very small ears. The several aspects of heat and heat flow required to understand these and other fascinating phenomena of biological heat regulation will be considered in this chapter.

14.2   Specific Heat Capacity

Heat is energy that can be transferred from one body to another because of temperature differences alone. The processes by which the heat is transferred will be considered in Section 14.5.

The quantity of heat, Q joules, required to change the temperature of m kilograms of a substance by ΔT kelvins or degrees Celsius[1] is:

$$Q = mC\Delta T \qquad \text{[14-1]}$$

---

1.   The numerical value of a temperature change is the same in kelvins and degrees Celsius.

where the *specific heat capacity, C* ($J \cdot kg^{-1} \cdot K^{-1}$), often called just *specific heat*, is characteristic of the substance and its state. Water, for instance, has a specific heat capacity of 4186 $J \cdot kg^{-1} \cdot K^{-1}$ and this is essentially constant over the temperature range of biological interest. However, ice has a specific heat capacity which is only about half of the value for liquid water. Moreover, the specific heat capacity of ice is not constant and decreases as the temperature decreases. A mixture of water vapour and air has even greater variations in specific heat capacity than does ice, and depends on temperature and pressure. All else being equal, it will require more energy to heat a quantity of gas at constant pressure P than at constant volume $\mathcal{V}$ because when pressure is constant, energy must be provided not only to increase the temperature but also to perform the work of expansion ($W = P\Delta\mathcal{V}$) that will simultaneously occur.

Not surprisingly, cells and tissues of living systems which have a high water content have specific heat capacities close to that of water. The complete body of a mouse, for instance, has a specific heat capacity of 3450 $J \cdot kg^{-1} \cdot K^{-1}$. Most substances have a specific heat capacity considerably less than that of water. Some specific heat capacities are given in Table 14-1.

Table 14-1   Specific Heat Capacities of Some Solids and Liquids

Substance	C ($J \cdot kg^{-1} \cdot K^{-1}$)
Aluminum	886
Crown glass	670
Lead	127
Ice (0°C)	2100
Water (20°C)	4186
Mercury (20°C)	140
Ethyl alcohol (20°C)	2500

If a hot object comes in contact with a cold substance, heat will be exchanged and, if no heat is gained or lost to the surroundings, a final intermediate equilibrium temperature will be achieved. This final temperature can be calculated by setting heat lost by the hot object equal to that gained by the cold substance, as in Example 14-1.

EXAMPLE 14-1

People lacking fireproof cooking utensils often heated liquids by placing hot stones in a container of the liquid. The native peoples of North America concentrated maple sap to syrup by doing this. If a 0.50 kg piece of granite ($C = 840$ J·kg^{-1}·K^{-1}) at an initial temperature of 500 K is placed in 2.0 kg of water at 300 K, what will be the final temperature T_F of the liquid?

SOLUTION

Heat lost by stone = Heat gained by water

Using Eq. [14-1], and subscripts "s" for stone and "w" for water, and writing each ΔT so that it is a positive quantity,

$$m_s C_s (T_s - T_F) = m_w C_w (T_F - T_w)$$

$$0.50 \text{ kg} \times 840 \text{ J·kg}^{-1}\text{·K}^{-1} \times (500 - T_F) \text{ K}$$

$$= 2.0 \text{ kg} \times 4186 \text{ J·kg}^{-1}\text{·K}^{-1} \times (T_F - 300) \text{ K}$$

$$2.10 \times 10^5 - 420 \, T_F = 8.37 \times 10^3 \, T_F - 2.51 \times 10^6$$

$$T_F = 3.1 \times 10^2 \text{ K}$$

14.3 Phase Changes

When phase diagrams were discussed in Section 11.6, it was shown that elements and compounds can exist in three phases: solids, liquids and gases. Considerable *heats of transformation* are required to be added or removed when there is a change of phase (or change of state). In a change from solid to liquid with no temperature change, the heat added per kilogram is called the *(latent) heat of fusion*, L_f, (J/kg). For the change from liquid to gas there is a corresponding *(latent) heat of vaporization*, L_v, (J/kg). If a mass m kg is to undergo a change of phase, then the heat Q that must be added or removed is given by

$$Q = mL_f \qquad \text{[14-2]}$$

or

$$Q = mL_v \qquad\qquad \textbf{[14-3]}$$

The heat removed when a gas is cooled to form a liquid is sometimes referred to as the *(latent) heat of condensation* and has the same magnitude as the heat of vaporization. Similarly a *(latent) heat of solidification* is removed on going from liquid to solid phase and has the same magnitude as the heat of fusion. Table 14-2 presents some typical values for heats of transformation at atmospheric pressure.

Table 14-2   Heats of Transformation

Substance	Melting Point (K)	L_f (J/kg)	Boiling Point (K)	L_v (J/kg)
Water	273	3.34×10^5	373	2.26×10^6
Gold	1336	0.67×10^5	2873	1.58×10^6
Ethyl alcohol	159	1.04×10^5	351	0.85×10^6
Copper	1356	2.05×10^5	2609	5.07×10^6
Oxygen	55	0.14×10^5	90	0.21×10^6

Under the right conditions, some substances can change directly from solid to gas in a process known as *sublimation*.

Why are these large amounts of heat required to bring about a change of phase even though the temperature will not change until the transformation is complete? Consider the changes that occur in water structure. In the form of ice it exists as a rather open crystalline structure with molecules held in place by hydrogen bonds, two to every molecule. The energy added in melting the ice breaks enough bonds to allow the structure to collapse into a denser fluid phase. (This makes water different from other substances which expand and become less dense on melting.) The surprising thing is that, for water, the energy added is sufficient to break only about 15% of the hydrogen bonds. If more heat is added to the resultant liquid it will, of course, warm up to 100°C as more bonds are broken. Even at this point some 65% of the bonds still must be broken for the molecules to separate and exist as a gas. The heat of vaporization provides the required energy. It must not be imagined that only certain bonds break in the transformation from solid to liquid and that the rest always remain intact. Bonds are constantly made and broken so that at any instant the majority are intact, but the molecules that are bonded are constantly changing.

EXAMPLE 14-2

Ice (0.50 kg) at 0 °C is placed in an open container on a 1500-W burner on a stove. Assuming all the heat energy is transmitted to the ice and resultant water, how much time will elapse until the container boils dry?

SOLUTION

The process must be treated in three steps: the transition from ice at 0 °C to water at 0 °C, the heating of the water to 100 °C, and finally the transition to water vapour at 100 °C.

The heat Q_1 required to melt the ice is given by Eq. [14-2], and L_f is given in Table 14-2.

$$Q_1 = mL_f = 0.50 \text{ kg} \times 3.34 \times 10^5 \text{ J/kg} = 1.67 \times 10^5 \text{ J}$$

Eq. [14-1] is used to calculate the heat Q_2 required to warm the water from 0 °C to 100 °C:

$$Q_2 = mC\Delta T = 0.50 \text{ kg} \times 4186 \text{ J·kg}^{-1}\text{·K}^{-1} \times 100 \text{ K} = 2.09 \times 10^5 \text{ J}$$

The heat Q_3 required to boil the resulting water is given by Eq. [14-3]:

$$Q_3 = mL_v = 0.50 \text{ kg} \times 2.26 \times 10^6 \text{ J/kg} = 11.3 \times 10^5 \text{ J}$$

Total heat required then is:

$$Q = (1.68 + 2.09 + 11.3) \times 10^5 \text{ J}$$

$$= 1.51 \times 10^6 \text{ J}$$

Heat is provided at 1500 J/s. Time required then will be:

$$t = \frac{1.51 \times 10^6 \text{ J}}{1500 \text{ J/s}} \approx 1.00 \times 10^3 \text{s} = 17 \text{min}$$

EXAMPLE 14-3

(a) 2.00 kg of ice at 0°C is added to 4.00 kg of water at 50°C. What is the final temperature?

(b) 2.00 kg of ice at 0°C is added to 2.00 kg of water at 50°C. What is the final temperature? Does all the ice melt? If not, how much is left?

In both (a) and (b), assume no heat lost to or gained from the environment.

SOLUTION

(a) The total heat gained by the ice equals the total heat lost by the water. The heat gained by the ice consists of two parts: that needed for the change of state as given by Eq. [14-2], where L_f is obtained from Table 14-2; and that given by Eq. [14-1], during an increase to a final temperature T_F. The heat lost by the water is also given by Eq. [14-1] as the water cools to T_F.

$(2.00 \text{ kg} \times 3.34 \times 10^5 \text{ J/kg}) + (2.00 \text{ kg} \times 4186 \text{ J·kg}^{-1}\text{·K}^{-1} \times (T_F - 0°C))$

$= (4.00 \text{ kg} \times 4186 \text{ J·kg}^{-1}\text{·K}^{-1} (50°C - T_F°C))$

$6.68 \times 10^5 \text{ J} + 8372 \, T_F \text{ J} = 8.37 \times 10^5 \text{ J} - 1.67 \times 10^4 \, T_F \text{ J}$

$$T_F = 6.7°C$$

(b) If part (b) is attempted using the same method as in part (a), a final temperature of −16.1°C is obtained. This is impossible!

To melt all the ice will require 6.68×10^5 J as in part (a). The water can cool down to only 0°C and in so doing can give up only (using Eq. [14-1] again):

$Q = mC\Delta T = 2.00 \text{ kg} \times 4186 \text{ J·kg}^{-1}\text{·K}^{-1} \times 50°C = 4.19 \times 10^5 \text{ J}$

This not sufficient to melt all the ice. The final result will be some ice floating in water at 0°C. The ratio of the heat available from the water to the heat needed to melt all the ice will be the fraction of the ice that melts.

$$\frac{4.19 \times 10^5 \, \text{J}}{6.68 \times 10^5 \, \text{J}} \times 2.00 \ \text{kg} = 1.25 \ \text{kg of ice melts}$$

Ice remaining = 2.00 kg – 1.25 kg = 0.75 kg

14.4 **Thermal Expansion**

If an object is heated, its length will increase by an amount that depends on the length of the object and the change in temperature. As the temperature increases, atomic vibrations in the object increase. This increases the distance between atoms and causes an increase in all dimensions of the object. Different substances will expand by different amounts. For the same temperature difference, for instance, a brass rod will increase in length considerably more than will a steel rod. Experiments show that the change in length, ΔL, is proportional to the original length L and the change in temperature ΔT. Therefore,

$$\Delta L = \alpha L \Delta T \qquad\qquad \textbf{[14-4]}$$

where the constant of proportionality α (K^{-1}) is the *coefficient of thermal expansion* of the substance being heated. For most substances the coefficient will depend only slightly on the temperature, and so α is a constant. Note that if two rods of the same material but of different lengths are heated by the same amount, the longer rod will expand more.

Figure 14-1

Increase in area of a heated surface.

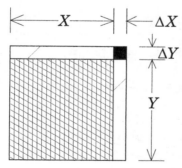

Real objects, of course, have three dimensions, and when heated all dimensions will increase, giving rise to an increase in the area of any face of the object and in its total volume.

Consider the area of the object in Fig. 14-1. Here the initial area was XY and the final area is:

$$(X + \Delta X)(Y + \Delta Y) = XY + X\Delta Y + Y\Delta X + \Delta X\Delta Y.$$

The increase in area is $X\Delta Y + Y\Delta X + \Delta X\Delta Y$ or just $X\Delta Y + Y\Delta X$ if the small area ($\Delta X\Delta Y$) in the top right corner of the figure is ignored.

Now, from Eq. [14-4], $\Delta Y = \alpha Y\Delta T$ and $\Delta X = \alpha X\Delta T$. Therefore $X\Delta Y = \alpha XY\Delta T$ and $Y\Delta X = \alpha XY\Delta T$. Consequently the total increase in area is:

$$\Delta \mathcal{A} = 2\alpha XY \Delta T = 2\alpha \mathcal{A} \Delta T \qquad \textbf{[14-5]}$$

where $\mathcal{A}$ is the initial area (XY).

A similar argument will show that if three dimensions are involved, the increase in volume can be approximated by:

$$\Delta \mathcal{V} = 3\alpha \mathcal{V} \Delta T = \beta \mathcal{V} \Delta T \qquad \textbf{[14-6]}$$

where $\beta = 3\alpha$ is the *coefficient of volume expansion* (units, K^{-1}).

A solid object and a hollow object of the same material and dimensions will have identical increases in volume when heated by the same amount. If a container such as a drinking glass is completely filled with some fluid, then the fluid volume will also increase according to the previous equation. If the fluid has a greater coefficient of thermal expansion than the material of the container, heating will result in some of the fluid overflowing. On the other hand, if the coefficient for the container is greater, then it will not be quite full after being heated.

Just as the interior volume of a container increases on heating, so the area of a hole in a flat object, contrary to what most people would expect, will also increase on heating. The spacing of the atoms lining the hole increases as the object is heated, and hence the hole's circumference increases.

The preceding argument has assumed no change in pressure. If sufficient pressure is used, the expansion can be decreased or even prevented, but this would be a very unusual occurrence.

Water is unusual in that its greatest density occurs at $4\,°C$; thus, it will increase in volume if heated above this temperature or if cooled below it. The volume increase from $4\,°C$ to $0\,°C$ is very slight.

Linear and volume expansion coefficients for several substances are provided in Tables 14-3 and 14-4.

Table 14-3   Linear Expansion Coefficients (α) near Room Temperature

Substance	α (K^{-1})
Steel	1.2×10^{-5}
Ice	5.1×10^{-5}
Copper	1.7×10^{-5}
Aluminum	2.4×10^{-5}

Table 14-4   **Volume Expansion Coefficients (β) near Room Temperature**

Substance	β (K⁻¹)
Water	20×10^{-5}
Mercury	18×10^{-5}
Gasoline	95×10^{-5}
Air	3.7×10^{-3}

EXAMPLE 14-4

A circular steel washer with a hole 9.500 mm in diameter at 293 K (20°C) is to be heated until it will fit over a circular bolt 9.550 mm in diameter. To what temperature must the washer be heated?

SOLUTION

Use Eq. [14-5]: $\Delta L = \alpha L \Delta T$ or $\Delta d = \alpha d \Delta T$

where d is diameter; α is given in Table 14-3.

Thus,

$$0.050 = 1.2 \times 10^{-5} \, (9.500) \, \Delta T$$

$$\Delta T = 440 \text{K} = 440°\text{C}$$

$$\text{Final } T = 20°\text{C} + 440°\text{C} = 460°\text{C}$$

14.5   Heat Exchange

Heat energy is transferred from one place to another by three physical processes: conduction, convection and radiation.

Conduction

Conduction is a transfer of heat through direct physical contact between two objects at different temperatures, or between different-temperature regions of one object. The actual agent responsible for the energy transfer varies depending on the nature of the material. In fluids (gases and liquids), where the molecules are free to move, adding heat at one location increases the average kinetic energy of motion of the molecules (or atoms) there, and by collision with other molecules the kinetic energy is passed on. In solids, adding heat at one location increases the vibrational energy of the molecules at their site, and by intermolecular interactions the energy of vibration is passed on to nearest neighbours.

A special case is that of metals. The conduction electrons (which account for the high electrical conductivity of metals) are free to move rapidly over long distances in the metal. These electrons take up most of the added heat energy and distribute it rapidly throughout the metal. It is for this reason that good electrical conductors are invariably good heat conductors.

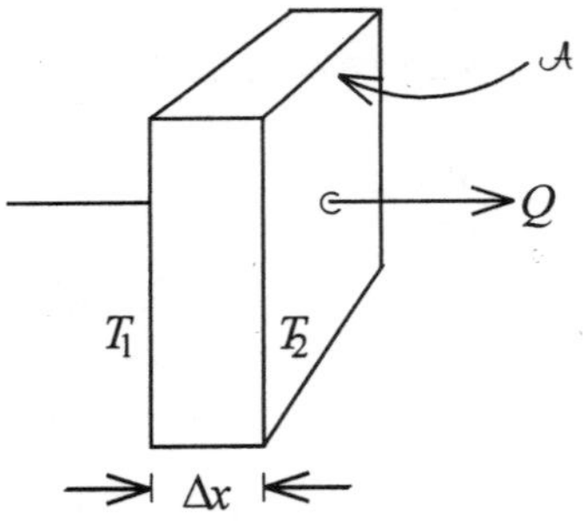

Figure 14-2

Heat flow through an insulating slab.

Fig. 14-2 shows a slab of heat-conducting material of thickness Δx, cross-sectional area $\mathcal{A}$, and temperature difference $T_1 - T_2 = \Delta T$ across the two faces. Experiments show that the amount of heat Q transferred in time t is proportional to the area $\mathcal{A}$, the temperature difference ΔT, and is inversely proportional to the thickness Δx. That is, the rate of heat flow, P watts (J/s) is given by:

$$\frac{Q}{t} = P = k\mathcal{A}\frac{\Delta T}{\Delta x} \qquad [14\text{-}7]$$

The constant of proportionality k is the thermal conductivity and has units of $W \cdot m^{-1} \cdot K^{-1}$. [2]

Some values of the coefficient of thermal conductivity are listed in Table 14-5. Notice the large value for the metal aluminum and the very small value for air. Air is an excellent thermal insulator but that property is not always apparent because heat transport by convection usually dominates (see subsection on convection in this chapter). Convection can be suppressed if the air is divided up into small pockets; this accounts for the low conductivity of cork as seen in Table 14-5. This is the way all home insulation works; the insulating effect is not brought about by the low conductivity of the glass in the fibreglass (it is rather high in fact), but by the suppression of the convection of the trapped air. The fur on animals acts in the same manner, and the hair on your skin rises when you are cold in order to suppress the motion of the air next to your skin and to conserve heat.

Table 14-5   Thermal Conductivities for a Number of Substances

Substance	k ($W \cdot m^{-1} \cdot K^{-1}$)
Aluminum	201
Ice	2.1
Cork	0.05
Fat	0.2
Muscle	0.4
Skin	0.3
Air (10°C)	0.024

---

2. This is just one example of a whole class of phenomena where the gradient of some potential drives the flow of some measured quantity. Ohm's law $\Delta V = IR$ (see Chapter 15) can be written as $q/(At) = (1/\rho)(\Delta V/\Delta x)$ where ρ is the resistivity, ΔV is the voltage and q is the electric charge. In diffusion, $n/(At) = D(\Delta C/\Delta x)$ is Fick's law where D is the diffusion coefficient, n is the number of moles transferred, and ΔC is the concentration difference of the diffusing molecules across the distance Δx.

EXAMPLE 14-5

A building's concrete wall 10 m × 2.5 m is 20 cm thick. The outside temperature is –10°C and inside is 20°C. The thermal conductivity of the concrete is 1.0 W·m⁻¹·K⁻¹.

(a) How much heat is transferred through the wall in one hour?

(b) 2.5 cm of wood ($k = 0.11$ W·m⁻¹·K⁻¹) is added to the warmer face. How much heat is now transferred in an hour?

SOLUTION

(a) Using Eq. [14-7], the rate of energy loss can be calculated:

$$P = k \, \mathcal{A} \, \Delta T / \Delta x$$

$$= 1.0 \text{ W·m}^{-1}\text{·K}^{-1} \times 10 \text{ m} \times 2.5 \text{ m} \times (20°\text{C} - (-10°\text{C}))/0.20 \text{ m}$$

$$= 3750 \text{ W}$$

And in one hour: $Q = P \times t = 3750$ J/s × 3600 s = 1.4×10^7 J

(b) The heat flow through the wood must equal the heat flow through the concrete once a steady state has been achieved. First, find the temperature T_i at the interface. Again using Eq. [14-7]:

Power transferred through wood
= Power transferred through concrete

0.11 W·m⁻¹·K⁻¹ × 10 m × 2.5 m × (20°C – T_i)/0.025 m
= 1.0 W·m⁻¹·K⁻¹ × 10 m × 2.5 m × (T_i–(–10°C))/0.20 m

$T_i = 4.0°$C

Now, for the wood layer, substitute $T_i = 4.0°$C into the left-hand side of the above equation:

$$P = 1760 \text{ W} = 1760 \text{ J/s}$$

and the total heat loss in one hour will be:

$$Q = P \times t = 1760 \text{ J/s} \times 3600 \text{ s} = 6.3 \times 10^6 \text{ J}$$

Notice that adding the wood has reduced the heat loss by more than a factor of two.

On flat surfaces if the thickness of insulation material is doubled or tripled, the heat loss will be decreased by a factor of two or three, as might be expected. The results, however, are quite different when curved surfaces such as pipes are covered with insulating layers of increasing thickness. If, for instance, two 1-cm diameter pipes are covered with 1- and 10-cm thicknesses respectively of the same material, there will be not only an increase in thickness but also a significant increase of the surface area which can lose heat to the surroundings. It will require forty times as much material for the thicker layer, but the losses will be reduced by a factor of only three-and-one-half.

Box 14-1   The *R* Value

Building materials, especially insulating materials, are frequently given an R value. This is a thermal resistance and is the thickness of the material times the reciprocal of thermal conductivity. However, the common units as used in the building trade are not SI units. Few people know what the units are. They just know that the bigger the number, the better. The actual units are: (ft.²)(degree Fahrenheit)(British Thermal Unit)⁻¹(hr). The R values are convenient because they add up. That is, if a piece of a material with $R = 5$ is placed over a piece of some other material with $R = 3$ then the total value for the combination will be 8. An R value of 6 is equivalent to the insulating value of 6 inches of (still) air.

Convection

Convection is a process by which the physical movement of a fluid (liquid or gas) can transfer heat energy from one location to another. On a global scale this is most evident in many characteristic air and sea currents, such as the Gulf Stream, that prevail in different areas. Local heating of a mass of fluid, as at the equator, will cause a decrease in density of the fluid which will rise and be displaced away from the equator by colder, denser fluids moving from below. The rotation of the earth causes a deviation to the right in the Northern Hemisphere and to the left in the Southern Hemisphere as the fluids leave the region of the equator, giving rise to prevailing wind and sea current patterns.

Convective flows may be passive, as when caused by changes in densities, or forced, as when a pump or fan moves the fluids faster than might ordinarily occur. The circulation of the blood is a forced convection and carries deep body heat to our extremities. This can help prevent the extremities from freezing in cold weather, or serve as an aid in cooling the body in situations of heat stress. The thermal conductivity values in Table 14-5 for fat, muscle and skin are approximate values when there is no blood flow through these tissues. Heat losses can be much greater than these figures would predict when there is a generous flow of blood through skin and subcutaneous fat. The maximum flow rate of blood in the skin is about 100 times the minimum flow rate.

There is no single equation to calculate heat losses by convection. The convective losses from two identical cylindrical objects will be quite different if one object is horizontal and the other vertical. Engineers have empirical equations that work moderately well for many different shapes in different orientations.

Regardless of the object and its orientation, there will still be a linear dependence of heat-loss rate on area as in all other forms of heat loss. As well, there will be a dependence on the temperature difference between the fluid and the surface in contact with the fluid, but this dependence is not necessarily linear.

Radiation

Heat energy can also be transmitted or received by the process of *radiation*, in which electromagnetic energy is radiated from an object and may travel through a vacuum or transparent medium. In Chapter 4 it was pointed out that certain compounds (e.g., the chlorophylls) could absorb such radiation in preferred wavelength bands, or emit it in preferred portions of the spectrum such as in fluorescence. In those instances the concern was with the visible and near-visible portions of the spectrum. As far as heat energy is concerned, the vast bulk of the radiation is in the infrared portion of the spectrum. Most animals, including humans, are *blackbodies* as far as such radiation is concerned: this means that they are perfect radiators or absorbers of all wavelengths within the spectral region of interest. There is some variation away from the behaviour of a perfect blackbody but the deviations are not of great significance.

Planck determined that, for a blackbody, the *spectral emittance*, W watts per square metre per unit range of wavelength (i.e., W/m^3 or $J \cdot m^{-3} \cdot s^{-1}$), is given by:

$$W = \frac{2\pi h c^2}{\lambda^5} \frac{1}{e^{\frac{hc}{\lambda k_B T}} - 1}$$

[14-8]

where h is Planck's constant (6.626×10^{-34} J·s), c the speed of light (2.998×10^8 m/s), T the absolute temperature, k_B Boltzmann's constant (1.381×10^{-23} J/K), and λ the wavelength of the radiation. This relationship is shown in Fig. 14-3 where it can be seen that the total power radiated per square metre, as represented by the area under the curve, increases rapidly with temperature, and that the wavelength λ_m at which maximum spectral emittance occurs increases with decreasing temperature.

Figure 14-3

Blackbody radiation curves.

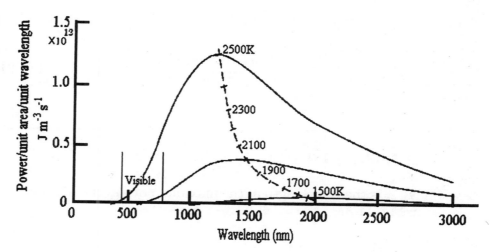

EXAMPLE 14-6

What is the spectral emittance for a perfect blackbody at a temperature of 2500 K at the peak of the radiation curve? The wavelength is 1.16 μm (as shown in Example 14-7).

SOLUTION

Using Eq. [14-8], and omitting units (all SI) for brevity,

$$W = \frac{2\pi hc^2}{\lambda^5} \frac{1}{e^{\frac{hc}{\lambda k_B T}} - 1}$$

$$= \frac{2\pi \left(6.626 \times 10^{-34}\right)\left(2.998 \times 10^8\right)^2}{\left(1.16 \times 10^{-6}\right)^5} \frac{1}{e^{\left(6.626 \times 10^{-34}\right)\left(2.998 \times 10^8\right)/\left(1.16 \times 10^{-6}\right)\left(1.381 \times 10^{-23}\right)(2500)} - 1}$$

$$= 1.782 \times 10^{14} \frac{1}{142.6 - 1}$$

$$= 1.26 \times 10^{12} \, \text{W/m}^3$$

This corresponds to the peak of the 2500-K curve in Fig. 14-3.

The wavelength of maximum spectral emittance is given by *Wien's law*:[3]

$$\lambda_m = \frac{2.897 \times 10^6 \, (\text{nm·K})}{T(\text{K})} \qquad \textbf{[14-9]}$$

EXAMPLE 14-7

What is the wavelength of the maximum of the 2500-K curve in Fig. 14-3?

SOLUTION

From Eq. [14-9],

$$\lambda_m = 2.897 \times 10^6 \, \text{nm·K} / 2500 \, \text{K} = 1159 \, \text{nm} = 1.159 \, \mu\text{m}$$

This result can be checked in Fig 14-3.

The shift of wavelength λ_m with increasing temperature explains why an object on heating will become red hot, then white hot and finally blue hot. At low temperatures, λ_m will be so far into the infrared portion of the spectrum that even the highest frequencies present will not extend into the visible region of the spectrum. As the temperature

---

3.   Wilhelm Wien (1864–1928), German physicist.

increases, the graph will shift towards higher frequencies and start to extend into the red end of the visible spectrum. This shift with increasing temperatures will eventually include all portions of the visible spectrum, and the resulting mixture of emitted wavelengths will create the sensation of whiteness.

The power density or intensity, I (W/m^2), radiated from the surface of a blackbody is the area under the Planck curve in Fig. 14-3. It is given by *Stefan's law:*[4]

$$I = \sigma T^4 \qquad \textbf{[14-10]}$$

where σ is the Stefan–Boltzmann constant,

$$\sigma = 5.671 \times 10^{-8} \text{ W·m}^{-2}\text{·K}^{-4}$$

For a non-blackbody, the right-hand side of the equation must be multiplied by a positive factor which approaches a maximum value of 1 as the object approaches the behaviour of a blackbody.

EXAMPLE 14-8

What is the power density radiated by a human body whose surface temperature is 30°C (303 K)?

SOLUTION

Using Stefan's law, Eq. [14-10]:

$$I = \sigma T^4 = 5.671 \times 10^{-8} \text{ W·m}^{-2}\text{·K}^{-4} (303 \text{ K})^4 = 478 \text{ W/m}^2$$

The power radiated per unit area in the previous example seems very large indeed. In practice the net radiation from a blackbody will rarely approach values of these magnitudes since the same blackbody will be receiving radiant energy from surrounding bodies according to the same fourth-power law. Thus, if a body is at one temperature, T_1, which is greater than that of the surroundings at T_2, the <u>net</u> intensity of radiation is:

$$I_{net} = \sigma(T_1^4 - T_2^4) \qquad \textbf{[14-11]}$$

---

4. Josef Stefan (1835–1893), Austrian physicist.

EXAMPLE 14-9

An absolute minimum metabolic rate might be that required to just maintain a body's surface temperature of 30°C in a room at 20°C. What is this rate if the body's surface area is 1.5 m²?

SOLUTION

The rate of heat loss P (in watts) is the product of the net intensity I_{net} (in watts per square metre) and the surface area $\mathcal{A}$ (in square metres):

$$P = I_{net}\,\mathcal{A}$$

Using $I_{net} = \sigma(T_1^4 - T_2^4)$, Eq. [14-11]:

$$P = \sigma\mathcal{A}(T_1^4 - T_2^4)$$
$$= 5.671\times10^{-8}\ \text{W·m}^{-2}\text{·K}^{-4}(1.5\ \text{m}^2)[(303\ \text{K})^4 - (293\ \text{K})^4]$$
$$= 90\ \text{W}$$

Newton's Law of Cooling

In real situations, heat exchange usually involves all or some combination of conduction, convection (free and forced) and radiation. Under any given set of conditions, the heat exchanged will be proportional to the area and depends in some way on the temperature difference. If the hotter body is that of a warm-blooded animal then metabolic activity may allow it to maintain a constant temperature, with a constant rate of heat loss to its environment if the environment itself does not change. On the other hand, a hot body that contains no internal source of heat will usually cool off according to *Newton's law of cooling*. This is an empirical relation which states that the rate of change of temperature will be proportional to the temperature difference between the body and its surroundings, i.e.,

$$\frac{dT}{dt} = -\kappa\left(T - T_s\right)$$

where T_s is the temperature of the surroundings and t is time. The κ is a constant of proportionality (decay constant) which will be different for each cooling situation. This equation can be rearranged and integrated over appropriate limits:

$$\int_{T_0}^{T} \frac{dT}{T - T_s} = -\kappa \int_{0}^{t} dt$$

The solution of this equation is:

$$T = T_s + (T_0 - T_s)e^{-\kappa t} \qquad \textbf{[14-12]}$$

This "law" is only an approximation, which works quite well in many cases. It does tend to break down if the temperature differences are large, especially if, at the same time, convection in still air is the major means of heat loss. It would work better for a cup of coffee in a breezy location than in still air.

EXAMPLE 14-10

A hot object cools from 100°C to 80°C in 10 minutes. How much longer will it take to cool to 50°C if the surroundings are at 20°C ?

SOLUTION

First, Eq. [14-12] is used with the information for the initial cooling period in order to find the decay constant, κ, for this particular situation. Note that temperatures either in degrees Celsius or in kelvins can be used with this equation.

$$T = T_s + (T_0 - T_s)\, e^{-\kappa t}$$

$$80°C = 20°C + (100°C - 20°C)\, e^{-\kappa t}$$

$$60°C = 80°C \times e^{-\kappa \times 10 \text{ min}}$$

Taking natural logarithms:

$$-\kappa \times 10 \text{ min} = \ln(0.75) = -0.288$$

$$\kappa = 0.0288 \text{ min}^{-1}$$

Now, again using Eq. [14-12], calculate the time to cool from 80°C to 50°C.

$$50°C = 20°C + (80°C - 20°C)\, e^{-0.0288/\text{min} \times t}$$

$$30°C = 60°C \times e^{-0.0288/\text{min} \times t}$$

$$-0.0288 \text{ min}^{-1} \times t = \ln(0.5) = -0.693$$

$$t = 24 \text{ min}$$

14.6 Biological Heat Regulation

The dependence of heat loss on surface area becomes relatively more important for smaller animals since they have a larger surface-to-volume ratio than do large animals. A 0.16-kg pigeon, for instance, loses in a day about 5×10^5 J/kg while the corresponding figure for a 680-kg steer is only about one-tenth of this. An isolated baby mouse in most situations would quickly die of hypothermia when the mother is away from the nest, but the babies will bunch up and become a single mass with a surface-to-volume ratio considerably smaller than for individuals.

Obviously, there are situations in which it is desirable for an animal to be able to lose heat, such as during exercise on a hot day, and other cases where the animal should conserve heat, as on a cold day, or in the case of the fish mentioned in the introduction to this chapter. Some animals have been more successful in developing appropriate mechanisms than have others.

The rabbit uses its ears extensively for heat regulation. The ears have a large surface for good heat exchange with the environment, and specialized connections, called arteriovenous anastomoses, between small arteries and veins. These can open up when the animal is overheated, allowing for greater blood flow than could be obtained through the capillaries alone. In extreme cold, the arteries themselves will constrict and the arteriovenous anastomoses will close down, markedly restricting blood flow and consequent heat loss. The northern rabbit does not have as great an overheating problem as does its cousin to the south, and there is a gradual increase in rabbit ear size to facilitate heat loss as one goes south. Such evolutionary developments are not only recent: there is now good evidence that the double row of bony plates along the back of the stegosaurus was, like the rabbit ears, a radiator to assist in heat regulation.

In the rat, the tail has developed as a heat exchanger which functions in essentially the same manner as the rabbit ear. There is a case on record where rats living in the loft above a henhouse in hot summer weather would sit with their tails hanging through knotholes in the loft floor into the cooler environment below, thus increasing their heat loss.

In the bluefin tuna, heat is retained within the body and consequently, muscle temperatures are higher; this may permit greater swimming speeds. If the blood passed directly from the heart to the gills, it would quickly lose to the environment any deep body heat it had acquired. Instead, the vessels leading to the gills break up into a network of fine vessels which intertwine with a similar network of fine vessels returning from the gills. This is called the rete mirabile (literally: the marvellous net) and acts as a heat exchanger. The cold blood coming from the gills is warmed by the blood going to the gills and returns much of the heat to the deep swimming muscles, rather than letting it proceed to the gills to be lost to the environment.

Heat exchangers can be built as parallel flow systems in which the hottest and coldest fluids are brought together at one end and both emerge as warm fluids at the other end, or as counterflow systems in which the hot and cold fluids enter at opposite ends and pass each other. Engineers have found that this second system is the more efficient. The bluefin tuna apparently "discovered" the same thing millions of years earlier. Still another example of nature's use of counterflow heat exchange is found in the flippers of whales, where the outgoing major arteries and the returning veins are in very close proximity. Thus some of the heat energy going into the cooler flipper is used to warm the returning, slightly chilled, blood. The same situation prevails in the case of wading birds whose feet are immersed in cold water most of the time.

Large nocturnal sphinx moths can fly best when their flight muscle temperature is about 40°C. At rest, the moth will be at the temperature of its environment, which is frequently low enough to make immediate flight impossible. If the resting moth is aroused, its flight muscles begin a violent shivering motion and warm up to the optimum value at about 10°C per minute. The surface-to-volume ratio of the body is very large and this would normally result in so much heat loss that the moth would have difficulty generating enough heat to both raise its temperature and compensate for losses to its surroundings. However, the thorax (which must be warmed because it contains the flight muscles) is separated from the large abdomen by an insulating air space. Thus, the abdomen surface does not contribute to heat loss when the moth is warming up.

Once in flight, however, the temperature buildup cannot continue much beyond 40°C or thermal denaturation of the proteins would begin. When a desirable maximum temperature is reached in the flight muscles, the heart rate suddenly increases markedly. This causes a sudden increase in the exchange of blood from the warm thorax to the cool abdomen, where it loses heat and is pumped back into the thorax.

Most insects lack the elaborate temperature-controlling mechanism of the sphinx moth and are very sluggish at low temperatures. However, the sphinx moth is not unique, as bees can also maintain elevated thorax temperatures of about 37°C even when the ambient temperature is approaching zero. Bees also use a flight-muscle-shivering type of mechanism and thermally isolate the abdomen from the thorax while warming up. While the bee rests on a flower, the shivering usually continues so that the bee is ready for instant flight. In cool fall weather, the nectar sources consist of flowers like goldenrod, in which many small florets in a panicle each provide minute amounts of the energy-supplying nectar. Here it becomes energetically impractical to maintain an elevated temperature for the considerable time it takes the bee to climb all over the panicle; consequently the bee stops shivering and conserves energy until a last-minute warm-up is required for flight to the next panicle.

14.7   **First Law of Thermodynamics**

When a gas increases in temperature, the molecules experience an increase in energy. In the case of a monatomic gas such as helium, only the average translational kinetic energy of the molecules will increase. To heat a diatomic gas molecule such as oxygen, which can have not only translational kinetic energy, but also vibrational and rotational energy, more energy is required for a comparable increase in temperature. More complex molecules can have even more vibrational modes and require even greater energy input for the same rise in temperature. The energy contained in the gas as a result of these motions is known as its *internal energy, U*. At the same time as heat energy is added (or removed), the gas may expand and do work (as in an internal combustion engine) or the gas may be compressed as work is done on it.

In any process involving a gas, there must be a careful balancing of the changes in heat exchange, internal energy, and work done on the gas. This balance, which is one way of expressing the law of conservation of energy, is represented by the equation:

$$\Delta U = \Delta Q + \Delta W \qquad\qquad \textbf{[14-13]}$$

which is known as the *first law of thermodynamics*. In this equation, ΔQ is positive when heat is added and negative when heat is removed. The work done <u>on</u> the gas, ΔW, is positive if the gas is compressed, and negative if the gas expands. In essence, Eq. [14-13] states that there are two ways of changing the internal energy of a gas: add (or subtract) heat, or do work on the gas.

EXAMPLE 14-11

A quantity of gas is heated by the addition of 6000 J of heat energy and, in expanding, does 3500 J of work. What will be the increase of internal energy of this gas?

SOLUTION

Since heat has been added to the gas, $\Delta Q = +6000$ J. As the gas expands, it does positive work on its environment (perhaps by forcing a piston down in an engine), and hence negative work has been done on the gas itself by the environment. Thus, $\Delta W = -3500$ J.

Using Eq. [14-13],

$$\begin{aligned} \Delta U &= \Delta Q + \Delta W \\ &= 6000 - 3500 \text{ J} \\ &= 2500 \text{ J} \end{aligned}$$

Exercises

14-1   Approximately how much heat would be required to raise the temperature of a 3-kg roast of beef from room temperature to fully cooked (80°C)?

14-2   On British sailing ships, barrels of tar for caulking seams were melted by plunging a hot iron ball on the end of a rod (called a "loggerhead") into the cold tar. A barrel has a volume of 35 gal. and holds 160 kg of tar which has a specific heat capacity of 2100 J·kg⁻¹·K⁻¹. The iron ball (density 7870 kg/m³, specific heat capacity 502 J·kg⁻¹·K⁻¹) was 20 cm in diameter and heated in an oven to 800°C (dull red). The tar starts at 20°C and is a liquid (although very viscous) at that temperature. What is the final temperature of the tar?

14-3   A 70-kg man when working hard might generate 230 W of power over and above the power generated while resting. Assuming all of this excess power generated by the hard work is removed by the evaporation of water (i.e., sweat), how much water would be evaporated in one hour? Take body temperature (the initial temperature of the water) as 37°C.

14-4   While a person is sitting and resting, metabolic processes generate heat at a rate of about 60 W/m². If all of the heat from a person could be transferred to a kilogram of water initially at 20°C, how long would it take to bring the water to the boiling point? The surface area of a person is about 1.5 m².

14-5   3.0 kg of crushed ice at 0°C is mixed with 3.0 kg of water at 20°C in an insulated container.

(a) What is the temperature at equilibrium?

(b) How much ice remains at equilibrium?

14-6   A steel-girder bridge 30 m long spans a river. What minimum gap must be left between one end of the bridge and the pavement if the bridge is put in place in the winter when the temperature is 0°C, and the maximum temperature envisaged in the summer is 38°C? The coefficient of expansion of steel is $11 \times 10^{-6} \, K^{-1}$.

14-7   The oceans of the world cover three-quarters of the earth's surface and have an average depth of 3.8 km. If the temperature of the oceans were to rise by 1.0 K, how much would the sea level rise on average due to thermal expansion alone? (Assume that the surface area of the oceans would not change. Write the volume of the oceans as area times depth, and consider a volume change with area held constant.)

14-8   The door of a refrigerator is 1.5 m high, 0.80 m wide and 6.0 cm thick. Its thermal conductivity is 0.21 W·m⁻¹·°C⁻¹. The inner surface is at 0°C and the outer is at 28°C. What is the heat loss per hour through the door? (Neglect convection effects.)

14-9   The mouth of a thermos bottle is closed with a cork 3.0 cm long and 3.0 cm in diameter. If there is an ice-water mixture in the thermos which sits in a room at 25°C, how much heat energy would be lost through the cork in 6.0 hours? The thermal conductivity of cork is 0.050 W·m⁻¹·K⁻¹.

14-10   It has been found experimentally that the heat lost per second by convection from a window, when the outside window surface is 20°C warmer than the outside air and when there is no wind, is:

$$P = 3.75 \, \mathcal{A} \, \Delta T^{1/4}$$

where P is in watts when $\mathcal{A}$ is in square metres and ΔT in kelvins represents the temperature difference between the outside window surface and the outside air. The coefficient of thermal conductivity of the glass is 0.84 W·m⁻¹·K⁻¹. What is the temperature difference across a window 3.0×10^{-3} m thick when the outside surface is at a temperature of 0°C and the outside air at a temperature of –20°C?

14-11   A mare and her colt are in a pasture on a cold day. The mare is three times as high as the colt. What is the ratio (for heat convection) of the heat lost by the mare to that lost by the colt?

14-12   A blackbody has a temperature of 5000 K. Compute the ratio of its spectral emittance at $\lambda = 550$ nm (in the visible) to its spectral emittance at $\lambda = 2000$ nm (in the infrared).

14-13   At what wavelength does your body have its maximum spectral emittance? (Assume a body surface temperature of 30°C.) In which region of the electromagnetic spectrum is this wavelength? (Refer to Fig. 3-2.)

14-14   A blackbody at 1000 K sits in a room at 27.0°C.

   (a) What is the wavelength of maximal spectral emittance of the body?

   (b) What is the intensity radiated by the body?

   (c) What is the net intensity radiated by the body?

14-15   What must be the minimum metabolic rate for a human exposed to still air at 0°C? Assume that the person's surface area is 1.5 m². Compare with Example 14-9.

14-16   A pathologist is investigating a fresh murder and records the core temperature of the corpse as 3.0°C at 10:00 pm. The body has been found outside where the temperature is 0.0°C. The pathologist knows that a fresh corpse in such an environment cools 10°C in the first half hour. When was the murder committed?

14-17   An object initially at 30°C in a room at 20°C cools off to 25°C in 15 minutes and to 22°C in an additional 25 minutes. Is Newton's law of cooling obeyed?

Problem

14-18   Two copper plates each 0.50 cm thick have a 0.10-mm sheet of glass sandwiched between them. The outer surface of one copper plate is kept in contact with flowing ice water, and the outer surface of the other is in contact with steam at 100°C. What are the temperatures of the two copper–glass interfaces, and what power is transferred through a 10 cm × 10 cm area? (Thermal conductivities: Cu, 384 W·m⁻¹·K⁻¹; glass, 1.05 W·m⁻¹·K⁻¹)

Answers

14-1   approx. $6–7\times10^5$ J

14-2   57°C

14-3   0.33 kg

14-4   1.0 hr

14-5   (a) 0°C (b) 2.2 kg

14-6   1¼ cm

14-7   0.76 m

14-8 4.2×10^5 J

14-9 6.4×10^2 J

14-10   0.028°C

14-11   9:1

14-12   11:1

14-13 9.6×10^3 nm, infrared

14-14   (a) 2.9×10^{-6} m (b) 5.67×10^4 W/m^2 (c) 5.63×10^4 W/m^2

14-15 2.4×10^2 W

14-16   6:00 pm

14-17   No. The decay constant for the first interval is 4.6×10^2 min^{-1}, and for the second interval is 3.7×10^{-2} min^{-1}.

14-18   11°C, 89°C, 8.2×10^3 W

15 Introduction to Electricity

15.1 Introduction

The relative positions of organisms on the evolutionary ladder are often correlated with their abilities to receive and process information from their environment. These abilities are directly related to the complexity of their sensory and nervous systems. Basic to these nervous systems is their function: the controlled flow of information from one part of the organism to another. The information is in the form of electrical signals, called nerve impulses, which arise from precisely determined movements of ions (current) through membranes of specially designed cells. While the fields of electricity and electronics are primarily concerned with the flow of electrons, not ions, certain aspects of these subjects are of great value in helping us understand the mechanisms involved in the production and propagation of nerve impulses. This chapter will provide the foundation for the study of any electrical phenomenon.

15.2 Electric Fields and Forces

Before proceeding with the discussion of currents, it must be realized that the motion of charged particles is influenced by electrical forces arising from the presence of other electrical charges nearby. The SI unit for a quantity of electrical charge is the *coulomb* (C).[1] The smallest elementary charge (i.e., that on an electron) has an absolute value of 1.602×10^{-19} C. The symbol for a general electrical charge is q. Since separated charges exert forces on each other, we imagine that about any isolated point charge there is a radially symmetric *electric field* ($\vec{E}$) of magnitude:

$$E = \frac{k|q|}{Kr^2}$$

[15-1]

_____

1.   Charles Augustin de Coulomb (1736–1806), French physicist.

Here the SI unit of the electric field is newtons per coulomb (N/C), r is the distance in metres from the charge q to the point where the electric field is to be measured, and K is the *dielectric constant*, which is a dimensionless quantity that is a property of the material where the field E is measured. The *Coulomb constant*, k, equals 8.988×10^9 N·m²/C². Note that if K is very large, then the field due to the charge is relatively small.

Table 15-1 contains dielectric constants for some common biological and non-biological substances. Note also the very high value for water, a property which in aqueous solution makes the electric fields due to electric charges much reduced relative to other media.

Table 15-1   Dielectric Constants of Selected Materials near 20°C

Substance	K
Air	1.0005
Transformer oil	2.10
*Palmitic acid	2.30
*Stearic acid	2.30
Beeswax	2.80
Wood (approximate)	5.0
Membrane (approximate)	6.0
Casein	6.5
Ethanol	26.8
Water	80.4

* Biological membrane components

Electric field is a vector quantity and has a direction defined by the direction that a small <u>positive</u> *test charge* would move when placed in the field.[2] Like charges (++ or −−) repel and unlike charges (+−) attract; consequently, the positive test charge would be repelled from any positive charge and attracted to any negative charge. Thus the electric field points away from a positive, or toward a negative, charge as shown in Fig. 15-1a,b. When two or more charges are close to each other, the resulting electric field is the vector sum of the fields due to

---

2.   In electric theory, all equations are written with positive charges in mind. When negative charges are encountered, there will be a sign change in the equations.

Figure 15-1

The electric field surrounding (a) an isolated positive charge; (b) an isolated negative charge; (c) two positive charges close to each other; (d) a positive and a negative charge close to each other in an arrangement known as an electric dipole.

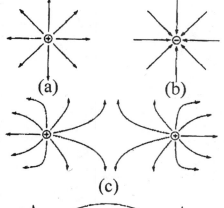

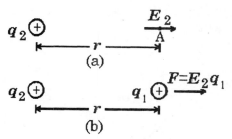

Figure 15-2

(a) Electric field at A due to charge q_2; (b) Force felt by a positive charge q_1 placed at A.

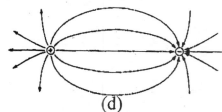

the individual charges, and may become quite complex as shown in Fig. 15-1c,d.

Consider a positive charge q_2. The magnitude of its electric field E_2 at point A at a distance r will be $E_2 = kq_2/Kr^2$ as shown in Fig. 15-2a. In general, when a charge q is placed at a point where there is an electric field $\vec{E}$, the force $\vec{F}$ experienced by the charge is given by

$$\vec{F} = q\vec{E} \qquad \text{[15-2a]}$$

Along the line joining the two particles, the magnitude of this force is given by

$$F = |q|E \qquad \text{[15-2b]}$$

In Fig. 15-2b, suppose that a charge q_1 is now placed at A. This charge will experience a force of magnitude $F = |q_1|E_2$ (from Eq. [15-2b]). Combining this with the expression for the magnitude of E_2 (using Eq. [15-1]) leads to *Coulomb's law* for the magnitude of the force:

$$F = \frac{k|q_1 q_2|}{Kr^2} \qquad \text{[15-3]}$$

The charge q_2 would feel an equal and opposite force, by Newton's third law of motion. If the product $q_1 q_2$ is positive, the force will be repulsive; whereas if $q_1 q_2$ is negative, the force will be attractive. It is interesting to compare Eq. [15-3] with Eq. [8-6], which describes the gravitational force between two masses. Although the origins of the interactions are different, the mathematical form of the equations is almost identical.

Equation [15-3] gives the magnitude of the force on a charged particle (such as an ion) due to a single nearby charge. This charge could be on a biopolymer or on another ion. The microscopic environment of biological membranes contains large numbers of ions and charged polymers. The net force on any one ion is the vector sum of the electric forces due to each of the surrounding charges. However, each individual force is a vector quantity and has a direction (due to the signs of the two charges involved) and a magnitude (due to the proximity and sizes of the charges). To illustrate this, suppose there is a distribution of ions as shown in Fig. l5-3a, and the magnitude and direction of the force on a positive charge at point x is to be determined. The magnitudes and directions of the forces at x due to each charge are shown and then added vectorially in Fig. 15-3b to produce the resultant force of magnitude F_T.

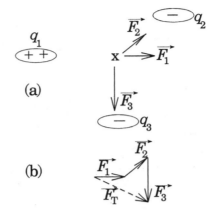

Figure 15-3

The electric force due to several charges: (a) the distribution of ions and the magnitude and direction of forces on a charge at point x; (b) the addition of the force vectors to give F_T.

EXAMPLE 15-1

Find the magnitude and direction of the resultant force on the sodium ion (Na^+) in the environment of calcium ions (Ca^{2+}) and chloride ions (Cl^-) shown in Fig. 15-4. All ions lie in the same plane, and the solvent is water.

SOLUTION

Start by determining the magnitudes of the individual fields created at the location of the sodium ion by each of the other three charges (using Eq. [15-1]):

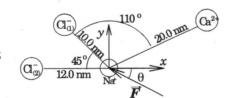

Figure 15-4

Example 15-1: Force on a sodium ion.

$$E_{Ca^{2+}} = \frac{k|q|}{Kr^2} = \frac{8.988\times10^9 \text{ N·m}^2/\text{C}^2\left(2\times1.602\times10^{-19}\text{ C}\right)}{80.4\left(20.0\times10^{-9}\text{ m}\right)^2}$$

$$= 8.95\times10^4 \text{ N/C}$$

$$E_{Cl^-_{(1)}} = \frac{8.988\times10^9 \text{ N·m}^2/\text{C}^2\left(1.602\times10^{-19}\text{ C}\right)}{80.4\left(10.0\times10^{-9}\text{ m}\right)^2}$$

$$= 17.9\times10^4 \text{ N/C}$$

$$E_{Cl^-_{(2)}} = \frac{8.988\times10^9\,\text{N}\,\text{m}^2/\text{C}^2\left(1.602\times10^{-19}\,\text{C}\right)}{80.4\left(12.0\times10^{-9}\,\text{m}\right)^2}$$

$$= 12.4\times10^4\,\text{N/C}$$

These fields can then be added vectorially using geometric methods or, more accurately, by components. If components are used, x- and y-axes are required. For convenience, let the line joining the second chloride ion and the sodium ion be the x-axis with origin at the sodium ion and the $+x$-direction to the right. Choose the $+y$-direction to be toward the top of the page. The x- and y-components of the three electric fields will then be as in Table 15-2 which also shows the net component along each of the axes.

Table 15-2   Calculation of Field Components

	x-component	**y-component**
2nd Chloride ion	-12.4×10^4 N/C	0
1st Chloride ion	-17.9×10^4 N/C cos(45.0°) $=-12.7\times10^4$ N/C	$+17.9\times10^4$ N/C sin(45.0°) $=+12.7\times10^4$ N/C
Calcium ion	-8.95×10^4 N/C cos(25.0°) $=-8.11\times10^4$ N/C	-8.95×10^4 N/C sin(25.0°) $=-3.78\times10^4$ N/C
Sum of components	-33.2×10^4 N/C	$+8.9\times10^4$ N/C

The magnitude of the total electric field is then:

$$E = 10^4\times\sqrt{\left(-33.2\,\text{N/C}\right)^2 + \left(8.9\,\text{N/C}\right)^2}$$

$$= 34.4\times10^4\,\text{N/C}$$

Equation [15-2b] ($F = |q|E$) can then be used to calculate the magnitude of the force on the sodium ion:

$$F = 1.602\times10^{-19}\,\text{C} \times 34.4\times10^4\,\text{N/C} = 5.51\times10^{-14}\,\text{N}$$

The direction of this force is at an angle θ with respect to the x-axis where

$$\tan(\theta)= (8.9/33.2)=0.268$$

$$\theta = 15°$$

15.3 Electric Potential Energy and Potential

Work *(W)* must be done to move any charged particle in an electric field. Hence, any charged species in an electric field has a potential energy equal to the work done placing it there. Consider for example the work done in bringing a single charge (such as on an ion of charge q_1) into the electric field, $\vec{E}$, of a fixed charge, q. Just as work must be done to lift an object from the surface of Earth against the force of gravity with a resultant increase in the gravitational potential energy of the object, so must work be done to move one electric charge closer to another of the same sign against the electrical force of repulsion, with a resultant increase in the *electric potential energy* of the charge. If q_1 were initially at infinity and then brought to some closer distance r from q, then the work done would be given by:

$$W = - q_1 \int_{\infty}^{r} E \, dr \qquad \text{[15-4]}$$

The minus sign in Eq. [15-4] arises because the force required to move q_1 is opposite to the force exerted by the electric field. The work done on the charge is also its increase in electric potential energy and W can be replaced by U, the symbol for potential energy. The integration must be performed because the electric field due to the fixed charge varies with position (i.e., it gets larger as the moving charge gets closer to the fixed charge). If both q and q_1 are of the same sign, then the potential energy at r would be positive and q_1 would accelerate away from q if released. When the charges are opposite, the potential energy at r would be negative and q_1 would accelerate toward q upon release. If q_1 is released when at infinity, it will not move because it has no potential energy there.

$$\frac{dy}{dx}$$

Inserting Eq. [15-1] for E into Eq. [15-4] and integrating yields:

$$U = \frac{k \, q \, q_1}{K r} \qquad \text{[15-5]}$$

This equation is often written:

$$U = V q_1 \qquad \text{[15-6]}$$

where V, the *electric potential*, is:

$$V = \frac{kq}{K r} \qquad \text{[15-7]}$$

Electric potential has units of joules per coulomb (J/C), which is given the name *volt*[3] (abbreviated as V). Note that electric potential is a property of the electric field due to charge q and is independent of the presence or absence of a charge such as q_1. Equation [15-7] indicates that the electric potential at an infinite distance from a charge q will be zero.

The electric potential at a particular point in a collection of charged particles is not a vector quantity as was the electric field. It is the scalar sum of all the potentials V_i due to the particles at distances r_i from the point of interest. Thus,

$$V = \sum V_i \qquad [15\text{-}8]$$

Experimentally, only *voltage differences* represented by ΔV can be measured. A scientist might measure a voltage difference across a cell membrane very accurately but would have no idea what the absolute value is on one side or the other. The *electric potential difference* (which is the same as voltage difference) between two points is defined as one volt when the amount of work done while moving one coulomb of charge from one point to another is one joule.

EXAMPLE 15-2

(a) Find the electric potential at each of the points A, B and C in Fig. 15-5. Assume $K = 1.00$.

(b) Find the work required to bring a sodium ion (Na^+) to A, B and C from infinity.

(c) Find the electric potential energy of a sodium ion at each of the three points.

(d) Redo (c) with a chloride ion (Cl^-) substituted for the sodium.

(e) How much work would be required to move the sodium ion from A to B to C and then back to A?

Figure 15-5

Example 15-2: Electric potential at A, B, C.

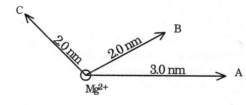

3.   Count Alessandro Volta (1745–1827), Italian physicist.

SOLUTION

(a) Using Eq. [15-7],

$$V = \frac{kq}{Kr}$$

$$V_A = \frac{8.988 \times 10^9 \, \text{N·m}^2/\text{C}^2 \left(2 \times 1.602 \times 10^{-19} \, \text{C}\right)}{1.00 \times \left(3.00 \times 10^{-9} \, \text{m}\right)} = 0.960 \text{ V}$$

$$V_B = V_C = \frac{8.988 \times 10^9 \, \text{N·m}^2/\text{C}^2 \left(2 \times 1.602 \times 10^{-19} \, \text{C}\right)}{1.00 \times \left(2.00 \times 10^{-9} \, \text{m}\right)} = 1.44 \text{ V}$$

(b) Since the potential energy at infinity is zero, the work done on the charge, which is equal to its change in potential energy, will equal its final potential energy. Using Eq. [15-6] ($U = V q_1$),

$$W_A = U_A = V_A q_1 = 0.960 \text{ V} \times 1.602 \times 10^{-19} \text{ C}$$
$$= 1.54 \times 10^{-19} \text{ J}$$

Similarly:

$$W_B = W_C = V_B q_1 = 1.44 \text{ V} \times 1.602 \times 10^{-19} \text{ C}$$
$$= 2.31 \times 10^{-19} \text{ J}$$

(c) This is a different way of wording (b); the electric potential energy of a charge at a particular point is the same as the work done in moving the charge from infinity to that point. Hence, the answers for (b) and (c) are the same.

(d) If the sodium ion were replaced by a chloride ion, the magnitudes of the answers would be the same but the energies would be negative, indicating a spontaneous movement from infinity with a resultant loss of energy. This is because the opposite charges are attracting each other and electric potential energy is lost as they move closer to each other, just as gravitational potential energy is lost when an object falls to Earth.

(e) Consider the first step from A to B. The potential energy at A is 1.54×10^{-19} J, and the potential energy at B is 2.31×10^{-19} J. Since the potential energy at B is greater than at A, work must be done that is equal to the increase in potential energy.

$$W_B - W_A = 2.31 \times 10^{-19} \text{ J} - 1.54 \times 10^{-19} \text{ J} = 0.77 \times 10^{-19} \text{ J}$$

Alternatively:

$$
\begin{aligned}
W_B - W_A &= W_{BA} \\
&= V_B q_1 - V_A q_1 \\
&= (V_B - V_A) q_1 \\
&= V_{BA} q_1 \\
&= (1.44 \text{ V} - 0.960 \text{ V})(1.602 \times 10^{-19} \text{C}) \\
&= 0.77 \times 10^{-19} \text{ J}
\end{aligned}
$$

Similarly:

$$W_C - W_B = W_{CB} = (V_C - V_B) q_1 = 0 \text{ V} \times 1.602 \times 10^{-19} \text{ C} = 0 \text{ J}$$

and:

$$
\begin{aligned}
W_A - W_C = W_{AC} &= (V_A - V_C) q_1 \\
&= (0.960 \text{V} - 1.44 \text{V})(1.602 \times 10^{-19} \text{ C}) \\
&= -0.77 \times 10^{-19} \text{ J}
\end{aligned}
$$

The total work =

$$
\begin{aligned}
W_{BA} &+ W_{CB} + W_{AC} \\
&= 0.77 \times 10^{-19} \text{J} + 0 \text{ J} - 0.77 \times 10^{-19} \text{J} \\
&= 0 \text{ J}
\end{aligned}
$$

Example 15-2(e) demonstrates a principle that can be proven generally. The work done in moving a charge from point A to point B in an electric field is independent of the path followed and depends only on the coordinates of the end points. A corollary of this says that the work done in carrying a charge around a closed path in the presence of an electric field is zero. Such fields are called *conservative fields*. The gravitational field is another conservative field.

15.4   Capacitance

A device consisting of a pair of parallel conducting plates separated by a thin layer of insulating material is frequently encountered in electrical circuits where there are changing currents. Such a component is called a *capacitor*. It is very useful as a means of storing electric charges. Positive charges collect on one plate and negative on the other, and the charges distribute uniformly across each plate. A uniform electric field occurs between the plates, and the electric field lines are parallel straight lines from charges on the positive plate to those on the negative.

A biological cell membrane can have charges on both the exterior and interior surfaces, which are analogous to the conducting plates. The surfaces are separated by the fatty acid tails of lipid molecules, and these have dielectric constants comparable to those of the materials between the plates of conventional capacitors. The membrane can therefore function as a capacitor.

As might be expected, the charge q that can be stored in a capacitor depends on the potential difference ΔV across it; in fact, it depends directly on ΔV. Thus a given capacitor can store a charge q given by

$$q = C\Delta V \qquad \textbf{[15-9]}$$

where q is the charge on the positive plate (and $-q$ is the charge on the negative plate). ΔV is the potential difference between the plates, and the constant of proportionality, C, is called the *capacitance* of the device. The capacitance is a function of the area $\mathcal{A}$ of each plate, the plate separation Δx, and the dielectric constant K of the material between the plates. From Eq. [15-9], the capacitance has units of coulombs per volt (C/V) and this combination is called the *farad*,[4] F. Many real capacitors have capacitances in the microfarad (μF) range.

For a parallel-plate capacitor with a material of dielectric constant K between the plates, it can be shown that

$$C = \frac{K\,\mathcal{A}}{4\pi\,k\Delta x} \qquad \textbf{[15-10]}$$

where k is the Coulomb constant.

The capacitance per unit area is:

$$\frac{C}{\mathcal{A}} = \frac{K}{4\pi\,k\Delta x} \qquad \textbf{[15-11]}$$

---

4.   Michael Faraday (1791–1867), English physicist and chemist.

EXAMPLE 15-3

A capacitor has plates separated by 1.0 mm of material with a dielectric constant of 6.0.

(a) What is the capacitance per unit area?

(b) If the capacitor has an area of 0.010 m², and a potential difference of 100 V is applied across the plates, what charge would be stored on it?

SOLUTION

(a) Using Eq. [15-11],

$$\frac{C}{\mathcal{A}} = \frac{K}{4\pi k \Delta x}$$

$$\frac{C}{\mathcal{A}} = \frac{6.0}{4\pi \times 8.988 \times 10^{9} \text{ N} \cdot \text{m}^{2}/\text{C}^{2} \times 1.0 \times 10^{-3} \text{ m}}$$
$$= 5.3 \times 10^{-8} \text{ F}/\text{m}^{2}$$

(b) Since the capacitance per unit area ($C/\mathcal{A}$) has been calculated, the capacitance can be determined by multiplying $C/\mathcal{A}$ by the area $\mathcal{A}$:

$$C = 5.3 \times 10^{-8} \text{ F}/\text{m}^{2} \times 1.0 \times 10^{-2} \text{ m}^{2} = 5.3 \times 10^{-10} \text{ F}$$

The charge stored on the capacitor is then given by Eq. [15-9]:

$$q = C \, \Delta V = 5.3 \times 10^{-10} \text{ F} \times 100 \text{ V} = 5.3 \times 10^{-8} \text{ C}$$

There would be this much positive charge on one plate, and an equal negative charge on the other.

15.5   Current, Resistance and Ohm's Law

Current

Electric current (represented by the symbol *I* and measured in coulombs per second or *amperes*, A)[5] is one of the fundamental quantities in physics (see Box 15-1). It is most properly defined in terms of the force developed between parallel wires carrying a specific current. For most purposes it is more useful to define electric current as the charge *q* that flows divided by the time *t* taken for the flow; that is, a current is one ampere when one coulomb of charge passes a given cross-sectional area in one second.

$$I = \frac{q}{t}$$ [15-12]

Sometimes when working with cell membranes it is especially convenient to use *current density* instead of current. Current density, *J*, is just the current per unit area flowing across a surface that is at right angles to the flow. The surface could be, for example, a biological membrane.

$$J = \frac{I}{\mathcal{A}} = \frac{q}{\mathcal{A}t}$$ [15-13]

The usual convention is to consider current direction to be in the direction of flow of positive charges, even though the actual charge carriers might be electrons or other negatively charged particles flowing in the opposite direction. If positively charged ions were flowing into a living cell across its membrane, the current would be considered to be inwards; if negatively charged ions were entering the cell, the current direction would be said to be outwards.

Box 15-1   Current Affairs—A Shocking Story!

Everyone has walked across a carpet on a dry day and been surprised by being "zapped" when reaching out to a metal door handle. The resultant shock may be slightly painful in the worst of cases. What sort of situation will result in a very painful, perhaps even lethal, shock?

(cont'd.)

---

5.   André Marie Ampère (1775–1836), French mathematician and physicist.

Obviously controlled experiments are not practical, but all the evidence suggests that the crucial factors are the magnitude of the current passing through the body and the part of the body through which the current passes. There are significant variations in person-to-person susceptibility, depending on such factors as skin moisture and thickness. External factors, primarily the extent to which the body is grounded (i.e., connected to the infinite capacity of the earth), are also obviously very important.

As little as one milliampere (10^{-3} A) of current flowing through the body is usually sufficient to cause a tingling. As current increases, muscles start to respond, usually by contracting. If the palm of a hand is in contact with the source of current, the muscle contraction may cause the hand to grasp the source more tightly and prevent it from releasing the source. The current at which this happens varies from person to person, but seems to be about 15 mA for most men, 10 mA for most women and as little as 5 mA for many children. If an alternating current source is encountered, the frequency is also a factor in this failure to release. It appears that the frequency to which we are most sensitive is the common 60-Hz frequency of our usual sources.

By 50 mA, severe damage may occur with symptoms such as pain, muscle damage, fainting, and difficulty in breathing. If the current increases to around 100 mA, ventricular fibrillation (a random beating of the heart) and death may result if the current actually passes through the heart; again, children are more sensitive. For shocks of very brief duration (i.e., about one-fifth of the duration of a heartbeat) there may very well be survival as the heart will fibrillate only if the shock occurs in a particular one-fifth of the heartbeat (i.e., at the time of what is known as the "t pulse"). Thus there is an 80% chance of survival for such very short shocks—a new version of Russian roulette! At very high currents such as 3 A there will be serious damage, but rather surprisingly the chances of survival are better than at 100 mA.

To receive a 100-mA current from an ordinary 115-V household source, the resistance of the body from source to ground must be about 1150 Ω. This low value implies that the person is making a strong contact with the source and is well grounded as in a bath or swimming pool.

Ohm's Law and Resistance

Experiments show that for most materials, the current through the material is proportional to the potential difference across the material. *Ohm's law*[6] states this relationship by introducing a constant of proportionality called *electric resistance, R*, through the equation:

$$\Delta V = IR \qquad \qquad \textbf{[15-14]}$$

Equation [15-14] (rewritten as $R = \Delta V/I$) actually defines what is meant by the resistance of an object. The ratio $\Delta V/I$ has units of volt/ampere (V/A), but this is given the special name *ohm* (Ω), which is the SI unit of resistance.

If for a given object the resistance is constant (at room temperature) regardless of the size of the applied voltage ΔV, then the material is said to be *ohmic*. For instance, metal wires at normal temperatures are ohmic; the temperature must be raised by a large amount to detect a dependence of the resistance on temperature. Examples of non-ohmic objects include semiconductors and transistors. Such dependencies will be ignored here.

Series and Parallel Resistors

In Fig. 15-6, portions of electric circuits are shown. The straight lines represent wires of negligible resistance and the sawtooth lines represent resistors. Currents would flow as indicated if a potential difference ΔV were applied from A to D in each case such that V_A is greater than V_D. This difference can be represented by the notation $V_{AD} = V_A - V_D$, which means the voltage at A relative to that at D.

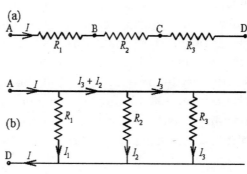

Figure 15-6

Resistors R_1, R_2 and R_3 (represented schematically by sawtooth lines) joined by wires (represented by solid lines) which are assumed to be perfect conductors with $R = 0$: (a) a series network, (b) a parallel network.

In Fig. 15-6a there is only one path for current from A to D. There is no opportunity to gain or lose current[7] in between, and consequently the current I entering at A is the same as the current through each

---

6.   Georg Simon Ohm (1787–1854), German physicist.

7.   This follows from the fact that charge cannot be created or destroyed, i.e., charge is conserved.

resistor, i.e., $I = I_1 = I_2 = I_3$. If the same current flows through a number of resistors, the resistors are said to be connected in "series."

In Fig. 15-6b, current I entering at A divides up into three smaller currents, I_1, I_2, and I_3, which then rejoin and exit at D. In this case, $I = I_1 + I_2 + I_3$. Note that the same potential difference is applied to each of the three resistors. If resistors are connected so that the current splits up into separate branches, as in Fig. 15-6b, so that the same potential difference (voltage) exists across each resistor, the resistors are said to be connected in "parallel." Note that this type of circuit has a number of branching points. At any branch point the total current entering must equal the total current leaving.

Ohm's law (Eq. [15-14]) can be used to examine the series circuit of Fig. 15-6a:

$$V_{AB} = IR_1 \qquad V_{BC} = IR_2 \qquad V_{CD} = IR_3$$

Since the potential difference across the whole chain is the sum of the potential differences across the individual resistors then:

$$V_{AD} = V_{AB} + V_{BC} + V_{CD} = IR_1 + IR_2 + IR_3$$

Therefore,

$$V_{AD} = I\,(\,R_1 + R_2 + R_3\,) = IR$$

where R is the single equivalent resistance that could replace the series combination, and the rule for combining series resistances is:

(Series) $\qquad\qquad R = R_1 + R_2 + R_3, \text{ i.e., } R = \Sigma R_i$ **[15-15]**

In Fig. 15-6b, the same voltage V_{AD} is being applied to each resistor, and Ohm's law can be rearranged as:

$$I_1 = V_{AD}/R_1 \qquad I_2 = V_{AD}/R_2 \qquad I_3 = V_{AD}/R_3$$

The total current entering at A equals the sum of the currents through the resistors:

$$I = I_1 + I_2 + I_3$$

Therefore,

$$I = V_{AD}\left(\frac{1}{R_1} + \frac{1}{R_2} + \frac{1}{R_3}\right) = V_{AD}\left(\frac{1}{R}\right)$$

where R is the single equivalent resistance that could replace the parallel combination, and the rule for combining resistances in parallel is:

(Parallel)

$$\frac{1}{R} = \left(\frac{1}{R_1} + \frac{1}{R_2} + \frac{1}{R_3} \right)$$

[15-16]

$$\text{i.e.,} \quad \frac{1}{R} = \sum \frac{1}{R_i}$$

EXAMPLE 15-4

Find I_1, I_2 and I_3 in the network of Fig. 15-7, if $R_1 = 10.0\ \Omega$, $R_2 = 8.00\ \Omega$, $R_3 = 5.00\ \Omega$, and $V_{AD} = 10.00$ V.

SOLUTION

First find the equivalent resistance R_4 that can replace the parallel combination of R_2 and R_3. Using Eq. [15-16],

$$\frac{1}{R_4} = \left(\frac{1}{R_2} + \frac{1}{R_3} \right) = \left(\frac{1}{8.00\ \Omega} + \frac{1}{5.00\ \Omega} \right) = 0.325\ \Omega^{-1}$$

Hence,

$$R_4 = 1/(0.325\ \Omega^{-1}) = 3.08\ \Omega$$

Then the network can be considered as a series combination of R_1 and R_4 which can be replaced by R using Eq. [15-15]:

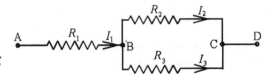

Figure 15-7

Example 15-4: Series-parallel resistor chain.

$$R = R_1 + R_4 = 10.0\ \Omega + 3.08\ \Omega = 13.1\ \Omega$$

The total current I flowing between A and D is given by Ohm's law:

$$I = V_{AD}/R = 10.00\ \text{V}/13.1\ \Omega = 0.763\ \text{A}$$

$$I = I_1 \quad \text{so } V_{AB} = I_1 R_1 = 0.763\ \text{A} \times 10.0\ \Omega = 7.63\ \text{V}$$

$$V_{BC} = V_{AD} - V_{AB} = 10.00\ \text{V} - 7.63\ \text{V} = 2.37\ \text{V}$$

$$I_2 = V_{BC}/R_2 = 2.37\ \text{V}/8.00\ \Omega = 0.296\ \text{A}$$

$$I_3 = V_{BC} / R_3 = 2.37 \text{ V}/ 5.00 \text{ }\Omega = 0.474 \text{ A}$$

Note that as a check, $I_1 = I_2 + I_3$ within the limits imposed by rounding off answers.

Resistivity

The resistance of a wire or any other object depends upon the material of which it is made and upon the size and shape of the object. *Resistivity*, ρ, is a property of the material alone and has units of $\Omega \cdot m$. The resistance, R, of a length of material decreases with the cross-sectional area, $\mathcal{A}$, and increases with the length, L. Therefore

$$R = \frac{\rho L}{\mathcal{A}} \qquad \textbf{[15-17]}$$

Materials with a very small resistivity are called conductors. Most metals fall into this category. Copper, for example, has a resistivity of $1.72 \times 10^{-8} \text{ }\Omega \cdot m$. Insulators have very large resistivities; teflon has a resistivity greater than $10^{13} \text{ }\Omega \cdot m$, more than 10^{20} times that of copper! Semiconductors have intermediate values, e.g., silicon has a resistivity of $2300 \text{ }\Omega \cdot m$ (see Box 15-2).

Box 15-2   Conductance and Conductivity

Biologists frequently find it more convenient, especially when considering the electrical properties of cell membranes, to use *conductance* instead of resistance when considering the flow of current due to a potential difference.

Conductance, g, is the reciprocal of resistance and is measured in siemens (S); i.e., $1 \text{ S} = 1 \text{ }\Omega^{-1}$. (Some very old books use "mho" in place of siemen.) Then Ohm's law $(\Delta V = IR)$ is written:

$$\Delta V = \frac{I}{g} \text{ or } I = g\Delta V \qquad \textbf{[1]}$$

Eq. [15-15] then becomes:

$$(Series) \quad \frac{1}{g} = \sum \frac{1}{g_i} \qquad \textbf{[2]}$$

(cont'd.)

And Eq. [15-16] becomes:

$$(Parallel) \quad g = \sum g_i \qquad [3]$$

In a similar fashion *conductivity*, σ, with units $(\Omega \cdot m)^{-1}$ is the reciprocal of resistivity, ρ $(\Omega \cdot m)$.

EXAMPLE 15-5

The resistivity of a material is 0.854 $\Omega \cdot m$. The material is formed into a cylindrical resistor of length 0.150 m and placed in an electrical circuit where it is found to have a resistance of 50.0 Ω. Determine the radius of the resistor.

SOLUTION

Using Eq. [15-17],

$$R = \rho \, L / \mathcal{A}$$

For a cylinder, $\mathcal{A} = \pi r^2$, where r is the radius, and hence,

$$50.0 \ \Omega = 0.854 \ \Omega \cdot m \times 0.150 \ m/(\pi r^2)$$

$$r = 0.0286 \ m$$

Kirchhoff's Rules

Sometimes circuits contain voltage sources such as batteries in more than one location and have more than one loop around which currents can flow. In such situations Ohm's law is not sufficient for an analysis of the circuit, and two rules known as *Kirchhoff's rules*[8] are of great assistance:

---

8. Gustav Kirchhoff (1824–1887), German physicist.

1. Kirchhoff's point rule: The total current entering a point in a circuit must equal the total current leaving that point. (This was used earlier but was not identified as one of Kirchhoff's rules.)

2. Kirchhoff's loop rule: The total change in electric potential around any closed loop in a circuit is zero, i.e., $\Delta V = 0$ for a closed loop.

Some sign conventions are required if these rules are to be used successfully.

1. If a current direction is not known, just pick a direction. If the assumed direction is correct, the answer calculated for this current will be positive. If the answer calculated is negative, then the direction is opposite to that initially assumed. However, the magnitude of the current will be correct even if the assumed direction is incorrect. Use Kirchhoff's point rule to obtain a first equation. You will eventually have to have as many equations as there are unknowns in the problem. Choose any loop and arbitrarily decide whether to go around it in a clockwise or counterclockwise direction.

2. Travel around the loop in the chosen direction, adding up potential differences as each battery or resistor in the loop is traversed. These terms will be battery voltages or *IR* products for resistors. Remember that a positive potential difference corresponds to an increase in electric potential. A battery voltage is considered positive if it is traversed from negative to positive terminal, and is considered negative if traversed from positive to negative. An *IR* term is considered negative if the resistor is traversed in the same direction as the current and positive when traversed in the opposite direction to the current.

EXAMPLE 15-6

Use Kirchhoff's rules to calculate the magnitude of the current in the circuit of Fig. 15-8 and determine if the assumed current direction (as indicated by the arrowhead) is correct.

SOLUTION

First an arbitrary decision is made to go around the loop in a counterclockwise direction as indicated by the curved arrow inside the loop. Since there is only one unknown, only one equation is required and that is obtained by applying the loop rule, i.e., $\Delta V = 0$.

The battery is traversed from negative to positive, and therefore its voltage term will be +10 V. The resistor is traversed in the direction of the current, so its IR term is negative and will be $-I \times 1000\ \Omega$. The loop-rule equation, $\Delta V = 0$, then is:

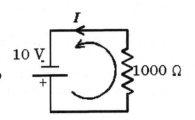

Figure 15-8

Example 15-6: Circuit diagram.

$$+10\ \text{V} - I \times 1000\ \Omega = 0$$

from which $I = 0.010$ A

Since I is positive, the assumed current direction is correct.

Suppose that the current direction had initially been assumed to be in the opposite direction to that in Fig. 15-8, but that the counterclockwise direction was retained for going around the loop. The sign conventions will still have the term for the battery as +10 V, but now the passage through the resistor is against the current and hence the IR term becomes positive. The above equation then becomes:

$$+10\ \text{V} + I \times 1000\ \Omega = 0$$

from which $I = -0.010$ A

The negative sign indicates that the assumed current direction was not correct but note that the magnitude is still correct.

EXAMPLE 15-7

Apply Kirchhoff's rules to the circuit in Fig. 15-9a to find:

(a) the current through the middle (50-Ω) resistor

(b) V_{BA}

Assume three significant digits in all data shown in the figure.

SOLUTION

Figure 15-9

Example 15-7: (a) Circuit diagram; (b) Assumed directions for currents and for going around loops.

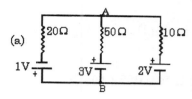

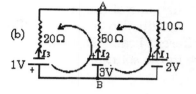

(a) Assume that the currents in the three branches are, from right to left, I_1, I_2, and I_3, and that all three are directed upwards. Assume also that both loops are circled in a counterclockwise direction in applying the loop rule. Figure 15-9b shows these directions.

Consider point A above the 50-Ω resistor. The point rule says:

$$I_1 + I_2 + I_3 = 0$$

This equation says three currents are flowing into the point and no current is flowing out. This is impossible. At least one of the assumed current directions is not correct. Continue with the solution. Remember, if a negative current is calculated it means only that the assumed direction for that current is wrong.

Next apply the loop rule ($\Delta V = 0$) to the left-hand and right-hand loops.

For the left loop, starting at the top left corner:

$$(20\ \Omega \times I_3)\ + 1\ \text{V} + 3\ \text{V} - (50\ \Omega \times I_2) = 0\ \text{V}$$

For the right loop, starting at the top left corner:

$$(50\ \Omega \times I_2)\ - 3\ \text{V} + 2\ \text{V}\ - (10\ \Omega \times I_1) = 0\ \text{V}$$

The point and loop rules have provided three equations which can be solved to give:

$$I_1 = 0.0765\ \text{A} = 76.5\ \text{mA};\ I_2 = 35.3\ \text{mA};\ I_3 = -112\ \text{mA}$$

Note that, within the limits imposed by rounding, these three currents do add up to zero, as required by the point rule.

The negative sign for I_3 indicates that the direction shown on the figure is incorrect but the magnitude is correct.

(b) V_{BA} represents $V_B - V_A$, i.e., the electric potential at B relative to that at A. It can be calculated by adding the changes in electric potentials in going <u>from</u> A <u>to</u> B through any of the three paths available.

Consider the central path through the 3-V battery and the 50-Ω resistor:

$$V_{BA} = (50\ \Omega \times I_2) - 3\ V = (50\ \Omega \times 0.0353\ A) - 3\ V$$
$$= -1.24\ V$$

As a check, calculate V_{BA} through one of the other branches. The answer of course will also be –1.24 V. Two different answers for two different branches would indicate an error somewhere in the solution.

Exercises

15-1   An electric charge of 2.0×10^{-15} C is placed in the vicinity of another electric charge where it experiences a force of magnitude 5.0×10^{-13} N. What is the magnitude of the electric field at the site of the first charge?

15-2   The charge in Exercise 15-1 moves a distance of 1.0 mm in the direction of the field. If the field is constant over that distance, how much work is done on the charge?

15-3   Through what potential difference has the charge of Exercise 15-2 moved?

15-4   A particular biological cell membrane can be considered as a capacitor with a plate separation of 4.0 nm and a dielectric constant of 6.0.

   (a) What is the capacitance per unit area of the membrane?

   (b) If the voltage across the membrane is 60 mV and the area of each surface of the membrane is 80 μm^2, how much charge is stored on each membrane surface?

15-5   In a chemical process, an average current of 25 A was maintained for 3.0 h. What electric charge was transferred?

15-6   If the chemical process in the previous question was an electroplating process and a singly charged metal ion was involved, how many moles of the metal were plated?

15-7   A battery of voltage 1.1 V is connected across a resistor of 2.2 Ω. What current is drawn?

15-8   A copper bar 0.50 cm in diameter is 5.0 m long. What is its resistance? The resistivity of copper is 1.72×10^{-8} Ω·m.

15-9   Two resistors each of 0.50 Ω are connected in series. What is the total resistance?

15-10   The resistors of Exercise 15-9 are connected in parallel; what is the resultant resistance?

15-11   In a particular biological cell, Na^+ ions are pumped from the inside to the outside at a rate of 1.0×10^5 per second. Calculate the corresponding electrical current in amperes.

15-12   In a time of 10 s, 8.0×10^5 Na^+ ions pass through a section of membrane having a total area of 50 μm^2. What are (a) the electrical current and (b) the current density?

15-13   When a neural voltage spike is generated, about 4.3×10^{-12} mol/cm^2 of sodium ions enter a nerve axon in 1.0 ms. Find the current density associated with this flow of ions. What is the electrical current if the ions flow through a membrane area of 5.0 μm^2?

15-14   For the purpose of calculating electrical current through the human body, the body fluid can be considered as a solution of conductivity 0.65 $(\Omega \cdot m)^{-1}$ at normal body temperature. Calculate the resistance of one arm, considering it to be 75 cm long and of a uniform cross-sectional area of radius 5.0 cm. Similarly, calculate the resistance of one finger, considering it to be 10 cm long and of a cross-sectional area of 4.0 cm^2.

Problems

15-15   Two negative charges of -1.0×10^{-6} C each are placed in water and separated by a distance of 0.10 m. A positive charge of 1.0×10^{-8} C is placed exactly midway between the two negative charges. Determine:

(a) the electric field (magnitude and direction) and electric potential at the position of the positive charge;

(b) the electric field (magnitude and direction) and electric potential at the position of either negative charge;

(c) the electric force (magnitude and direction) experienced by the positive charge;

(d) the electric force (magnitude and direction) experienced by either negative charge.

15-16 Find the force (magnitude and direction) on the chloride ion (Cl_1^-) in the following figure. All the ions lie in the same plane, and the solvent is water. Assume two significant digits in all data.

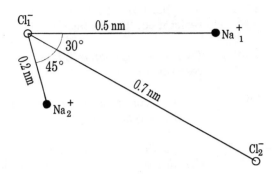

15-17 If the charge in Exercises 15-1 and 15-2 was initially at rest and was carried by a dust mote of mass 5.0 μg, what is its speed after moving the 1.0 mm?

15-18 The resistors in the network in the following figure are:

$R_1 = 6.0\ \Omega$ $R_4 = 7.5\ \Omega$
$R_2 = 3.0\ \Omega$ $R_5 = 5.0\ \Omega$
$R_3 = 15\ \Omega$ $R_6 = 4.0\ \Omega$

If the current I is to be 14 A, what is the voltage of the battery that must be connected across a and b? What is the voltage across R_1 and R_2? What is the voltage across the three-resistor section?

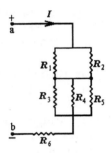

15-19 In the circuit of the accompanying figure, determine I_1, I_2, and I_3, and the voltages across the 1-Ω, 2-Ω, and 3-Ω resistors. Assume two significant digits in all data.

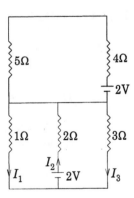

Answers

15-1	2.5×10^2 N/C	**15-14**	$1.5 \times 10^2\ \Omega$, $1.2 \times 10^2\ \Omega$

15-2 5.0×10^{-16} J

15-3   0.25 V

15-4   (a) 0.013 F/m^2
         (b) 6.4×10^{-14} C

15-5 2.7×10^5 C

15-6   2.8 mol

15-7 $0.50\ \Omega$

15-8 $4.4 \times 10^{-3}\ \Omega$

15-9 $1.0\ \Omega$

15-10 $0.25\ \Omega$

15-11 1.6×10^{-14} A

15-12   (a) 1.3×10^{-14} A
          (b) 2.6×10^{-4} A/m^2

15-13   4.1 A/m^2, 2.1×10^{-11} A

15-15   (a) $\vec{E} = 0$, $V = -4.5 \times 10^3$ V
          (b) $\vec{E} = 1.1 \times 10^4$ N/C
            toward other charges,
            $V = -1.1 \times 10^3$ V
          (c) $\vec{F} = 0$
          (d) $\vec{F} = 1.1 \times 10^{-2}$ N
            away from other charges

15-16 $\vec{F} = 7.1 \times 10^{-11}$ N at 69°
          below Cl_1–Na_1 line

15-17 1.4×10^{-5} m/s

15-18 $V_{ab} = 1.2 \times 10^2$ V;
          $V_{1,2} = 28$ V;
          $V_{3,4,5} = 35$ V

15-19 $I_1 = 6/11$ A $= 0.55$ A;
          $I_2 = 8/11$ A $= 0.73$ A;
          $I_3 = 2/11$ A $= 0.18$ A
          $V_1 = 6/11$ V $= 0.55$ V;
          $V_2 = 16/11$ V $= 1.5$ V;
          $V_3 = 6/11$ V $= 0.55$ V

16 Magnetism

16.1 Introduction

In Chapter 15, some properties of direct current (DC) electricity were examined and several important relationships were derived. One of the effects of flowing electricity is the production of a magnetic field. As we will see, magnetic fields are always a result of electric currents either at a macroscopic level or at the atomic level. Magnetism has increasing relevance in the life sciences, particularly with the development of Magnetic Resonance Imaging (MRI) and concerns about the health effects of magnetic fields.

16.2 Basic Magnetism

The magnetic effect of natural magnets such as lodestone have been known for centuries. This magnetic effect causes a compass needle or filings of iron to align themselves up along invisible lines of the *magnetic field* surrounding a natural magnet. Figure 16-1 is a picture of iron filings sprinkled on a sheet of paper placed over a bar magnet. The lines, as delineated by the filings, seem to point toward two locations (or at least small regions) near the ends of the magnets called the *poles*. They are designated "north" (N) and "south" (S) according to which pole of the Earth they point to if the magnet is allowed to swing freely about its centre.

Figure 16-1

The magnetic field of a bar magnet.

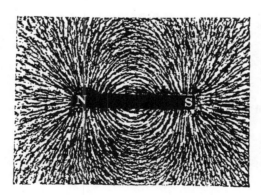

A further observation about the poles is that unlike poles (N-S) attract and like poles (N-N, S-S) repel each other, much like positive and negative electric charges.

A basic difference between electric and magnetic fields is that whereas electric field lines originate on positive charges and terminate on negative charges, the magnetic field lines occur only in closed loops. In Fig. 16-1 the lines, by convention, are imagined to extend <u>from</u> the N <u>to</u> the S pole and return to close the loop within the magnet itself as shown in Fig. 16-2.

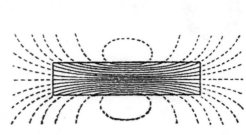

Figure 16-2

The magnetic field inside a bar magnet.

16.3   Electromagnetism

For a long time electricity and magnetism were thought to be unrelated phenomena until, in 1820, Oersted[1] discovered that a compass needle, placed near a current-carrying wire, was deflected. Following this discovery of *electromagnetism* it was realized that magnetic fields are a result of moving charges. In the case of the wire, the moving charges are the flowing electrons; in the case of a permanent magnet they are the electrons orbiting in their individual atoms. (It is important not to confuse electric and magnetic fields: electric fields exist between electric charges whether stationary or moving; magnetic fields accompany only moving charges.)

Oersted's investigations showed that the magnetic field that accompanied a current-carrying wire was in the form of circular lines about the wire as shown in Fig. 16-3. The direction of the field can be determined from the following *right-hand thread rule*:

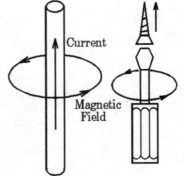

Current

Magnetic Field

Figure 16-3

The magnetic field about a current-carrying wire.

---

1.   Hans Christian Oersted (1777-1851), Danish physicist.

To advance a right-hand threaded screw (the ordinary kind) in the direction of the conventional current (i.e., flow of positive charge), the screwdriver would have to be rotated in the direction of the magnetic field.

Examine Fig. 16-3 in light of this rule.

Figure 16-4

The interaction of a moving charge with a magnetic field.

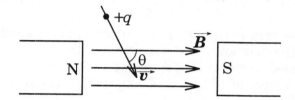

Figure 16-4 shows a uniform magnetic field $\vec{B}$ established between the poles of two large magnets. Note that, by convention, the field points from N to S. If a charge $+q$ is injected at a velocity $\vec{v}$, at angle θ, into this field, it will experience a force that is at right angles to both the magnetic field lines <u>and</u> the direction of the velocity vector.[2] This means that in Fig. 16-4 the force is directed either into, or out of, the page. Whether it is into or out of the page is determined by another *right-hand thread rule* shown in Fig. 16-5:

A screwdriver turned in the direction of vector $\vec{v}$ rotated into vector $\vec{B}$ would advance a right-hand threaded screw in the direction of the force.

In the situation shown in Fig. 16-4, the force would be out of the page.

Figure 16-5

The direction of the electromagnetic force.

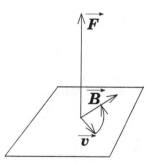

This rule, like all rules in electricity, has been given for a positive charge. Of course, if the charge is <u>negative</u> the force will be in the <u>opposite</u> direction.

---

2. In all that follows in this chapter, it will be assumed that the speed with which charges move are small enough compared to the speed of light that relativistic methods are not needed. Speeds less than 10% of c will give results correct to ½%.

Experiments show that the force experienced by the charge is proportional to the magnitude of the charge, the speed (magnitude of the velocity), and the sine of the angle θ. From these observations, the *magnitude of the magnetic field B* is given by

$$B = F/|q|v \sin(\theta)$$

or

$$F = |q|vB \sin(\theta) \qquad \textbf{[16-1]}$$

In Eq. [16-1] when the orce is in newtons (N), the charge in coulombs (C), and the speed in m/s, the magnetic field strength has units $N \cdot s \cdot C^{-1} \cdot m^{-1}$; this unit is called the *tesla* (T).[3] A magnetic field has a strength of 1 T if it exerts a force of 1 N on a 1 C charge that is moving at 1 m/s at right angles to it. An older, but still current, unit is the *gauss* (G)[4] where 10^4 G = 1 T.

EXAMPLE 16-1

A He^{2+} ion travels at 45° to a magnetic field of 0.80 T with a speed of 4.0×10^5 m/s. Find the magnitude of the force on the ion.

SOLUTION

The charge on a He^{2+} ion is twice the elementary charge or $2 \times 1.602 \times 10^{-19}$ C.

From Eq. [16-1],

$$
\begin{aligned}
F &= |q|vB \sin(\theta) \\
&= 2(1.602 \times 10^{-19} \, C)(4.0 \times 10^5 \, m/s)(0.80 \, T)\sin(45°) \\
&= 7.2 \times 10^{-14} \, N
\end{aligned}
$$

---

3. Nikola Tesla (1856–1943), a brilliant but eccentric Croatian-American electrical engineer who invented the induction motor and the transformer.
4. Johann Karl Friedrich Gauss (1777–1855), German mathematician who first established the units of magnetism.

Path of a Charged Particle in a Uniform Magnetic Field

Figure 16-6 shows a charged particle $q > 0$ moving with speed v at right angles to a uniform magnetic field $\vec{B}$. Since the magnetic force is always at right angles to both $\vec{v}$ and $\vec{B}$ then it acts toward a point C and is the centripetal force for motion in a circle with radius r. Using Eq. [16-1] and Eq. [9-8] we have (with $\theta = \pi/2$)

Figure 16-6

Circular motion of a charged particle moving at constant speed in a uniform magnetic field.

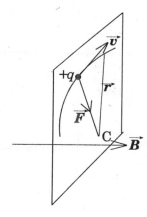

$$F = qvB = mv^2/r$$

or

$$r = \frac{mv}{qB} \qquad \textbf{[16-2]}$$

The time to make one circuit of the circle (the period) is $T = (2\pi r)/v$ and the frequency f is the reciprocal of the period. Therefore, using Eq. [16-2],

$$f = \frac{1}{2\pi}\frac{qB}{m} \qquad \textbf{[16-3]}$$

The quantity f given by Eq. [16-3] is called the *cyclotron frequency*. If $q < 0$, Eq. [16-2] and [16-3] are still valid with q replaced by $|q|$.

EXAMPLE 16-2

A proton ($q = +1.602 \times 10^{-19}$ C, $m = 1.673 \times 10^{-27}$ kg) is injected at a speed of 3.0×10^6 m/s into a region of uniform magnetic field of strength 4.0×10^{-2} T.

(a) What is the radius of the orbit of the proton?

(b) What is the frequency of its revolution?

SOLUTION

(a) Using Eq. [16-2]

$r = (mv)/(qB)$
$ = (1.673 \times 10^{-27}$ kg $\times 3.0 \times 10^6$ m/s)/
$ (1.602 \times 10^{-19}$ C $\times 4.0 \times 10^{-2}$ T)
$ = 0.78$ m

(b) Using Eq. [16-3]

$f = (1/2\pi)(qB/m)$
$\quad = (1/2\pi)(1.602\times10^{-19}\,\text{C} \times 4.0\times10^{-2}\,\text{T})/1.673\times10^{-27}\,\text{kg}$
$\quad = 0.61\times10^{6}\,\text{Hz} = 0.61\,\text{MHz}$

The circular motion described here is used in a number of physical and medical instruments and machines including high-energy particle accelerators like the *cyclotron* and the *mass spectrometer* (see Box 16-1).

Box 16-1   The Mass Spectrometer

The mass spectrometer is an instrument used for the accurate determination of the mass of atoms and molecules. Modern mass spectrometers have become essential tools for the analysis of protein structures. The particles must be ionized to give them a predetermined charge—usually positive. The beam of ions from some source S is collimated by slits and accelerated by a voltage V to a speed v. They pass into an evacuated chamber in a uniform magnetic field $\vec{B}$ (shown into the page in Fig. 16-7). In the chamber, the ions travel in a circle of radius r and are recorded by a detector D. The work done on each ion by the electric potential V is qV (Eq. [15-6]) and appears as the kinetic energy $\frac{1}{2}mv^2$ of the ions:

Figure 16-7

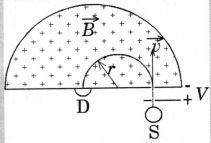

A schematic diagram of a mass spectrometer.

$$\tfrac{1}{2}mv^2 = qV$$

From Eq. [16-2],

$$v = (q/m)Br$$

Eliminating v between these two equations we have

$$\frac{q}{m} = \frac{2V}{B^2r^2}$$

If the charge q is known then the mass can be found, since the position of D determines r. To find r, either the magnetic field can be altered to put the ions on the detector, or the detector can be moved with B constant to find the ion beam.

Force on a Current-Carrying Conductor in a Magnetic Field

Figure 16-8

Force on a current-carrying conductor in a magnetic field.

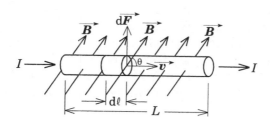

In Fig. 16-8, a conductor of length L carrying a current I is in a uniform magnetic field $\vec{B}$, making an angle θ with the direction of the current. The velocity of the moving charges is $\vec{v}$. The magnitude of the magnetic force acting on the conductor can be determined as follows:

The charge within the volume element of length $d\ell$ moves a distance $d\ell$ in a time dt. Therefore $d\ell = v\, dt$. Since $I = q/dt$, then $I = qv/d\ell$ or $qv = I\, d\ell$. Therefore, for the element $d\ell$, the contribution dF to the total force is, from Eq. [16-1]:

$$dF = IB \sin(\theta)\, d\ell$$

which must be integrated over the length of the conductor from 0 to L giving

$$F = IBL \sin(\theta) \qquad \textbf{[16-4]}$$

Torque on a Current Loop in a Magnetic Field

Figure 16-9

The torque on a current loop in a magnetic field.

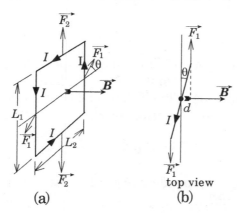

(a)　(b)

Figure 16-9a shows a rectangular loop of dimensions $L_1 \times L_2$ carrying a current I in a magnetic field of magnitude B. The loop makes an angle θ with the direction perpendicular to $\vec{B}$ as shown in Fig. 16-9b. Equation [16-4] can be used to evaluate the forces acting on the four sides of the loop. On the sides of length L_2, forces of magnitude $F_2 = IBL_2 \sin(\theta)$ act but in opposite directions, so the net effect on the loop is zero.

On each side of length L_1 the force has magnitude $F_1 = IBL_1$ acting in such a manner as to produce a torque on the loop attempting to rotate it to a transverse position in the field. The lever arm of the torque (τ) is $d = \frac{1}{2}L_2 \sin(\theta)$ and the total torque from the two forces F_1 is

$$\tau = 2 F_1 d = 2IBL_1\frac{1}{2} L_2 \sin(\theta) = IBL_1 L_2 \sin(\theta) = IB\mathcal{A} \sin(\theta) \qquad \textbf{[16-5]}$$

where $\mathcal{A} = L_1 L_2$ is the area of the loop. The expression has been worked out for a rectangular loop but, in fact, it applies to a loop of any shape since it can be thought of as infinitesimal rectangular elements in the horizontal or vertical directions. This torque is reponsible for the motion of the needle attached to a coil in most electrical measuring instruments. The basic device is called a galvanometer and is discussed in Box 16-2.

If the loop is a coil of N turns then the result is simply multiplied by N and

$$\tau = NIB\mathcal{A} \sin(\theta) \qquad \textbf{[16-6]}$$

The *magnetic moment* ($\mathcal{M}$) is defined as:

$$\mathcal{M} = NI\mathcal{A} \qquad \textbf{[16-7]}$$

so Eq. [16-6] becomes

$$\tau = \mathcal{M} B \sin(\theta) \qquad \textbf{[16-8]}$$

The magnetic moment can be thought of as a vector that points in the direction perpendicular to the plane of the coil. The concept of a magnetic moment will be important in later considerations in this chapter.

EXAMPLE 16-3

A square coil has sides of length 10 cm and consists of 20 turns of wire. The coil carries a current of 0.10 A and is in a magnetic field $\vec{B}$ of magnitude 0.200 T. The coil is oriented so that its plane is in the same direction as $\vec{B}$.

(a) What is the torque acting on the coil?

(b) If the magnetic field is perpendicular to the plane of the coil, what is the torque?

SOLUTION

(a) The area of the coil $\mathcal{A} = 0.10$ m $\times$ 0.10 m $= 1.0 \times 10^{-2}$ m²

The magnetic moment of the coil $= \mathcal{M} = NI\mathcal{A}$
$= 20(0.10$ A$)(1.0 \times 10^{-2}$ m²$) = 0.020$ A·m²

From Eq. [16-8],

$\tau = \mathcal{M}B \sin(\theta)$
$= (0.020$ A·m²$)(0.200$ T$) \sin(90°) = 4.0 \times 10^{-3}$ N·m

A note on the units:

A $=$ C·s^{-1} and T $=$ N·s·C^{-1}·m^{-1} (see Eq. [16-1]).

Therefore A·m²·T $=$ C·s^{-1}·m²·N·s·C^{-1}·m^{-1} $=$ N·m as required.

(b) When the coil is perpendicular to the field, $\theta = 0°$
and $\tau = 0$.

Box 16-2   The Galvanometer

Figure 16-10 shows the structure of a *galvanometer*
which is the basic analog (non-digital) instrument
for making electrical measurements.

Figure 16-10

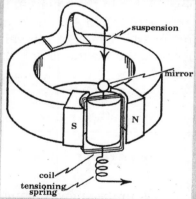

The galvanometer.

A small coil of N turns is suspended between the
poles of a magnet. The poles are shaped and there
is a cylindrical iron core in the centre of the coil.
This ensures that the magnetic field is always in
the plane of the coil no matter what its orientation,
i.e., $\theta = \pi/2$ and $\sin(\theta) = 1$.

When current flows through the coil, the torque
$N\mathcal{A}IB$ turns the coil against the restoring torque of
the suspension. Mechanical equilibrium is achieved
at some angle of deflection which can be measured by reflecting a

(cont'd.)

light beam from the mirror. Since $\sin(\theta) = 1$ for any orientation of the coil, then the stationary deflection of the coil is proportional to the current I.

If the suspension is replaced by bearings, the restoring torque supplied by a hairspring (as in a watch), and the mirror and light beam replaced with a pointer and dial, then the common analog electrical meter is the result.

Magnetic Field of a Moving Charge

Figure 16-11 shows a charge $q > 0$ moving with a velocity $\vec{v}$. At the point P located at a distance r from the charge and a distance R from the direction the charge is moving, and inclined at an angle θ to the direction of $\vec{v}$, there is an electric field $\vec{E}$ directed away from q (according to the discussion of Chapter 15) and a magnetic field $\vec{B}$ directed at right angles to both $\vec{v}$ and $\vec{E}$.

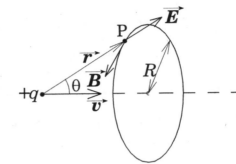

Figure 16-11

The magnetic field of a moving electric charge.

Experiments show that the magnitude of $\vec{B}$ is proportional to q and $v\sin(\theta)$, and inversely proportional to r^2, that is

$$B = \text{const}\ \frac{qv\ \sin(\theta)}{r^2}$$

The constant in the equation is defined as $\mu_0/4\pi$ where $\mu_0 = 4\pi \times 10^{-7}$ T·m·A^{-1} (or m·kg·C^{-2}) so that the constant is exactly equal to 10^{-7} T·m·A^{-1}; μ_0 is called the *permeability of free space*.[5] Therefore the magnetic field of a moving charge is given by

$$B = \frac{\mu_0}{4\pi}\ \frac{qv\sin(\theta)}{r^2} = 10^{-7}\ \frac{qv\sin(\theta)}{r^2}\ \text{T} \qquad \textbf{[16-9]}$$

EXAMPLE 16-4

A proton moves along the x-axis at a speed 10% of the speed of light. What is the magnitude of the magnetic field 15 cm ahead of the proton and 10 cm away from the x-axis?

---

5. The reasons for these units of magnetic properties are complicated and beyond the present discussion. Justifications can be found in standard texts on electricity and magnetism.

SOLUTION

$q = 1.602\times10^{-19}$ C and $v = c/10 = 2.998\times10^{7}$ m/s
$\theta = \tan^{-1}(10/15) = 33.7°$ and therefore $r = 18$ cm or 0.18 m

$B = (10^{-7}$ T·m·A$^{-1})(1.602\times10^{-19}$ C$)(2.998\times10^{7}$ m/s$)\sin(33.7°)/$
$\quad (0.18$ m$)^2$

$= 8.2\times10^{-18}$ T

Magnetic Field of a Straight Conductor Carrying a Current

Figure 16-12

The magnetic field from a current-carrying conductor.

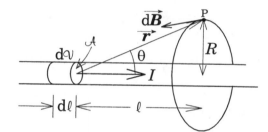

Figure 16-12 illustrates the same geometry as that in Fig. 16-11 except that the single moving charge q is replaced by a continuous current I (of positive charges). The magnitude of $\vec{B}$ at a perpendicular distance R from the current-carrying conductor can be derived as follows.

$\dfrac{dy}{dx}$

The result given in Eq. [16-9] can now be applied to the moving charge element in the volume $d\mathcal{V}$ to evaluate the field dB at r and then integrated over all the elements in the wire from $-\infty$ to $+\infty$.

The charge contained in the volume element $d\mathcal{V}$ is dq. Since all the charges contained in the element $d\mathcal{V}$ flow out through area $\mathcal{A}$ in the time dt to provide the current I, the length of the element is, as before, $d\ell = v dt$, Therefore

$$dq = dq(d\ell/d\ell) = dq \; v \; dt/d\ell$$

and

$$I = dq/dt = dq \; v/d\ell$$

which gives

$$I d\ell = v \; dq$$

This expression is substituted for qv in Eq. [16-9] for the part of the magnetic field dB at the point P contributed by the current element $Id\ell$.

$$dB = \left(\frac{\mu_0}{4\pi}\right)\frac{Id\ell}{r^2}\sin(\theta)$$

The total field is obtained by integrating this from $\ell = -\infty$ to $\ell = +\infty$. From symmetry, the contributions of the right-hand and left-hand portions of the conductor are equal so that the integration can be from $\ell = -\infty$ to $\ell = 0$ (or $\theta = 0$ to $\theta = \pi/2$) by multiplying by a factor of 2.

$$B = \left(\frac{\mu_0}{2\pi}\right)I \int_{-\infty}^{0}\frac{d\ell}{r^2}\sin(\theta) \qquad \textbf{[16-10]}$$

The variable ℓ can be eliminated in favour of θ by using the relations $\sin(\theta) = R/r$ and $R/\ell = \tan(\theta)$, from which $r = R/\sin(\theta)$ and $d\ell = -(R/\sin^2(\theta))\,d\theta$. Substituting these into Eq. [16-10] gives

$$B = \left(\frac{\mu_0}{2\pi}\right)\frac{I}{R}\int_{0}^{\pi/2}\sin(\theta)d\theta$$

which upon integration gives

$$B = \left(\frac{\mu_0}{2\pi}\right)\frac{I}{R} = 2\times10^{-7}\,\frac{I}{R}\,\text{T} \qquad \textbf{[16-11]}$$

With the result of Eq. [16-11] we can evaluate the force on a current-carrying conductor in the vicinity of another parallel current-carrying conductor as shown in Fig. 16-13. The centre-to-centre distance of the two conductors is d and they carry currents I_1 and I_2. The field at the centre of the second conductor due to the current in the first is given by Eq. [16-11] with $R = d$:

$$B_1 = \left(\frac{\mu_0}{2\pi}\right)\frac{I_1}{d}$$

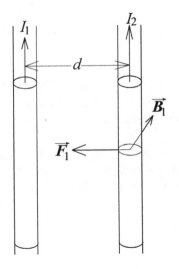

Figure 16-13

The force between two current-carrying conductors.

This field interacts with the current I_2 according to Eq. [16-4], $F = IBL\sin(\theta)$, with $\theta = 90°$

$$F_1 = I_2 B_1 L = I_2 \frac{\mu_0}{2\pi}\frac{I_1}{d}L$$

$$\frac{F_1}{L} = \frac{\mu_0}{2\pi}\frac{I_1 I_2}{d} \qquad \textbf{[16-12]}$$

where L is the total length of the conductors and so F_1/L is the force per unit length. Of course there is an equal and opposite force F_2 acting on conductor 1 to the right. When the currents are in the same direction as shown, the two conductors are attracted to each other; when the currents are in the opposite direction, the forces are repulsive.[6]

If the same current ($I_1 = I_2 = I$) passes through the conductors in opposite directions, and the force, length, and separation are measured,

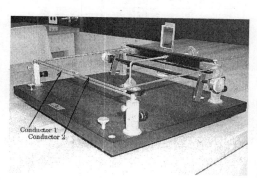

Figure 16-14

A student current balance: conductor 1 is fixed and conductor 2 is balanced on knife edges.

Conductor 1
Conductor 2

then from Eq. [16-12] the square of the current (I^2) can be calculated. A device to do this is called a "current balance"; it is the fundamental way in which electric current (and thus electric charge) is defined in terms of force and distance. A student version of the current balance is shown in Fig. 16-14.

EXAMPLE 16-5

The high voltage cables on the top of a pylon carry a current of 1.0×10^4 A and are separated by a distance of 2.0 m. What is the magnitude of the force between a 5.0 m length of a pair of them? Is it attractive or repulsive?

SOLUTION

From Eq. [16-12]:

$$F/L = (2 \times 10^{-7}\,\text{T·m·A}^{-1})(I)^2/d$$

$$= (2 \times 10^{-7}\,\text{T·m·A}^{-1})(1.0 \times 10^4\,\text{A})^2/(5.0\,\text{m}) = 4.0\,\text{N/m}$$

The force is repulsive since currents in two parallel wires in a closed circuit flow in opposite directions.

---

6. The student should check the directions of B and F using the rules given earlier in the chapter.

Magnetic Field of a Current Loop

The magnetic field pattern of a current loop is shown in Fig. 16-15. Note the similarity with the field pattern for a bar magnet shown in Fig. 16-2. Since that pattern arises from a N-S pair of magnetic poles, the field pattern is said to be that of a *magnetic dipole*. We see that the field of a current loop is also that of a magnetic dipole. The mathematical expression for the magnitude of the field B at any point is quite complicated. The evaluation of the magnitude of the field at any point on the axis at a distance R from the centre is simpler but beyond what we wish to do here. It is given by

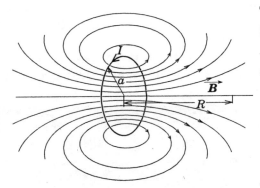

Figure 16-15

The magnetic field of a current loop.

$$B = \frac{\mu_0 I a^2}{2(a^2 + R^2)^{3/2}} \qquad \textbf{[16-13]}$$

where a is the radius of the loop. The field points along the symmetry axis.

At the centre of the loop ($R = 0$), the magnitude of the field is

$$B = \frac{\mu_0 I}{2a} \qquad \textbf{[16-14]}$$

16.4 Magnetization of Matter

Magnetic fields arise as a result of the motion of electric charges. It is not a surprise, therefore, to discover that the motion of charge on the atomic level has magnetic effects. The easiest case to visualize is that of the magnetic moment produced by the orbital motion of the electron around the nucleus of an atom as depicted in Fig. 16-16a. The motion of the charge in its orbit is a microscopic version of the current loop depicted in Fig. 16-15. What is not so obvious is that there is also a magnetic moment produced by the spin of the electron. The spin of the nuclear particle—neutron and

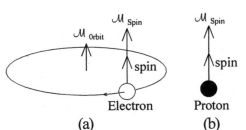

Figure 16-16

The magnetic moment at the atomic level: (a) the orbit and spin moment of the electron; (b) the spin moment of the proton.

proton—also produce magnetic moments as illustrated in Fig. 16-16b. This fact will be important later.

Figure 16-17

The magnetization of matter: (a) in the absence of an external field; (b) in the presence of an external field B.

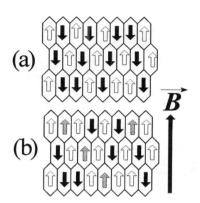

These moments do not all point in the same direction in a multi-electron atom, but largely cancel each other out (think of the "up" and "down" spins discussed in Chapter 4). Materials made of these atoms are called *diamagnetic*. Moderate external magnetic fields have no effect on such materials as there is no magnetic moment for the field to interact with. In some atoms, however, there is one or more unpaired electrons and the atom is left with a residual magnetic moment; examples of this are oxygen and iron. As a result, when placed in a magnetic field, they exhibit magnetic effects that are different from non-magnetic materials. In the absence of an external magnetic field, oxygen does not have an overall magnetic moment since the atoms themselves are randomly oriented and the individual atomic magnetic moments cancel out. This is illustrated symbolically in Fig. 16-17a where the magnetic moments are depicted half pointing in one direction and half in another.[7]

When oxygen is placed in a magnetic field of magnitude B, some of the moments change direction as shown in Fig. 16-17b, and the material as a whole acquires a magnetic moment which disappears when the field is removed; such material is called *paramagnetic*.

If iron is placed in an external field, the atomic magnetization is sufficiently strong as to force the alignment of neighbouring atoms, and so small regions, called *domains*, acquire a permanent magnetic moment; such materials are called *ferromagnetic*. In the absence of an external field it is not the individual atoms that are randomly oriented to produce zero magnetization but the domains themselves. The domains are smaller than 1/10 mm in size, so particles of iron smaller than a single domain can be permanent magnets.

If bulk iron is placed in an external magnetic field, some of the domains switch their alignment as in Fig. 16-17b, and the iron acquires a magnetic moment. In this case, however, when the field is removed, not all of the domains switch back to their initial direction. It takes considerable energy to switch a domain, and so the iron retains some

7.   In reality, of course, the moments point equally in all directions.

permanent magnetization and becomes a permanent magnet. If sufficient energy is supplied, for example by heating, the permanent magnetization can be removed.

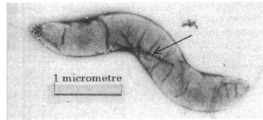

Figure 16-18

A magnetotactic spirillum: the arrow points to the line of single crystal magnets of iron oxide.

A striking example of a biological application of the permanent magnetization of iron domains occurs in the case of the magnetotactic[8] spirillum shown in Fig. 16-18. About 20 single crystals of iron oxide are each smaller than a single domain and are permanent magnets. The bacterium uses the interaction of the magnets with Earth's magnetic field to navigate.[9]

An important medical application of magnetic fields and magnetization is that of *Magnetic Resonance Imaging* or MRI, which uses the magnetization of living matter to provide a three-dimensional picture of living tissues with very high spatial resolution. The technique is described in more detail in Box 16-3.

Box 16-3   Magnetic Resonance Imaging (MRI)

Figure 16-19

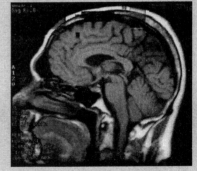

An MRI scan of the mid-section through a male adult head.

Since nuclear particles (protons and neutrons) also have magnetic moments, they also can be aligned in an external magnetic field. When water is placed in a strong external magnetic field, the protons which are the nuclei of the hydrogen atoms have their magnetic moments aligned in the direction of $\vec{B}$. In fact there are two alignments which are possible: the magnetic

(cont'd.)

8.   Magnetotactic means "responding to magnetic fields."
9.   Iron oxide is also found in the brains of some birds that use the Earth's magnetic field to navigate.

moment <u>in</u> the direction of the applied field or <u>opposite</u> to the direction of the field. We might call these two configurations "parallel" (p) and "anti-parallel" (p′).

There is an energy difference between these two configurations: the p configuration has the lower energy. The energy difference between the two states is proportional to the applied field magnitude B and so

$$E_{pp'} = \gamma B$$ [16-15]

where γ is the *gyromagnetic ratio* of the nucleus (usually measured in units of MHz/T). Table 16-1 gives a list of the number of unpaired protons and neutrons along with γ for a selection of nuclei.

Table 16-1. Number of Unpaired Protons and Neutrons and Gyromagnetic Ratio for a Selection of Nuclei

Nucleus	Number of Unpaired Protons	Number of Unpaired Neutrons	Gyromagnetic Ratio γ (MHz/T)
^{1}H	1	0	42.58
^{2}H	1	1	6.54
^{13}C	0	1	10.71
^{14}N	1	1	3.08
^{23}Na	2	1	11.27
^{31}P	0	1	17.25

Since there is an energy difference between the p and p′ states, a photon of the correct energy will convert a low-energy magnetization state into a high one, i.e., will invert the magnetic moment. For example, for protons in a 2.0 T magnetic field, a photon with frequency $2.0 \times 42.58 = 85.16$ MHz will be absorbed.

If a sample containing water (protons) is placed in a magnetic field which varies slightly in a known way with position, then the resonant frequency will also vary with position in a predetermined way. In this way the presence of water can be mapped in the sample. MRI employs other more elaborate techniques to increase the sensitivity and resolution of the mapping which are beyond the present discussion.

Modern MRI techniques can easily achieve spatial resolution of 1 mm, which is sufficient for many diagnostic purposes in medicine.

16.5   Electromagnetic Induction

Oersted's discovery that an electric current produced a magnetic field raised the question as to whether or not a magnetic field would produce an electric current or, more precisely, an electric potential that could, in turn, drive a current in a circuit. This question was first seriously investigated by Michael Faraday[10] who, after lengthy investigation, discovered *electromagnetic induction*.[11]

In Fig. 16-20a, a conductor of length ℓ, such as a wire, is being pulled by an externally applied force, at a constant speed v through a magnetic field of magnitude B; for simplicity we choose $\vec{v}$, ℓ, and $\vec{B}$ to be at right angles to each other.

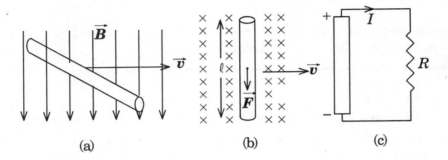

(a)   (b)   (c)

Figure 16-20

Three configurations of a conductor moving in a magnetic field.

In Fig. 16-20b the situation is viewed from above where the crosses represent the tails of the arrows representing $\vec{B}$ which are pointing into the page. Imagine a conduction electron (i.e., free to move) in the wire being carried along with the wire. From the right-hand thread rule it will experience a force $\vec{F}$ in the direction shown. (Remember that an electron carries a negative charge.) Since the electron, like all the others, is free to move in the wire, it will move toward one end making it negatively charged with respect to the other end. An electric field $\vec{E}$ is therefore produced in the wire pointing from the negative end to the positive end. When sufficient charge has been separated, the electric force will counterbalance the magnetic force and the process will come into equilibrium so long as the wire continues to move in the magnetic field. A voltage called the *induced voltage* has been developed along the wire. Balancing the electric and magnetic forces of Eq. [15-2a] and [16-1] ($\theta = 90°$), we have

$$eE = evB$$

---

10.   Michael Faraday (1791–1867), English physicist and chemist.

11.   Electromagnetic induction was also discovered independently by the American physicist Joseph Henry (1797–1878), who was also the inventor of the electromagnet and the electric telegraph.

or

$$E = vB$$

From Chapter 15, Eq. [15-5] can be written (with $r = \ell$) as $V =$ $kq/K\ell$. Upon dividing both sides by ℓ we have $V/\ell = kq/K\ell^2$ which is simply the electric field E as in Eq. [15-1]. Thus an alternative way of writing the electric field is $E = V/\ell$, and so

$$V = vB\ell \qquad \textbf{[16-16]}$$

Thus, the motion of the wire through the field induces a voltage across it. The wire can be considered as a voltage source for some external circuit containing a resistance, and so the moving wire would act like the battery in a circuit, driving current, I, through the external resistance, R, as shown in Fig. 16-20c. The current would continue to flow as long as the wire continued to move. The current is given by

Figure 16-21

Magnetic flux.

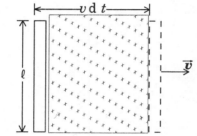

$$I = V/R = vB\ell/R \qquad \textbf{[16-17]}$$

In a time dt the wire moves a distance vdt, tracing out an area $d\mathcal{A} = v\ell dt$ as shown in Fig. 16-21, thus

$$d\mathcal{A}/dt = v\ell$$

Using Eq. [16-16],

$$V = B(d\mathcal{A}/dt) = (d/dt)(B\mathcal{A}) = (d/dt)\phi \qquad \textbf{[16-18]}$$

The quantity $\phi = B\mathcal{A}$ is called the *magnetic flux*.[12] Eq. [16-18] is Faraday's *law of electromagnetic induction* which states that

The voltage induced in a closed circuit is equal to the time rate of change of the magnetic flux.

Obviously the flux in a circuit can change in two ways: the circuit can move in the magnetic field, changing the area through which the magnetic field acts; or the magnetic field itself could change in a circuit

12.   The word *flux* is often encountered in science where some physical quantity flows or passes through some area. For example, a sound flux would be the total sound energy falling on some area; a fluid flux is the flow from a pipe of a given cross-sectional area.

of fixed size. The operation of a generator is an example of the former and a transformer is an example of the latter. Of course in practical cases, both processes might occur simultaneously.

If the closed circuit is a coil of N turns, then the magnetic flux through the coil is just N times the flux through one coil or

$$\phi = NB\mathcal{A} \qquad\qquad \text{[16-19]}$$

EXAMPLE 16-6

A circular loop of wire 5.0 cm in diameter is placed between the poles of a magnet perpendicular to the lines of a uniform magnetic field of strength 0.20 T (Fig. 16-22). The leads from the loop are fed through a tube and are connected to a resistor of 393 Ω. A steady pull on the leads causes the loop to shrink until its area is halved after a time of 2.0 s. What is the <u>average</u> current in the resistor during this process?

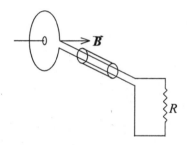

Figure 16-22

Example 16-6: A circular loop in a magnetic field.

SOLUTION

The initial flux through the loop is, from Eq. [16-19],

$\phi = NB\mathcal{A} = 1(0.2 \text{ T})\pi(2.5\times10^{-2} \text{ m})^2 = 3.93\times10^{-4} \text{ T·m}^2$

The final flux is half this amount since the area is reduced to one half. Therefore the change in the flux is $\Delta\phi$
$= 3.93\times10^{-4} \text{ T·m}^2/2 = 1.96\times10^{-4} \text{ T·m}^2$.

From Eq. [16-18], $V = d\phi/dt = 1.96\times10^{-4} \text{ T·m}^2/2 \text{ s}$
$= 9.82\times10^{-5} \text{ V}$

From Ohm's Law $I = V/R = 9.82\times10^{-5} \text{ V}/393 \text{ Ω} = 2.5\times10^{-7} \text{A}$
$= 25 \text{ μA}$

Exercises

16-1   If a horizontal wire carries a DC current from east to west, and is located in a magnetic field that is vertically downward, what is the direction of the magnetic force on the moving charges in the wire?

16-2   A charge is moving with a velocity v in a magnetic field B. The non-zero components of v and B are given by the subscripts. Find the direction of the resulting force in each diagram.

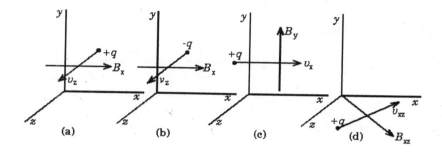

(a)   (b)   (c)   (d)

16-3   Earth's magnetic field at Toronto is 57 μT in magnitude and the dip angle (angle of $\vec{B}$ below the horizontal) is 75°. A vertical wire 3.0 m long in the wall of a house carries a current of 10 A. What force acts on the wire and in what direction?

16-4   The wire used in making superconducting magnets has a diameter of 1.00 mm. This wire will remain superconducting so long as the magnetic field at its surface is less than 0.10 T. What is the maximum current that can be passed through the wire?

16-5   A wire of length 0.20 m moves at a speed of 15 m/s at right angles to a magnetic field of strength 0.0040 T. The wire is also at right angles to the field. What is the induced voltage in the wire?

16-6   Consider a rotating rectangular generator coil as shown in the diagram. When the area is perpendicular to the field $\vec{B}$, as in part (a) on the diagram, consider the direction of the force on a negative charge in each of the four sides of the coil by answering the following questions:

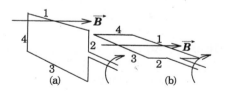

(a) What is the magnitude of the force on a charge in any portion of the coil in part (a) on the diagram?

(b) What is the instantaneous current in the wire?

(c) What is the direction of the force on a negative charge in portions 1 and 3 in part (b) on the diagram? Does this force contribute to a current in the coil?

(d) What is the direction of the force on a negative charge in portions 2 and 4 in part (b) on the diagram? Do they contribute to the current flowing in the coil?

(e) In what direction is the conventional current in the coil in part (b) on the diagram?

Problems

16-7   One of the ways to separate the isotopes of uranium ^{235}U and ^{238}U is to use a device similar to a mass spectrometer. Such a device produces singly ionized uranium atoms (a mixture of ^{235}U and ^{238}U), accelerates them through a potential difference of 815 kV, and injects them into a region of uniform magnetic field of strength 2.00 T.

(a) What is the radius of the orbit of the ^{235}U beam?

(b) How far apart are the separated ^{235}U and ^{238}U atoms after traversing a semicircle (see Fig. 16-7). (This part has a simple solution without repeating the calculation of part (a).)

16-8   A beam of electrons in an evacuated glass tube can be rendered visible by having a small pressure of mercury atoms present which fluoresce upon collision with a few of the electrons. In an undergraduate experiment to measure the charge-to-mass ratio (q/m) of the electron, the apparatus is placed in a magnetic field of 10.0 gauss $(1.00 \times 10^{-3}$ T$)$. The trajectory of the electrons is seen to be a circle of diameter 10.0 cm.

(a) With what speed were the electrons injected into the field?

(b) Through what voltage were the electrons accelerated to achieve this speed?

16-9   An electron, e, is moving in a straight line with a speed of 10% the speed of light. What is the magnitude and direction of the magnetic field produced by the electron at the points A, B and C (see diagram)?

16-10   The #14 cable used in house wiring has a cross-section as shown in the diagram. Embedded in a plastic matrix 10 mm by 6.0 mm are two current-carrying copper wires 1.0 mm in diameter and separated by 4.0 mm (centre-to-centre). (There is also a ground wire offset

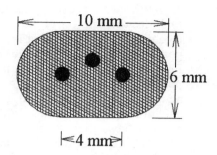

from the current-carrying pair.) The circuit containing the cable is carrying a current of 10 A.

(a) What is the magnitude of the magnetic field midway between the wires?

(b) What is the magnetic field at the centre of one wire due to the current in the other?

(c) What is the magnitude of the force per metre of length between the wires, and is it attractive or repulsive?

16-11   The smallest orbit of the electron in the hydrogen atom has a radius of 0.053 nm (the Bohr radius). The centripetal force holding the electron in this circular orbit is the Coulomb attraction between the nucleus (a proton) and the electron.

(a) What is the speed of the electron in the orbit?

(b) Treated as a circular one-turn coil, what is the equivalent current?

(c) What is the magnetic field at the position of the proton produced by the motion of the electron?

16-12   An object which is known to contain water in it somewhere is placed in a magnetic field that varies linearly from 1.00 T to 1.01 T over a distance of 10.0 cm. A tunable radio frequency oscillator irradiates the sample with high frequency EM waves, and absorption is observed at a frequency of 42.72 MHz. Where is the water in the sample?

16-13   A horizontal wire 0.75 m long is falling at a speed of 4.0 m/s perpendicular to a magnetic field of strength 2.0 T. The field is directed from north to south. What is the induced voltage of the wire? Which end of the wire is positive?

16-14   A copper bar 1.0 cm long is dropped from rest at a height of 3.0 m and falls in a horizontal orientation through the gap of a magnet which produces a uniform field of 0.20 T. What voltage appears across the ends of the bar when it is in the gap?

16-15 A metal rod rests on two conducting rails completing a circuit which includes the resistor of 3.0 Ω (see diagram). The circuit is perpendicular to a magnetic field of strength 0.15 T that is pointing into the diagram as shown. The rod is pulled at a constant speed of 2.0 m/s. What current flows through the resistor? In what direction does the current flow?

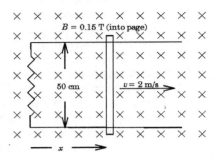

Answers

16-1 south

16-2 (a) $+y$ (b) $-y$ (c) $+z$ (d) $-y$

16-3 1.5×10^{-4} N, horizontal

16-4 500 A

16-5 0.012 V

16-6 (a) zero (b) zero (c) 1) toward 4, 3) toward 2, Yes (d) Perpendicular to the length of the wire, No (e) $1 \to 2 \to 3 \to 4$

16-7 (a) 1.00 m (b) 4 mm

16-8 (a) 8.8×10^6 m/s (b) 220 V

16-9 A, zero; B, 6.0×10^{-18} T, 30° down to the left; C, 12.0×10^{-18} T down

16-10 (a) zero (currents in opposite direction) (b) 5.0×10^{-4} T (c) 5.0×10^{-3} N, repulsive

16-11 (a) 2.2×10^6 m/s (b) 1.06×10^{-3} A (c) 12.5 T

16-12 3.3 cm from the 1 T position

16-13 6.0 V, west

16-14 15 mV

16-15 0.050 A, counterclockwise

1 Appendix: The Poisson Distribution

Many processes in biological and physical systems depend on random distributions of events. Examples include the number of photons incident on a light-sensitive cell in an eye, the number of visits of bees to a particular flower, the killing effect of high-energy radiation in living cells, and the emission of radioactive particles from unstable nuclei. The *Poisson distribution* describes things that happen randomly and rarely.

Imagine the following situation. Suppose that we have observed buses passing a particular bus stop for a year, and have found that during the hours when buses are running, an average of three buses pass in an hour. The bus company has an unusual philosophy of scheduling: the drivers maintain random schedules. There are many buses and many drivers, and they start their routes whenever they feel so inclined. Hence, buses come by the stop randomly, and we have observed that three go by per hour, on average.

How many buses will go by in a given hour? There is no answer to that question, since it might be any number; no bus may pass, or one, or two, or any number. But what we can do is tell what the chances are that a specified number of buses will pass the stop in the given hour. We can do this because the random arrival of the buses is governed by the Poisson distribution. The probability $\mathcal{P}(n;a)$ that n buses will go by in a given hour, if the average number of buses passing in an hour is a, is

$$\mathcal{P}(n;a) = \frac{e^{-a}a^n}{n!}, \qquad n = 0, 1, 2, \ldots \qquad \textbf{[A1-1]}$$

where the symbol $n!$ (n factorial) is the integer that equals the product of n with the preceding positive integers, i.e., $n! = 1 \times 2 \times 3 \times \ldots \times n$. For example, $4! = 1 \times 2 \times 3 \times 4 = 24$. A special case arises when $n = 0$: zero factorial is defined to be one, i.e., $0! = 1$.

EXAMPLE A1-1

In the bus situation described above, $a = 3$.

(a) What is the probability that four buses will go by in a given hour?

(b) What is the probability that no buses will go by in a given hour?

SOLUTION

(a) Since $a = 3$ and $n = 4$, what is required is $\mathcal{P}(4;3)$. Using Eq. [A1-1],

$$\mathcal{P}(n;a) = \frac{e^{-a}a^n}{n!}$$

$$\mathcal{P}(4;3) = \frac{e^{-3}3^4}{4!} = \frac{e^{-3} \times 81}{4 \times 3 \times 2 \times 1} = 0.17$$

This result indicates that the probability that four buses will pass in a given hour is 17 percent.

(b) The probability that no bus will pass is $\mathcal{P}(0;3)$:

$$\mathcal{P}(0;3) = \frac{e^{-3}3^0}{0!} = \frac{e^{-3} \times 1}{1} = 0.050$$

Thus, the probability that not a single bus will pass in a given hour is 5.0 percent.

If the value of $\mathcal{P}$ is known for a specific a and n, and another value of $\mathcal{P}$ is required for the same a but a different n, there is a useful recursion relation that avoids doing the whole calculation of Eq. [A1-1]. This relation, which allows $\mathcal{P}(n;a)$ to be determined easily if $\mathcal{P}(n-1;a)$ is known, is developed by first writing the ratio of $\mathcal{P}(n;a)$ to $\mathcal{P}(n-1;a)$:

$$\frac{\mathcal{P}(n;a)}{\mathcal{P}(n-1;a)} = \frac{e^{-a}a^n}{n!}\frac{(n-1)!}{e^{-a}a^{n-1}} = \frac{a^n}{a^{n-1}}\frac{(n-1)!}{n!}$$

This expression can be simplified by noting that $a^n/a^{n-1} = a$. As well, since $(n-1)! = 1 \times 2 \times 3 \times \ldots \times (n-1)$, and $n! = 1 \times 2 \times 3 \times \ldots \times (n-1) \times n$, then $(n-1)!/n! = 1/n$.

Thus,

$$\frac{\mathcal{P}(n;a)}{\mathcal{P}(n-1;a)} = \frac{a}{n}$$

Rearranging to give $\mathcal{P}(n;a)$:

$$\mathcal{P}(n;a) = \frac{a}{n}\mathcal{P}(n-1;a) \qquad \textbf{[A1-2]}$$

Equation [A1-2] shows that $\mathcal{P}(n;a)$ can be calculated easily by multiplying $\mathcal{P}(n-1;a)$ by a/n.

EXAMPLE A1-2

Continuing with the bus situation of Example A1-1,

(a) what is the probability that two or fewer buses go by in an hour?

(b) what is the probability that three or more buses go by in an hour?

SOLUTION

(a) The probability that two or fewer buses go by is just the sum of the probabilities that zero buses, one bus, or two buses go by, that is,

$$\mathcal{P}(0;3) + \mathcal{P}(1;3) + \mathcal{P}(2;3)$$

$\mathcal{P}(0;3)$ was already determined to be 0.050 in Example A1-1. The recursion relation [A1-2] can be used to determine $\mathcal{P}(1;3)$ and $\mathcal{P}(2;3)$:

$$\mathcal{P}(n;a) = \frac{a}{n}\mathcal{P}(n-1;a)$$

$$\mathcal{P}(1;3) = (3/1)\,\mathcal{P}(0;3) = 3(0.050) = 0.15$$

$$\mathcal{P}(2;3) = (3/2)\,\mathcal{P}(1;3) = 1.5(0.15) = 0.23$$

Thus, the required probability that two or fewer buses go by is

$$\mathcal{P}(0;3) + \mathcal{P}(1;3) + \mathcal{P}(2;3) = 0.05 + 0.15 + 0.23 = 0.43$$

Hence, the probability that fewer than three buses go by in an hour is 43 percent.

(b) One way to determine the probability that three or more buses go by is to calculate $\mathcal{P}(3;3)$ and $\mathcal{P}(4;3)$ and $\mathcal{P}(5;3)$, and so on *ad infinitum*, and then add these probabilities. That would certainly do it, but there is a better way. It is 100 percent certain that some number of buses will go by, that is, it is certain that either zero buses, or one bus, or two buses, etc., will go by. Thus, the sum of the probabilities of zero buses, one bus, two buses, etc., going by is 1 (i.e., 100 percent). That is,

$$\mathcal{P}(0;3) + \mathcal{P}(1;3) + \mathcal{P}(2;3) + \mathcal{P}(3;3) + \mathcal{P}(4;3) + \mathcal{P}(5;3) + \ldots = 1$$

Moving the first three terms on the left over to the right,

$$\mathcal{P}(3;3) + \mathcal{P}(4;3) + \mathcal{P}(5;3) + \ldots = 1 - [\mathcal{P}(0;3) + \mathcal{P}(1;3) + \mathcal{P}(2;3)]$$

Thus, the probability of three or more going by is just 1 minus the probability that two or fewer go by. Using the result from part (a) for the probability that two or fewer go by:

$$\mathcal{P}(3;3) + \mathcal{P}(4;3) + \mathcal{P}(5;3) + \ldots = 1 - 0.43 = 0.57$$

Hence, the probability that three or more buses go by in an hour is 57 percent.

The Beer–Lambert Law

In Section 4.10 the Beer–Lambert law was derived in a traditional way by adding up (integrating) contributions to the absorption of a beam of light from a succession of thin slices of a light-absorbing sample. Since the encounter between the photons in the incident beam and the absorbing molecules is a random process, the law can also be derived using the Poisson distribution.

The average number of photon-absorbing interactions encountered along a straight-line path traveled by a photon depends on the concentration of the absorbers (C), the path length of the sample (x) and the effective absorption cross-section (σ) of each absorber. Therefore,

$$\text{average number of photon-absorbing interactions} = a = \sigma C x$$

For a photon to survive its passage through the sample, it must encounter zero photon-absorbing interactions. The probability that this will occur is $\mathcal{P}(0;a)$ of the Poisson distribution:

$$\mathcal{P}(0;a) = \frac{e^{-a}a^{0}}{0!} = e^{-a} = e^{-\sigma Cx}$$

This probability is also equal to the ratio of the number (N) of photons that survive the passage through the sample to the number (N_0) that are incident on the sample. For example, if the probability of a photon surviving its passage is 25%, then $N/N_0 = 25\% = 0.25$. Hence,

$$\mathcal{P}(0;a) = \frac{N}{N_0} = e^{-\sigma Cx}$$

As well, this probability equals the ratio of the transmitted light intensity to the incident light intensity. Writing the transmitted intensity of light (having wavelength λ) as $I(\lambda)$, and the incident intensity as $I_0(\lambda)$:

$$\frac{I(\lambda)}{I_0(\lambda)} = e^{-\sigma Cx}$$

$$\text{or} \quad I(\lambda) = I_0(\lambda)e^{-\sigma Cx}$$

This last equation is identical to Eq. [4-19] in the discussion of the Beer–Lambert Law in Chapter 4.

Exercises

A1-1 The average number of mice caught by a bobcat is 4.3 per day. What is the probability that on a particular day the bobcat will catch more than 2?

A1-2 Pierre is portaging with his canoe and finds he gets 300 blackfly bites in a portage that takes 1 hour and 40 minutes. For how many one-minute intervals was he free of blackfly bites?

A1-3 A radioactive sample placed in front of a Geiger counter records 540 counts in 2 hours and 20 minutes. During a one-hour period, during how many one-minute intervals will 5 counts be recorded? (Assume that the average number of counts per hour is constant.)

A1-4 While waiting in an airport during a thunderstorm, a statistician decides to count lightning flashes. He notes the flashes in each one-minute interval during the storm and gets the following data:

# of flashes in 1 min.	Frequency
0	2
1	7
2	7
3	4
4	2
5	1

(a)   Calculate the average number of flashes per one-minute interval.

(b)   Assuming the data fit a Poisson distribution, calculate the expected frequency of 0, 1, 2, 3, 4, or 5 lightning flashes per one-minute interval in a similar 23 minutes.

(c)   Is the assumption of a Poisson distribution a good one?

Answers

A1-1 $0.80 = 80\%$

A1-2 5

A1-3 9

A1-4 (a) 2.0

(b) 3, 6, 6, 4, 2, 1

(c) Reasonably good

2 Appendix: Web-Based Resources

The advent of the World Wide Web (WWW) has made available many excellent and often amusing resources to aid the teaching of physics in an interactive and dynamic way. The difficulty in including the addresses of these resources in a published book is that the sites are often ephemeral. By the time this book is published, many sites active during the writing period will have disappeared or been altered.

We have chosen, therefore, to list only one site which will be maintained in an active state by the authors and the Department of Physics at the University of Guelph.

http://www.physics.uoguelph.ca/biophysics/

On this site will be found links to excellent tutorials suitable for study using this book. The available material includes several *remedial tutorials* such as the one on Dimensional Analysis referred to in Chapter 1, and interactive activities suitable for virtually every chapter in the book. At this site they are grouped by chapter after the remedial tutorials, and include a brief description of the activity and what the student may expect to see. Many of the sites are "applets" which is a term applied to interactive applications which can be accessed and manipulated on the Web. These applet activities introduce the student to a new form of tutorial and are highly recommended as a strong reinforcement of many basic concepts. The authors are always looking for new Web-based tutorials and welcome suggestions or comments. These should be e-mailed to biophysics@physics.uoguelph.ca.

3 Appendix: Symbols

α	alpha-particle	λ_m	wavelength of maximum emittance
α	angular acceleration	μ	linear absorption coefficient
α	smallest resolvable angle	μ	linear attenuation coefficient
α	thermal expansion coefficient, linear	μ_A, μ_B	chemical potentials
β	beta-particle	μ_k	coefficient of kinetic friction
β	thermal expansion coefficient, volume	μ_m	mass attenuation coefficient
γ	gamma ray	μ_s	coefficient of static friction
γ	gyromagnetic ratio	ν	neutrino
γ	surface tension	$\bar{\nu}$	antineutrino
δ	phase angle	ξ	molecular displacement
Δ	change (finite)	Π	osmotic pressure
ϵ	extinction coefficient	ρ	density
ϵ	strain, tensile	ρ	resistivity
ϵ_S	strain, shear	σ	Stefan–Boltzmann constant
η	viscosity coefficient	σ	stress, tensile
θ	angular displacement	σ	conductivity
θ	contact angle	σ	absorption cross-section
κ	decay constant	σ_S	stress, shear
λ	decay constant	τ	torque
λ	wavelength	τ	time constant

Φ	work function		e	electron
Ψ	wave function, electron		e	elementary charge
ω	angular frequency		e^+	positron
ω	angular velocity		$\vec{E}$	electric field vector
Ω	ohm		E	electric field magnitude
$\vec{a}$	acceleration vector		E	energy
a	acceleration (magnitude)		E_V	energy density
a	aperture size		f	focal length
$\vec{a}_{av}$	average acceleration vector		f	frequency
a_{av}	average acceleration (magnitude)		f_0	threshold frequency
a_c	centripetal acceleration		$\vec{F}$	force vector
a_t	tangential acceleration		F	force
A	absorbance		$\mathcal{F}$	friction factor
A	activity		F_c	centripetal force
A	amplitude		g	acceleration due to gravity
A	baryon number		g	conductance
$\mathcal{A}$	area		G	elastic bulk modulus
$\vec{B}$	magnetic field vector		G	shear modulus
C	concentration		G	universal gravitational constant
c	speed of light		h	Planck's constant
C	capacitance		H	equivalent dose
C	specific heat capacity		I	electric current
$\mathcal{C}$	mass density of solute		I	intensity
C_{os}	osmolality		I	moment of inertia
D	diffusion coefficient		I	number of inactivating events
D	dose		I	power density
d	depth		$I(\lambda)$	intensity at wavelength λ

IL	intensity level	$\vec{p}$	linear momentum vector
J	current density	p	linear momentum (magnitude)
J	particle flux	p	number of targets
k	Coulomb constant	p	object distance
k	force constant	P	power
k	thermal conductivity coefficient	P	pressure
K	dielectric constant	P_{abs}	absolute pressure
K	kinetic energy	P_{atm}	atmospheric pressure
k_B	Boltzmann's constant	P_{gauge}	gauge pressure
L, ℓ	length	$P(x)$	probability density
L	angular momentum	$\mathcal{P}(n)$	probability of n events
L_f	latent heat of fusion	$\mathcal{P}_E$	probability
L_v	latent heat of vaporization	q	image distance
m	magnification (spherical surface)	Q, q	charge
m	mass	Q	heat quantity
m_e	mass of electron	Q	flow rate
M	magnification (lens)	$\vec{r}$	position vector
M	molar mass	R, r	radius
$\mathcal{M}$	magnetic moment	R	range
n	electron quantum number	R	resistance
n	mode (or node) number	R	molar gas constant
n	number of moles	R_e	Reynolds number
n	refractive index	$s, \Delta s$	arc length
$\vec{N}$	normal force	s	electron spin quantum number
N	number of nuclei	s	sedimentation coefficient
N	number of neutrons	S	siemens
N_A	Avogadro's number	t	thickness

t	time		V	electric potential
T	period		$\mathbf{V}$	voltage
T	tension		V_m	partial molar volume (solvent)
T	temperature		V_s	stopping potential
T	transmittance		V_s	sensitive volume
$\%T$	percent transmittance		$\mathcal{V}$	volume
$T_{1/2}$	half-life		w	weight
U	elastic potential energy		w_R	radiation weighting factor
U	electric potential energy		W	spectral emittance
U	internal energy		W	work
U	gravitational potential energy		x	position
U	potential energy		x_0	initial position
$\vec{v}$	velocity		x_{rms}	root mean square displacement
v	speed		X_v, X_u	mole fraction (solvent and solute)
v	vibrational quantum number		y	position
v_{av}	average speed		y	object height
v_0	initial speed		y_0	initial position
v_{rms}	root mean square speed		y'	image height
v_t	terminal speed		Y	Young's modulus
			Z	atomic number

4 Appendix: Constants, Conversions, Prefixes, & Greek

A4-1 Physical Constants

Name	Symbol	Value
magnitude of electronic charge	e	1.602×10^{-19} C
speed of light in vacuum	c	2.998×10^{8} m/s
Planck's constant	h	6.626×10^{-34} J·s
universal gravitational constant	G	6.673×10^{-11} N·m^2/kg^2
Avogadro's number	N_A	6.022×10^{23} mol^{-1}
Coulomb constant	k	8.988×10^{9} N·m^2/C^2
molar gas constant	R	8.315 J·mol^{-1}·K^{-1}
Stefan–Boltzmann constant	σ	5.671×10^{-8} W·m^{-2}·K^{-4}
Boltzmann's constant	k_B	1.381×10^{-23} J·K^{-1}
mass of electron	m_e	9.109×10^{-31} kg
mass of proton	m_p	1.673×10^{-27} kg
mass of neutron	m_n	1.675×10^{-27} kg
atomic mass unit	u	1.661×10^{-27} kg

speed of sound (dry air, 17°C, standard pressure)	340 m/s
speed of sound (water, 20°C)	1.41×10^{3} m/s

A4-2 Conversion Factors

Name	Equivalent
electron volt (eV)	1 eV = 1.602×10^{-19} J
litre (L)	1 L = 1.00×10^{-3} m^3 = 1.00 dm^3
absolute zero (temperature)	0 K = –273.15°C
calorie (cal)	1 cal = 4.186 J

A4-3 SI Prefixes

Factor	Prefix	Symbol	Factor	Prefix	Symbol
10^{24}	yotta	Y	10^{-1}	deci	d
10^{21}	zetta	Z	10^{-2}	centi	c
10^{18}	exa	E	10^{-3}	milli	m
10^{15}	peta	P	10^{-6}	micro	μ
10^{12}	tera	T	10^{-9}	nano	n
10^{9}	giga	G	10^{-12}	pico	p
10^{6}	mega	M	10^{-15}	femto	f
10^{3}	kilo	k	10^{-18}	atto	a
10^{2}	hecto	h	10^{-21}	zepto	z
10^{1}	deka	da	10^{-24}	yocto	y

A4-4 Greek Alphabet

A	α	alpha		N	ν	nu
B	β	beta		Ξ	ξ	xi
Γ	γ	gamma		O	o	omicron
Δ	δ	delta		Π	π	pi
E	ϵ	epsilon		P	ρ	rho
Z	ζ	zeta		Σ	σ	sigma
H	η	eta		T	τ	tau
Θ	θ	theta		Υ	υ	upsilon
I	ι	iota		Φ	ϕ	phi
K	κ	kappa		X	χ	chi
Λ	λ	lambda		Ψ	ψ	psi
M	μ	mu		Ω	ω	omega

Index

Q

R